CERAMICS IN ADVANCED ENERGY TECHNOLOGIES

Colloquium sponsored by:

- Commission of the European Communities, Directorate General: Science, Research and Development, Directorate General: Industrial Affairs and Internal Market
- Committee of Senior COST Officials

Co-sponsors:
- CERAME UNIE
- British Ceramic Society
- Dutch Ceramic Society
- French Ceramic Society
- German Ceramic Society
- Institute of Industrial Research and Standards, Dublin, Irl.

CERAMICS IN ADVANCED ENERGY TECHNOLOGIES

Proceedings of the European Colloquium
held at the Joint Research Centre,
Petten Establishment, Petten, The Netherlands,
20-22 September 1982

edited by

H. KRÖCKEL
M. MERZ
O. VAN DER BIEST

CEC, JRC Petten Establishment, Petten, The Netherlands

D. Reidel Publishing Company

A MEMBER OF THE KLUWER ACADEMIC PUBLISHERS GROUP

Dordrecht / Boston / Lancaster

for the Commission of the European Communities

Library of Congress Cataloging in Publication Data
Main entry under title:
Ceramics in advanced energy technologies.

Proceedings of the European Colloquium on Ceramics in Advanced Energy Technology.
1. Ceramic materials–Congresses. 2. Energy development–Congresses.
3. Power (Mechanics)–Congresses. I. Kröckel, H. II. Merz, M. III. Biest, O. van der.
IV. European Colloquium on Ceramics in Advanced Energy Technology (1982 : Joint
Research Centre, Petten Establishment) V. Commission of the European Communities.
Directorate-General for Science, Research and Development. VI. Commission of the
European Communities.
TA455.C43C47   1984   620.1′4′024621   84–8212

ISBN-13: 978-94-009-6426-6   e-ISBN-13: 978-94-009-6424-2
DOI: 10.1007/978-94-009-6424-2

Lay-out: Reproduction service J.R.C. Petten
Design frontcover: J. Wells, J.R.C. Ispra

Publication arrangements by
Commission of the European Communities
Directorate-General Information Market and Innovation, Luxembourg

EUR 9210
☺ 1984 ECSC, EEC, EAEC, Brussels and Luxembourg
Softcover reprint of the hardcover 1st edition 1984

LEGAL NOTICE
Neither the Commission of the European Communities nor any person acting on behalf of the
Commission is responsible for the use which might be made of the following information.

Published by D. Reidel Publishing Company
P.O. Box 17, 3300 AA Dordrecht, Holland

Sold and distributed in the U.S.A. and Canada
by Kluwer Academic Publishers,
190 Old Derby Street, Hingham, MA 02043, U.S.A.

In all other countries, sold and distributed
by Kluwer Academic Publishers Group,
P.O. Box 322, 3300 AH Dordrecht, Holland

LIST OF CONTENTS

Session 3: Chemical Features.
Chairman: H. Hausner, Technische Universität, Berlin, FRG.

Session 4: Electrical and Electronics Features.
Chairman: C. Palmonari, Centro Ceramico, Bologna, I.

Final session:
Chairman: A. Baudran, Société Française de Céramique, Paris, F.

ORGANIZING COMMITTEE

PREFACE

The European Colloquium on Ceramics in Advanced Energy Technology dealt with structural, mechanical, thermo-physical, chemical and electrical/electronics aspects of ceramics, as candidates materials in advanced energy conversion systems. The technical programme included the presentation of 22 invited papers. The Colloquium was concluded by a panel discussion which was preceeded by a presentation on the Ceramic Fabrication.

For convenience, papers and discussions are arranged in the same way as they were presented at the Colloquium, followed by the conclusions drawn by the panel members. The editors wish to express their gratitude to the authors for the preparation and the presentation of their papers and to the Colloquium participants for providing their discussion contributions. They also acknowledge the valuable contributions to the planning and realisation of the Colloquium made by the representatives of the various sponsoring and co-sponsoring organisations. The excellent guidance of the Colloquium discussion periods exercised by the session chairmen is highly appreciated as well as the efforts of the panel members to prepare the conclusions from the Colloquium sessions. The contributions of all others which assured a successful Colloquium from which these proceedings evolved are gratefully appreciated.

In a Post-Colloquium Workshop a number of experts discussed results obtained from the Colloquium and identified areas warranting future R & D efforts. The conclusions drawn from the Workshop are annexed to these proceedings.

The Editors.

Welcome and opening

P.J. van Westen,
C.E.C., D.G. XII, J.R.C., Petten Establishment.

Ladies and Gentlemen,

It is my duty and above all my pleasure, as Director of the
Petten Establishment, to welcome you here to this place.
I should like to add to my welcome a few words on the
connection between this colloquium and our work at Petten,
which is one of the four establishments of the Joint Research
Centre. We are, in fact, in charge of three different
research activities and among them, it is the High
Temperature Materials programme which has brought us together
at this colloquium.

The experimental research of this programme is concerned with
mechanical properties of H.T.M.'s in aggressive environments.
Around this experimental research, some other activities are
placed, including a data bank on H.T.M. properties and an
Information Centre. This Information Centre establishes
communication with industries, research institutes and other
organisations, carries out surveys and inquiries and
organises courses, conferences, workshops and meetings. It is
in this context that this colloquium has been organised.

In opening this colloquium I should like to stress that high
temperatures are fundamental to advanced energy conversion
and utilization technologies. We all realise that development
and technology in this field is of the utmost importance now
and will, very likely, remain as important in decades to
come. The present energy supply problems force us to reduce
our dependence on gas and oil by strong efforts in two areas:
-    conserving energy both by using it more effectively and
     by avoiding wastage, and
-    supplementing these fuels by coal and nuclear energy and
     by other primary energy sources such as biomass, wind
     power, solar energy, wherever technically and
     economically feasible (and environmentally acceptable).
Since the thermal efficiency of energy conversion processes
generally shows a strong temperature dependence, most of
these systems imply advanced technologies operating at high
temperatures. This has led to an increasing demand for
structural materials of higher temperature capability.

Industries like steel, cement and electricity production,
which operate with high temperatures, provide the greatest

1

potential for energy saving. However, energy is also conserved by using it more efficiently and numerous options for improved thermal efficiency resulting from higher working temperatures are presently being considered for commercialization. These include:
- Advanced cycles for combined, power generation;
- Advanced Automotive Power Plants;
- Electrochemical Energy Conversion Devices.

Features like availability and reliability, which are essential for the economic and technical feasibility of any process and the limitations imposed by the structural materials employed in high temperature components, present one of the main obstacles to present progress. In addition to the technical factors, economic or political aspects complicate the situation. For example, the availability of chromium and cobalt, essential elements for alloying high temperature materials, is strategically vulnerable due to the geographical location of their sources. Substitution and saving of these metals has recently provided an additional incentive for research, particularly on the wider application of ceramics for high temperature components.

Finally, energy technologies, high temperatures and materials are frequently inseparable items. Studies on materials, and specifically ceramics, for new energy applications are receiving growing attention both in Europe and elsewhere in the world.

Current applications of high-strength ceramics are still restricted to rather simple applications in the sense that, in general, the forms of the components are not intricate and the real constructional use of advanced ceramics is limited to relatively small items. Furthermore, the scope for the application of materials to replace the currently used metallic alloys is very wide. It embraces fields of large-scale energy production and conversion processes, energy usage in transport and a range of high temperature industrial processes. Thus there is considerable variation in the form of components required in such applications.

The particular mechanical and physical properties required of ceramic components are also extremely wide-ranging. The need to design and develop plant or machines to utilize such materials to the best advantage in the light of achievable characteristics should provide a stimulating challenge to design and development engineers.

Many possible applications of ceramics in high temperature

technology are being currently explored by engineers and
these may be exemplified in the first place by inlet guide
vanes, rotor blades and other high temperature components for
gas turbines. An increase in turbine inlet temperature
without the need for blade cooling would give immediate
benefits in both specific power output and in fuel
consumption.
The feasibility of ceramics in gas turbines is rather well
established and the practical use of ceramics for commercial
components is approaching slowly but steadily. The maximum
temperatures involved range from about 1100-1200°C in
stationary power generation turbines to 1500-1600°C in
advanced aero engines.
The use of ceramics in Diesel engines is also envisaged. The
main advantage is to minimize heat losses and to convert a
greater proportion of the thermal energy to mechanical
energy.

Other fields of energy production or conversion are also
potentially very important users of ceramics. One is magneto-
hydrodynamic generation of electrical power using fossil
fuels. Ceramics are likely constructional materials for the
walls of the combustion channel and for the electrodes.
Another is related to coal conversion systems requiring
operation at high pressures and temperatures ranging up to
1500°C. Research on heat exchangers is also related to this
field with use in most other energy conversion systems.

Application to the field of solar energy is very wide.
Although temperatures in battery and fuel cell systems are
not very high, chemical corrosion and electrical conductance
properties have much in common with fields like MHD. Related
is the use of ceramics in the production of sensors for
oxygen-monitoring.
The application of ceramics in the metallurgical industry and
in industrial furnaces is well described, where energy
savings are most often based on ceramic fibre insulation.

In conclusion I should like to point out that these major
industrial branches generate economic requirements to
investigate all problem areas open to a greater use of
ceramics, and where the reliable and reproducible production
of high performance ceramic components needs further
improvement.

The Community has recently launched its first R & D action in
the field of Technical Ceramics and I hope that this
colloquium will provide indicators for future R & D efforts
in Europe.   Thank you for your attention.

The Impact of the High Temperature Ceramics in Industrial Growth in the Community.

A.A. van Rhijn, Deputy Director General,
CEC, Directorate General: Internal Market and Industrial Affairs.

Throughout history the technological progress of mankind has been determined by the interplay between materials and ideas. Advancing from the stone age, where natural ceramics were used to make tools, through the bronze, iron and super-alloy ages we are now entering an era where ceramics once more hold the key to progress.

Modern technical ceramics are, however, not natural artifacts nor even simply processed natural materials but are carefully prepared with certain properties in mind, and components made from technical ceramics can be found in almost all industrial sectors.

Energy is a sector where in particular the progress of the dialectic between material properties and design concept is determined by the properties of materials. Since the time of Carnot it has been possible to conceive of a heat engine which could function at relatively high efficiency, provided it could stand the high temperatures of operation which would be necessary. From its conception the internal combustion engine has been improved mechanically, but its overall thermal efficiency has been imprisoned within a range broadly dictated by the materials used in its construction.

The advent of the gas turbine enabled higher efficiency to be achieved for a number of uses, but it was necessary for those gas turbines to develop alloys that retained their strength at the higher operating temperatures and some further advances in gas turbine technology have only been possible due to the incorporation of ceramic elements where the conditions are most severe.

Up to 1973 we were living in an era of relatively cheap energy so that the economic pressures to improve heat engine efficiency were weak. Now, however, the problem of limitation of resources cannot be left to the next generation. We have immediate economic reasons for making the best of the fuels we have available, even if at the moment we seem to have a temporary respite in the upward pressure of energy prices – but let us not be fooled on this temporary element. The reduction of thermal and combustion product pollution is also an important environmental benefit which flows from a greater energetic efficiency, since less fuel is burned to achieve the same output.

4

The role of advanced ceramics in heat engines will have to be reassessed, so that they may not become minor, but major components. It is no secret that Japan is consecrating large sums into research in this field and the Community must be prepared to make a similar if not even a larger effort in order that its industry can remain competitive.

It would be impossible for me and useless for you, because you know it all much better than I do, to enumerate all of the instances in which technical ceramics are used in other industries, but let me stress the point that they are frequently a key element in equipment whilst representing a small part of the total capital cost.

Because of its disperse but key position the sector of technical ceramics can be seen to be a vital link in development and technical innovation in many other sectors. At this point I would like to indicate the relevance of your work in the field of ceramics against the broader background of things that are going on in Europe and in the Community. Your work is not work being done in isolation as far as its substance and its intentions are concerned; your work is on the contrary highly relevant, specially also to what the Commission sees as essential developments. Over the last period the Commission has published three papers of major importance. One is the so-called industrial strategy paper, the second one is a paper about the R & D objectives and the third is a competitiveness-report of the European industry. In that last report an analysis of the competitive position of European industry over the last ten years has been made, and the results are most worrying. Nearly all sectors in all the countries of the Community are experiencing a decrease in investment rates; though at different levels in the various countries, nevertheless a similar development overall. Today in a number of countries industry is not investing at all, or is even disinvesting at a national level. Of course here the economic crisis plays a role, but all the same the natural structural tendency I just indicated is playing its role as well. And why is this so worrying? Because even though Japan and the United States are faced with similar economic difficulties of a business cycle nature, nevertheless in their countries they do not have the same experience as far as the investment trends are concerned. Not to speak of the newly industrialised countries that are moving ahead rapidly. There is a very strong need to reverse the trend, because the relatively low investment levels in Europe have led to an ever increasing gap in competitiveness with the countries just mentioned. The reasons are, decreasing profitability on the one hand through cost increases in the sense of taxes, wages,

energy prices and interest costs, on the other hand because of
a lack of confidence in future profitability of new
activities.
A second point that is closely related to the developments as
we see them in the European industry over the last ten years
is a decreasing part of own risk capital in the financing of
the enterprise, which makes the enterprise of course much more
vulnerable to difficulties and periods of losses because they
then immediately have to turn to the outside, to the
government, to the banks, in order to overcome the
difficulties. We know all in our countries enough of these
cases.

What has to be done? Here the industrial strategy report and
the R & D objectives papers give the answer. In close relation
to the policy that is needed to overcome the economic crisis,
special attention has to be given to those things that at the
same time would cope with the competitiveness-problem, and so
it is advocated that a strong stimulant should be given to
investments, and specially in those areas where the government
is playing a major role, because there the governments are
able to influence the investment decisions in a direct way and
they do not depend on the confidence of industrialists to make
new investments. Energy production and energy saving being in
this respect of particular importance, but other sectors could
be mentioned.

Closely related to stimulating investment is stimulation of
R & D. In the past the Community has done relatively little in
the field of industrial research and development, but a few
years ago the tendencies have changed and with even more
emphasis the Commission is now pursuing a policy in which it
tries to elaborate programmes in close cooperation with
industries, to a large extent also being executed by
industries, for R & D in various sectors and new technologies.
One of these sectors, outside your direct concern, but
nevertheless you will certainly know about it, is the
information technology in which field the Commission is now,
in cooperation with industry, developing a major R & D plan in
order to be able to cope with the lagging behind of Europe vis
à vis the Japanese above all, but also the Americans in the
field of information technologies. Other similar programmes of
a bigger or smaller size are also under consideration or on
their way.

Your sector of interest, ceramics, is in this respect also
playing a major role, at the same time fitting in very well
with another part of the approach of the European Commission
as far as the R & D is concerned, that is application of new

technologies in traditional sectors. I think this is also an element that is of extreme importance, that one is not just only considering the development of new technologies as such, but also trying to work up programmes which promote the application of those new technologies in the sectors concerned; here again I think your sector is a very important one if we relate that to the automotive industry, the industry of gas turbines etc. which are just pure examples. But what is the reason that the Commission is going for such a policy? It is because there is an evergrowing conviction that industries cannot do it alone anymore and cannot even in many cases do it on a national level effectively. This should be done at a broader level, i.e., for the countries of the European Communities, on the level of the E.E.C.

Earlier in my introduction I referred to an element that seems to be very important, that is the role of advanced ceramics in heat engines which will have to be reassessed so that they may not become minor but major components, and it's no secret that Japan is consecrating large sums into research in this field, and that the Community must be prepared to make a similar effort. It is of no use to delude ourselves that this can be done on a national scale. Major technological advances of this kind demand large financial commitments which can, if we are to match our competitors, only be achieved on a Community scale through cooperation. Such cooperation would not spell the end of technical competition within the Community industry, as there would still be plenty of scope for this in manufacturing and marketing new concepts. I think also that in this respect Japan is giving very interesting illustrations of on the one hand a close cooperation in the research domain , and then a very strict competition between the Japanese firms in other areas.
A problem that is also pointing in this direction is that budgetary problems in the Member States have led them to reduce public funded research and development, and that just at a time when the need for an increased effort has become more acute. Just as our Community industry is having to restructure organizationally on a Community dimension to gain the advantages offered to it by the Common Market, so research will have to be carried out on a Community dimension as well, so that we can deploy the resources necessary to the task in hand. The elimination of duplication and cross-fertilisation of our ideas engendered by cooperation, will give greater cost effectiveness in our research efforts and respond both to budgetary and technological needs.

There is another point which is close at heart to the Community which I would like to mention. Even with an active

innovative industrial base the Community, which is largely
deficitary in raw materials, is often strategically vulnerable
to a disruption of supplies or manipulation of raw material
markets. The raw materials for most technical ceramics are
indigenous to the Community. However, in many cases the metals
they replace are made from raw materials coming from
non-European, non-Community countries. Technical ceramics have
therefore a strategic role in our industrial policy and this
is one of the principal reasons why the Community has included
them in the multiannual research programme on raw materials
adopted by the Council last May.

A last point. There are many processes which can be used to
manufacture technical ceramic components. The products are
frequently required to individual specifications and in
relatively small numbers, so that small and medium sized
industries and enterprises can be highly effective, and this
fits very well into the philosophy of the European Commission
in stimulating innovation through new small and medium size
firms.

If you will allow me a few concluding remarks: the disperse
nature of the technical ceramics industry and the multiplicity
of its products make it an ideal candidate for collaborative
effort in both research and the diffusion of the results. We
have many products looking for outlets and many problems
looking for a solution, but at the same time this is a sector
where the spin-off potential is very large. All cooperation
demands some sacrifice of independence, but the long-term
gains far outweigh any imaginary short-time losses and we are
convinced that we should be prepared to work together if we
are not to succumb in splendid isolation.

Most of you participating in this colloquium are technical
experts and by your advanced and detailed knowledge of your
subject you are in a privileged position to assess its
potential. I would urge you to seize every opportunity which
may arise during the next two days to lay the foundations for
on-going cooperation, and when you return home, to impress the
advantages of working together on your firms. Only in this way
we will be able, as Europeans, to restore the competitiveness
we are losing, and to keep it, where we just have not lost it
yet. The Commission has with respect to all this set its
targets and the Commission will also do all that is in its, I
accept, limited powers to see that the means are provided to
achieve them. In this respect I would like to refer to the
programme I just mentioned in which also a not inconsiderable
sum is set aside for the ceramics sector and I would suggest
that Mr. Bourdeau says some more about these programmes,

because the period for making offers for participating in this programme is open up to the 15th of October; this is your last chance in trying to get part of the money made available.

May I finish with the remark: it depends on you in the end, and on European industry in general, whether Europe will be successful in overcoming its actual problems.

CERAMICS FOR FUTURE AUTOMOTIVE POWER PLANTS

Peter Walzer
Research Division
Volkswagenwerk AG

Wolfsburg - West Germany

Synopsis:

Ceramic materials can improve automotive gas turbines, turbo
chargers and Diesel engines. The paper describes the status of
component development in the various applications. Further
improvement of the ceramic materials and component fabrication
technologies with respect to less variation of strength properties and to longer time thermal stability will be necessary.

1. Introduction

The components of alternative power plants with continuous combustion are subject to high, continuously acting gas temperatures. New highly heat-resistant materials will have to be found before gas turbines or Stirling engines can be seriously considered as potential drive units.

Further development of conventional automotive engines, too, demands for materials with better heat stability, heat-insulation properties and resistance to erosion. Where the conventional metallic materials are unable to fully meet such higher demands, super alloys have to be used.

New ceramic materials could better fullfill these future requirements. For high stress applications this could be silicon nitride and silicon carbide. For other applications aluminum silicate, zirconium oxide or aluminum titanate are of particular interest.

In the following, the properties of these materials are outlined at first. Then the state of development of components for vehicle gas turbines, turbo chargers and piston engines is demonstrated. It will be shown that future work has to concentrate on increasing the survival probability in mass production and on improvement of the long time thermal stability of these materials.

2. Properties of ceramic materials

In Figure 1 properties of some ceramics are compiled.

Non oxide ceramics like silicon nitride (Si_3N_4) and silicon carbide (SiC) attain higher strengths than metallic super alloys at temperatures above 135o K, which is somewhat above 1o5o° C. The strength of the material increases with the density attained in the production process. Various production processes lead to materials of different densities and properties. Highest densities and highest strengths yield hot pressing and pressureless sintering processes. In the case of silicon nitride, for example, porous components with the advantage of being formed to complicated net shapes can be made by reaction bonding. As forming operations classic methods like injection moulding or slipcasting can be used. Methods for a subsequent densification of porous bodies are under development. In addition to their high temperature strength properties these ceramics have only one third of the density of ferrous superr alloys, which offers low inertia masses in rotating or oscillating parts.

Magnesium aluminum silicate (MAS) attains only moderate strength, its thermal expansion, however, is almost zero. With this property it is useful in applications where a component with low mechanical loads is exposed to high temperature gradients.

The main advantage of ceramics like zirconium oxide (ZrO_2) or aluminum titanate ($Al_2O_3-TiO_2$) is their low thermal conductivity, which is only one fifth of that of metals. Effective insulation can be achieved, the low strength of the latter material, however, demands for a design where the ceramic is used as a heat shield and the mechanical load is carried by the metallic structure.

3. Ceramics in vehicle gas turbines

We shall now go on to consider first how important the use of these new ceramic materials could be in the development of a non-conventional drive unit, i.e. a gas turbine.

Figure 2 is a design study of a vehicle gas turbine. It contains a compressor which inducts air from the atmosphere and compresses it. The compressed air is then preheated in a heat exchanger. The combustion of fuel in the combustion chamber produces so much energy that the released combustion gases can drive the compressor via a first turbine stage and the vehicle via a second. A large portion of the heat energy still contained in the exhaust gases is removed in the heat exchanger.

Gas turbines work with continuous combustion and with high air excess. The advantages are that the exhaust gases are very clean and that various types of fuel can be used. Indeed, 80 % of all crude oil derivatives can be burnt in a gas turbine, probably even coal dust. The disadvantage of this process has hitherto been that in contrast to the piston engine, the components of a gas turbine are continuously exposed to the high temperature of the combustion gases and that cooling of the small components is impossible.

For this reason, the maximum gas temperature permissible with metallic materials was 1300 K. At these working temperatures gas turbines cannot reach higher thermal efficiency or fuel economy than today's spark ignition engines. This is particularly true in the case of low output gas turbines. However, if metallic gas turbine components can successfully be replaced by high-temperature ceramic parts, it should be possible to increase the process temperatures by approximately

3oo K. Apart from the clean exhaust gas and multi-fuel capability which I have already mentioned the gas turbine then yields fuel consumption figures which are lower than those of today's reciprocating piston engines.

Because of the inviting prospects held out by the high-temperature gas turbine, work has been underway now for several years, especially in the USA, with the intent to putting silicon nitride and silicon carbide to work. Since 1974, the German Federal Ministry for Research and Technology has been supporting such work in Germany. Engaged in this development program along with Volkswagen are other power unit manufacturers such as Motoren-Turbinen-Union and Daimler Benz. Companies such as Annawerk, ESK, Rosenthal and Feldmühle represent the ceramics manufacturers in this program. University institutes play a part in the basic research into these materials.

Figure 3 shows the components which have to be made from ceramics in a high temperature gas turbine. These are the combustion chambers, the turbine nozzle rings, the turbine rotors and the heat exchanger.

In the development work done so far it has been proven that components with low mechanical loads like the combustion chamber or the nozzle can survive cyclic loads at gas temperatures as high as 165o K for at least 1oo hours. In case of the combustion chamber the material is Si-infiltrated SiC and in case of the nozzle rings injection moulded reaction bonded Si_3N_4.

In the rear a rotary heat exchanger disc made from AS. This component was developed in the United States ceramic program and has shown survival capability of several 1ooo hours at hot operating conditions.

The most difficult task, however, is the development of the mechanically and thermally highly loaded turbine rotors. Figure 4 shows monolithic rotors which are machined at Daimler Benz out of hot pressed Si_3N_4 billets. These wheels have been cold spintested at 79.ooo rpms. In the meantime one of the rotors has accumulated 115 hours hot test including 2o hours engine test. For these tests the speed was up to 6o.ooo rpms, the blade tip speed being 365 m/s and the temperature up to 15oo K.

Figure 5 shows the instant of blade failure of a rotor with a metallic hub and ceramic blades. Those rotor concepts are being investigated at VW and MTU. At VW we use blades made from injection moulded reaction bonded Si_3N_4 in the as fired condition with a layer between the blade root and the wheel. In cold spin proof tests we go up to a blade tip speed of 475 m/s. Blades made from hot pressed Si_3N_4 at MTU accumulated 2oo hours test with blade tip speeds of 35o m/s at 132o K and 75 hours with 1525 K.

4. Ceramic components for turbochargers

The exhaust gas turbocharger could be improved by the use of ceramic components, too. In particular, when there are frequent changes of load, as typical in passenger car operation, the response behaviour of today's turbochargers is still unsatisfactory. If the metallic rotor is replaced by a ceramic rotor because of its lower densitiy leading to a smaller mass moment of inertia, the response delay can be shortened. Apart from this, today's turbochargers are still too expensive, partly because of the high-temperature alloys necessary for the turbine rotor. The use of ceramic should bring improvements in both instances.

Figure 6 shows a prototype ceramic turbocharger turbine rotor. The material used was sintered SiC. This rotor has survived a 4o hour test rig loadcycle with speeds as high as 11o.ooo rpms at gas temperatures of 13oo K. Presently the rotor is tested in a turbocharged Diesel engine under actual driving conditions.

5. The use of ceramics in conventional engines

Finally we shall consider whether, and if so, where, ceramic materials could be used to advantage in the combustion chambers of conventional engines.

The advantageous use of ceramics is most easily envisaged in those applications where either the materials used at present are insufficiently resistant to heat and erosion or where expensive alloys are needed. Those components are prechambers for stratified charge engines as well as swirl chambers with connecting passage in indirect injection diesel engines and the bowl in piston crown in direct injection diesel engines. For these applications it might be possible, that more suitable but previously thermally non-viable designs could be developed where there is promise of more efficient combustion.

Let us now discuss whether a higher thermodynamic efficiency can be expected from the use of ceramics in conventional engines. The lower thermal conductivity of some ceramic materials leads to higher temperatures in the combustion

14

chamber wall. In spark ignition engines this will very soon lead to the danger of knocking combustion. For this reason, our further discussions on this subject will be confined to the Diesel engines.

Calculations indicate that a ceramic insulation of the combustion chamber of a naturally aspirated Diesel engine means no great benefit in terms of thermodynamics. It is, however, possible that the hotter combustion chamber walls could have an positive effect in terms of CO and HC emissions as well as in terms of smoke formation. It is also worth investigating how far the hotter walls could contribute to a shortening of ignition lag and thus to a reduction in combustion noise. It is also possible that the thermally insulated combustion chamber could have a positive effect in terms of cold start behaviour. Figure 7 shows an engine which has been specially prepared for the study of these very effects. The piston and cylinder head of this engine have been coated with ZrO_2 and the swirl chamber is made of $Al_2O_3-TiO_2$.

The use of ceramics in a Diesel engine decreases the heat transfer to the wall and to the cooling system and increases the amount of heat removed with the exhaust gas. If the heat of the exhaust gases can be utilized, for instance by means of a turbocharger, considerable thermodynamic improvement can be expected. Thermodynamically it would even be more suitable to insert a steam power device. Figure 8 shows a system of this kind. The exhaust gas turbo charger is followed by a steam power device whose turbine drives the drive shaft via a reduction gear. Computational analysis which has been published by the Cummins Engine Company, Inc., show that 61 % overall efficiency is attainable. This is far more than the 35 % which can be achieved with today's direct injection diesel engines.

6. Priorities for future work

As it has been shown, promising results have been obtained in the development of ceramic components. Nevertheless, considerable further progress has to be made before ceramic components advantageously can be put into production engines. According to my opinion the priorities of the future work have to be in the following areas:

Design of ceramic components:

Ceramic materials are brittle. Generally speaking, they cannot dissipate stress concentrations by plastic deformation. Therefore, the stresses resulting from mechanical and thermal loading and in particular stress concentrations resulting from notch effects, point load or steep local temperature gradients must be under control when designing the components. On one hand, more experience has to be accumulated by component tests. On the other hand test results must be utilized to improve existing simulation models of the components and comprehensive computing programs for determining 3-dimensional temperature and stress distributions. As an example Figure 9 shows the results of an analytical simulation of a ceramic turbocharger turbine rotor.

Connections between ceramic and metal components:

Ceramics and metals have very different thermal expansion coefficients. This makes it extremely difficult to design suitable connections between both materials. Figure 10 shows examples for mechanical connections, for glueing or brazing, as well as for casting metal around ceramic shells. Since all these methods are not really satisfactory so far, research for better solutions is necessary.

Scatter of material strength

The state of development reported so far has been related to individual pieces of hard ware. These pieces have been carefully selected and pretested before the load cycles. Ceramics are brittle materials whose strength is highly related to impurities and defects in the microstructure of the material and to production deviations in the component. One of the most important goals of future work is to reduce the scatter in the strength properties of the materials. This will raise the probability of survival for mass produced components.

The scatter in strength of a material is expressed in the Weibull modulus m. Smaller variations correspond to higher m-values. Figure 11 shows, how material strength and Weibull modulus influence the probability of failure of an axial turbine rotor. Today we think that such rotors can be built with material strength of 350 MN/m² on a Weibull factor m = 16. This would indicate we could expect that 199 from 200 mass produced rotors would survive the critical start conditions.

Thermal stability of ceramics:

The component tests and the survival probabilities mentioned so far are related to short time material properties. In order to investigate the long time properties of ceramics, more than 10.000 ceramic samples have been exposed for more than 1.000 hours to thermal cycles by Volkswagen researchers.

Figure 12 shows different development steps of reaction bonded

Si_3N_4. In the diagram on the left the materials degrade heavily with the time of exposure to hot air either of one or the other test temperature. The diagram on the right shows the results obtained from a material with considerable thermal stability after all applied cycles. However, much further work has to be done in this area.

7. Conclusion

Promising results have been accomplished in the development of ceramic components. Future work should concentrate on the design side on improving the stress simulation methods and on investigating the most suitable connection methods between the ceramics and the metallic surrounding. On the material side the scatter in strength properties inside and between components has to be reduced and the long time thermal stability to be improved. Progress in these areas will need long-term and steady step-by-step development. So far the work does not give a guarantee of success. If this work is successfull, however, there will be a new material available for great improvements in heat engines and this material will be available in almost unlimited quantities all over the world.

Properties\n\nMaterial	Bending Strength (at 1350 K)\n$N/mm^2$	Density\n$\times 10^3\,kg/m^3$	E-Modul (at 1260 K)\n$\times 10^3\,N/m^2$	Thermal Expansion α (300..1260K)\n$\times 10^{-6}\,m/mK$	Thermal Conductivity λ (at 1260 K)\n$W/mK$
Si_3N_4, hot pressed	800	3.2	290	3.1	18
Si_3N_4, reaction bonded	350	2.6	180	3.0	9
SiC , Si infiltrated	450	3.1	380	4.4	50
MAS	20	2.2	12	0.6	1
ZrO_2 with MgO/CaO	300	5.7	200	9.8	2.5
Al_2O_3-TiO_2	40	3.2	23	3.0	2
Inco 713 C	200	7.9	170	15.0	25

Fig. 1   Properties of Ceramic Materials

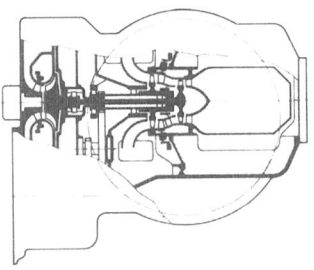

Fig. 2   High Temperature Gas Turbine

Fig. 3   Ceramic Components for a High Temperature Gas Turbine

Fig. 4   Hot Pressed Si_3N_4 Turbine Rotors

Fig. 5   Hybrid Turbine Rotor with Metallic Hub and Injection Moulded Si_3N_4 Blades.

Fracture Speed 58.000 rpm

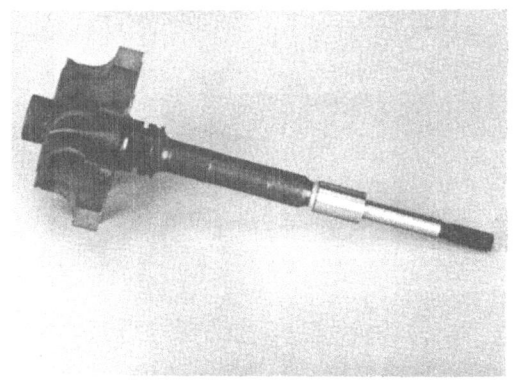

Fig. 6   SiC Turbo Charger Rotor

Fig. 7   Diesel Engine with Ceramic Insulation. Swirl
           Chamber in Al_2O_3-TiO_2,
           Cylinder Head and Piston Crown Coating in ZrO_2.

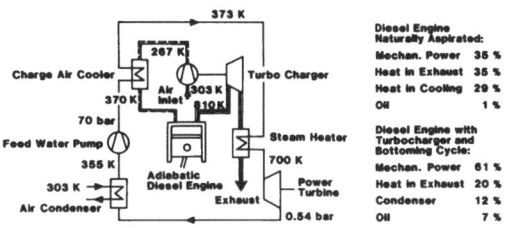

Fig. 8   Adiabatic Diesel Compound Engine

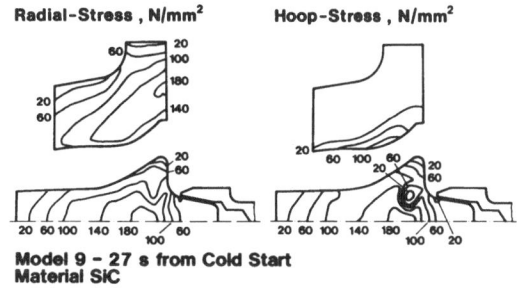

Fig. 9   Radial and Hoop Stress Distribution in a
Ceramic Radial Turbine

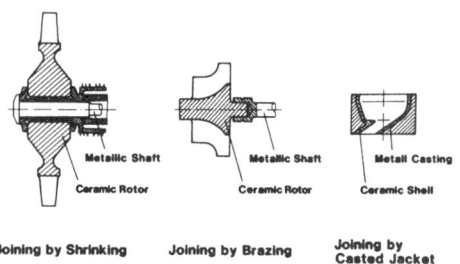

Fig. 1o   Joining Methods of Ceramic and Metal
Components

21

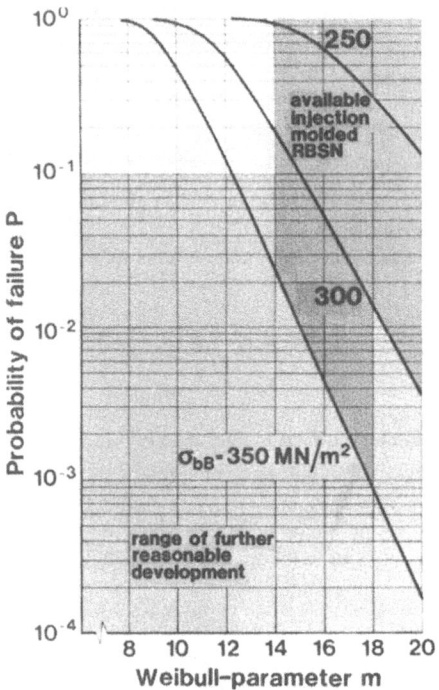

Fig. 11   Probability of Failure of a Reaction
           Bonded Axial Si_3N_4 Turbine Rotor

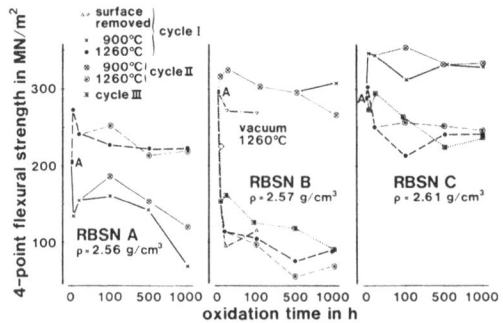

Fig. 12   Strength Behaviour of Reaction Bonded
           Si_3N_4 after Annealing

NON OXIDE CERAMICS

(Silicon Carbide, Silicon Nitride, Sialon)

E. Gugel

Annawerk Keramische Betriebe GmbH

D-8633 Roedental

SYNOPSIS

The non-oxide ceramic materials belong to the latest developments in the field of materials technology and this development has certainly not come to an end yet. They find increasing use in applications especially to solve materials problems. Because of their most interesting profile of mechanical, thermal and chemical properties we can consider them as an already settled new group of materials that stand at the beginning of a bright future.

Silicon carbide and silicon nitride ceramics are considered as candidate materials for energy production and energy conversion, especially in high temperature engines. They have already reached a distinct place in abrasive and chemically attacking environmental applications.

All this was possible only by intensive research and development work to understand their basic chemistry and microstructure. Since the technological aspects that include the powder preparation, different ceramic forming methods, sintering and machining have also been experienced and have advanced to a relatively high level, it is now possible to produce even complicated shapes in satisfactory homogeneity with a high degree of performance ability. All this needs further improvement to make possible a mass production at acceptable costs.

The different kinds of material - reaction-bonded, pressureless sintered, hot-pressed - will be characterized with regard to the present level of their properties, shape and dimension capabilities, as well as experiences in already realized applications mainly as wear and corrosion resisting
structural components.

In spite of this high level of development that is already
reached, more research is necessary to understand completely
the behaviour of these materials under the conditions they
are planned for. So among others, new methods of e.g. fracture mechanics, life time prediction are necessary for more
understanding and to guide the material development itself
towards the best composition for each application.

1. <u>INTRODUCTION</u>

The non-oxide ceramic materials do in fact belong to
the latest developments in the field of material technology - and this development has certainly not come
to an end yet. They are increasingly being used in
practice . One of the first achievements was the
use of reaction-sintered silicon nitride in aluminium
melting metallurgy [1], but very soon it was also used for
manufacturing machine components, particularly for the
seals of the rotary piston engine [2]. The important impulse to the development of these materials was given
by the desire to increase the working temperature of
vehicular engines, which leads to extensive work in several countries to realize a ceramic gasturbine [3] and
also a piston engine.

Much earlier than the now very often reported ceramic
demonstration piston engines in Japan, it was shown in
Great Britain with a small diesel engine [4] that ceramics
can work in heat engines.

For this most spectacular development intensive research
in the field of material evaluation, testing and technology was necessary and is still important. The results
are reported in the proceedings of numerous meetings and
governmental reports in the past ten years, the last
ones are mentioned in reference [5].

Though the history of the non-oxide materials is relatively short, the literature published up to now is
already difficult to overlook. So the references fill only
a small section in this paper.

The main applications today are not in the high temperature area but in the field of wear resistant parts, in
the seals- and valve technology. Besides there seems to
be almost no area in which no tests are on the way in
order to realize the interesting combination of properties
of the non-oxide ceramic products. The production has now
been developed so far that shapes in a wide scope of
dimensions can be provided and today existing materials limits
can be overcome.

2. MATERIAL

Non-oxide ceramics include those on the basis of the
compounds that the light elements boron, carbon, nitrogen and silicon form together. This group includes silicon carbide SiC, silicon nitride Si_3N_4, boron nitride
BN, boron carbide B_4C, the boron silicides SiB_4 and SiB_6
and, properly seen, also carbon in the form of graphite
and diamond (carbon carbide!).

These compounds share the characteristics of strongly
covalent bonding with short atomic distances - which
means high theoretical strength and hardness, low thermal
expansion and, advanced thermal and chemical stability
(without regard to their oxidation sensibility). The mechanical properties, however, are restricting their use because
of their incapability of relieving stress concentrations by
local yielding. They are brittle as all ceramics are and the
designer and engineer has to get acquainted with this
intrinsic feature.

Silicon nitride appears to have an astonishing ability
to form solid solutions with a variety of oxides. Of
most interest are the Si_3N_4-Al_2O_3-solid solutions, known
as sialons. The basic system has been thoroughly investigated
and the results show that apart form the sialon-ss $Si_{6-x}O_x$
$N_{8-x} (x = 4)$ there exist a series of other compounds. Besides even more compounds can be formed by introducing
other oxides like Y_2O_3, BeO, Li_2O, MgO to Si_3N_4 and/or
Al_2O_3 [6].

For high temperature application silicon carbide and silicon nitride are the candidate material because of their
resistance to oxidation. That is due to the formation of
a self-healing coating of silicon dioxide on the surface
of the compounds which are themselves thermodynamically
not resistant to oxidation. Continued oxidation is
controlled by oxygen diffusion [7], which shows that
the composition of the material (contaminations or sin-

tering additives) are of fundamental influence.

Due to this property together with some advantageous
thermal and mechanical properties, the interest in the
application of ceramics was focussed on materials based
on SiC and Si_3N_4, which already led to the development
of several very interesting engineering materials.

The sialons are less explored although in many institutes large-scale R & D activities are on the way. They
represent a potential for the future that can not yet
be judged. At present also the technical realisation of
the sialons is not as far advanced as that of SiC- and Si_3N_4-
materials.

3. MANUFACTURING

The stability of these crystals complicates sintering
by volume diffusion, the process ceramics are normally
subjected to. In order to be able to utilize the favourable properties of a component, other methods for manufacturing had to be developed, namely reaction sintering and hot-pressing.

Meanwhile also the pressureless sintering is so far advanced that it becomes an interesting and possible approach for commercial and reasonable production. Table 1
gives a survey of the different procedures to densify
silicon nitride and silicon carbide.

3.1 Reaction sintering

By this process the material is developed in the already
shaped part as a result of a chemical reaction between
the constituents. The shaping itself is done by any
method usual in ceramic processing, depending on dimensions, shape and quantity [8]. For first test parts isostatic pressing and machining in a prefired stage is recommended. For a medium number of pieces slip casting is
a good method, because tooling is relatively cheap and
a change in design causes only little problems. For more
or less mass production injection molding or automatic
dry-pressing is the answer.

3.1.1. Reaction sintered silicon nitride

Since the early work has been done in Great Britain [9]
and Germany [10], the base of the sintering procedure is

well-known and relatively well established for the production of commercial ware. A lot of research work has
been published which has increased the understanding to a
high level, especially the work in Great Britain [11] ,
in USA at Ford [12] , Airresearch and Norton and in Germany [14] should be mentioned here.

But nevertheless there are still problems in the technical scale production, especially of parts with bigger
wall thickness, higher density (and strength) and reproducible homogeneity in a reasonable short sintering time.
The main parameters controlling the process are listed
in table 2.

The key problem appears in controlling the exothermic
reaction, which makes necessary a relatively long sintering time. When silicon is converted to silicon nitride, an increase in weight and in volume will result.
The latter takes place overall within the
body into the porous area and consequently has practically no effect on the total volume [15] . This proves to
be a significant advantage of RBSN, i.e. its dimensional
accuracy referred to the raw body is much below $\pm$ 1 %.
Thus little or no machining is required. Certain additions or impurities such as iron accelerate the nitriding process or make it possible only with parts having
thicker walls. The material so produced is porous, but
because of the very fine grain structure of the silicon
nitride crystals in the pores, the pore size is remarkably small and is normally below 0.2 μm, which is essential for the oxidation behaviour. Usually densities are
ranging from 2.4 - 2.6 g/cm^3 [16] , but it is doubtful,
whether such a material can be produced homogeneously
and reproducibly in a mass production component.

The microstructure of RBSN (Fig. 1) consists of pores
and silicon nitride crystals in a needle-like, fine mat.
It is essential to note that the strength is governed
by the always existing single pores that appear in various quantities up to a size of 10 μm or even more. The
main task is therefore to minimize the quantity and size
of these pores in order to get a material with a high
strength quality.

3.1.2. Reaction sintered silicon carbide (SiSiC)

This is an already well established material with some
sales volume. It is commercially available for almost 30
years, starting in USA by a vapour infiltration process
[17] , later in Great Britain as well as in Germany[19] by a
liquid infiltration method [18].

The production of this material starts with a mixture of carbon, silicon carbide and a suitable plasticizer - depending on the forming process - which is shaped by any ceramic shaping method and then siliconized in a special furnace with molten or gaseous silicon.

This process can be controlled so that a porous, but pure silicon carbide body is obtained, or that its porous area is filled with non-reacted silicon (Fig. 2). This is possible because of the good wettability between the two materials, yet cannot be done with reaction sintered silicon nitride [20].

Since the silicon-containing, dense product is much more advantageous with regard to strength, it is the most-fabricated product. Here, too, the outer change of volume of the component is neglegible, but this can only be achieved by a co-ordination of the relative amount and grain size of primary silicon carbide, carbon or graphite, pore space as well as by controlling the temperature schedule in sintering because the siliconizing reaction is exothermic and also causes a volume increase in the body. The main parameters that influence the production of SiSiC are listed in Table 3.

The production of parts up to a wall thickness of about 20 mm and dimensions in rings to 450 mm and tubes up to 2 m is relatively well established. Problems exist with high volume parts with thicker walls (like gas turbine rotors), where cracks may appear.

Although it is basically possible to produce a microstructure with as low as 4 % silicon, it is not easy to achieve this in a technical scale production. Since for some applications a low silicon content is favoured, there is some demand to develop this. On the other hand a material with a higher silicon content - this means more than 10-12 % by weight - has a slightly higher strength level with lower scatter.

3.2 Hot Pressing

This process starts with silicon nitride, respectively silicon carbide-powder with special additives allowing densification up to practically theoretical density. Normally this is done between graphite dies axially pressed together. The limited flowing capability of the material gives an explanation for the fact that only comparatively simply shaped bodies can be manufactured.

Moreover it is remarkable that approaches have been developed to press HPSN-blanks of various wall thicknesses rather close to the finally required dimensions with a remarkable homogeneity, shown by a Weibull modulus of more than 25 [21]. A very interesting procedure is isostatic hot-pressing, where the pressing media is a gas, so that the pressure is applied uniformly from all sides [22]. Theoretically more complex shapes can be pressed, but only when the very serious question of sealing the porous green body towards the pressing fluid can really be solved, which is an extremely difficult research goal. It seems that the work by ASEA is highly advanced in this field [23].

The additional isostatic compression of already pre-sintered components seems less problematic. The value of that additional treatment is still quite unclear, as up to now the tests have not absolutely shown an improvement of the mechanical properties with this relatively expensive process.

3.2.1. Hot Pressed Silicon Nitride (HPSN)

A silicon nitride powder in the α-modification is necessary to obtain the needle-like linked microstructure (Fig. 3) by transformation to β-Si_3N_4 [24]. The additive – normally MgO or Y_2O_3 – reacts with the SiO_2 in the α-Si_3N_4-lattice as well as at the surface to form a liquid silicate glass, in which the α-Si_3N_4 dissolves and from which it reprecipitates as elongated β-Si_3N_4-grains. The so developed microstructure is the key for the unique mechanical properties of HPSN that has the extraordinary strength level of 700 N/mm^2 MOR and more at room temperature. In Table 4 the essential fields to understand HPSN and to produce a good material are shown.

With the aid of the high resolution electron microscope the grain boundry phase was intensively investigated for MgO-HPSN [25] as well as for Y_2O_3-HPSN [26], which leads to the expression of "grain boundary engineering" [27]. Furtheron the phase relations and stabilities are all-important in the development of the microstructure and thus for the mechanical properties [28]. Crystallisation can lead to an increase in the stability of the grain boundary phase [29].

The principles mentioned here are generally the same as for pressureless-sintered silicon nitride (SSN). And so this work and knowledge is of high value even when the HPSN will lose importance and will be replaced by SSN.

For the production of complicated shapes by hot-pressing, the approach of post hot-pressing of RBSN, either by axial [30] or by isostatic hot-pressing [31], has to be mentioned. The material must be doped with the necessary fluxing aids. The advantage is the much less shrinkage of only about 6 lin. %. This method can be favourable for the realisation of components with different densities, e.g. the duo density gas turbine rotor [32].

3.22 Hot-Pressed Silicon Carbide (HPSC)

Fine α-SiC powder can be densified by hot-pressing with a relatively small amount of additives like Al, B, Fe and compounds of them. Though some success could be achieved in developing a good strength, especially with HPSC, where no degradation can be determined up to 1600°C [33], hot-pressed silicon carbide has rather little impact in application, since it has been experienced that hot-pressed silicon nitride is more suitable in most cases, which is caused by its lower thermal expansion and its lower Young's modulus; so it is somewhat tougher.

3.3 Pressureless Sintering

Pressureless sintering of covalent compounds to full density is extremely difficult, even with the use of additives, especially when they dissociate rather than melt at normal pressures, such as SiC and Si_3N_4. At temperatures, where the atomic mobility is high enough, decomposition is already a severe problem. So it can be understood that the sintered non-oxide ceramics belong to the youngest materials of this group.

3.31 (Pressureless) Sintered Silicon Nitride (SSN)

The main work for the development of this material was done only in the last five years in USA [34,35], probably in Japan and also in Germany [36].

For the pressureless densification of silicon nitride a higher amount of oxide additions of 10 % and more is necessary. So the result is more a composite material but with interesting properties that offers basically the chance to replace the more sophisticated hot-pressed silicon nitride, as far as the high temperature strength is satisfactory [37]. The sintering itself causes some difficulties because of the low decomposition temperature in relation to the necessary sintering temperature of 1800°C (Table 4).

Especially for the purpose of satisfying the demand for good high temperature properties the amount and kind of the oxide phase is of extreme importance. On the one hand it should make possible the sintering of Si_3N_4 at the lowest possible temperature and on the other hand it should realize good mechanical strength at high temperatures. This is difficult to achieve and therefore it is not surprising that numerous approaches have been made with different additives, e.g.: MgO, Y_2O_3, Al_2O_3, Fe_2O_3, ThO_2, ZrO_2, CeO_2, Cr_2O_3, BeO, $ZrSiO_4$, $MgAl_2O_4$, TiO_2, SiO_2, rare earth oxides, separate or in mixtures, sintered under gas pressure or/and in a powder bed in different time schedules. The grain size distribution, morphology, crystallinity of the silicon nitride powder and additives are all-important to achieve an optimum microstructure.

The oxygen content of silicon nitride powder which is mainly responsible for the forming of the second phase in sintering, is difficult to control. So it should be considered to handle the powder without exposing it to the air, as it is done with special metal powders.

So the work already starts in preparing the best powder. Normally a powder is used that is prepared by nitriding elemental silicon [38]. A cheaper way of production would be by reduction of silicon dioxyde by carbon and nitridation in the same process, but this powder has the disadvantage that it is difficult to achieve the desired low concentration of carbon and oxygen.

The pyrolysis of a silicon- and nitrogen containing gas phase has to be mentioned too. This leads to the purest and finest-grained powder, but comprises other problems in mixing and homogeneous densification of the batch; besides this is a very expensive process. This cannot be considered as basic process for technical scope production.

The shaping of sintered products can again be done by any ceramic forming method, using temporary binders, so the shaping capability is not limited as with hot-pressing (prior to grinding). A disadvantage is certainly the high shrinkage - linearly up to 20 % - which decreases the accuracy of net shaped components by warping, especially in comparison with the reaction-sintered products.

Nevertheless good results could be achieved by slip casting [39] and injection molding [40], the two most promising methods for complex shapes and/or mass production.

Considering mass production especially of bigger-sized components the handling of the fine sintering powder will cause some problems. Aggregations appear very easily, which leads to pores, inhomogeneities in density and last not least to cracks. New ways of powder preparation like filtration or freeze drying have to be explored.

A very interesting approach to SSN is the production and post-sintering of a RBSN-part, doped with an appropriate sintering aid [41]. This process combines the advantages of the well established forming technology of RBSN with a relatively low shrinkage of lin.~ 6 % to the final dense product. The sintering takes place at about 1800°C in nitrogen atmosphere. The problems arising are similar to those with the sintering of silicon nitride powder compounds (Table 4) and also reaction sintering of silicon nitride (Table 2).

3.32 (Pressureless) Sintered Silicon Carbide (SSiC)

It was not before 1973 that silicon carbide could be pressureless sintered for the first time, using submicron powder with simultaneous addition of boron and carbon [42]. It is remarkable that amounts of additives of only a few percent are sufficient for a complete densification, which is, however, influenced by a range of parameters like it is shown in Table 5. An extremely fine powder is necessary that can be of the α - or of the β -modification. The sintering may be governed by energy ratios, but is certainly basically also a solution - reprecipitation process by a very thin, glassy phase between the SiC grains [43].

The result of this sintering process is a material with a homogeneous, fine microstructure (Fig. 4). Its strength level of about 400 N/mm^2 MOR is not yet satisfying. It is remarkably low for a material with such a fine microstructure and almost no second phase. Thus defects like pores or inclusions seem to make up a major part and further development work is necessary in this respect.

3.4 Sialons

To complete the survey of material processing, the sialons have to be mentioned. This is a special class of
materials comprising solid solutions of metal oxides -
originally Al_2O_3 and later also MgO, Y_2O_3, BeO and others
have been used - in the β-Si_3N_4 crystal structure [44].
They can be hot-pressed or pressureless sintered. Although known for ten years they have no importance in
applications up to now. Only recently some success was
reported as cutting tool tips [45], which is an Y_2O_3-Al_2O_3-
Si_3N_4-material.

The pressureless sintering of sialons is one of the few
examples of ceramic alloying. For the first time here
a way is shown how a special material with excellent properties out of the group of sintered silicon nitride materials can be produced with regard to the exact adherence to the necessary production parameters and knowledge of
the phase diagrams. For further progress it is necessary to define exactly the required parameters for
reproducible production and to let this idea of alloying
establish itself in the general field of ceramics. The
fundamental research work is done in several countries,
like Great Britain, France, USA and Germany.

3.5. Recrystallized Silicon Carbide

For the sake of completeness the recrystallized silicon
carbide has to be mentioned. Here a silicon carbide compact is fired at temperatures so high that even a certain evaporation-condensation takes place that leads to
a recrystallisation possible by differences of the surface energies and herewith to a porous product with a
moderate strength level, which is used as superior kiln
furniture in the china ware industry.

3.6 Machining

Machining by means of diamond tools to the final shape
is normally necessary for the dense products, especially
HPSN. The procedures have to be developed as all other
technologies for the non-oxide ceramics [46]. The significance of grinding, however, also lies in shaping the
surface in such a way that optimum mechanical properties
are achieved. The direction of the machining procedure
in relation to the direction of the load in application
has an essential effect on the service value of the component being used [47]. Should this not be considered, a
loss of strength of as much as 40 % can result.

Thus the optimum surface treatment of any component has to be specially considered. The method of free abrasive grinding might be successful for the treatment of curved surfaces [21].

4. PROPERTIES

In Table 6 a short survey of some properties is shown to characterize the just discussed materials. Here the mechanical strength is of main interest. It shows realistic values for pilot and technical scale production. Sometimes in laboratories much higher strength values have been achieved, but a long and hard development work will be necessary to realize a component with reliable reproduction possibilities.

For the mechanical testing a statistical base is necessary to measure the failure probability. So the meanwhile well-known Weibull statistic is a very valuable approach to the demand of the designer. Thus the designing with brittle material [48] is another key for the success of non-oxide ceramics in the field of engineering.

In Table 6 no data is given for sialons, because there is no known material that can be technically realized, maybe with the exception of the cutting tool tips [45]. Their flexural strength can be between that of RBSN and SSN.

The strength behaviour at high temperatures is of main interest. Here the existence of a second oxide phase is of most influence and weakens the material. Furtheron oxidation can be very important in influencing the strength behaviour. Subcritical crack growth appears [49], which governs the life time. It is obvious that this behaviour depends on the microstructure, which is the result of numerous parameters. Silicon nitride is especially sensitive, because of its second phase, i.e. glass or oxides in HPSN and SSN and pores in RBSN.

In parallel with the material development, testing techniques have to be developed as well in order to get the approval for application in technical engineering.

The significance of non-oxide ceramics lies in the combination of their favourable properties. Together with their manufacturing potential, they present a most important factor in the material sector.

34

It should be mentioned here that every year a report is published in which the very interesting results of intensive measurements of all important properties of commercially available non-oxide materials are reported [50].

5.   APPLICATION

Practically every process in the modern technique is limited by the properties of the used materials. Therefore the appearance of a new material - or even a group of materials - leads always to an increasing optimism to be able to solve existing material problems.

The first step to reach this aim is to develop a material with good appropriate properties. In order to achieve a technical progress, the material has to be realized in a product that is working in a system and that can be manufactured in a reproducible and economical way. The application side is another step of the development which has to be done by mechanical engineers together with material specialists.

For any application the problems of attaching a ceramic part to the metallic system have to be solved, concerning the mismatch of properties, especially thermal expansion that hinders material joining. Failures of ceramic components in an engineering equipment or in a test rig have mostly been the result of attachment problems. This remains a critical development area for any application including basic research for joining technologies.

All these aspects imply not only the necessary technical application tests, but also adapted material developments as well as possibly special control techniques, i.e. a non-destructive testing [51,52].

These fields can hardly be separated and therefore it seems reasonable to mention some typical applications and to describe their stage of development.

5.1 Application in Metallurgy

RBSN has excellent resistance to aluminum and other light metals as well as to copper and its alloys. Therefore it may be used for melting transport equipments, stirrers, crucibles, pump components, valves and nozzles for foundry equipment, as well as for ascending pipes for low-pressure casting and for thermocouple protection tube.

A severe obstacle is the fragility of the material. This field is by far not yet fully explored, which is due to the application development as well as to the basic understanding of chemical attack.

5.2 Application as Drawing Tool

Hardness, abrasion resistance, exact dimensioning and surface quality are the most significant properties for the use of hot-pressed silicon nitride for tools to draw tubes and wires. Especially its high abrasion resistance together with the surface quality remaining unaffected by this are responsible for the fact that, compared with other drawing materials, the tolerance of the drawn body can be maintained even after an unusually prolonged drawing period, with the quality of the surface being notably good.

The production of 40 km steel tubes in one hot-pressed silicon nitride nozzle and of 10,000 stainless steel bushes in a tool out of the same material with extremely low wear shows [53] the potential and the reason to develop this further.

5.3 Ball Bearings

A potential field of application is that of bearings, which can only break through if mass production could lead to reasonable prices.

Modern processing of dense ceramic materials allows the manufacturing of complete ball bearings out of dense silicon nitride with an accuracy equal to that reached with metals. They are suitable for temperatures up to approx. 700°C in oxidizing atmosphere and for operation under extremely corrosive conditions and without lubricants. The low specific weight combined with the high compressive strength of the material allow high speeds. Life has proved to be longer than that of any comparable steel bearings [54]. Contrary to expectations, failure is not caused by catastrophic fracture, but by formation of pits. The behaviour of ceramic balls thus damaged is judged favourably for further use. Hot-pressed silicon nitride ball bearings are in service without lubrication in aggressive liquids and vapors at approx. 300°C and 2000 RPM for three years now without failures, whereas any other bearings had to be removed after several hours or days.

5.4 Valve Components

A typical application for HPSN are components for high pressure valves in which aggressive and corrosive fluids are controlled at high pressures and temperatures. This is of special interest in the coal gasification and liquefaction field. Such high performance valves are applied for instance in the chemical process technique for pressure release in one or more steps. On this occasion the pressure energy of the medium in the slot between the seat and the plug is transformed into kinetic energy. Solid particles carried along with the medium are strongly accellerated on to constricted areas causing considerable erosion. The abrasive wear caused, for example, by grains of sand is increased in most cases by the simultaneous chemical attack of the aggressive medium; at the same time corrosion occurs. In some extreme cases pressure shocks caused by vapour bubbles lead to cavitation.

HPSN-valve plugs and valve seats (Fig. 5) have led to life times that have never been achieved before. The application of these valve components improves the functioning of the valve and increases its durability. Enormous reduction in shut down time of the plant and low cost of repair lead to a cost reduction in large plants.

5.5 Mechanical Seal Rings

A most interesting field of application is the utilization for face seal rings and packing rings. The improved corrosion resistance compared with hard metal and better resistance to thermal shocks in comparison with aluminum oxide distinguishes silicon carbide/silicon as a preferred material. Fig. 6 shows only some of the shapes and dimensions used.

In power plant seals, for instance, the change-over to silicon carbide rings led to a considerable improvement, as this material has a higher chemical resistance, a far higher erosion resistance, better anti-seizure properties and also a higher thermal stress resistance factor and it is most wear-resistant. On the basis of the high wear resistance of the couple silicon carbide/ antimony-impregnated hard carbon, pump seals of pipelines today are produced with only one seal.

Furthermore seal rings made of silicon carbide are predominantly used in high duty seals for drilling and percussion equipment, for power plant seals and seals in coal refinement and in oil production. They have also been proved for centrifugal pumps for chemical industries and acid pumps.

5.6 Brazing Fixtures

Low thermal conductivity together with high resistance to thermal shocks and chemical as well as erosive resistance and the possibility to manufacture complex shapes, distinguish reaction-sintered silicon nitride – in some cases hot-pressed silicon nitride – as an excellent material for supports, placing fixtures, spacers, mountings in numerous different shapes and dimensions for soldering and other thermal treatments. The high electric resistivity offers the additional advantage that the material does not heat up in case of inductive heating.

5.7 High-Temperature Testing Equipment

Especially the range of high-temperature materials requires the possibility of testing mechanical properties at high temperatures. The linkage, punches, supports, etc. are components exposed to high mechanical load and made of non-oxide ceramics, especially reaction-sintered silicon nitride and hot-pressed or pressureless-sintered silicon carbide. They show clear advantages as compared to oxide ceramics that hitherto were the only material available for this purpose.

5.8 Cutting Tools

Silicon nitride used as cutting bar for plastics proved to be a material with long life time. This seems to be an open field for more, similar possibilities. For cutting tips first very promising results have been achieved in machining gray iron and super-alloys with remarkably high cutting speed. It appears that the composition of the material seems to be very sensitive for this application, so more investigations are necessary.

5.9 Automotive Engine Components

The most spectacular potential application of these materials is in the field of energy conversion, mainly in automotive engines.

For more than 40 years designers were looking for ceramic materials, but only since the non-oxide class of ceramic materials have been developed in the last decade, there is a realistic chance to reach the goal.

A second prerequisite is responsible for this: The now developed computer capability to calculate the stresses in a component with the degree of refinement required in brittle materials design.

A large number of technical demonstration programmes are on the way which should encourage to use ceramics as engineering material. The results are very promising. The basical applicability in heat engines has been demonstrated at operating temperatures that are about $300^{\circ}C$ higher than are possible with uncooled super-alloys.

The most spectacular appearance of vehicular engines on the roads that are partly or almost fully made of ceramic, is a major step that may possibly lead to a "ceramification" of heat engines step by step in the course of the next decades.

5.10 Other Applications

There are many other potential applications for non-oxide ceramics. One of them should be mentioned here, because of its importance in energy saving: Heat exchangers are in development for house heating burners [55] and for the receiver in solar power plants [56], made of RBSN, respectively SiSiC.

6. CONCLUSION

The high-performance non-oxide ceramics have shown their feasibility in many applications. They solved material problems that up to now could hardly be overcome.

For future development the question appears to what degree of reliability ceramic components can be realized in an affordable mass production. Net shaping, rather than machining the sintered parts has to be applied and the experience with mass-production of oxide ceramics gives an optimistic outlook.

Finally the importance of non-destructive testing has to be pointed out, which is a further key for structural ceramics application in order to insure that only reliable components are put in use.

Ceramic engineering components are on the way to penetrate in almost every area of technology to resist better to the specific environmental conditions. To achieve this, a lot of problems have still to be solved, and it will certainly take time and effort. But it seems sure that the ceramic penetration cannot be stopped any more and ceramics will be used in applications unthinkable before.

TABLE 1 Manufacturing Possibilities of Silicon Nitride and Silicon Carbide Ceramics (Additives/Temperature.

Process \ Material	Silicon Nitride	Silicon Carbide
REACTIONSINTERING	(+ Fe, Fe_2O_3) 1300 - 1500 ^oC	(with Si-surplus) 1400 - 1700 ^oC
HOT PRESSING	+ MgO, Y_2O_3,... 1700 - 1800 ^oC	+ Al_2O_3, B 2000 - 2100 ^oC
PRESSURELESS SINTERING	+ Al_2O_3, AlN $\longrightarrow$ Sialon + Y_2O_3, Al_2O_3, CeO 1700 - 1900 ^oC	+ B, C 2100 - 2200 ^oC
RECRYSTALLIZING	———	2300 - 2500 ^oC

TABLE 2 Main Factors for Production of Reaction-Bonded Silicon Nitride (RBSN).

ORIGIN: (Chemical/physical background)	EFFECT: (Product/Properties)	INFLUENCES: (Raw Material/Technology)
Reaction kinetics	Microstructure	Grain size
Exothermic heat	Density/Porosity	Contaminations
Volume increase	Pore size distribution	Additives
Si $\rightarrow$ Si_3N_4	Homogeneity	Shaping method
	Local inhomogeneities	Green density
	Residual silicon	Wall thickness
	Unreacted cores	Sintering technology
	Single great pores	Reaction gas
	Strength / Oxidation	Additives
	Dimensional stability	Flow rate

41

TABLE 3 Main Factors for Production of Reaction-Sintered Silicon Carbide (SiSiC).

ORIGIN: (Chemical/physical background)	EFFECT: (Product/Properties)	INFLUENCES: (Raw Material/Technology)
Volume increase	Microstructure	Silicon composition
$Si+C \longrightarrow SiC$	Silicon content	Carbon source
$Si_l \longrightarrow Si_s$	Dimensional stability	Relation of SiC, C and pores
Oriented crystal growth	Homogeneity	Green density
Cappilarity	Internal stresses	Pore size distribution
Exothermic heat	Cracks, streaks	Shaping method
	Non-reacted cones	Wall thickness
		Sintering technology

TABLE 4 Main Factors for Production of Hot-Pressed and Sintered Silicon Nitride (HPSN/SSN).

ORIGIN: (Chemical/physical background)	EFFECT: (Product/Properties)	INFLUENCES: (Raw Material/Technology)
Crystal growth	Microstructure	Additives
Solution-reprecipi-tation	Strength $<$ room temp. / high temp.	Contaminations Oxygen
$\alpha - \beta - Si_3N_4$ trans-formation	(Dimensional stability, shrinkage)	Grain size
Phase equilibria	Homogeneity	Crystallinity
(Decomposition)	Local inhomogeneities: Pores Inclusions Aggregations	$\alpha/\beta - Si_3N_4$-ratio
Devitrification		Technology Temperature Pressure Time
		Post heat treatment
		Size of compact
		(Shaping method)

TABLE 5   Main Factors for Production of Pressureless Sintered Silicon Carbide (SSiC).

ORIGIN: (Chemical/physical background)	EFFECT: (Product/Properties)	INFLUENCES: (Raw Material/Technology)

Kinetics of crystal growth:
  Equiaxially
  elongated
 	
Sintering mechanism:
  Diffusion
  Surface energy
  Grain boundry
        energy
	Microstructure
	Shrinkage
	Dimensional stability
	Homogeneity
	Local inhomogeneities:
	      Pores
	      Cracks
	      Inclusions
	Strength
		SiC-Modification
		Grain size
		Contaminations
		Sintering aids
		Shaping method
		Green density
		Sintering technology
		Atmosphere

TABLE 6   Characteristic Properties of Silicon Nitride and Silicon Carbide Ceramics.

PRODUCT N = Silicon Nitride C = Silicon Carbide	Bulk Density	Porosity	Modulus of Rupture (4-point)			Young's Modulus	Thermal Expansion
			20 °C	1000 °C	1400 °C	20 °C	20-1400 °C
	g/cm^3	%	MN/m^2	MN/m^2	MN/m^2	$10^3 MN/m^2$	$10^{-6} K^{-1}$
N - Reaction Bonded (1)	1,9 - 2,2	25 - 40	150	150	150	120	3,0
N - Reaction Bonded (2)	2,4 - 2,7	15 - 25	250	250	300	150	3,0
N - Pressureless Sintered	3,3	< 2	600	550	300	310	3,3
N - Hot Pressed	3,2	0	700	650	400	320	3,2
C - Reaction Bonded	2,7	16	250	250	250	280	4,5
C - Silicon Impregnated	3,1	0	350	400	200	360	4,3
C - Recrystallized	2,6	20	100	100	100	200	4,5
C - Pressureless Sintered	3,15	< 2	400	450	400	400	4,4
C - Hot Pressed	3,2	0	550	550	450	420	4,5

(1) = Axial dry pressed; extruded
(2) = Isopressed, slip cast; injection and transfer molded

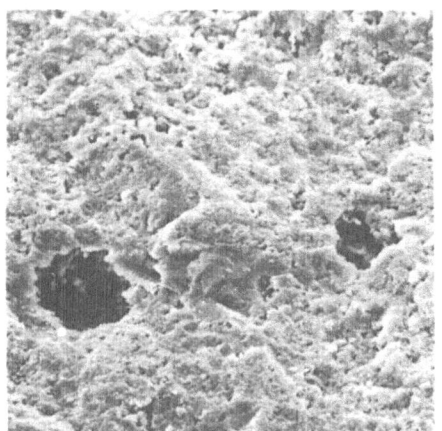

Fig. 1: Microstructure (SEM) of RBSN. ⊢——⊣ = 10 μm

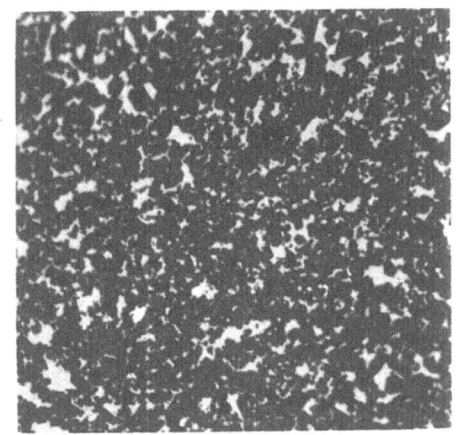

Fig. 2: Microstructure of SiSiC. ⊢——⊣ = 25 μm

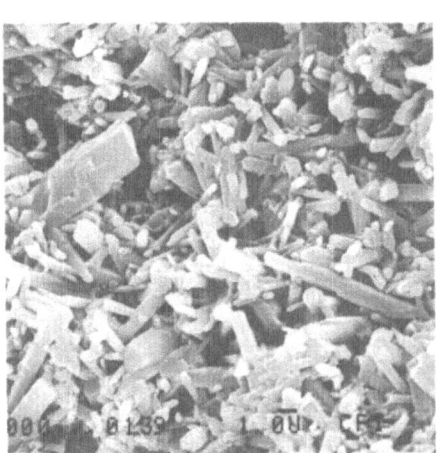

Fig. 3: Microstructure of MgO-HPSN. ⊢——⊣ = 4 μm

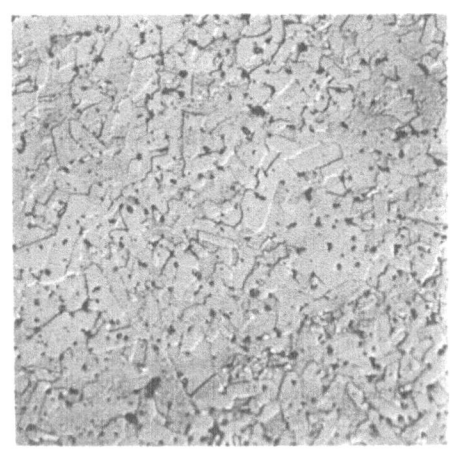

Fig. 4: Microstructure of SSiC. ⊢——⊣ = 25 μm

Fig. 5: Valve Components Made
of Hot-Pressed Silicon Nitride
and Reaction-Sintered Silicon
Carbide.

Fig. 6: Mechanical Seal Rings
Made of Reaction-Sintered Silicon Carbide.

R E F E R E N C E S

1   H. Feld: Sprechsaal 102, 1969, Nr. 24, 1098-1101.

2   N. N.: Auto, Motor und Zubehör, 1972, Nr. 24.

3   E. Gugel, G. Leimer: Ber. Dt. Keram. Ges. 50, 1973, Nr. 5, 151-155.

4   D. J. Godfrey, P. G. Taylor: Eng. Mat. Design, 1969, Sept., 1339-1342.

5   a) 5th International Symposium on Automotive Propulsion Systems, 1980.

    b) 18th Automotive Technology Development Contractors Coordination Meeting, 1980.

    c) AGARD Conference Proceeding No. 276, 1980.

    d) Nitrogen Ceramics I and II, 1977/1982, F. L. Riley.

    e) W. Bunk, B. Böhmer: Keramische Komponenten für Fahrzeug-Gasturbinen I und II, 1978/1981.

    f) Papers of ASME, Gasturbine Conference, every year.

    g) Ceramics for High Performance Applications III, Reliability, 1979, Orcas Island.

6   Gauckler, G. Petzow: Representation of Multicomponent Silicon Nitride Based Systems, 41-62; 1977, Leyden, Holland, F. L. Riley Noordhoff Intern. Publ..

7   E. Gugel, H. W. Hennicke, P. Schuster: Ber. Dt. Keram. Ges. 46, 1969, 481-485.

8   E. Gugel: AGARD Conf. Proceedings, Advanced Fabrication Processes, 256, 1978, 14-1/14-16.

9   N. L. Parr: Si_3N_4 - A new ceramic material for high temperature engineering and other applications, Research, 1960, 13, 261-269.

10 H. Feld, E. Gugel, H. G. Nitzsche: Werkstoffe und Korrosion, 1969, 20, 571-574.

11 a) A. Atkinson, A. J. Moulson, E. W. Roberts: Journ. Am. Cer. Soc., 1976, 59, 7/8, 285-289.

    b) H. J. Merkin, G. S. Hughes, C. McGreavy: Journal Mat. Sci., 1980, 15, 9, 2345-2353.

12   J. A. Mangels: Nitrogen Ceramics, 569-576; 1977, Leyden, Holland, F. L. Riley Noordhoff Intern. Publ..

13   J. A. Mangels: Journal of Materials Science, 1980, 15, 8, 2132-2135.

14   E. Gugel, G. Leimer: Ber. Dt. Keram. Ges., 1977, 54, 4, 116-117.

15   E. Gugel, N. Hauck, O. W. Flörke: Ber. Dt. Keram. Ges., 1979, 56, 1, 10-14.

16   J. A. Mangels: Automotive Propulsion Systems, 1980, 5, 500-514.

17   K. M. Taylor: Materials and Methods, 1956, 44, 92-95.

18   C. W. Forrest, P. Kennedy, J. V. Shennan: Special Ceramics, 1972, 5, 99.

19   E. Gugel, G. Leimer, H. Colert, G. Gratwohl, F. Thümmler: Statusseminar: Keramische Komponenten für Fahrzeug-Gasturbinen II, 1981, 473-498.

20   G. Leimer, E. Gugel: Zeitschrift für Metallkunde, 1975, 66, H. 10, 570-576.

21   H. Kessel: Ind.-Diam.-Rundschau, 1981, 15, 4, 226-233.

22   M. Böhmer, J. Heinrich: BMFT-Statusseminar, 1981, 517-534.

23   H. T. Larber: AGARD Conf. Proc., 1980, 276, 18/1-4.

24   E. Gugel, A. Fickel, H. Kessel: Powder Metallurgy International, 1974, 6, 3, 136-140.

25   D. R. Clarke, G. Thomas: Journal Am. Ceram. Soc., 1977, 60, 491-495.

26   D. R. Clarke, G. Thomas: Journal Am Ceram. Soc., 1978, 61, 114-118.

27   R. N. Katz, G. E. Gazza: Materials Science Res., 1978, 11, 547-560.

28   F. F. Lange: Nitrogen Ceramics, 491-509; 1977, F.L. Riley.

29   J. E. Weston, P. L. Pratt: Journal Materials Science, 1978, 13, 2147-2156.

30   E. Gugel, H. Kessel: Ceramics for High Performance, Applications-II, Proceedings of the 5th Army Materials Technology Conf., Newport, Rhode Isl., 1977, March, 515-526.

31   J. Heinrich, M. Böhmer: Science of Ceramics 11, 1981, 439-440.

32   E. Gugel, H. Kessel, N. Müller, Lange: Herstellung von Siliziumnitrid-Verbundbauteilen für die Gasturbine, BMFT-Statusseminar I, 1978, 141-160.

33   K. Hunold, W. Grellner: BMFT-Statusseminar II, 1981, 499-516.

34   R. N. Katz, G. E. Gazza: 5th International Symposium on Automotive Propulsion Systems, 1980, 470-481.

35   Th. Smith, Q. L. Quackenbush, V. Nehring: 5th Intern. Symp. on Autom. Prop. Systems, 1980, 482-499.

36   H. Hausner, H. Landfernmann, G. Wötting, E. Gugel, G. Leimer: BMFT-Statusseminar II, 1981, 265-296.

37   D. Steinmann, E. Gugel: Science of Ceramics 11, 1981, 113-124.

38   G. Schwier: BMFT-Statusseminar II, 1981, 255-264.

39   A. Novotny, E. Gugel, G. Leimer: Schlickergegossene Gasturbinen-Bauteile aus Siliziumnitrid und Siliziumkarbid, BMFT-Statusseminar I, 1978, 162-182.

40   J. T. Neil, K. W. French, C. L. Quackenbush, J. T. Smith: ASME-Paper, 1982, 82-GT-252.

41   A. Giachello, P. Popper: Ceramurgia International 5, 1979, 110-114.

42   S. Prochaska: General Electric Report, 1973, 73CRD325.

43   E. Gugel, G. Leimer: AGARD, Conf. Proc. 276, 1979.

44   K. M. Jack, W. J. Wilson: Nature Physics Science (London), 1972, 238, July, 28-29.

45   K. M. Jack: Science of Ceramics II, 1981, 125-142.

46   H. Kessel: Industrie-Diamanten-Rundschau, 1976, 3, 128-133.

47   H. Kessel, E. Gugel: Industrie-Diamanten-Rundschau 12, 1978, 3.

48   J. J. Mueller, A. S. Kobayashi, W. D. Scott: Design with brittle material, Textbook University of Washington, 1979.

49   D. Steinmann: Science of Ceramics, 1980, 10, 637-644.

50   D. C. Larsen: Property Screening and Evaluation of Ceramic Turbine Engine Materials, Techn. Rep. AFML-TR

51   K. Goebbels, H. Reiter: Zerstörungsfreie Prüfung Keramischer Bauteile, Science of Ceramics 11, 1981, 361-374.

52   A. G. Evans: Non-destructive failure prediction in ceramics. AGARD Conf. Proc., 1980, 276, 18/1-4.

53   K. Haarscheidt, Seminar Forschungsgesellschaft Umformtechnik, 1981/82.

54   H. K. Lorösch, J. Vay, R. Weigand, E. Gugel, H. Kessel: Fatigue strength of silicon nitride for high-speed rolling bearings, Trans. ASME, 1980, 102, 128-131.

55   St. R. Schindler, A. Krauth: Beitrag der Keramik zur besseren Energieausnutzung, CFI-Ber. DKG, 1981, 58, 75-84.

56   W. Gehrke, M. Kuczera, G. Willmann: Heat Exchanger, Science of Ceramics 11, 1981, 71-78.

DISCUSSION

A.J. Burggraaf: Which of the two materials families (Si_3N_4 or SiC) will become the dominant one in applications where thermal conductivity does not play a role and where mechanical properties are most important?

E. Gugel: Which material should be used for a special application is not only dependent on the properties, but also of the shaping possibilities. This means, in what degree can the desired component be produced in a reliable and affordable way? This is the one side.

The other side is, that at today's state of the art at lower temperatures dense Si_3N_4 has some advantages regarding to toughness, but at higher temperatures – say above $1200^\circ C$ – dense silicon carbide has to be considered better in long term strength. Therefore we are glad to have both materials for mechanical applications, and one will not beat the other.

K.H. Lim: As far as I know the problem of SiC, in order to be able to be applied in coal gasification, is its bonding agent, which is quite susceptible to the attack of the gaseous mixture in the reactor. Do you think there is any new development or breakthrough in this field?

E. Gugel: If silicon carbide itself stands the chemical attack in coal gasification, the only answer is to use "self bonded" products. If a porosity is allowed, recrystallised silicon carbide, or maybe silicon nitride bonded silicon carbide, can be considered; otherwise sintered silicon carbide is certainly the best but most expensive answer.

P. Popper: Somebody asked whether Si_3N_4 or SiC were preferable for certain engine applications from the thermal stress resistance point of view because of their different thermal expansions and conductivities. I asked the same question at a gas turbine conference and the answer I got may be of interest. It was said that a variety of components were investigated theoretically by computer and in all of them Si_3N_4 came out better because of its lower modulus of elasticity.

E. Gugel: Yes, the low Young's modulus is the big advantage for silicon nitride.

TRANSFORMATION-TOUGHENED CERAMICS (TTC)

Nils Claussen

Max-Planck-Institut für Metallforschung

7000 Stuttgart 80, F.R.G.

SYNOPSIS

The paper defines transformation toughening of ceramics in a rather broad sense: "Toughening" encompasses all types of mechanical-property enhancement which can be achieved by utilizing the tetragonal-to-monoclinic phase transformation of zirconia particles dispersed in a ceramic matrix. Hence, the necessary conditions are outlined to improve the fracture toughness, the strength and the thermal shock properties as well as to enable stable crack growth. A schematic grouping of the great variety of TTC is presented based on characteristic microstructural features; the three main groups being a) partially-stabilized zirconia ceramics, b) dispersed zirconia ceramics and c) complex zirconia systems. This grouping allows a survey-type discussion of the state-of-the-art in science and technology, the deficiences in research and development and the future trends of TTC. It is reported on applications already commercially available and presently being tested in pilot runs.

INTRODUCTION

Brittle ceramic materials can be toughened and strengthened by utilizing the tetragonal-to-monoclinic phase transformation of ZrO_2 particles dispersed or precipitated in a ceramic matrix[1-8]. The toughening originates from the volume and shape change associated with the transformation.*) Even though this fact has been well recognized, the exact mechanisms, i.e., the micro-mechanics of the toughness increase, still remain under active discussion. This is especially due to the fact that at least two types of toughness enhancement have been found; in one case, the martensitic transformation of ZrO_2 particles near the advancing crack tip is directly involved in the energy absorption (stress-induced transformation) and, in the other case, nucleation and extension of matrix microcracks which are due to particles transformed on cooling before the specimen is loaded, are responsible for increased energy absorption during crack propagation (microcrack nucleation and extension).

Transformation toughening, which was previously thought to be a phenomenon inherent only in partially stabilized ZrO_2 (PSZ) in which tetragonal particles are coherently precipitated within the cubic-stabilized matrix, has been shown to be also applicable to Al_2O_3 and other ceramic matrices in which ZrO_2 can be incorporated. Furthermore, fine-grained, fully tetragonal ZrO_2 has been developed with extremely high toughness. Another important aspect of transformation toughening is the generation of compressive surface stresses (surface toughening) resulting in considerable strength increases. Although all of the work on TTC has been centered on ZrO_2 or HfO_2-alloyed ZrO_2 as "toughening agent", it is likely that phase transformations in other materials, not necessarily martensitic in nature, are suitable for toughening of ceramics.

The objective of the present paper is to review the state-of-the-art in science and technology of TTC and to point out the deficiencies and the trends in this new field of ceramics. Like with many comparable developments, theory and model experiments are rather advanced, while technology and field tests are lacking, especially information on high-temperature strength as well as on the static, dynamic and thermal fatigue behavior. However, fracture mechanical studies at high temperature on single crystalline partially-stabilized ZrO_2 give rise to an optimistic

---

*) In the following, the definition of TTC should also include the rare cases where the direct transformation from cubic to monoclinic symmetry also leads to some sort of toughening, e.g., transformation of the surface region of cubic stabilized zirconia by a certain heat-treatment or the complete conversion of cubic to fine-grained monoclinic material by sub-eutectoid aging, etc.

picture as far as toughening in the temperature region of the stable tetragonal form is concerned[9].

For basic understanding, the following chapter will briefly discuss the transformation of ZrO_2 particles constrained in a ceramic matrix and the consequent toughening. For a more detailed treatment of this aspect, recent review papers are available[10-13].

2. TRANSFORMATION AND TOUGHENING

The martensitic start temperature (M_s: start of the tetragonal monoclinic transformation on cooling) of single crystals and poly-crystalline bulk ZrO_2 occurs between 950° and 850°C. ZrO_2 particles embedded in ceramic matrices are usually retained to lower temperatures due to the constraint of the transformational volume expansion and shape change by the surrounding matrix. External application of stress and/or relief of this constraint by matrix microcracking triggers the transformation to the stable monoclinic symmetry. In some cases, an intermediate orthorhombic phase has been detected[14] which may be always associated with the tetragonal → monoclinic transformation.

A convenient way of representing and characterizing the transformation behavior of ZrO_2 particles in a dispersion composite is by a dilatation-temperature curve, as shown schematically in the upper left of Fig. 1. One important magnitude which is related to the toughening mechanisms is the M_s temperature because it determines which type of toughening may be dominant. M_s depends in a very complex way on various parameters such as particle size, chemical composition, shape, site (within the grains or along grain boundaries), elastic properties, thermal expansion coefficients, etc.[15,16] In composites with a ZrO_2 particle size distribution, M_s, relates only to the martensite start temperature of a few, usually the largest particles in a composite. Correspondingly, the completion of the transformation is associated with the smallest particles. However, in most cases the smallest fraction is retained in the tetragonal symmetry to temperatures far below room temperature, often to 0 K, making M_f difficult to define. The reverse transformation starting at A_s has been found to be much less influenced by the various parameters, although a tendency similar to that for M_s exists, e.g., A_s decreases slightly with smaller particle size. Additions such as HfO_2 which increase the chemical driving force can drastically increase A_s[17]. As for M_s, A_s refers to only one group of particles, probably the smallest particles.

Toughening, as used in the following, will be understood in a rather broad sense. As indicated in the right hand side of Fig. 1 it implies not only an increase in the critical stress intensity factor, K_{Ic}, i.e., an increase in critical stress or strain (hatched area), but also improvement of the breaking load as well as the resistance to mechanical and thermal shock. This, under certain circumstances, requires the material to undergo stable crack growth (dashed curve in Fig. 1). Although, K_{Ic} usually parallels these other mechanical properties, the broad interpretation of toughening must be stressed because the ZrO_2 transformation offers the opportunity to improve the properties according to the applicational need. For instance, resistance to severe thermal shock[18,19] or extreme strength under bending conditions do not necessarely require optimum K_{Ic} values, but rather ability for intensive crack branching or high residual compressive stresses in the surface region[15]. Hence, TTC can be tailored to various applications, with the operative factor often being the relationship between M_s and the application temperature.

Transformation toughening addresses at least two mechanisms, stress-induced transformation and nucleation and extension of microcracks. It is yet unclear whether and to which extent these mechanisms are additive or interacting. Stress-induced transformation is possible for matrix/particle systems in which M_s is below the test temperature, i.e., where the particles are retained in tetragonal symmetry. Tensile or shear stress, when reaching a critical level, triggers the transformation to the monoclinic form. Transformation in a zone ahead of a crack tip causes the material to "yield", i.e., to dissipate elastic strain energy. Such "crack blunting" allows further loading until K_{Ic} is reached. Another form of toughening by stress-induced transformation is achieved by introducing compressive surface stresses, as is state-of-the-art in glass technology. Surface transformation can be obtained by grinding or sand-blasting,[19-21] processes which usually cause strength degradation in conventional ceramics. The transformation of tetragonal ZrO_2 particles in spinel and mullite, as induced by in situ TEM straining,[22,23] is shown in Fig. 2.

Microcrack nucleation adjacent to transformed monoclinic particles may cause a K_{Ic} increase by crack-tip energy dissipation during the process of microcrack extension[24,25]. In this case, the ZrO_2 particles, which may be monoclinic or transform in the stress field, essentially provide a higher energy density (more microcrack nuclei sites) than ZrO_2-free ceramics would exhibit. Controlled microcracking can be achieved by controlling the monoclinic particle size. Large particles, for example, can be utilized to induce crack branching, an important factor in strength retention under severe thermal shock conditions[26].

The type of toughening depends on the ceramic system and the specific microstructure and can, to a certain extent, be designed to meet to demands.

3. CHARACTERIZATION OF TTC BY THE MICROSTRUCTURE

Research and development is steadily increasing the variety of TTC types. In order to present an overview on the different types, microstructural features are used to classify the materials. The schematic microstructures in Fig. 3 are grouped in three classes: A) Ceramics based on partially-stabilized zirconia (PSZ); B) dispersed zirconia-containing ceramics and C) complex zirconia systems encompassing all other composites not fitting into either group A or B. In all cases ZrO_2 can partially or fully be substituted with HfO_2. The microstructural classification (Fig. 3) is not thought to be complete, but seems to be more useful than other systems, e.g., based on materials, applications etc., to characterize the various properties, the potential applications, the technological state, etc. Most of the materials have only been prepared on a very small research scale, some are more advanced and have found commercial applications, some are being studied in pilot tests, and others just indicate future possibilities. Most of the microstructural types presented in Fig. 3 and their combinations can be prepared by different techniques. Some, however, are more typical for particular preparation methods.

3.1. Ceramics Based On Partially-Stabilized Zirconia (PSZ)

A1:   Conventional zirconia ceramics, partially stabilized by addition of MgO, CaO, Y_2O_3 or rare-earth oxides are usually sintered in the cubic solid solution range, i.e., at temperatures between 1700 and 1900°C [27,28]. Upon cooling, the microstructure contains large (50-100 μm) cubic grains with coherent precipitates of tetragonal symmetry. The precipitates are then coarsened by aging at temperatures between 1300 and 1500°C in order to optimize their transformability under stress, i.e., the stress field of the approaching crack front must be able to transform large numbers of tetragonal particles in order to optimize the toughness of the two phase material. Sintering and aging can also be combined in a single-step heat-treatment in the cubic/tetragonal phase region yielding similar microstructures and properties.

A2:   Fine-grained PSZ ceramics, also consisting of cubic grains with tetragonal precipitates, have so far found little attention, mainly because of the technological difficulties. The preparation of fine-grained (< 10 μm) PSZ should, however, be feasible through more refined powder technology. For instance, starting with highly active PSZ powders prepared by evaporation-decomposition of solutions (EDS), water atomization [29], solid-solution annealing and re-grinding, etc., would allow the densification at temperatures (1300-1500°C) where grain growth is rather limited while allowing the optimization of the tetragonal precipitate size. Furthermore, grain growth inhibition, e.g., by Al_2O_3 additions, or HIPping could be used to achieve the same goal. The great advantage of fine-grained PSZ would lie in improved strength properties, especially after introducing compressive surface stresses.

A3: Tetragonal Y-PSZ is a fine-grained, single-phase material in which the retention of the tetragonal form is due to the constraint imposed by neighboring grains on one another and Y_2O_3 doping[5,30]. It can be thought of as a material which only consists of tetragonal precipitates. The critical grain size is between 0.1 - 1 µm depending on density and amount of Y_2O_3 content (Fig. 4). Similar preparation conditions prevail as in A2 materials, though far less critical and sintering can be carried out in the tetragonal single phase field (1.5 to 3.5 mole% Y_2O_3) at reasonable temperatures ($\sim 1450°C$). Small additions of Al_2O_3 help in limiting the grain growth[37]. This type can presently be considered the toughest and strongest of all PSZ, probably of all polycrystalline ceramics made so far. The reasons are twofold: The extremely fine grain size is far below the critical flaw size, and the "active", transformable phase may reach ~ 100 %.

A4: Fine-grained, monoclinic zirconia can be prepared by hydrothermal synthesis[32], the technique, however, is rather limited for the making of useful parts. In a more recent technique[33], coarse-grained cubic Mg-stabilized (~ 15 mole%) material is decomposed by aging at temperatures < 1100°C. The resulting microstructure consists of 1-5 µm monoclinic grains with highly dispersed MgO-rich pipes. The decomposition may not involve a transitional tetragonal phase. The extremely high thermal-expansion anisotropy causes enhanced grain-boundary microcracking and crack branching. Even though, the strength is relatively low, the thermal-shock properties are good, which is, to some extent, also due to the reduced Young's modulus, hardness and effective coefficient of thermal expansion compared to other PSZ types.

A5: Overaged conventional Mg-PSZ is essentially a three-phase material where at least part of the tetragonal precipitates are transformed to monoclinic symmetry by aging at temperatures >1100°C[34]. At this low temperature little precipitate coarsening takes place, however, the tendency to loose coherency increases with annealing time. Furthermore, monoclinic material is formed at the cubic grain boundaries. This complex microstructure, offering a variety of modifications between the A1 and A4 types, combines various toughening mechanisms.

A6: Single crystalline PSZ provides the opportunity to utilize the intrinsic properties of PSZ ceramics without the limitations encountered in polycrystals, i.e., without the influence of grain-boundary phases, flaws, etc. Y-PSZ single crystals (up to ~ 60 mm length) have been prepared by skull melting with similar tetragonal precipitates as in Y-PSZ (A1)[9]. The growth of Mg- and Ca-PSZ single crystals seems to be technologically more difficult due to fast overaging of the less stable precipitates. Typical precipitate morphologies of Ca-, Mg- and Y-PSZ single crystals are shown in Fig. 5. The great advantage of single crystal PSZ lies in the high strength ($\sim$1400 MPa). Even at temperatures of 1500°C, values of up to $\sim$700 MPa have been measured[9]. Hot forging of these materials offers the possibility of shaping parts without loosing the good properties inherent to the single crystal material.

3.2. Ceramics Containing Dispersed Zirconia

This group contains all systems in which zirconia particles are dispersed in a ceramic matrix other than ZrO_2.

B1: The intercrystalline dispersion of tetragonal ZrO_2 particles in a fine-grained matrix usually leads to very high strengths compared to the matrix alone. This is due to an increased fracture toughness (K_{Ic}) and to grinding-induced compressive surface stresses. The latter case is especially effective for high Young's modulus materials (e.g., Al_2O_3). The intercrystalline dispersion normally results from homogenizing techniques, i.e. fine grinding and mixing, applied to fine-grained ceramics[3,21]. The retention of all ZrO_2 particles in the tetragonal form is, however, difficult to achieve unless great care is taken to keep a narrow size distribution. For ZrO_2 volume fractions >0.15 small amounts of stabilizers (1-2 mole% Y_2O_3) are usually required to retain most particles in tetragonal symmetry at room temperature[30].

B2: Fine-grained ceramics with an intercrystalline dispersion of monoclinic (at RT) particles are made in the same way as B1 and with a slightly larger ZrO_2 particle size (or without stabilizer at higher volume fraction)[25]. Ceramics with this microstructure exhibit lower strengths than the B1-type, however, the toughness (K_{Ic}) may be higher. Their advantage lies in the ability of microcrack-induced stable crack growth (R-curve behavior), and in improved thermal shock resistance under severe conditions.

B3: The intercrystalline dispersion of both tetragonal and monoclinic particles is actually the most common microstructure of the dispersion-type TTC (Fig. 6). This type is usually obtained through conventional ceramic mixing techniques. As the microstructure so are the properties intermediate between B1 and B2.

B4: Intracrystalline tetragonal ZrO_2 particles can, mainly for geometrical reasons, be best dispersed in coarse-grained ceramics. This type of microstructure has so far only been prepared for the TEM-analysis of the transformation behavior of ZrO_2 in ceramic matrices[22,35] so that the anisotropy effects associated with the grain-boundary location are excluded. It is yet unclear whether the B4 type offers any advantages over B1 to B3, especially since intragranular particles are far less sensitive to stress-induced transformation[16], i.e., they are more stable and may therefore contribute only little to the toughening. This material is an analog of the A1 PSZ material and may be expected to exhibit useful properties only when most features of A1 PSZ are met.

B5: A medium-grain size ceramic with intracrystalline tetragonal particles and monoclinic particles at the grain boundaries should exhibit comparable properties to A5 PSZ. Grinding-induced surface compressive stresses would cause high initial strength and enhanced intercrystalline microcracking may create crack branching. B5-type microstructures (Al_2O_3-ZrO_2) have been obtained through sol-gel techniques. The possitive effect of surface grinding, however, has not yet been confirmed[36].

B6: In situ-reacted microstructures seem to be characteristic for the manufacturing process[37-39]. They range somewhat between the B-type microstructures and the more complex zirconia systems (C-type), however, the technological simplicity based on the dispersion of zircon or other Zr-compounds with oxide reactants, may justify this grouping. The typical feature of this microstructure are rounded ZrO_2 particles which are randomly located within the grains and at grain boundaries (Fig. 7). The zirconia phase (monoclinic or tetragonal) present at RT is determined by the size (e.g., the critical size in mullite is ~ 1 µm) which again can be controlled by the sintering and reaction conditions[40,41].

3.3. Complex Zirconia Systems

Most of the microstructures presented in this group are in a very early state of investigation, some may never be successfully realized. It is obvious that the following microstructure types only represent examples of many more to be developed in the future.

C1: A typical microstructure of mixed TTC systems can clearly not be described in a schematic form. Evidently, the variety would be quite large. However, for the present purpose, the C1 type of mixed systems is defined to represent any microstructure which may be obtained by simple combination of A or B types. For instance, a combination of A3 and B1 has been prepared by mixing high volume fractions of zirconia ($\div \sim 2$ mole% Y_2O_3) with alumina[30]; the resulting properties are correspondingly intermediate, i.e., the material exhibits a higher Young's modulus and hardness than A3 and a higher fracture toughness than B1.

C2: The analog of PSZ type materials (A1) based on ceramic matrices other than zirconia is a highly attractive goal because of the nearly ideal distribution of the tetragonal (or monoclinic) ZrO_2 phase. This microstructure, in which the ZrO_2 "precipitates" may be optimized by aging as in A1, has not yet been realized. Efforts are being made, for instance, by the use of rapidly quenched near-eutectic Al_2O_3-ZrO_2 compositions which, in the form of amorphous powders, may be densified and crystallized to the desired microstructure (cf Fig. 8)[29].

C3: The crystallization of tetragonal (or monoclinic) ZrO_2 in a glass or glass ceramic may be a useful tool to improve the toughness of this class of materials. It may be technologically even more feasable than C2 microstructures. A first step towards this type has been indicated by plasma dissociation of zircon: Small ($\sim$10 nm) tetragonal and larger monoclinic ZrO_2 particles have crystallized in the SiO_2- glass matrix during cooling after the dissociation process[42].

C4: This microstructure is obtained by adding ZrO_2 to a ceramic matrix in which a third phase can be precipitated. ZrO_2 in either monoclinic or tetragonal form would function as low-temperature ($T<A_s$) toughening agent, and the (transgranular) precipitates act in the conventional way of precipitation strengthening at high temperatures ($T>A_s$). An example is Al_2O_3-rich spinel with dispersed tetragonal ZrO_2 particles which, after annealing at $T > 1000°C$, contains Al_2O_3 precipitates without having changed the morphology of the ZrO_2 particles[43].

C5: The duplex microstructure consists of a high-strength matrix (e.g., B1 type) with relatively large ($\sim$100 μm), spherical inclusions consisting of a material containing a high volume fraction of monoclinic particles (high M_s). The intent of this microstructure is to combine high resistance for crack initiation with the ability for crack branching, i.e., increasing resistance with crack propagation. This behavior is shown in Fig. 9 in an idealized diagram. Although very few experiments have been performed with this type of microstructure, it seems to offer a good chance for the development of thermal and mechanical shock resistant materials[44,45].

C6: Directionally solidified eutectics involving ZrO_2 as one phase have been prepared at a time when the effect of transformation toughening was not yet discovered. Since these aligned eutectic microstructures have been shown to exhibit good high-temperature strength properties, it seems to be worthwhile to investigate the possible addition of ZrO_2 precipitates with analogies to the $\gamma/\gamma'-\alpha$ superalloys.

4. EXPERIMENTAL FACTS AND THEIR INTERPRETATION

Great differences in the mechanical behavior exists between the various TTC types, especially between PSZ and the B- and C-type ceramics. This is partly associated with the inherent system properties. However, controlled microstructural design within the frame of the various groups allows for rather specific, application-oriented properties (see also Tables 1 and 2).

Some general experimental facts yet will be presented using typical data. The interpretation provided is usually not yet, however, generally accepted. Further difficulties arise from the fact that practically all variables have interacting effects.

4.1. Size-Dependency of the Transformation Temperature (M_s)

In all systems studied so far it has been found that M_s decreases with decreasing ZrO_2 particle size. Since M_s determines the type of toughening or whether toughening is possible at all, it is of great technological interest to precisely control the precipitate or particle size. This is usually done by alteration of the sintering and aging schedule (A1, A2, A5) and milling duration (B1-B3), etc. Furthermore, additions of stabilizers will shift M_s to lower temperature for a given particle size.

It is convenient to define a critical size, d_c, above which, at room temperature, spontaneous transformation will occur and below which will not. Table 3 gives d_c values for some systems, mostly obtained in TEM.

There are essentially four theoretical explanations for the size effect*):

a) As in metallic systems where similar critical size effects are attributed to nucleation, the size dependency of M_s is considered in terms of different propensities for nucleating the transformation in large and small particles. Experimental evidence suggests that the nucleation argument is the most important[76]. One indication is the fact that round transgranular ZrO_2 particles in an Al_2O_3 matrix are far more stable than irregularly-shaped (faceted), intergranular particles; the facet edges probably act as sites for nucleation. The stability of ZrO_2 particles in an Al_2O_3 matrix increased with annealing time at 1500°C, indicating that embryo-like defects at the interface or within the particles were annealed out.

---

*) Early explanation[48] for the critical size in unconstrained particles, based upon the difference in surface energy between monoclinic and tetragonal phases, seem inadequate to explain the differences encountered in constrained systems.

b) The solubility of impurities may be size dependent; for this case increased solute content with decreasing particle size would clearly decrease M_s. EELS and EDAX measurements have shown that Al_2O_3 is soluble to a few percent in ZrO_2 particles [46], whereas the Al_2O_3 solubility in bulk ZrO_2 is usually taken to be significantly less. Hence, the chemical argument cannot be discarded.

c) In twinned ZrO_2 particles long-range shear stresses may cancel when they are opposed in neighboring twins or variants. In this case, only short-range stresses which are localized at the particle-matrix interface, will remain, and hence will scale as the particle area (d^2). Since all other energy terms vary with volume (d^3), the expected size dependency of M_s will result [47].

d) Another argument [30] postulates that microcracking and/or twinning will relieve some of the matrix constraints, thus the strain energy associated with the transformation. In this case, the size dependency is thought to arise either from the energy per unit volume of transformed material associated with the crack surface, or the energy of the twin surface per unit volume of transformed material.

4.2. Temperature Dependence of the Stress-Induced Transformation

The magnitude of the toughening is clearly associated with the chemical driving force for the transformation (ΔF_o). Since ΔF_o decreases with temperature (becomes more negative), the toughness would also be expected to decrease; this has been confirmed in most experiments. Fig. 10 gives K_{Ic} versus temperature for various C1-type compositions [30] and Fig. 11 the strength for a single-crystalline material (A6) and A1 PSZ. It is interesting to note that the A6 material retains high strength to $\sim 1500°C$; this is interpreted as a direct interaction of the crack with the tetragonal precipitates (dispersion hardening) [9].

4.3. Optimum Amount of Stabilizer

The chemical driving force decreases with solute amount (especially stabilizing elements) which implies that, for PSZ and high-volume fraction ZrO_2 ceramics, an optimum addition, for which the retained volume fraction and the transformability are maxima, must exist. The position of this optimum depends very much on the type of microstructure, especially the homogeneity of the dispersion, i.e., with increasing size uniformity, the optimum K_{Ic} is shifted towards lower solute contents. Fig. 12 shows K_{Ic} of PSZ ceramics versus mole fraction of stabilizer [6,9,30,49] and Fig. 13 K_{Ic} of C1-type Al_2O_3-ZrO_2 compositions made by two different sol-gel routes [50]. It is obvious

that for some TTC types the technological development must be directed towards lowering the amount of solutes down to zero, e.g., it is
to be expected that ultra-fine grained A3 types may be fabricated
without any Y_2O_3 addition as predicted by theory at grain sizes
< 0.03 μm (Fig. 14)[57].

4.4. Optimum Volume Fraction of ZrO_2

All theoretical derivations for the fracture toughness predict increasing K_{Ic} with increasing volume fraction of the dispersed ZrO_2
phase. Since, however, all microstructures, except for A3, impose
a geometrical, homogeneity-dependent limit on the volume fraction,
an optimum develops, as is shown for Al_2O_3-ZrO_2 composites[7] in
Fig. 15. The same is true for A1 PSZ where the optimum volume fraction of tetragonal precipitates results from a combined optimization of stabilizer fraction and aging conditions.

4.5. Strengthening by Surface Grinding

The grinding-induced transformation of tetragonal ZrO_2 introduces
residual compressive stresses which can considerably increase the
bend strength of TTC. The magnitude of this surface strengthening,
which is state-of-the-art in glass technology, depends mainly on
the Young's modulus and the grain size[3,57]. Therefore, high Young's
modulus B1 materials (e.g., Al_2O_3-ZrO_2) or the fine-grained A3 PSZ
exhibit much greater strengthening than, for instance, A1 PSZ. The
influence of increasing grinding roughness, manifested by an increasing fraction of transformed ZrO_2 in the surface region, is illustrated in Fig. 16 for a B1 Al_2O_3. Although rough grinding
introduces deep surface flaws, the transformation depth (proportional to the compressive layer) obviously extends further. Some
transformational profiles measured in different TTC are shown in
Fig. 17[53-55]. From the curves it becomes evident that reducing the
transformability by stabilizing additives (e.g. Y_2O_3 in material ①)
decreases the transformation depth. In Fig. 18 the transformability
of the dispersed ZrO_2 particles is varied with the attrition-mill-
ing time, i.e., prolonged milling time reduces the particle size.[15,56]
The maximum strength (as-ground surface) at 10-12 h coincides
with the maximum in K_{Ic} and transformation zone depth.

Although grinding-induced transformation is comparable to the transformation in the crack tip stress field, the exact mechanisms have
hardly been investigated yet. There are a number of possible mechanisms which may be simultaneously active: a) transformation is
triggered by the high shear strains caused by the grinding process;
b) a high density of parallel flaws, induced by the abrasive media,
creates similar stress-field transformation as prevailing at single
sharp crack tips; c) the dislocations produced in the process pile

up at particles and consequently cause their transformation. From
various experiments with B1 ceramics using different matrices it
becomes obvious that surface grinding results in a much larger
transformed depth than that due to crack propagation.

4.6. Improvement of Thermal Shock Resistance

Appropriate selection of ceramics for the many different forms of
thermal shock is usually rather complex. However, the variety of
TTC microstructures (cf. Fig. 3) provides a unique opportunity to
design a suitable type. In a recent paper the various possibilities
of ZrO_2 toughening for improving thermal shock resistance of ceramics have been outlined[18]. In addition to the toughening described
in Chapter 2., the effective thermal coefficient of expansion can
be reduced by using ZrO_2 particles with a wide size distribution
(thus a wide M_s-temperature range), by changing the elastic behavior through introduction of a high density of spontaneous microcracks and by the occurence of transient compressive stresses
which oppose the thermal tensile stresses. Principally there are
two different directions of microstructural developments: a) pre-
vention of initiation of thermal stress failure and b) arresting
propagating cracks in thermal stress fields. Generally, the high-
strength types (e.g., A1-A3, B1, C1) are suitable for case a)
and those where microcracking can be induced (e.g., A4, A5, B3,
B5 and especially C5) are suitable for case b). This has been confirmed in very recent investigations of the thermal-shock behavior
of A5-type PSZ[34].

5. APPLICATION

The field of potential application for TTC   seems to be huge, because, for most mechanical engineers, the thought of using ceramics
instead of metals is still somewhat unusual, i.e., the menace of
brittleness still outweights the many specific advantages like corrosion and wear resistance, high-temperature strength, heat insulation and abundance of raw materials. Especially in the area of
mechanical and chemical engineering, the possibilities are numerous.
So far TTC, mainly PSZ types, are being used where there was a definite need, i.e., where metals were actually unable to fulfill the
requirements. Since very little has been done to design for high-
temperature applications, the TTC parts already commercially available are made for low or medium temperature applications. Some are
in the prototype stage, still others are in the experimental stage.
A few examples are listed in the following.

5.1. Commercial

Cold forming tools for all types of metals and alloys have been
produced for a number of years, for applications such as calibra-
ting dies, wire drawing cones (over 500 mm in diameter), etc. These
are mainly of the A1 type PSZ where toughness, combined with low
friction, results in long tool lives and high surface quality of the
formed metals[2]. Because of high oxygen conductivity, corrosion resistance and good thermal shock properties, A3 types are utilized
in automotive oxygen sensors. Al_2O_3-based cutting-tool inserts
having B3 microstructures are used for steel and cast iron machining; high fracture toughness is needed especially for interrupted
cutting.

5.2. Prototype

A variety of low-temperature engine components exposed to severe
service conditions are being tested in cars. There are cam fol-
lowers, push rods, connecting-rod pins, cams etc.[57], usually of
A1 or Al_2O_3-B1 types. Exaust compression brake bearings[57] or other
dry bearings for the mining industry[58], mainly made of A1 PSZ, show
excellent field test result. The same is true for warm extrusion
dies (A1)[33,59] which last several 100 extrusions without redressing
the surface. Pouring nozzles for continuous casting made of B2
zircon ($ZrSiO_4$ + m-ZrO_2) exhibit much better wear and thermal
shock resistance than zircon alone[58].

5.3. Experimental

Most engine components (e.g. cylinder liners, piston caps, hot
plates) in contact with combustion are still in an experimental
stage[60]. The A1 and A3 type PSZ tested so far have not shown sufficient thermal stability for long life. However, more stable PSZ
types and TTC based on other ceramics are presently being developed.
A2 PSZ prostheses seem to show good results in animal tests[58], i.e.,
they are biocompatible but tougher than the pure alumina parts.
Further recent developments are A3 tape cutters, fish knives,
scissors, golf club heads, etc.

6. PROBLEMS AND R+D REQUIREMENTS

The main problems, and consequently the most important requirements
for future testing and development, concern the high-temperature use
of TTC. Since the toughening mechanism of stress-induced transformation is only available when retained tetragonal can be transformed
to monoclinic ZrO_2, the microstructure must provide long-term thermal

stability at the maximum application temperatures. The system-in-
herent disadvantage of transformation toughening lies in the fact
that it is only useful at $T < A_s$. Since, however, most critical
stress situations occur at low to medium temperatures (start-up,
up and down shock, etc.) this handicap is reduced. Furthermore, A_s
can be increased by hafnia additons, and increasing plasticity with
incrasing temperature may further contribute to the toughness.
Some of the major problems and little-researched areas associated
with TTC are outlined in the following sections.

6.1. Microstructural Development

No attempts have been made so far to design TTC microstructures
which provide strength and toughness at temperatures in the
tetragonal stability range. The high-temperature strength results
obtained with PSZ single crystals (A6) indicate that the tetragonal precipitates cause a kind of dispersion hardening (cf. Fig. 11)[61].
Polycrystalline PSZ types with "clean" grain boundaries or with
dispersions preventing grain-boundary sliding should be developed
to avoid the typical strength break-down of PSZ.

High-temperature strength development in the B- and C-type TTC
should follow the general requirements for ceramics, i.e., preven-
tion of grain-boundary sliding (increased grain size), precipitation strengthening, etc. An example may be ZrO_2-toughened spinel
with corundum precipitates in the spinel grains. Also fibre re-
inforcement (e.g. SiC) should be included in the analysis.

6.2. Thermal Stability

The long-term thermal stability of PSZ seems to be little investigated and may give rise to major problems. Due to the high oxygen mobility, decomposition in many PSZ systems may occur even at moderate
temperatures. Fine-grained A3 type PSZ, for instance, looses its high
initial strength (~1000 MPa) after annealing for 500 h at only 300°C
(Fig. 19)[62]. Similar degradation of an A3-type cubic-tetragonal mixtures was found when annealed between 200 and 300°C; the presence of
water vapor may enhance the decomposition process[63]. A further indication of thermal instability is the fact that fully Mg-stabilized
PSZ decomposes to fine-grained monoclinic PSZ (A4) by annealing at
temperatures > 900°C [33]. Thus, extended investigations must be carried
out to find the exact causes and to look for ways to overcome this
severe handicap.

6.3. Slow Crack Growth

Static, dynamic and thermal fatigue properties of most TTC are practically unknown except for some low-temperature data with A1 Mg-PSZ. This material exhibits a power-law crack extension at $K_I < 4.6$ MPa$\sqrt{m}$.[64]

At higher stress concentrations, a region II plateau is supposed to be caused by stress-induced tetragonal$\rightarrow$monoclinic transformation, a feature which seems to be absent at low K_I-values. Careful examination of the fracture surfaces, however, indicates that, at low K_I values, transformation of tetragonal precipitates into an orthorhombic phase[14], which can be considered as intermediate phase between tetragonal and monoclinic, had occured.

The crack-growth behavior of a C1-type Al_2O_3-ZrO_2 composite is shown in Fig. 20[50]. The lower n value indicates an increased "ductility" with a general shift to higher K values compared to the Al_2O_3 alone. Preliminary measurements with B1 Al_2O_3-based material, however, indicated very high n values[65], hence general trends cannot yet be indicated. Major efforts must therefore be undertaken to examine the subcritical crack-growth behavior of TTC, especially at elevated temperatures.

6.4. Test Techniques

Most fracture mechanics test techniques have to be carefully analysed when applied to TTC. Both stress-induced transformation and microcrack nucleation and extension may yield unexpected and controversial results. Sawing or grinding induced compressive stresses strongly influence K_{Ic}-measurements, e.g., the maximum K_{Ic} value of a B2 Al_2O_3 at 15 vol.% ZrO_2 (Fig. 21)[66] is drastically overestimated in an NB test when samples have not been annealed at $T > A_f$. ISB (indentation-strength in bending) tests (Fig. 22) exhibit deviation from the normal 1/3-power law when unannealed (as-ground) surfaces are used[66]. Unusually high K_{Ic} data may be recorded with B2 Al_2O_3 when samples are measured directly after cooling from 1275°C (Fig. 23)[25]; in this case microcracks may have just nucleated, however, not yet extended. With increasing time at the test temperature (e.g. RT), the measured values drop to a steady-state value indicating that microcracks have subcritically propagated, hence "used up" potential energy sinks. ICL (indentation crack length) tests seem to be very difficult to evaluate with most TTC.

Inability to distinguish between cubic and tetragonal phases using X-ray peaks has also caused some difficulty in the literature. These problems may be overcome by using Raman microprobe spectroscopy[67].

6.5. Theoretical Aspects

The problem associated with theoretical interpretations of transformation and of toughening mechanisms are numerous, however, they will not be touched in the present paper. It is referred to the respective literature [10] and the upcoming Second Conference on the Science and Technology of Zirconia in June 1983 in Stuttgart.

7. FUTURE TRENDS

One of the major future objects is to tailor TTC microstructures to fit each specific application and to carry out actual component testing. PSZ types may be limited to low and medium temperature uses unless ways are found to overcome thermal instability problems. New, more perfect TTC types may be developed through crystallizing techniques, possibly based on classical ceramic compositons. One safe conclusion can be drawn for the future: The new field of TTC will further expand and a variety of surprising results must be expected.

8. ACKNOWLEDGEMENT

The author thanks G. Petzow and J.S. Wallace for helpful discussions.

9. REFERENCES

1. R.C. Garvie, R.H. Hannink, R.T. Pascoe, Nature 258 (1975) 703.
2. H.H Sturhahn, W. Dawihl, G. Thamerus, Ber. Dt. Ker. Ges. 52 (1975) 703.
3. N. Claussen, J. Am. Ceram. Soc. 59 (1976) 49.
4. D.L. Porter, A.H. Heuer, J. Am. Ceram. Soc. 60 (1977) 183.
5. T.K Gupta, F.F. Lange, J.H. Bechtold, J. Mater. Sci. 13 (1978) 1464.
6. U. Dworak, H. Olapinski, G. Thamerus, Science of Ceramics 9 (1978) 543.
7. P.F. Becher, J. Am. Ceram. Soc. 64 (1980) 37.
8. R.W. Rice, K.R. Mc Kinney, R.P. Ingel, J. Am. Ceram. Soc. 64 (1981) C175.
9. R.P. Ingel, Ph.D. Thesis, Catholic University of Washington, D.C., 1982.
10. First International Conference on the Science and Technology of Zirconia, Advances in Ceramics, Vol. 3, The American Ceramic Society, 1981.
11. A.G. Evans, A.H. Heuer, J. Am. Ceram. Soc. 63 (1980) 241.
12. N. Claussen, Z. Werkstofftechn. 13 (1982) 138 and 185.

13. R. Stevens, Trans. Brit. Ceram. Soc. 80 (1981) 81.
14. L.K. Lenz, A.H. Heuer, to be published in J. Am. Ceram. Soc.
15. N. Claussen, J. Jahn, Ber. Dt. Ker. Ges. 55 (1978) 487.
16. A.H. Heuer, N. Claussen, W.M. Kriven, M. Rühle, J. Am. Ceram. Soc. 12 (1982)
17. N. Claussen, F. Sigulinski, M. Rühle, Advances in Ceramics 3 (1981) 164.
18. N. Claussen, D.P.H. Hasselmann, In: Thermal Stresses in Severe Environments, Plenum Press, New York, (1980) 381.
19. M.V. Swain, to be published
20. F.F. Lange, J. Am. Ceram. Soc. 63 (1980) 38.
21. N. Claussen, M. Rühle, Advances in Ceramics 3 (1981) 137.
22. M. Rühle, B. Kraus, to be published in J. Am. Ceram. Soc.
23. M. Rühle, E. Bischoff, Proc. Int. Conf. Solid-Solid Phase Transformations, Pittsburgh, 1981.
24. N. Claussen, J. Steeb, R.F. Pabst, Am. Ceram. Soc. Bull. 56 (1977) 559.
25. N. Claussen, R. Cox, J.S. Wallace, J. Am. Ceram. Soc. 11 (1982)
26. G. Ziegler, this volume.
27. R.H.J. Hannink, R.T. Pascoe, K.A. Johnson, R.C. Garvie, Advances in Ceramics 3 (1981) 116.
28. A.H. Heuer, Advances in Ceramics 3 (1981) 98.
29. N. Claussen, G. Lindemann, G. Petzow, Proc. 5th CIMTEC, Lignano, Italy, 1982.
30. F.F. Lange, J. Mat. Sci. 17 (1982) 225.
31. H. Schubert, N. Claussen, unpublished research.
32. M. Yoshimura, S. Somiya, J. Am. Ceram. Soc. 59 (1980) 246.
33. M.V. Swain, R.C. Garvie, R.H.J. Hannink, to be published
34. M.V. Swain, R.C. Garvie, R.H.J. Hannink, R. Hughan, M. Marmack, to be published in Proc. Brit. Ceram. Soc., Dec. 1981.
35. N. Claussen, J. Am. Ceram. Soc. 61 (1978) 85.
36. P.F. Becher, G.C. Culbertson, B.A. Mac Farlane, C.Cm. Wu, Am. Ceram. Soc. Bull. 59 (1980) 357.
37. E. DiRupo, E. Gilbert, T.G. Carruthers, R.J. Brook, J. Mat. Sci. 14 (1979) 193.
38. N. Claussen, J. Jahn, J. Am. Ceram. Soc. 63 (1980) 228.
39. Sh. Yangyun, R.C. Brook, to be published in Ceramics Int.
40. J.S. Wallace, M. Rühle, G. Petzow, N. Claussen, Proc. of Int. Conf. on Surface and Interfaces, Berkeley 1980.
41. J.S. Wallace, Ph.D. work, MPI Stuttgart.
42. R. Mc Pherson, B.V. Shater, Mi Ming Wong, J. Am. Ceram. Soc. 4 (1982) C-57.
43. N. Claussen, unpublished research.
44. N. Claussen, J. Steeb J. Am. Ceram. Soc. 59 (1976) 457.
45. N. Claussen, M.V. Swain, unpublished research.
46. M. Rühle, L.H. Schoenlein, A.H. Heuer, unpublished research.
47. A.G. Evans, N. Burlingame, M. Drory, W.M. Kriven, Acta Met. 29 (1981) 447.
48. R.C. Garvie, J. Phys. Chem. 82 (1978) 218.

49. M.V. Swain, R.H.J. Hannink, Proc. Int. Conf. on Fracture V, Cannes, 1981.
50. P.F. Becher, V.J. Tennery, In: Fracture Mechanics of Ceramics 5/6, Penn State, 1981.
51. K. Haberko, R. Pampuch, Proc. 5th CIMTEC, Lignano, Italy, 1982.
52. M.V. Swain, J. Mat. Sci. 15 (1980) 1577.
53. R.T. Pascoe, R.C. Garvie, In: Ceramic Microstructures, West-view Press, Boulder, 1977.
54. R. Wagner, Ph.D. Thesis, University of Stuttgart, 1980.
55. T. Kosmac, J.S. Wallace, N. Claussen, J. Am. Ceram. Soc. 65 (1982) C66.
56. T. Kosmac, R. Wagner, N. Claussen, J. Am. Ceram. Soc. 64 (1981) C72.
57. C. Razim, Daimler-Benz AG, personal communication.
58. R.C. Garvie, CSIRO, personal communication.
59. W.A. Plummer, S.T. Gulati, In: Thermal Expansion 7, Plenum Publ. Co., 1982.
60. M.E. Woods, I. Oda, SAE Technical Paper Series 820429, 1982.
61. R.P. Ingel, D. Lewis, R.W. Rice, J. Am. Ceram. Soc. 65 (1982) C150.
62. O.T. Masaki, K. Kubayashi, Proc. Ann. Meeting, Jap. Ceram. Soc., 1981.
63. M. Matsui, T. Soma, I Oda, Am. Ceram. Soc. Bull. 60 (1981) 382.
64. L. Li, R.F. Pabst, J. Mat. Sci. 15 (1980) 2861.
65. R.F. Pabst, I. Bognar, N. Claussen, Science of Ceramics 10 (1980) 603.
66. M.V. Swain, N. Claussen, to be published in J. Am. Ceram. Soc.
67. D.R. Clarke, F. Adar, J. Am. Ceram. Soc. 65 (1982) 284.
68. M. Rühle, E. Bishoff, N. Claussen, Proc. Int. Conf. Solid-Solid Phase Transformations, Pittsburgh, 1981.
69. N. Claussen, J. Jahn, J. Am. Ceram. Soc. 61 (1978) 94.
70. N. Claussen, R. Pabst, C.P. Lahmann, Proc. Brit. Ceram. Soc. 25 (1975) 139.
71. U. Dworak, H. Olapinski, G. Thamerus, Ber. Dt. Keram. Ges. 55 (1978) 98.
72. R.C. Garvie, R.H.J. Hannink, C. Urbani, Ceramurgia Int. 6 (1980) 19.
73. H. Ruf, A.G. Evans, to be published.
74. R.M. Cannon, T.D. Ketchum, unpublished research.
75. R. Lupold, Ph.D. Thesis, University of Stuttgart, 1980.
76. L.J. Gauckler, J. Lorenz, J. Weiss, G. Petzow, Science of Ceramics 10 (1980) 577.

Table 7. Fracture toughness and strength of ZrO_2-containing ceramics (B-types).

Matrix	vol.% ZrO_2 (Additive)[1]	Tetrag. ZrO_2 at RT[1] %	K_{Ic} (Matrix) MPa$\sqrt{m}$	Test Techn.[3]	Strength (Matrix) MPa	Type of Micro-struc.(cf Fig.3)	Ref.
Al_2O_3	15	20	9.6 (5.4)	NB-a	480 (550)	B2	3
Al_2O_3	16	100	15.0 (5.2)	NB-a	1200 (400)	B1	15,54
Al_2O_3	7.5(0.12 w% MgO)	95	6.8 (5.2)	NB-b	600 (460)	B1	55
Al_2O_3	11.5	70	6.5 (4.2)	DCB	750 (480)	B3	7
Al_2O_3	29.5(2 m% Y_2O_3)	100	7.4 (4.9)	ICL	950 (600)	C1	30
ZnO	36	0	3.3 (1.6)	NB-b	n.d.	B2	73
ThO_2	13	n.d.	3.5 (1.5)	ICL	n.d.	B3	74
Spinel	17.5	50	4.6 (2)[2]	NB-b	500 (200)[2]	B3	43
Mullite	23	30	4.5 (3)[2]	NB-a	400 (270)[2]	B6	38
Si_3N_4	25(10 eq% Al)	30 cub.	8.5 (5.6)	NB-a	950 (670)	B2	69
Si_3N_4	10 (1 eq% Al)	30 cub.	7.2 (5.5)	NB-b	700 (610)	B2	75
SiC	15 ZrO_2	0	5.9 (4.7)	ICL	n.d.	B2	76

1: Rest monoclinic
2: assumed
3: NB: notched beam, a: as-notched, b: annealed at 1250°C, DCB: double-cantilever beam,
   ICL: indentation-crack length

Table 2. Fracture toughness and strength of partially-stabilized zirconia (PSZ).

Material	Composition		Tetrag. Phase Content[2] %	K_{Ic}[1] MPa$\sqrt{m}$	Test Technique[3]	Strength MPa	Grain Size μm	Type of Microstructure (cf Fig.3)	Ref.
Mg-PSZ	3	w% MgO	40	9.5[4]	NB	650	60	A1	70
Mg-PSZ	9	m% MgO	40	8.95	NB	600	80	A1	6
Ca,Mg-PSZ	10	m% (CaO+MgO)	10	5.6	NB	360	60	A1	71
Ca-PSZ	3.7	w% CaO	40	6.8[5]	NB	645	80	A1	72
Y-PSZ	6	w% Y_2O_3	30	5.9	NB	600	n.d.	A2	49
Y-PSZ	2.5	m% Y_2O_3	97	6.4	DCB	700	1	A3	5
Y-PSZ	3	m% Y_2O_3	30	9	ICL	1400	30	A6	9

1: The toughness of fullly-stabilized (cubic) ZrO_2 ranges between 2 and 3 MPa$\sqrt{m}$
2: Rest cubic, traces of monoclinic
3: NB: notched beam, DCB: double cantilever beam, ICL: indentation crack length
4: As-notched
5: Notched and etched; as-notched: 9.6 MPa$\sqrt{m}$

Table 3. Critical sizes, d_c, for M_s = room temperature observed in different TTC.

	d_c (μm)	ZrO_2 (vol.%)	Type (cf.Fig.3)	Ref.
Al_2O_3	0.52	16	B1	54
Al_2O_3	0.3	15	B3	17
		$Zr_{0.5}Hf_{0.5}O_2$		
Spinel	0.8-1.0	17.5	B1	68
Mullite	1	22	B6	40
Forsterite	0.5	32	B6	39
Si_3N_4	0.1	15	B2	69
SiO_2-Glass	0.01	20	C3	42
Zircon	0.1	20	B3/B6	42
Mg-PSZ	0.1-0.2	8.1 mol% MgO	A1	53
Y-PSZ	0.32	2.5 mol% Y_2O_3	A3	5

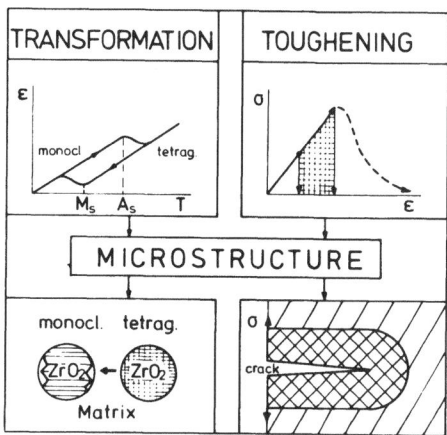

Fig. 1. Schematic representation of transformation toughening.

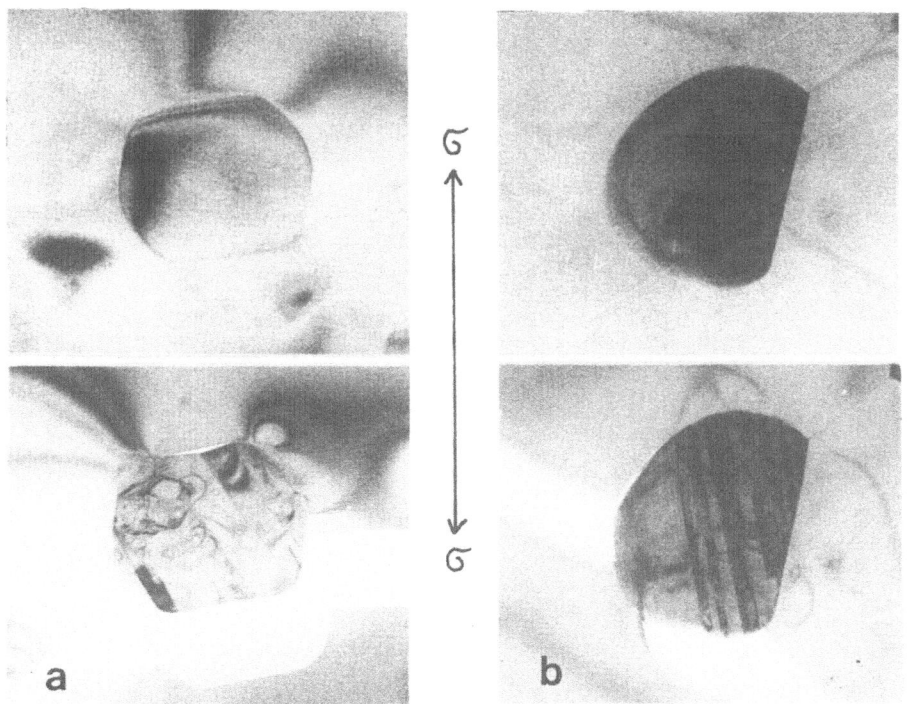

Fig. 2. Upper photographs: Tetragonal particle in spinel (a) and mullite (b) matrix. Lower photographs: After transformation to monoclinic symmetry on in situ straining (stress direction σ) in the TEM[22].

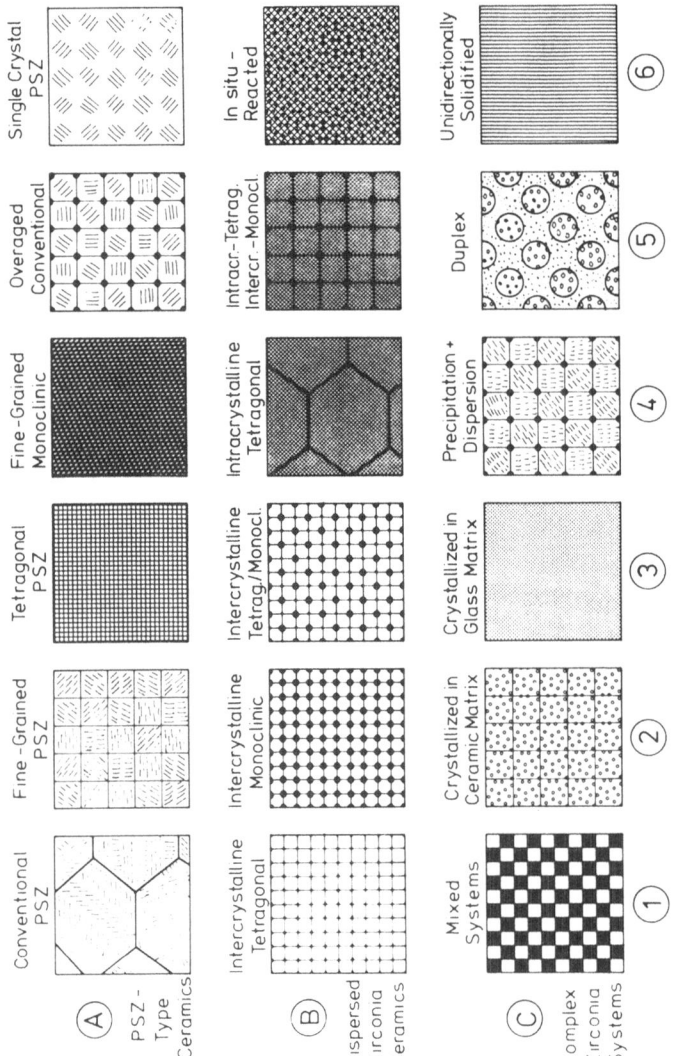

Fig. 3. Classification of transformation-toughened ceramics (TTC) based on typical microstructural features. Grouping example: A5: Overaged, conventional partially-stabilized zirconia (PSZ).

Fig. 5. Dark field TEM of ZrO_2 single crystals[9] (A6, cf. Fig. 3) with a) 4 w/o CaO, b) 2.8 w/o MgO and c) 8 w/o Y_2O_3.

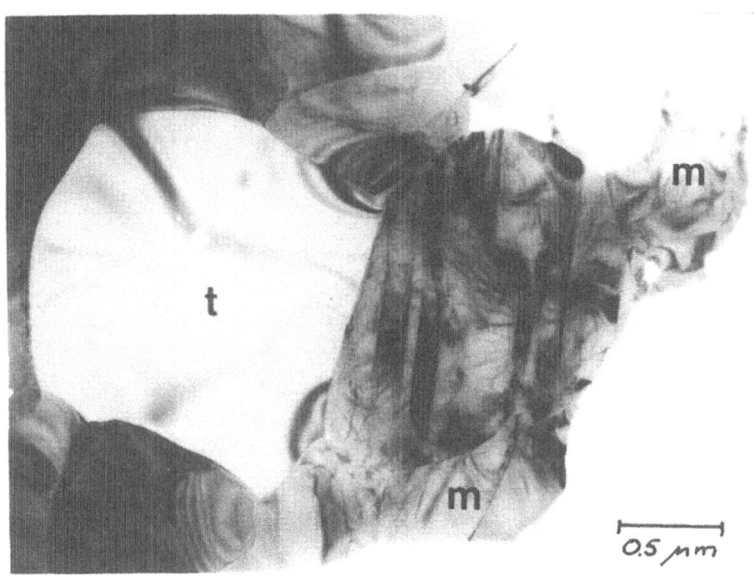

Fig. 4. Bright field TEM micrograph of fine-grained, tetragonal 2 m/o Y_2O_3-PSZ (A3, cf. Fig. 3). Some grains have transformed to monoclinic (m) after electron beam bombardment.[31]

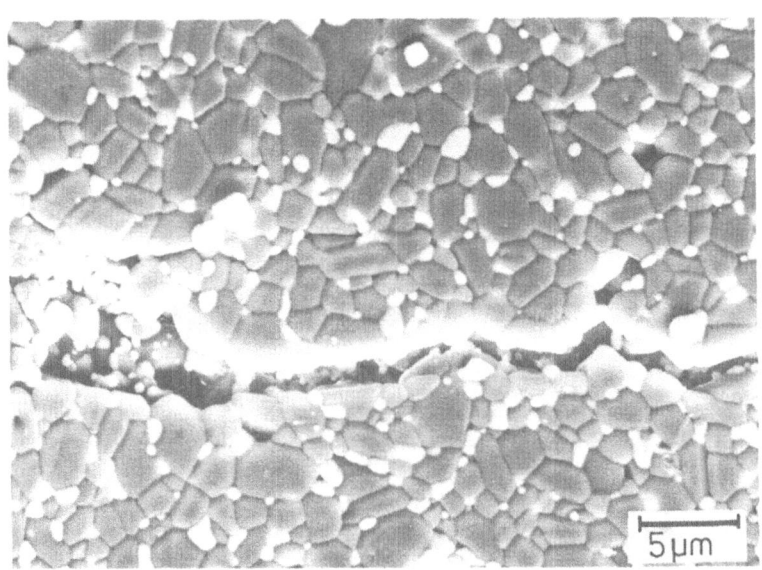

Fig. 6. Thermally etched surface of an Al_2O_3-15 v/o ZrO_2 composite (B1, cf. Fig. 3). Crack had been introduced by Vickers indent.

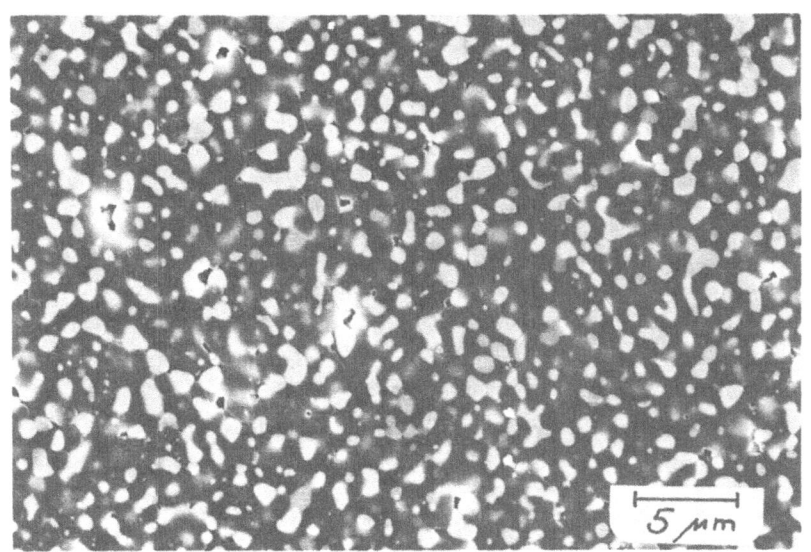

Fig. 7. In situ-reacted mullite 22 v/o ZrO$_2$ composite[40] (B6, cf. Fig. 3).

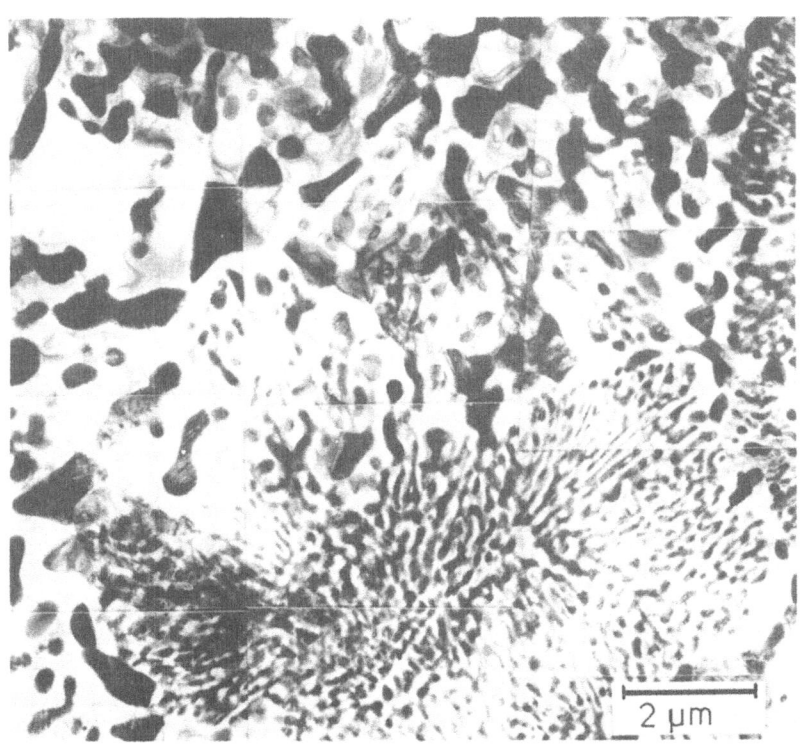

Fig. 8. Bright-field TEM micrograph of Al$_2$O$_3$-33 v/o ZrO$_2$[29] (C2, cf. Fig. 3), crystallized during hot-pressing of rapidly-quenched, amorphous particles; lower part tetragonal upper monoclinic ZrO$_2$ phase.

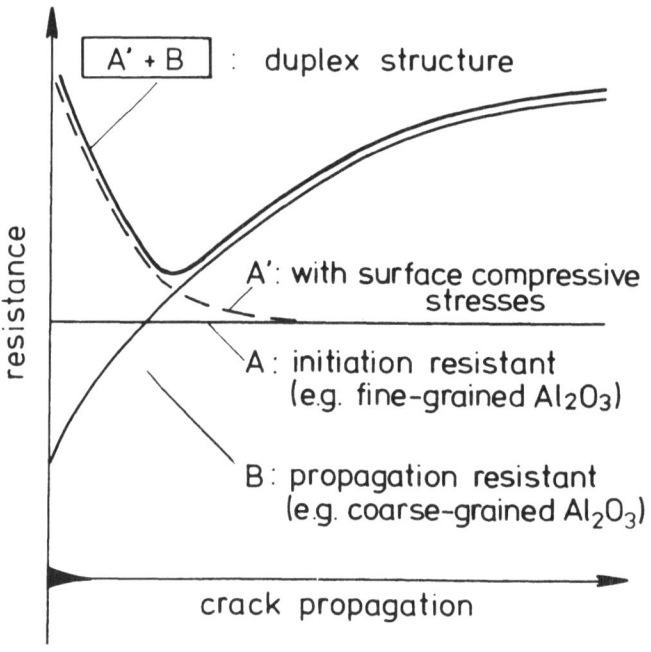

Fig. 9. Schematic R-curve of composite with duplex structure (C5, cf. Fig. 3) A' + B. A' may represent an Al_2O_3 matrix with tetragonal ZrO_2 particles (B1 type) with dispersed B areas also made of an Al_2O_3 matrix, however, with monoclinic ZrO_2 particles (B2 type).

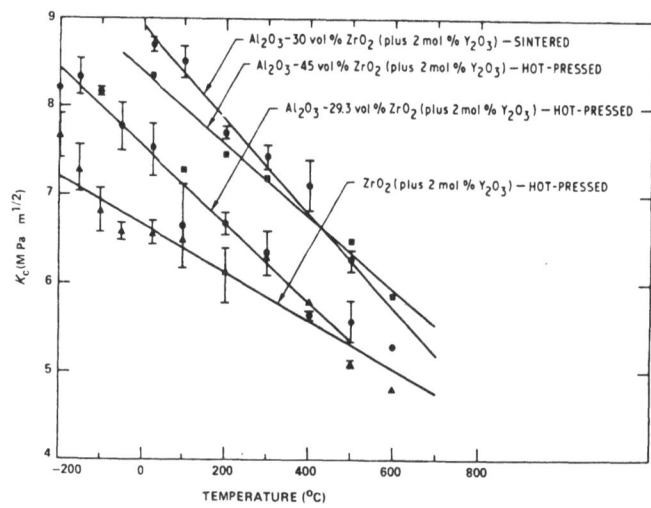

Fig. 10. Temperature dependence of fracture toughness of various C1 type Al_2O_3 and A3 type ZrO_2 ceramics.[30]

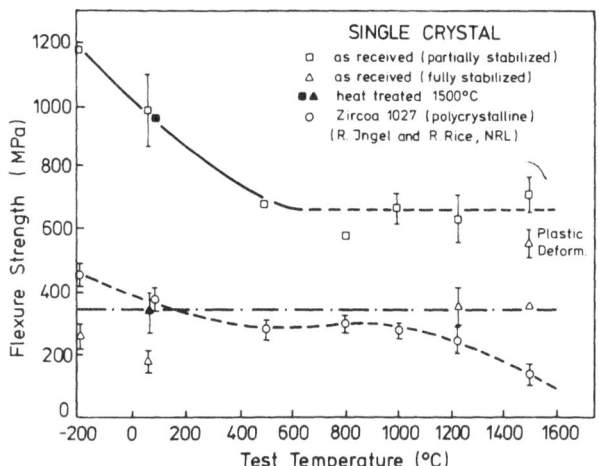

Fig. 11. Bend strength of single crystals (A6 cf. Fig. 3) with temperature.[9,61]

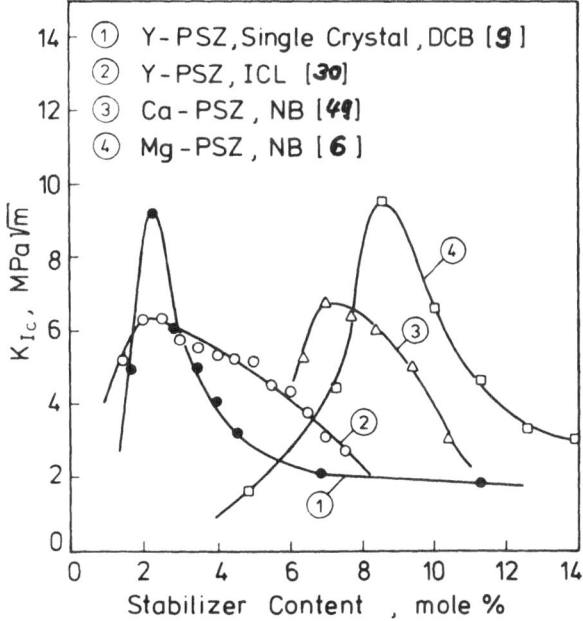

Fig. 12. Fracture toughness of various PSZ types with stabilizer content.

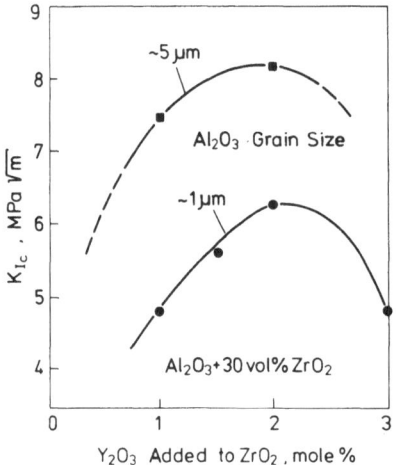

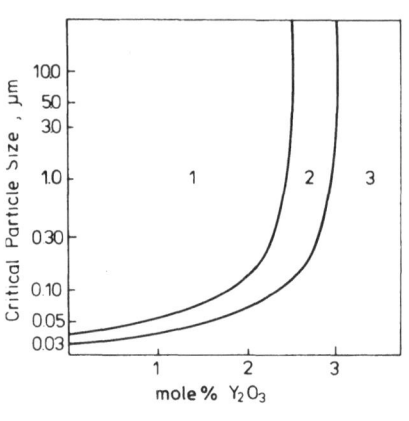

Fig. 13. Fracture toughness of Cl-type Al_2O_3-ZrO_2 composites made by two different sol-gel routes[50].

Fig. 14. Theoretical phase fields as a function of grain size and Y_2O_3 content: 1: monoclinic; 2: monoclinic + tetragonal; 3: tetragonal[51].

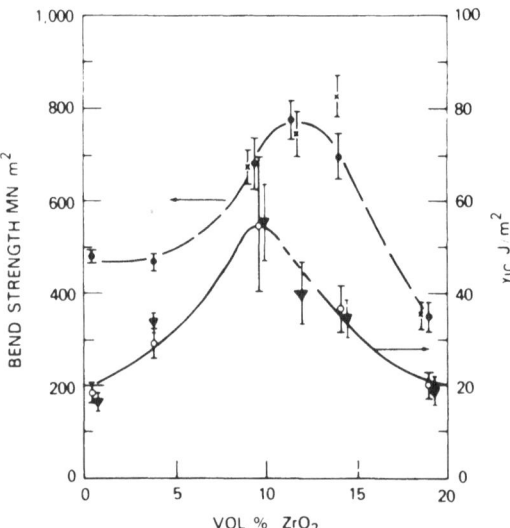

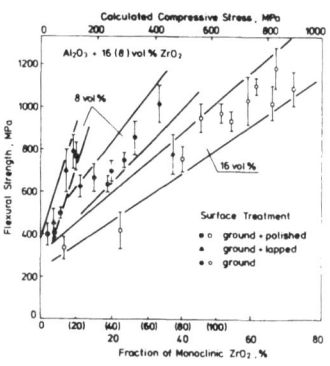

Fig. 15. Bend strength and fracture energy of Bl-type Al_2O_3-ZrO_2 versus volume fraction ZrO_2[7].

Fig. 16. Bend strength of Bl-type Al_2O_3-ZrO_2 as a function of transformed ZrO_2 in the surface region. Increasing monoclinic content was produced by increasing grinding roughness (pressure)[12].

80

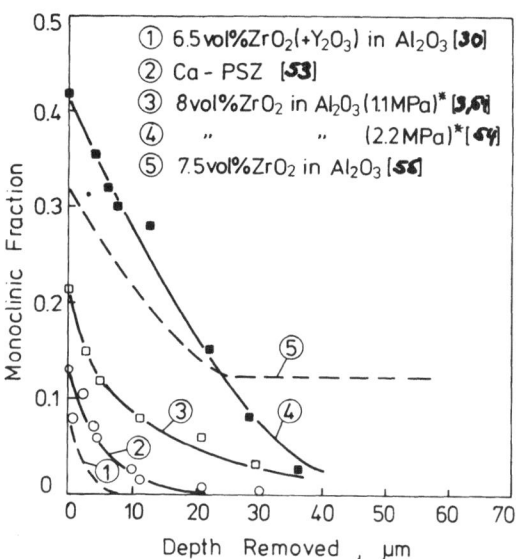

Fig. 17. Transformational depth profiles in different TTC produced by surface grinding.

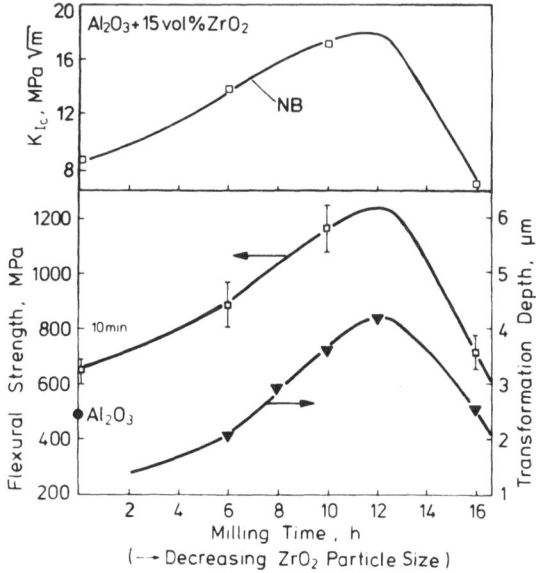

Fig. 18. Bend strength and fracture toughness of Al_2O_3-ZrO_2 composites compared with transformation zone size. ZrO_2 particle size was changed with milling time, i.e., short times yielded B2 type and long times B1 type microstructures.[56]

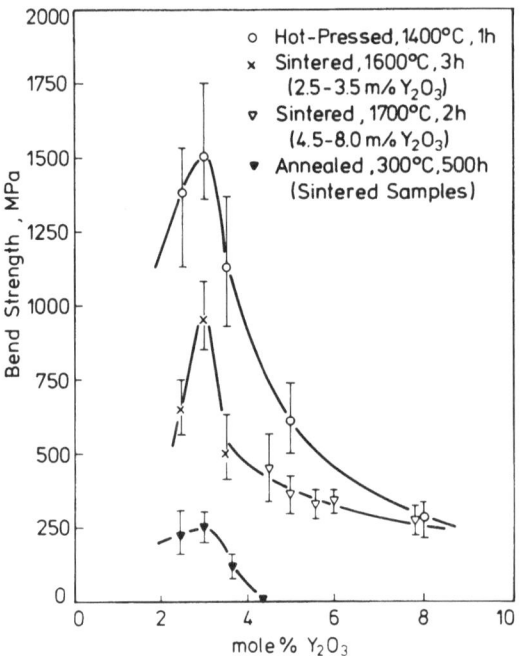

Fig. 19. Bend strength of A3-type PSZ versus Y_2O_3 content[62].
Note the drastic degradation on low-temperature annealing for
long periods of time.

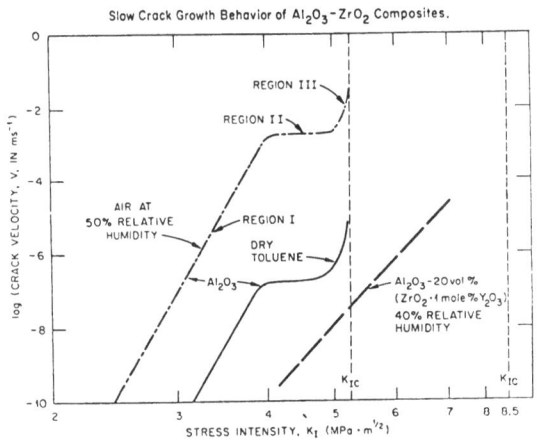

Fig. 20. Slow crack growth of C1-type Al_2O_3-ZrO_2 compared with
Al_2O_3 alone[50].

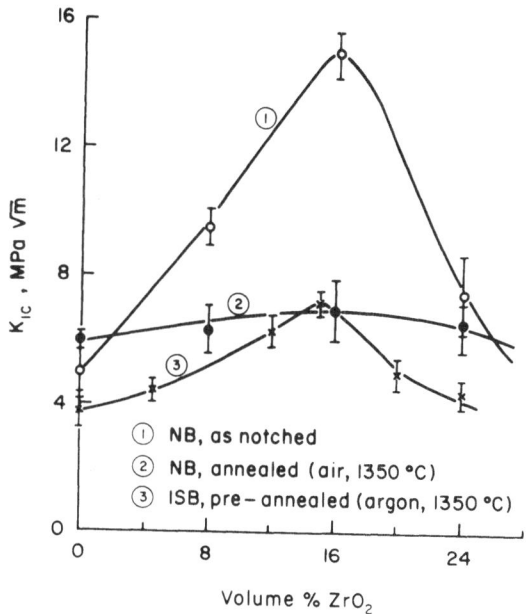

Fig. 21. Fracture toughness of B3-type Al_2O_3-ZrO_2 composites, obtained by different test techniques[66].

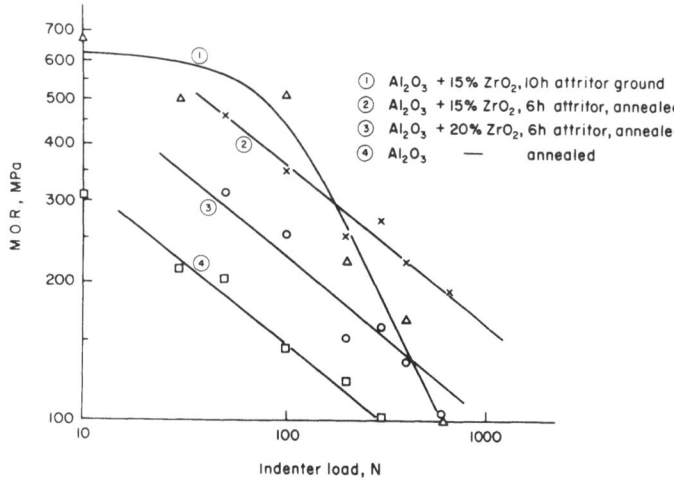

Fig. 22. Bend strength of B3 type Al_2O_3-ZrO_2 versus indenter load Note, material 1 contains compressive surface stresses after grinding.[66]

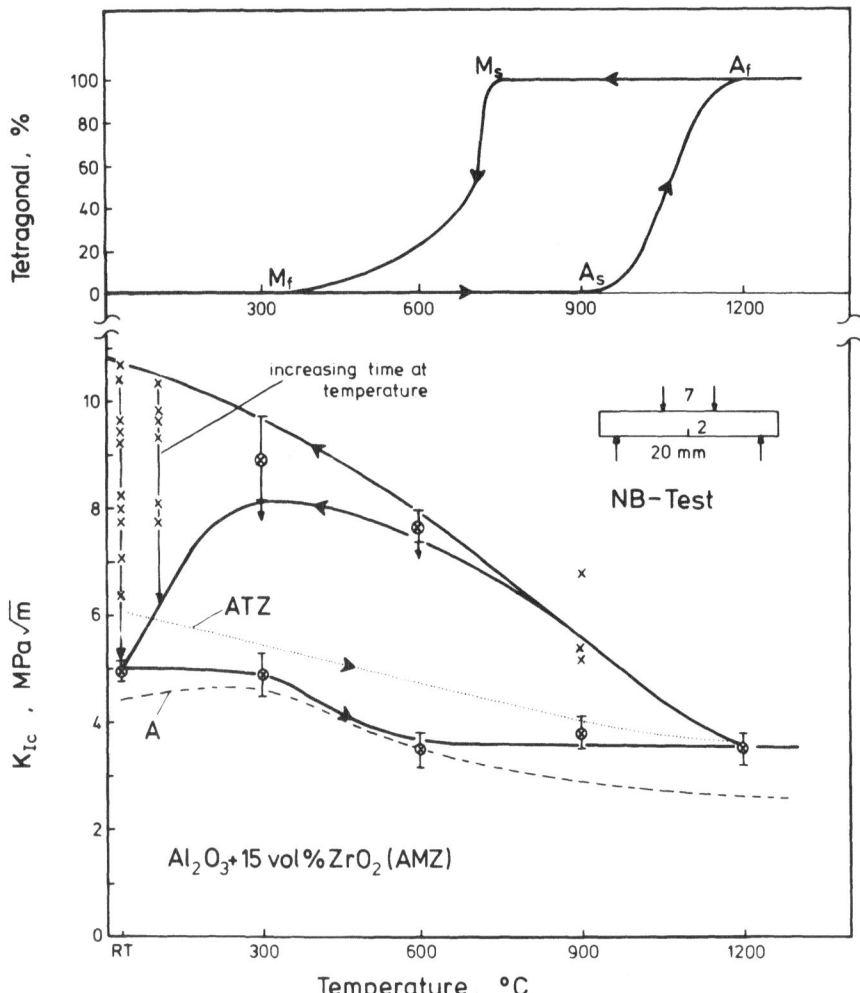

Fig. 23. Fracture toughness and tetragonal content of 15 vol.% ZrO_2-containing Al_2O_3 (B2, cf Fig. 3) as a function of temperature.[25] Lower curve represents K_{Ic} on heating, upper curve splitting on cooling depending on resting time at test temperature. For comparison, K_{Ic} values on heating of Al_2O_3 (A) and B1 type Al_2O_3-ZrO_2 (ATZ) are also indicated.

DISCUSSION

<u>H.T. Larker</u>: An advanced PSZ-material with a flexure strength of 1700 MPa has been announced by a Japanese company. Can you comment on what type of structure is typical for this material?

<u>N. Claussen</u>: The typical microstructure of this high-strength PSZ material consists of fine (< 1 μm) tetragonal, 2 to 4 mole% Y_2O_3 containing zirconia (see Fig. 3, type A3). With increasing Y_2O_3 content larger cubic grains will appear. This material, which is presently being made by several companies either by hot-pressing or sintering, degrades, however, on long-term annealing at relatively low temperatures (< 500°C).

<u>R. Webb</u>: As a manufacturer of zirconium raw materials for Zirconia Ceramics, we are naturally interested in the relationship between demands on purity and on cost, especially for ceramics with potential engineering applications. Can Dr. Claussen comment on the impurity limitations which will affect the mechanical properties - for example, is there is limiting silica level in partially stabilised zirconia?

<u>N. Claussen</u>: There are two primary effects of impurities on the properties of PSZ. The first effect is a complex chemical interaction between the stabilizer and the impurities which changes the effectivity of the stabilizing agents; these effects must be determined experimentally. The second effect is as source of inhomogeneity, either as a grain-boundary phase or as a source of mechanical flaws. The levels at which these impurities become unacceptable is dependent on the service conditions, i.e. for coarse-grained PSZ (A1, Fig. 3) silica levels of probably less than 0.2 w/o are desirable for materials which must resist creep at elevated temperatures.

<u>F. Starr</u>:   Dr. Claussen has presented a great deal of mechanical property data on zirconia-strengthened refractories and also shown a number of slides of load bearing components. Could he indicate how the level of minimum design stress can be determined from a knowledge of the raw data; in other words, what fraction of the tensile strength needs to be taken?

<u>N. Claussen</u>: We dont't design actual components, i.e. we leave this up to the engineers. However, the proof test concept along with Weibull statistical analysis could be

applied to determine the acceptable stress in the components. I can not speculate on the design strength based on a fraction of the strength obtained in laboratory tests.

K. Goebbels: One of the problems is the K_{Ic} determination for the material. Possibilities are given to measure this nondestructively, e.g. by acoustic emission or the combination of stress wave propagation parameters (velocity, attenuation). Are measurements known?

N. Claussen: Only comparative measurements have been carried out, e.g., during notched-beam K_{Ic} testing, acoustic emission was recorded. A direct measurement of K_{Ic} by AEA does not seem to be feasible.

CRACKING RESISTANCE OF ALUMINA-BASED POLYPHASE CERAMICS

H. LE DOUSSAL - G. BISSON

SOCIETE FRANCAISE DE CERAMIQUE
23, rue de Cronstadt
75015 PARIS (France)

ABSTRACT

The cracking resistance of four alumina-based products is investigated. They are the following : 1 sintered product Al_2O_3-Cr_2O_3, 1 sintered product Al_2O_3-mullite, 1 sintered product Al_2O_3-SiC, 1 molten-cast product Al_2O_3-ZrO_2-SiO_2. The cracking tests were conducted under 4-point bending on SENB* specimens of 230 x 50 x 12,5 mm³ with a notch depth of 13 mm. The appearance of the first cracks and the evolution of cracking during the test were followed by acoustic emission. The graph of the functions $K_I = f(a)$ and $V = f(K_I)$ of each product was plotted from tests at imposed cracking and delayed cracking rates (under relaxation and creep). The critical stress intensity factor for initiation K_{Ii} varies from 0,7 to 2,7 MPa$\sqrt{m}$ and that corresponding to the crack propagation plateau K_{Ip} from 3,2 to 6,5 MPa$\sqrt{m}$. The sintered products Al_2O_3-SiC and Al_2O_3-mullite exhibit the most favourable K_{Ip}/K_{Ii} ratio as well as the best delayed cracking range, the cracking rate being independent of K_I. Acoustic emission measurements show a very good correlation between K_I and the logarithm of the acoustic emission rate. The acoustic emission rate moreover makes it possible to determine the number of microfractures occurring per unit of crack length.

1 - INTRODUCTION

Advanced ceramics intended for applications which may be highly varied undergo aggression in service generally of two origins :

 . thermo-dynamic stresses due to expansion, thermal shock and erosion phenomena;

* SENB: single edge notched beam

. thermochemical stresses due to solid, gas or liquid corrosion phenomena.

Research and development on products which must have sometimes contradictory physical and chemical properties, involves problems relative to the choice of compound corresponding to the optimum product. As in most cases, all the conditions must be fulfilled simultaneously, the material ultimately adopted is a compromise in which one of the desired properties is privileged.

The present investigation deals with the search for a product having, on the one hand, good behaviour with respect to corrosion and to erosion by a liquid metal and, on the other, good resistance to thermal shock capable of reaching amplitudes of the order of 1400-1500°C in a few fractions of a second.

2 - MATERIALS INVESTIGATED

To cope with the corrosive environment imposed by the application and to meet thermal shock requirements, research was oriented toward products having alumina as the basic essential constituent and a second phase consisting of one or several other oxide and/ or non-oxide compounds.

The four basic products adopted are :

- a sintered product, $Al_2O_3-Cr_2O_3$ (reference C),
- a sintered product, Al_2O_3-mullite (reference Z),
- a sintered product, Al_2O_3-SiC (reference R),
- a molten-cast product $Al_2O_3-ZrO_2-SiO_2$ (reference E).

3 - TEST PHASES

The tests were conducted in two successive phases :

. During a first phase, the main physical and mechanical properties of the four products were measured. The characteristics taken into account are :

- bulk specific gravity,
- modulus of rupture by bending at ambient temperature,
- static and dynamic modulus of elasticity at ambient temperature,
- thermal expansion coefficient,
- Weibull's modulus.

This modulus, which allows a statistical analysis of flaws in products, was determined from prismatic specimens of 230 x 20 x 10 mm^3.

. During a second phase, the cracking tests were carried out by applying the concepts of linear fracture mechanics.

In order to take into account the texture of the products and the observations made during preliminary tests which made it possible to optimize the specimen format, the tests were conducted on SENB prisms measuring 230 x 50 x 12,5 mm^3. Moreover, the specimen has a notch depth of 13 mm and a notch root radius of 0,20 mm. To comply with these experimental data, the upper spacing of the 4-point bending test setup was placed at 100 mm and the lower spacing at 200 mm.

Under these conditions, the available cracking length below which the measurements are valid can be estimated at about 23 mm and remains compatible with the size of the damaged zone which may be encountered with this type of material.

All the mechanical tests were performed by means of an INSTRON 250-kN hydraulic press with the possibility of programming both the loading rate or the strain rate. During the test, a SEPEMA acoustic emission system makes it possible to detect the appearance of the first cracks and to follow the evolution of the crack during the plotting of the diagram $K_I = f(a)$ where :

K_I : stress intensity factor,

Δa : main crack length growth.

4 - RESULTS

4.1. Physical and mechanical characterization tests :

Table 1 groups the results obtained in the first part of the study. It should be noted that the values indicated in this table represent the average of 30 determinations (except for the expansion coefficient).

The examination of this table shows substantial differences between the four materials.

The products referenced Z and R have a relatively low modulus of rupture by bending and a relatively low modulus of elasticity compared with the two other products referenced C and E. It will be noted that the static modulus of elasticity, measured in a simple bending test, is lower than the dynamic modulus of elasticity déterminé on the basis of the propagation of an ultrasonic wave train in a bar of the material. This difference, which is valid for the four products investigated, was taken into account in the interpretation of the cracking measurements. A precise knowledge of the modulus of elasticity is in fact necessary because this property is used for determining the apparent length of the crack during the cracking tests.

On the other hand, no significant differences were noted as among the expansion coefficients.

The Weibull moduli indicate a great density of flaws for the product Z and a small density of flaws for the product R (see figure 1).

4.2. <u>Testing with imposed cracking rate</u> :

Table 2 groups the toughness parameters obtained during tests with a controlled cracking rate, having the advantage of allowing stable specimen fracture.

The following parameters are defined :

K_{Ii} : value of K_I corresponding to crack initiation, determined by acoustic emission,

$K_{I\ 0,5}$: value of K_I corresponding to a crack growth equal to 50 % of a_{Rp} (a_{Rp} is the crack length at the moment of the plateau),

K_{Ip} : value of K_I corresponding to the crack propagation plateau at constant K_I.

The graph of the functions $K_I = f(a)$ (R-curve) has different shapes depending on the material considered (see figure 2).

It should be noted that the flaw size a_{Rp} making it possible to reach K_{Rp} is independent of the cracking rate and of the modulus of elasticity.

On the other hand, the initiation value K_{Ii} is of a random nature because it depends on the surface condition of the crack root.

4.3. <u>Delayed cracking tests</u> (relaxation and creep) :

In this type of test, the logarithm of K_I is placed on the abscissa and the logarithm of the crack propagation rate V is placed on the ordinate.

In general, K_I and V are related by a relationship of the form $V = bK_I^n$ where n is a characteristic of the material and b is generally a constant.

In the present study, two procedures were followed :

. under relaxation : the displacement is controlled and the load is recorded,

. under creep : the load is controlled and the displacement is recorded.

4.3.1. Relaxation tests :

Table 3 gives the slopes n of the relationship $V = bK_I^n$ for the four products in the different crack propagation rate ranges.

In this type of test, the crack has an extension inversely proportional to the steepness of the slope for the same drop in the stress intensity factor K_I.

The products C, E and Z have slopes n which are close to each other in the range III (range in which $V = bK_I^n$) which is the one most easily reached in the present case. On the other hand, the product R has a clearly steeper slope. This difference in slope must be related to the nature of the bond at the level of the cracked zone, these products moreover having a porosity of the same order of magnitude.

A difference in behaviour is also noted with respect to the delayed cracking ranges (in particular range II in which the crack propagation rate is independent of K_I). Thus, the range II appears toward 10^{-2} m/s for the products Z and R, toward 10^{-3} m/s for the product C and toward 5.10^{-5} m/s for the product E. The products Z and R thus have a better cracking behaviour in the case of a stress developed by a controlled deformation.

This assessment is confirmed by the value of the ratio K_{Ip}/K_{Ii} of these two products which is higher than that of the products C and E (see Table 2).

4.3.2. Creep tests :

Figure 3 illustrates one of the curves $V = f(K_I)$ obtained with each of the products investigated.

Compared with the relaxation tests, the crack propagation rate has a tendency to increase during the test and the slopes n obtained are smaller (in this case, n varying from 10 to 40).

The ranges are themselves more delimited and it is especially the range of the high propagation rates which is explored, and the risks of catastrophic fracture are significant.

Based upon these limitations, the relaxation tests allowing a more stable fracture are easier to interpret.

4.3.3. Discussion :

Following these tests, it appears that, in the case of the creep tests, a damaged zone is created at the crack tip and there is an interaction between the extension of the damaged zone characterized by the R-curves and the extension of the main crack. On the other hand, in the case of the relaxation tests, it appears that the extension of the damaged zone remains small and that the essential part of the behaviour law is governed by the progression of the main crack.

4.4. Acoustic emission measurements :

These measurements make it possible to determine K_{Ii} which is the stress intensity factor for which there is the appearance of acoustic emission owing to the formation of microcracks within the material. This emission corresponds to the loss of the linearity in the load-deformation diagrams which is not always easy to show visually.

The appearance of acoustic emission also serves to fix the optimum loading during the flow tests.

In the controlled crack propagation rate tests, the acoustic emission rate T_{EA} can be related to the crack growth Δa and to the value of the stress intensity factor K_I.

It appears that there is a strong correlation between K_I and $\ln T_{EA}$. Thus, when K_I becomes constant (K_{Ip}), the acoustic emission rate T_{EA} also remains constant. Figures 4 and 5 illustrate these observations for the products R and Z. Thus, the correlation coefficient is 0,92 for the product R and 0,97 for the product Z.

The stress intensity factor can be expressed as a function of the acoustic emission rate in the following form :

$$K_I = b.\ln T_{EA} + \text{constant.}$$

The acoustic emission rate expressed in number of events makes it possible to determine the number of microfractures occurring per unit of crack length. Thus, for the product R under stable cracking conditions (K_{Rp}), there are 150 events per millimetre of cracking. Since this occurs on a front 25 mm wide, the "elaboration" zone, which relaxes at the tip of the main crack, probably has a small extension. On the other hand, for the product Z, where the stable cracking conditions have not yet become established, almost 500 events are observed per millimetre of cracking.

5 - CONCLUSIONS

In research and development on new products capable of withstanding thermomechanical stresses, the determination of toughness parameters is an important area. These measurements should be approached by previously defining a specimen format suited to the microstructure of the product in order to have test conditions in conformity with the concepts of linear fracture mechanics.

To obtain the maximum amount of information from these tests, it is necessary to simultaneously carry out tests under controlled cracking (K_I = f(t) and delayed cracked (V = f(K_I)).

Acoustic emission measurements also provide precious information as regards both the detection of initial cracks as well as the evaluation of the number of microfractures.

Among the four products studied, the products R and Z appear to give the best guarantees with respect to resistance to mechanical stresses.

These observations will be correlated in future investigations with the results of thermal shock tests conducted on specimens of 210 x 50 x 25 mm heated on one side to 1000°C.

TABLE 1 : MAIN PROPERTIES OF MATERIALS INVESTIGATED

Properties / Nature of product	Bulk density (kg.m⁻³)	Modulus of rupture by bending à 20°C σ_F (M.Pa)	Modulus of elasticity at 20°C (G.Pa)		Expansion coefficient α (10^{-6} °C⁻¹)	Weibull's modulus
			statique	dynamiqueUS		
Sintered Al$_2$O$_3$-Cr$_2$O$_3$ (C)	3330	26,3	109	131	7,5	13,2
Sintered Al$_2$O$_3$-mullite (Z)	3120	16,6	43	57	7,0	4,9
Sintered Al$_2$O$_3$-SiC (R)	2970	17,5	44	48	7,0	18,5
Molten-cast Al$_2$O$_3$-ZrO$_2$-SiO$_2$ (E)	3725	46,1	110	137	7,0	9,2

TABLE 2 : TOUGHNESS PARAMETERS OBTAINED DURING CONTROLLED PROPAGATION RATE TESTS.

Parameter / Nature of product	Crack length a_{Rp} (mm)	Stress intensity factor (MPa$\sqrt{m}$)			Ratio K_{Ip}/K_{Ii}
		at crack initiation K_{Ii}	at 50 % a_{Rp} $K_{I\,0,5}$	at crack propagation plateau K_{Ip}	
Sintered Al$_2$O$_3$-Cr$_2$O$_3$ (C)	15	1,4	3,5	4,3	3
Sintered Al$_2$O$_3$-mullite (Z)	25	1,0	3,0	3,8	3,8
Sintered Al$_2$O$_3$-SiC (R)	20	0,7	2,6	3,2	4,3
Molten-cast Al$_2$O$_3$-ZrO$_2$-SiO$_2$ (E)	18	2,7	5,7	6,5	2,4

TABLE 3 : DELAYED CRACKING TESTS UNDER RELAXATION - VALUES OF SLOPES n OF LINE SEGMENTS
$\ln K_1 = n \ln V$

slope n of line segments / Nature of product	Range IV	Range III	Range II	Range I	Remarks
Sintered Al$_2$O$_3$-Cr$_2$O$_3$ (C)	450	107	60		Range IV : V < 10^{-6} m/s Range II : V > 10^{-3} m/s
Sintered Al$_2$O$_3$-mullite (Z)		96			Range III : 10^{-7}<V<10^{-3} m/s
Sintered Al$_2$O$_3$-SiC (R)		240			Range II : V > 10^{-2} m/s
Molten-cast Al$_2$O$_3$-ZrO$_2$-SiO$_2$ (E)		97	33	57	Range II : V ≈ 10^{-5} m/s

FIGURE 1 : VARIATION OF FRACTURE PROBABILITY P_s % with modulus of rupture by bending σ_r for the products C, E, R and Z.

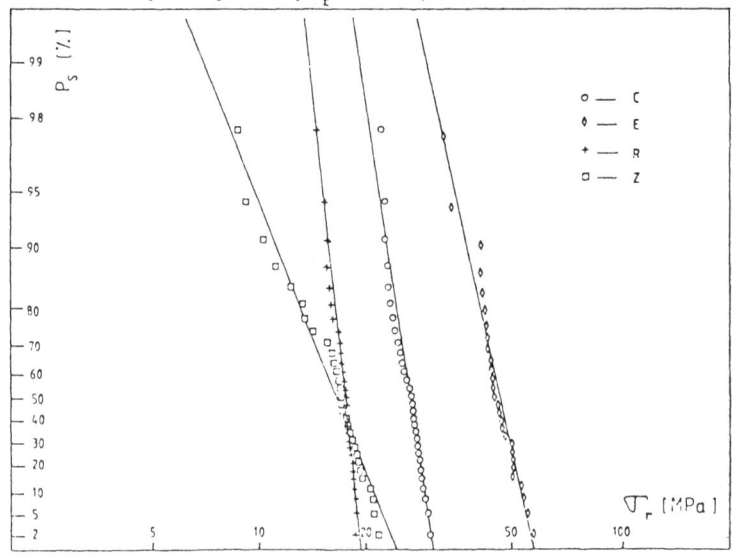

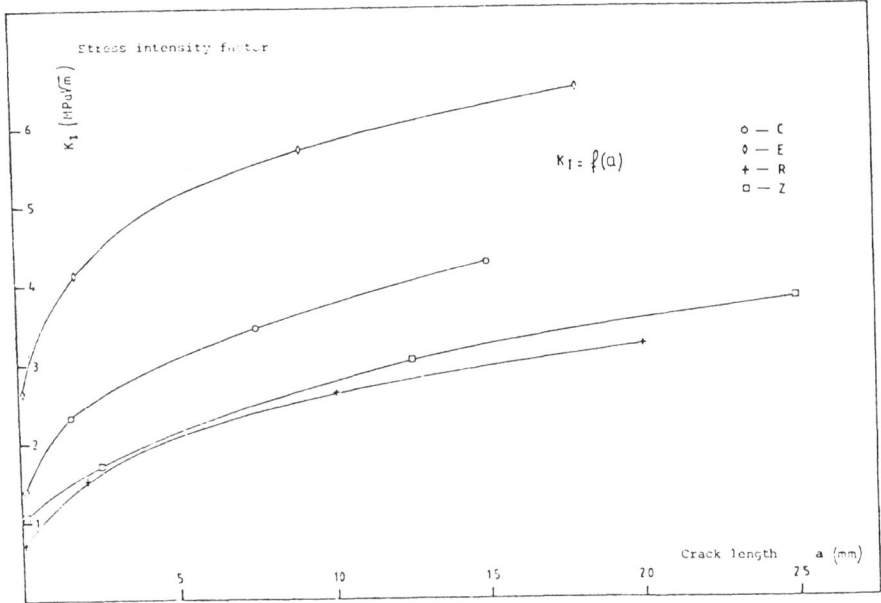

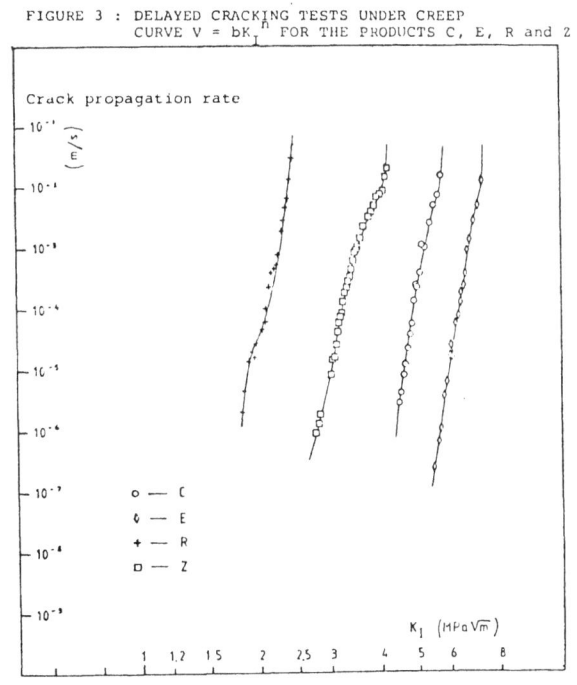

FIGURE 3 : DELAYED CRACKING TESTS UNDER CREEP
CURVE $V = bK_I^n$ FOR THE PRODUCTS C, E, R and Z

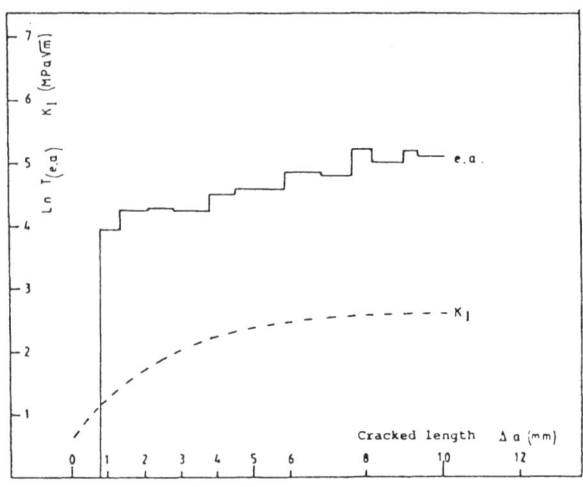

FIGURE 4 : ACOUSTIC EMISSION RATE AS A FUNCTION OF
CRACKED LENGTH Δa for the product R

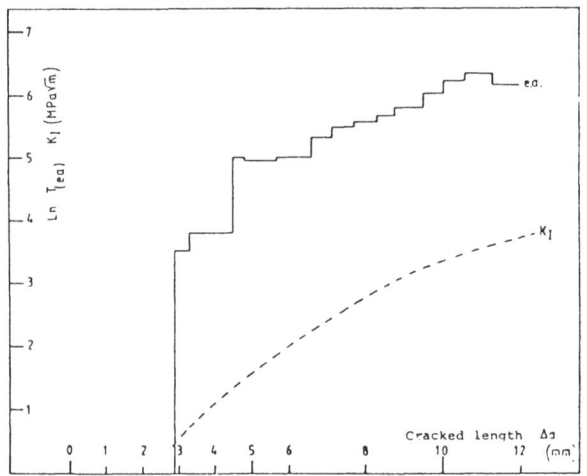

FIGURE 5 : ACOUSTIC EMISSION RATE AS A FUNCTION OF CRACKED
LENGTH Δa FOR THE PRODUCT Z

DISCUSSION

W.E. Buresch: In your paper you discuss the behaviour of materials with different internal stresses caused by the mismatch of thermal expansion coefficients between different phases. What is the contribution of the internal stress to your measured values, especially the K_{Ii}, $K_{I\,0,5}$ and K_{Ip} ?

H. Le Doussal: The measured values concern the entire product. Additional studies under way, notably on the product E, have shown that the tensile stresses developed in the corundum crystals from the zirconia segregations produced microcracking when they were located in the immediate vicinity of the main crack.

The internal stresses present in the material contribute to the preferential cracking of certain phases, having the effect of absorbing energy during cracking, thereby stopping the cracking process.

M.K. Halpin: You use acoustic emission, but is it not a fact that it is very difficult to make definite interpretations because of interferences, causing noise etc.

H. Le Doussal: Acoustic emission measurements are in fact disturbed by noises of mechanical origin and background noise of electronic origin.

Noises of mechanical origin are eliminated by two highpass filters of 50 kHz producing a damping of 80 dB.

Electronic background noise is eliminated by a signal processing system involving a single variable which adjusts automatically over random electronic noise.

The high frequency signals generated by the electronics are eliminated by a 2-MHz lowpass filter.

The natural frequency of the transducer, which is 200 kHz, is the one which is least absorbed by the products we have investigated.

MECHANICAL PROPERTIES AND ENGINEERING DESIGN WITH CERAMICS

R.W. DAVIDGE

Materials Development Division,
Building 552, AERE Harwell,
Didcot, Oxon, OX11 0RA, UK

SYNOPSIS

The ability to predict the mechanical performance of engineering ceramic components based on measurements of the mechanical properties of the ceramic is the theme of this paper. Ideally one would wish to perform a standard set of mechanical property measurements, and through various computational procedures predict the lifetime of a component subjected to realistic regimes of stress, temperature and environment. In principle, it is possible to do this in an entirely empirical way but there are obvious dangers if there is an absence of basic understanding. The emphasis of the paper is on the reliability of engineering design data for predicting mechanical performance, but the more fundamental considerations relating microstructure to mechanical properties will be considered where appropriate. Performance prediction is good only in an ideal case where fracture occurs in a perfectly brittle manner in inert conditions. In reality the situation is complicated by a number of destabilising factors including sub-critial crack growth which may be chemical in origin or related to limited crack tip plasticity; crack tip rounding associated with localised plastic deformation; microstructural changes such as phase transformations, creep or grain growth; and chemical changes, particularly oxidation. These factors are discussed in the context of engineering design with ceramics.

1. INTRODUCTION

The basic equation relating fracture strength (σ_f), a critical flaw size (C) and the critical stress intensity factor (K_{IC}) is crucial to understanding engineering design in ceramics[1]

$$\sigma_f Y \sqrt{C} = K_{IC}. \qquad \qquad \dots (1)$$

Fracture strength can be measured in a straightforward way. The critical stress intensity factor usually is a function of the failure load of a specimen, the specimen dimensions and an intentionally produced flaw or crack in the specimen. In some specimens however (for example the double torsion test) the failure load is not dependent on the crack size and thus the stress intensity factor may be obtained solely from the failure load and the specimen dimensions. In principle therefore one can measure strength and the critical stress intensity factor and use these for engineering design purposes.

Before considering some of the actual complexities it is illustrative to discuss a simple idealised situation as presented by Griffith[2,3]. A thin slab of material is subjected to a biaxial stress in the plane of the sheet at its outer edge, Fig. 1. The specimen contains a slit crack approximating in shape to a narrow ellipse. Under these conditions a number of ceramic specimens tested at ambient temperatures and under an inert environment would be expected to fail at a very similar fracture stress. Additionally, the critical stress intensity factor could then be calculated through eqn. (1). In this case one could use the fracture stress as a simple design criterion. Provided that the actual stress in service was less than the fracture stress then the component would operate satisfactorily.

In practice however there are a number of complicating factors. The remainder of this paper is devoted to a discussion of these and the implications for engineering design with ceramics. The discussion is under four main headings: State of Stress; Statistical Variations in Strength; Time Dependence of Strength; Effects of Microstructure and Chemical Changes.

2.   STATE OF STRESS

For simplicity the strength of ceramics is measured most commonly in bending (uniaxial tensile stress). For design purposes, where stresses in practice may be different from this, it is important to have information about the effects of the state of stress. Griffith, in his original work, suggested that the state of stress should not affect the strength of a specimen of the type in Fig. 1 for any biaxial tensile stress or a compressive/tensile stress for compression/tension ratios less than 3. Subsequent mathematical analyses have questioned the validity of this hypothesis that the second stress component has no effect on the tensile failure stress normal to the crack plane. Unfortunately

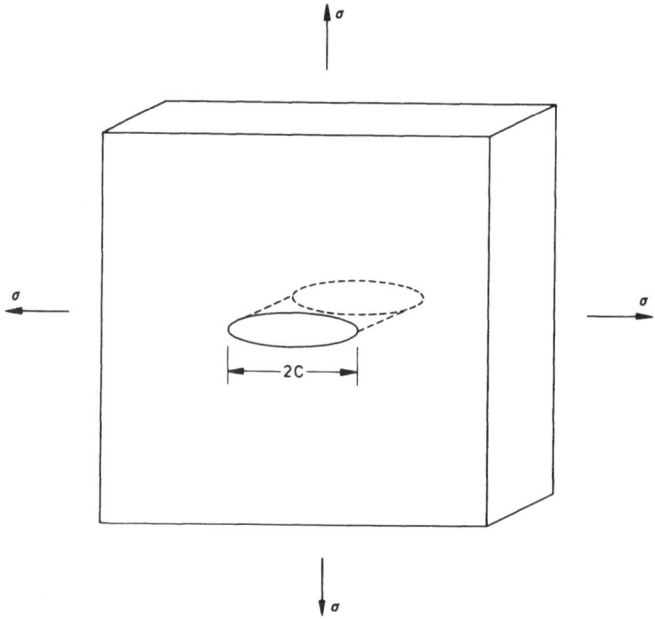

Fig. 1   A sharp elliptical crack in a thin sheet of material.

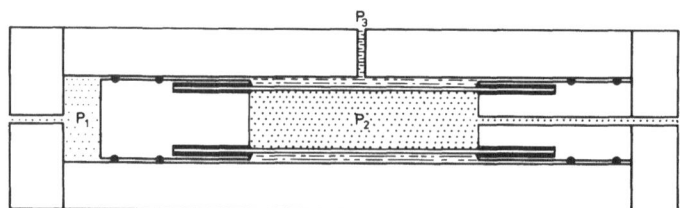

Fig. 2   Experimental rig for studying effects of biaxial stresses
          on the strength of ceramic tubes.

there is no current agreement on this point and for example in an
equibiaxial tensile stress condition the strength may be less than
equal to or greater than the uniaxial strength depending on the
value of Poisson's ratio[1].

Despite a voluminous literature on the effects of biaxial
stresses it is difficult to investigate effects unambiguously
because of the interference of other factors such as a random
source of cracks and the effects of specimen size (to be discussed
below). In an ideal experiment one would wish to keep the specimen
size and shape identical and ensure that each specimen has the same
sized single flaw. Such data have been reported for a tube
specimen geometry with flaws introduced by diamond indentation[4].
The biaxial stress ratio is controlled by the application of two
out of three independently controlled pressures on the tube, Fig.
2. Results showed that for silicon carbide tubes the equibiaxial
tensile strength was smaller by about 20% than for uniaixal tension
or compression/tension. More recent work however, comparing
uniaxial tensile strengths, measured in bending, and biaxial
strengths, measured in a ball on ring test, for glass specimens
with single indentation flaws showed that the two strengths were
not significantly different[5]. Clearly further experimental and
theoretical work is required in this field but the effects will
probably be of minor importance compared with those to be discussed
next.

3. STATISTICAL EFFECTS

The statistical treatments used to describe the strength of
ceramics are based on weakest link theory and on assumptions
referred to in statistical theory as the Poisson postulates. The
general theory of statistics of extremes is given in the book of
Gumbel[6], and the weakest link theory of strength, pioneered by
Weibull[7] is one aspect of this. In essence the problem is: given
a particular distribution of property values (e.g. strength), what
is the minimum value expected for a particular number of samples?
For example one might be required to estimate the highest operating
stress that could be applied to a component to give a failure rate
of less than 1 in 10^6 from a number of observations of strength
(typically < 100).

The theory assumes that the material can be divided into a
number of small regions (which could be volume dV or area dA
depending on whether volume or surface flaws are controlling) that
act independently, each containing a flaw associated with a
particular strength S. The strength of the whole is then
determined by the strength of the weakest region, as in the links
of a chain. The theory as applied to ceramics[8-12] considers a
function g(S) such that the number of flaws per unit volume (or
area) associated with a strength between S and S + dS is g(S)dS.

If $d\phi$ is the probability of failure of an element

$$d\phi(S) = dV \int_0^s g(S)dS \qquad \qquad \dots (2)$$

with a survival probability $1 - d\phi(S)$. In material of volume V there are V/dV volumes dV. The probability $(1-\phi)$ that no volume has a strength lower than S is the product of the separate probabilities and

$$1 - \phi = (1-d\phi)^{V/dV} = \left(1-dV \int_0^s g(S)dS\right)^{V/dV} \qquad \dots (3)$$

This reduces to, as $V/dV \to \infty$,

$$1-\phi = \exp -V \int_0^s g(S)dS \qquad \qquad \dots (4)$$

The theory of the statistics of extremes considers three kinds of asymptotic function for g(S): the first two refer to variates that are **unlimited** in extreme values; for distributions that are **bounded** at one extreme the so-called asymptotic function of the third kind holds. This third function is thus of most relevance to the strength of engineering ceramics which must have a tensile strength $\geqslant 0$. Analysis shows that the **simplest** function for g(S), relevant to a lower limit S_u, is given by

$$\int_0^s g(S) = \left(\frac{S-S_u}{S_o}\right)^m \qquad \qquad \dots (5)$$

where S_o is a 'scale' parameter and m the 'shape' parameter. This will be recognized as the well-known Weibull distribution[7]; m is usually known as the Weibull modulus and increases with decreasing variability in strength. There is no underlying reason (yet) why strength should follow the above equations.

Weibull statistics are widely applied to the strength of ceramics, and for many data there is good agreement between experimental results and the above function. In many cases the lower limit of strength S_u is taken as zero and thus, rearranging equations (4) and (5) gives a proportionality between survival probability $P_s = 1-\phi$ as

$$\ln \ln (1/P_s) \propto m \ln S \qquad \qquad \dots (6)$$

Typical data fitting this relation are shown in Fig. 3. The prediction of stress levels for high survival probabilities is possible but this should be treated with some caution as extrapolations way beyond the experimental data are involved and there is little evidence to indicate whether the predictions are accurate.

An alternative approach was taken by Evans and Jones[11] who extended the treatment of Matthews et al.[8] whereby the function of g(S) was not considered to have any specific form but is directly determined from experimental strength measurements. This is obviously a valuable and more general technique but it requires graphical or analytical evaluation of the strength data and again extrapolation outside the range of measurements.

An additional attraction of Weibull statistics is that, for materials that are consistent with the theory, it is possible to predict effects on strength of the state of stress (e.g. uniaxial, biaxial), the distribution of stress, and the specimen size. In this case however we are concerned more with the average rather than extreme values of strength. Both these effects are a manifestation of the fact that the greater the volume of material under stress, or the greater the number of flaws perpendicular to the tensile stress, the higher the chance of finding a large flaw.

These effects are quantified in the examples below for a range of values of Weibull modulus of 5, 10 and 20. For specimens under a tensile stress, assuming a volume distribution of flaws, the mean strengths σ_{V1} and σ_{V2} for specimens of volume V_1 and V_2 are

$$\frac{\sigma_{V1}}{\sigma_{V2}} = \left(\frac{V_2}{V_1}\right)^{1/m} .$$

Thus for various values of V_1/V_2 and m we have values for σ_{VI}/σ_{V2} of:

V_1/V_2 \ m	5	10	20
10	0.63	0.79	0.89
100	0.40	0.63	0.79
1000	0.25	0.50	0.71

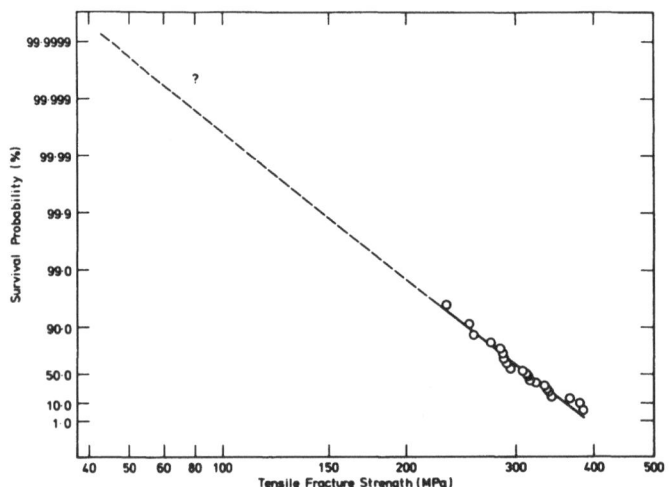

Fig. 3   Statistical data for the tensile strength of REFEL silicon
carbide according to eqn. (6)[1].

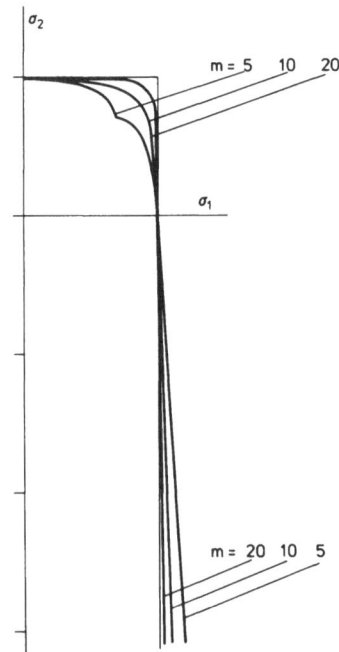

Fig. 4   Variation of biaxial fracture stress with Weibull
modulus[13,14].

For a bar under a tensile stress, three point, four point (quarter length knife edges) and pure bending, assuming a volume distribution of flaws, the relative mean strengths are:

$$1 \quad : \quad [2\,(m+1)^2]^{1/m} \quad : \quad \left[\frac{4(m+1)^2}{m+2}\right]^{1/m} \quad : \quad [2(m+1)]^{1/m},$$

or for particular values of m:

Stress state / m	5	10	20
Tension	1	1	1
3-point bend	2.35	1.73	1.40
4-point bend	1.83	1.45	1.25
Pure bend	1.64	1.36	1.21

Finally, Fig. 4 gives data for the effect of biaxial stress on strength (this ignores the possible effects mentioned in section 2) for various values of Weibull modulus.

The above discussion is based mainly on an empirical approach to statistical variations in strength but the methods are widely used with reasonable success. It is worth however considering some related microstructural aspects of the problem. The implication from eqn. (1) is that scatter in strength may be understood in terms of scatter in either the largest flaw size or the critical stress intensity factor, which should be a materials constant.

The determination of K_{IC} from standard fracture mechanics tests requires the introduction of a large crack into the specimen which is then fractured. The crack is thus very large compared with microstructural features. There are few experimental data on the statistical variation of the critical stress intensity factor but this has been measured by Pankow and Finnie[15] for polycrystalline alumina. Their results are shwon in Fig. 5. The K_{IC} data were generated from specimens with sharpened notches (radius 20 μm) and are compared with strength data for smooth beams broken in three-point bending. The scatter in K_{IC} is four times less than the scatter in strength. This suggests that for large sharp cracks K_{IC} may be essentially constant. This is intuitively reasonable as the crack is sampling a statistically large area of the specimen as it propagates.

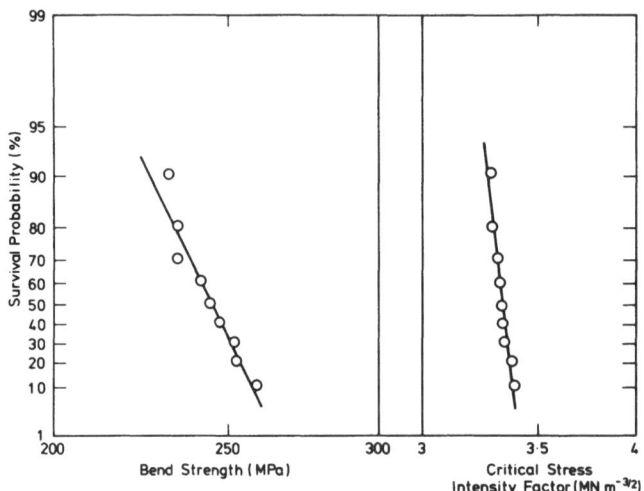

Fig. 5   Statistical variations in strength and stress intensity
         factor for alumina[15].

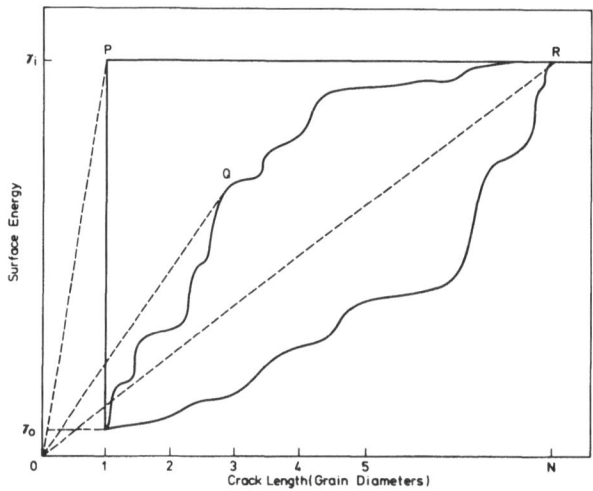

Fig. 6   Variation in effective surface energy for crack initially
         of grain size dimension with length of crack where γ_0 is
         value for grain boundary or cleavage fracture and γ_i
         value from fracture mechanics tests. Strength is
         controlled by maximum value of γ/C, the slope of lines OP,
         OQ, or OR.

The critical stress intensity factor ideally is a material constant but this is unlikely at flaw sizes comparable in size to microstructural features. The surface energy for a grain size crack should approximate to the cleavage or grain boundary surface energy (γ_0) which is usually at least an order of magnitude less than the effective surface energy for macroscopic fracture (γ_1). Further propagation of the crack beyond a grain size dimension has a greater surface energy requirement and this may be a simple step function, or more complicated as indicated in Fig. 6 where it is envisaged that γ rises to γ_1 over a length of a few grain dimensions. Strength is controlled by the maximum value of γ/C and in Fig. 6 the value at points P, Q and R for the three curves drawn. Fracture may thus be preceded by a period of subcritical crack growth based on microstructural considerations. (This is in addition to the environmental stress corrosion effects discussed in the next section.)

Thus for inherent flaws it is conceptually difficult to discuss strength in terms of an independent crack size and stress intensity factor in that these two parameters interact strongly.

A further complication arises from the density of inherent cracks. The theory assumes that these flaws do not interact mutually. However, when the crack spacing is similar to the crack size this is not true. Recent experiments of Okada and Sines[16] have identified a relatively high density of active microcracks in polycrystalline alumina. Under these conditions a number of close but small microcracks can coalesce and lead to failure. Fig. 7 shows three such arrays of cracks that could all propagate in a short time at the same stress as the single large crack at the top of the figure. Note that this has serious consequences for non-destructive testing techniques. The nil-observation of flaws above a particular size in a sample could lead to an optimistic prediction for strength if fracture was initiated from closely-spaced smaller flaws. An additional problem is that the times to failure under delayed fracture conditions for the various crack arrays varies by an order of magnitude.

In conclusion: although statistical variations in strength can be treated with reasonable success in an empirical manner the related microstructural features are far from clear and further work is warranted here to establish a sound scientific foundation.

4. TIME DEPENDENCE OF STRENGTH

The longer the time that a ceramic is subjected to a stress, the lower is its ultimate failure strength. This effect is manifested for example by a strain rate dependence of strength (the lower the strain rate the lower the strength) and delayed fracture effects where failure occurs in a finite time under a fixed stress.

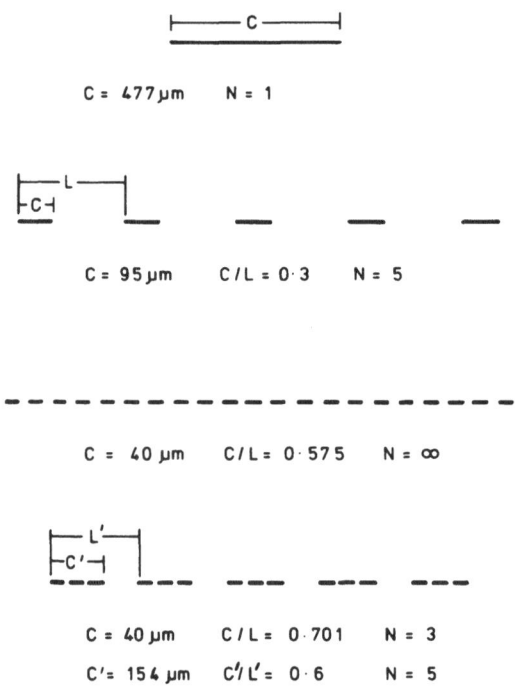

C = 477 μm   N = 1

C = 95 μm   C/L = 0·3   N = 5

C = 40 μm   C/L = 0·575   N = ∞

C = 40 μm   C/L = 0·701   N = 3
C'= 154 μm   C'/L'= 0·6   N = 5

Fig. 7   Various linear arrays of microcracks that would all propagate at the same stress[16].

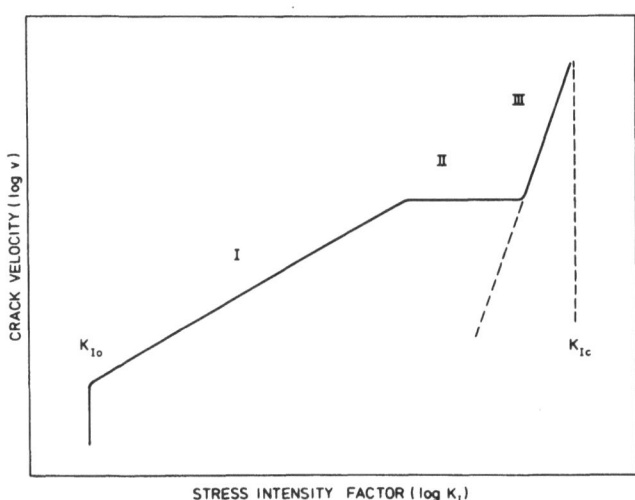

Fig. 8   The general stress intensity/crack velocity behaviour for ceramics.

For the moment we will restrict the discussion to low-to-medium temperatures and leave the additional complexities of high temperatures to the next section. The time dependence of strength is associated with sub-critical crack growth. Note that we have already alluded to other related effects in the previous section involving the variation of fracture surface energy with crack length for small cracks.

Data relating crack velocity to strain intensity factor have been obtained for a wide range of ceramic materials using standard fracture mechanics tests, particularly the double torsion test and the double cantilever beam test. The classical behaviour observed is as shown in Fig. 8, where there are three well defined regions. Above a threshold stress intensity factor the crack velocity in region I is proportional to the nth power of the stress intensity factor. In region II the crack velocity is independent of stress intensity factor and then in region III the behaviour is similar to that in region I but with a significantly higher slope. Fracture finally occurs at the critical stress intensity factor K_{IC}. The rate of crack growth in region I is reaction rate controlled, whereas in region II it depends on the diffusion of a corrosive species to the crack tip. In most cases the corrosive agent is water and region II occurs at higher crack velocities with increase in water content. Region III represents an inherent behaviour of the material and is the only stage observed when testing is done in vacuum.

The important effects relating to time dependence of strength are generally concerned almost solely with region I in that the behaviour associated with other regions occurs during very short times. This has led to the emergence of the parameter n as an all-important one in determining the time dependent behaviour of ceramics.

For simple systems such as glass, where the effects of micro-structure are eliminated, there is very good agreement between basic theory and experimental observation[17]. The theory is based on a stress enhanced chemical reaction between water and the highly stressed material near the crack tip. The basic rate equation from chemical reaction rate theory can be modified to reflect physical and chemical processes that occur at the crack tip. The theory is consistent with data for crack growth dependence on temperature, applied stress intensity factor, and the concentration of water in the environment.

One discrepancy between this theory and the general use of the data is that the theory predicts an exponential dependence between crack velocity and stress intensity factor. Referring to Fig. 8, this means that stress intensity factor scale for the theory is linear rather than logarithmic, which is the way most people use the data for engineering design. This perhaps illustrates the dangers of using double logarithmic plots where straight lines are commonly obtained.

A more serious problem however relates to the different behaviour of large cracks, as used to generate the above data, and small cracks from which the fracture originates in normal specimens. Values of n obtained from K/v diagrams should thus be treated with some caution, and it is often preferable to use data for n obtained from strain rate dependence of strength or delayed fracture measurements. For example in alumina it has been found that the value of n obtained from strain rate variations is approximately one half that obtained from double torsion tests[18]. The K/v data from macroscopic cracks would thus lead to a highly optimistic performance of components.

For engineering design purposes one can combine the statistical and time dependent effects which are usually expressed in the forms of the well known strength-probability-time diagrams[18,19] or proof test diagrams[20]. Examples are shown in Figs. 9 and 10 and generally one can obtain a good prediction of engineering performance using these mainly empirical approaches to the problem.

5. EFFECTS OF MICROSTRUCTURAL AND CHEMICAL CHANGES

In the discussion so far the only changes occurring in the material involve the subcritical growth of microcracks up to the point of catastrophic failure. Whereas this situation obtains for low-to-moderate temperatures, the situation is much more complicated at higher temperatures where additional effects can occur involving microstructural or chemical changes to the material. There are a number of important effects including:

(a) localised plastic flow which can induce crack healing or the generation of creep induced voids,

(b) chemical changes, particuarly oxidation, which again can produce crack healing but also the generation of new defects,

(c) microstructural changes such as grain growth, phase changes or surface evaporation.

Any of these changes may be either deleterious or advantageous with respect to strength.

The important consequence is that the original flaw population present at ambient temperature cannot necessarily be regarded as relevant in controling the mechanical response of a material to

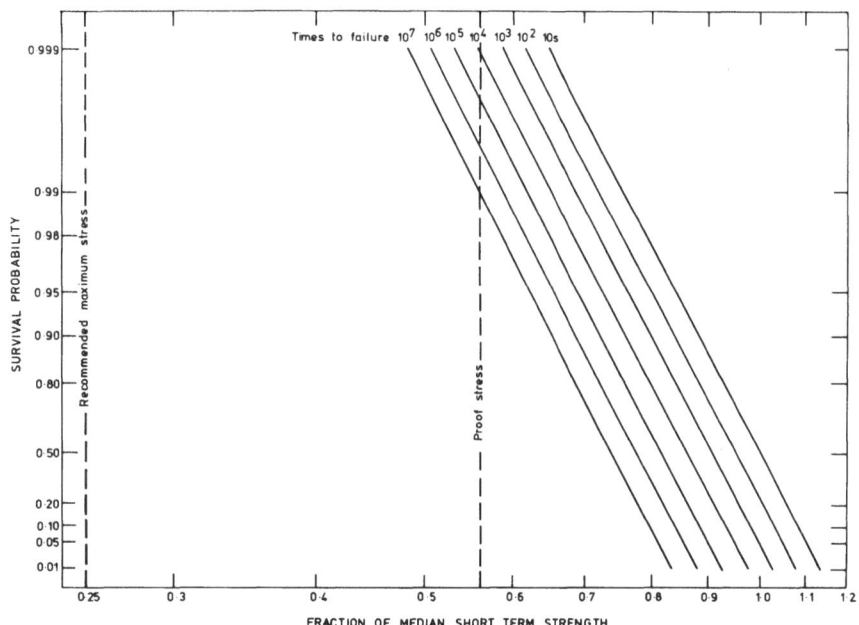

Fig. 9   Strength-probability-time diagram for vitreous bonded
          alumina ceramic[19].

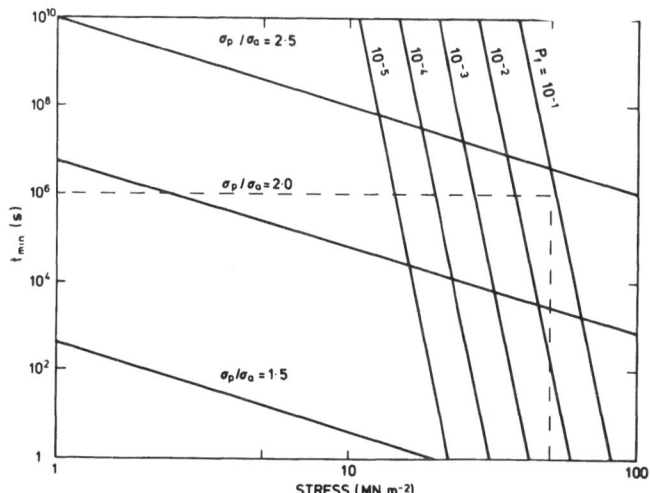

Fig. 10   Proof test failure diagram for glass[20].

stress at high temperatures. This means that the valuable techniques mentioned above such as proof testing or non-destructive testing have limited validity for high temperature performance prediction. This is because the ranking of the strength of a set of components is different at low temperature compared with high temperature. A dramatic example of this for hot pressed silicon nitride is shown in Fig. 11. Data for two sets of specimens are compared at 25 and 1200°C, before and after proof testing at 25°C. In the data at 25°C the strength of the proof-tested specimens is highly truncated, and in agreement with theory, indicating the value of the proof test. On the other hand, similar data for specimens tested at 1200°C, having first been annealed for 30 min at 1200°C, show virtually identical strength distributions irrespective of proof-testing. This indicates strongly that the original flaws are not relevant in controlling strength under high temperature conditions.

With regard to plastic flow, one of the definite benefits is that sharp cracks can become rounded at temperatures where limited plasticity in the material can occur. An example is shown in Fig. 12 which gives data for an alumina ceramic containing a small proportion of a glassy phase. At temperatures where the glassy phase begins to show significant ductility, $\sim 650°C$, the sharp cracks can be blunted under the action of stress due to plastic flow. This results in the increase in strength shown after ageing specimens under stress.

In contrast, for fine grain sized alumina deformed under conditions where creep can occur ($\sim 1500°C$) a number of voids occur at grain boundaries and coalescence of these eventually lead to failure. Under certain conditions the time for cavities to propagate across a grain facet is the dominant factor in the creep rupture process. Generally however the details may be too complex to enable satisfactory predictions to be made[23].

The most important chemical change is probably oxidation in connection with the high temperature engineering ceramics: carbides and nitrides. In these materials oxidation is usually significant at temperatures above $\sim 1000°C$ and results in a silica layer on the surface. This can lead to various effects including crack healing as described above. In other cases the surface oxidised layer may act in a similar way to a convitoanl glaze on a ceramic with a beneficial increase in strength, or alternatively the oxidised layer may crack and give a reduction in strength. These two effects are shown in reaction bonded silicon nitride specimens which have been heated in air for 24 h at 1400°C and then cooled directly to a test temperature and fractured[24], Fig. 13. The dramatic fall in strength at $\sim 300°C$ is concerned with a phase change in the oxidised layer involving the two crystal forms of cristobalite. In other cases such as hot pressed silicon nitride

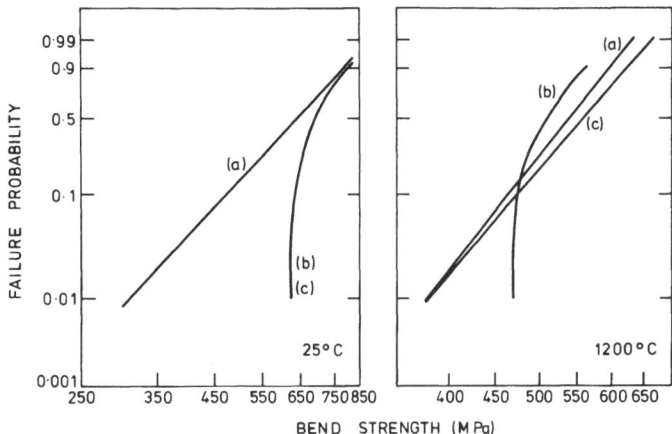

Fig. 11   Effects of proof testing at low temperature on the
strength of silicon nitride tested at 25°C and after 30
min at 1200°C[21].
(a)   Samples not proof tested.
(b)   Theoretically expected curve after proof testing.
(c)   Data for proof tested samples.

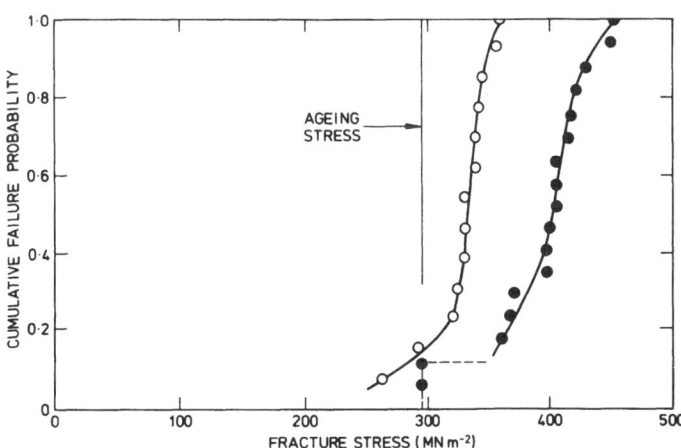

Fig. 12   Effects of ageing under stress at 650°C on the strength at
650°C of a 95% alumina ceramic[22]. Open circles,
original strength; closed circles, strength after ageing.

containing impurity inclusions, a pit generation mechanism on oxidation has been identified as the factor controlling the strength of samples[25].

One has thus a daunting list of possibilities for potential high temperature effects that control strength and it is thus not surprising that performance prediction techniques under these conditions are still in their infancy. Nevertheless, attempts have been made to structure the problem and one interesting approach as described by Wiederhorn[25] is the strength degradation diagram. An example is shown in Fig. 14, relevant to magnesia doped hot pressed silicon nitride where four failure regimes are mentioned. The initial distribution of strengths is indicated on the left side of the figure. Those specimens with a strength lower than the applied stress fail instantly. A second set of specimens with initial strengths above the applied stress but below a static fatigue limit fail by subcritical crack growth. The remaining, stronger specimens, were found by post fracture analysis to have failed following a number of mechanisms involving, sequentially, healing of surface machining damage, pit generation due to impurity inclusions on oxidation, and finally cavitation creep. In this case there is some possibility for the understanding of behaviour in terms of fundamental mechanisms and thus the derivation of the boundaries between the various regimes using calculations from basic data. This is however clearly a priority area for research if the reliability of structural components for high temperature applications is to be derived with any confidence.

6. CONCLUSIONS

In summary, the table below draws out the main conclusions of this paper with indications as to those areas where further research work is needed. Comment is made on three topics; time dependence, statistical effects, and life prediction, for both low and high temperature applications.

	Low Temperatures	High Temperatures
Time Dependence	Effects restricted to sub-critical crack growth. Effects are well understood theoretically and prediction is good providing that data obtained is relevant to growth of inherent cracks rather than large cracks	A multitude of effects occur including sub-critical crack growth and crack rounding, chemical pitting, oxide layer formation, creep void generation, and microstructural changes. Most effects are understood individually but there are complications when several occur simultaneously.

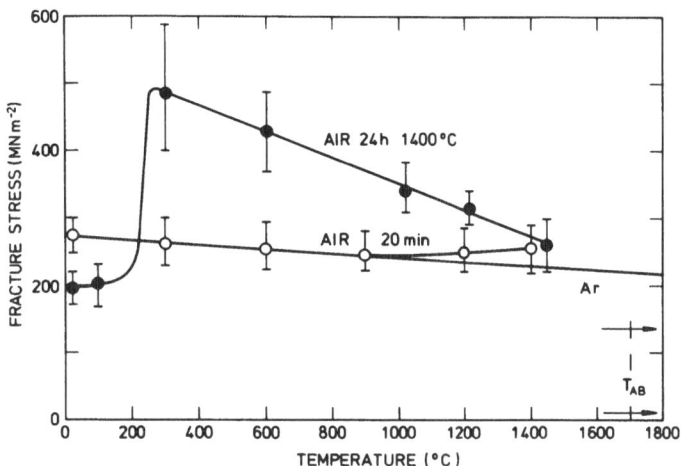

Fig. 13   Effects of oxidation on the temperature dependence of
strength of reaction bonded silicon nitride.   Open
circles, tested immediately; closed circles, tested after
heating for 24 h at 1400°C and cooled directly to test
temperature[24].

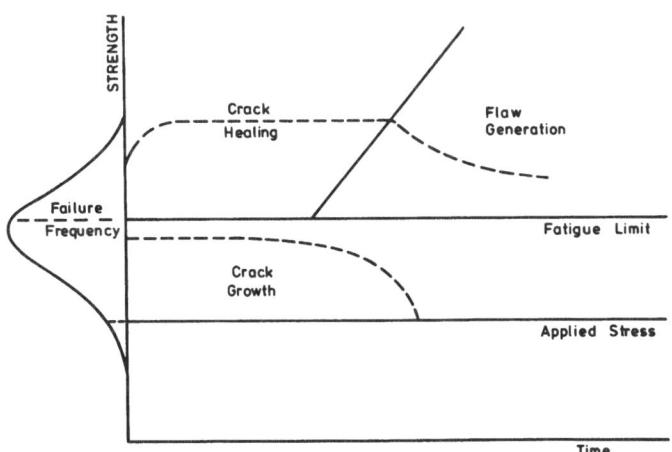

Fig. 14   Features of the strength degradation diagram relevant to
hot pressed silicon nitride at 1200°C[25].

	Low Temperatures	High Temperature
Statistical Effects	Weibull statistics are usually satisfactory and useful for prediction of effects of stress state and specimen volume. Alternatively an empirical statistical approach can be used.	A major problem is that the strength ranking of specimens can change during high temperature use which makes prediction very difficult.
Life Prediction	There is a reasonable empirical understanding, and life predictions can be made with some confidence. Weak specimens can be eliminated via proof testing procedures, and non-destructive testing techniques although increases in sensitivity here are required.	Understanding is restricted. Proof testing and non-destructive testing may be of very limited usefulness.

REFERENCES

1. R.W. Davidge, 'Mechanical behaviour of ceramics, (1979), Cambridge University Press.

2. A.A. Griffith, 'The phenomena of rupture and flow in solids', Phil. Trans. Roy. Soc. Lond. A221 (1920) 163.

3. A.A. Griffith, 'The theory of rupture', in Proceedings of the First International Congress on Applied Mechanics (ed. C.B. Biezeno and J.M. Burgers), p. 55, (1924), J. Waltman, Delft.

4. G. Tappin, R.W. Davidge and J.R. McLaren, in Fracture Mechanics of Ceramics (ed. R.C. Bradt, D.P.H. Hasselman and F.F. Lange), (1978), p. 435, Plenum, New York.

5. D.J. Campbell, B.R. Lawn and P.L. Kelly, J. Amer. Ceram. Soc. 65 (1982), C31.

6. E.J. Gumbel, 'Statistics of extremes', (1958), New York, Columbia University Press.

7.   W. Weibull, J. Appl. Mech., $\underline{18}$, (1951), 293.

8.   J.R. Matthews, F.A. McClintock and W.J. Shack, J. Amer. Ceram. Soc., $\underline{59}$, (1976), 304.

9.   F.A. McClintock, 'Fracture mechanics of ceramics', (Ed. R.C. Bradt et al), $\underline{1}$, (1974), 113; New York, Plenum Press.

10. F.A. McClintock and F. Zaverl, Int. J. Fract., $\underline{15}$, (1979), 107.

11. A.G. Evans and R.L. Jones, J. Amer. Ceram. Soc., $\underline{61}$, (1978), 156.

12. S. Batdorf, 'Fracture mechanics of ceramics', (ed. R.C. Bradt et al), $\underline{4}$, (1978), 1; New York, Plenum Press.

13. O.K. Salmassy, W.H. Duckworth and A.D. Schwope, 'Behaviour of brittle state materials', Technical Report 53-50, part 1, (1955), Wright Air Development Centre, Ohio.

14. R.J. Price and H.R.W. Cobb, 'Application of Weibull statistical theory to the strength of reactor graphite', in Proceedings of the Conference on Continuum Aspects of Graphite Design, (ed. W.L. Greenstreet and J.C. Battle, (1972), p. 547, Oak Ridge National Lab., Tenn.

15. D.H. Pankow and I. Finnie, 'Mechanical behaviour of materials', (ed. K.J. Miller and R.F. Smith), $\underline{3}$, (1979), 67; New York, Pergamon Press.

16. T. Okada and G. Sines, J. Amer. Ceram. Soc. 65 (1982), in press.

17. S.M. Wiederhorn, E.R. Fuller and R. Thomson, Metal Science, $\underline{14}$ (1980), 450.

18. R.W. Davidge, J.R. McLaren and G. Tappin, J. Mater. Sci. $\underline{8}$, (1973), 1699.

19. J.R. McLaren, R.W. Davidge, D.C.L. Cotton, S.A. Haywood and M.E. Robson, Proc. Br. Ceram. Soc., $\underline{26}$, (1978), 67.

20. S.M. Wiederhorn, 'Reliability, life prediction, and proof testing of ceramics', in Ceramics for High Performance Applications (ed. J.J. Burke, A.E. Gorum et al.), (1974), p. 633, Brook Hill, Chestnut Hill, Mass.

21. S.M. Wiederhorn and N.J. Tighe, J. Mat. Sci. <u>13</u>, (1978), 1781.

22. J.R. McLaren and R.W. Davidge, Proc. Br. Ceram. Soc. <u>25</u>, (1975), 151.

23. C.H. Hseuh and A.G. Evans, Acta Met. <u>29</u> (1981) 1907.

24. A.G. Evans and R.W. Davidge, J. Mater. Sci. <u>5</u> (1970), 314.

25. S.M. Wiederhorn, 'Fracture mechanics of ceramics', ed. R.C. Bradt et al, Plenum Press NY, <u>5</u> (1982), in press.

DISCUSSION

W.E. Buresch: In your Fig. 7 you show various linear arrays of microcracks that would all propagate at the same stress. I agree with you, because the strength increases to the inverse of the square root of the microcrack length. Conversely the strength decreases with increasing density of colinear microcracks $\frac{C}{L}$. Thus there are two reverse features. On the other hand the Young's modulus depends only on the density of colinear microcracks $\frac{C}{L}$ (see Koitor and Yokoberi e.g.) and therefore will decrease in the sequence of the first three lines of microcrack configurations of your Fig. 7.

R.W. Davidge: I agree with your comments about Young's modulus. But to quantify the effect we also need to know the width of the specimen (W) compared with the total length of the crack array (L_{tot}). For large W/L_{tot} the effects on Young's modulus would be small, and vice versa.

G. Grathwohl: Dr. Davidge mentioned statistics as a feature in mechanical behaviour of ceramics, which is fairly well understood. Is this an understanding in terms of microstructural items or does it just refer to the possibility of fitting the experimental strength distribution by a particular mathematical function, e.g. the Weibull distribution? What happens to the applicability of this statistical approach if the original flaw population is changed, e.g. after a period of application of load at higher temperatures?

R.W. Davidge: The statistical variation in the strength of

ceramics is understood mainly at an <u>empirical</u> level. It is
not possible generally to relate the size distribution of
inherent flaws to the strength variation. At high
temperatures there is a strong likelihood that entirely new
flaws will be generated (as opposed to extension of existing
flaws) - in this case any statistical strength measurements
on as-manufactured materials are not relevant.

ENGINEERING OF REFRACTORY STRUCTURES

PADGETT, G.C. and PALIN, F.T.

BRITISH CERAMIC RESEARCH ASSOCIATION LIMITED

QUEENS ROAD, PENKHULL, STOKE-ON-TRENT, ENGLAND

SYNOPSIS

The conditions pertaining to refractory structures are examined
for the purposes of specifying engineering data. A comparison between service conditions and those that are prescribed in National
Standards - traditionally used for measuring engineering data -
shows clear differences. The influence of these differences are
studied in simulative experiments on two of the most important properties, thermal expansion and modulus of elasticity. A theoretical model for cylindrical structures, modified to take account of
conditions that exist in service, is briefly delineated. This model is employed to illustrate the significance of data derived by
simulative and traditional procedures. Several examples are given
of how changes in structural design affect the stresses in refractory linings and retaining steelwork. A rig designed to verify
the validity of the measured engineering data and the theoretical
model is described. Areas where further research and development
should be directed are highlighted.

1. INTRODUCTION

The engineering design of refractory structures differs from that
of other load bearing systems because of the high temperatures involved.   Due to these high temperatures there is a much greater need to critically examine the chemical, thermal and mechanical
aspects involved in design.

Chemistry is the most important as is reflected in the abundance
of published papers, supplemented by phase diagrams, describing

120

and interpreting the degradation processes where the conditions are particularly corrosive or erosive. Some processes inevitably dictate a compromise solution because the slags involved have different basicities at different stages within a refining cycle; an example being the use of magnesite in a B.O.F. vessel where in the early stages of refining the slags being of low basicity are especially aggressive particularly with iron in the ferrous state.

Thermal design has two important but apparently opposing functions. In the important field of fuel efficiency the insulation of many refractory structures with fibre blanket has made a significant contribution to energy saving. There are, however, other processes where insulation would produce a disaster by raising the temperature and increasing the chemical reaction. In the case of known examples like blast furnaces, electric arc furnaces and slagging gasifiers, the opposite approach is adopted, and the successful operation depends on lowering the reaction temperatures by including water cooling.

Mechanical design together with satisfactory chemical and thermal conditions confers engineering stability to the structure. It is generally accepted that there is a dearth of engineering data that can be used directly in the design of refractory structures and experience remains a vital factor. It is therefore appropriate to select this as a suitable subject for a review of the 'state of the art' in engineering design.

2. MECHANICAL DESIGN

The requirements for mechanical design can be conveniently divided into two sections:

(i)   Engineering properties of all materials involved in the construction including refractory, joints and steelwork. Measurements must cover the range of values likely with each material.

(ii)   A theoretical model to define the thermal and mechanical loads with particular attention to those areas of high intensity.

2.1. Measurement of Engineering Properties

To formulate methods for measuring engineering properties it is necessary to understand the conditions that refractories are subjected to in service. The complexity of the two real situations is simplified in Figure 1.

In the case of a cylindrical structure the steel shell will be protected by an inner lining of refractory which could be bricks or a monolithic material such as refractory concrete. As the temperature in the centre increases the refractory expands and this expansion is restrained by the shell. It can be readily shown that the stresses are of the hoop and radial type. The hoop stresses act circumferentially around the vessel, compressive at the hot face and tensile at the cold face, and the radial stresses will be compressive but, being much lower than the hoop stresses, are less important. The stress levels developed will be influenced by numbers and types of joints which could be filled with cement and also by the thermal strain associated with the heating rate. A similar situation would exist in a furnace wall where stresses develop as the expansion is restrained, joints will be present and there will be a temperature gradient. The only additional factor which has to be considered is the gravitational stress acting vertically.

The important features to be considered when developing test procedures are thermal gradients, thermal strains, mechanical stresses and joints.

To theoretically predict stresses in these systems, as has already been stated, certain engineering data are required and these are given in Table 1. Density has been included because of its importance to other properties. Existing procedures for measuring these properties are given in National and International Standards. A comparison between the standard procedures and the service conditions shows that, at best, only one of the important service features is accounted for in those procedures. Of the properties given in Table 1, thermal expansion, modulus of elasticity, thermal conductivity and Poisson's Ratio are used directly for predicting stresses in refractory structures. Thermal expansion and modulus of elasticity are known to have the greatest influence on stress development and, as shown in the flow diagram (Figure 2), were selected for special attention. Separate and combined experimentation evaluated the influence of a temperature gradient, stress and joints on these two properties. A range of data was provided which when combined with thermal conductivity and Poisson's Ratio permitted theoretical predictions of stress to be made. Clearly, detail of such experimentation is not required other than that necessary to understand the implications and limitations of the measured data.

2.1.1. Thermal expansion

An apparatus for the measurement of thermal expansion under the combined influence of a stress and a temperature gradient is shown in Figure 3. This shows a cross-sectional view with the test specimen placed in the centre heated on one face only. Above and below the test specimen are three sets of loading blocks in which a stress gradient can be applied. A similar loading system exists at right angles to those shown in the figure giving an additional stress gradient in the other plane producing a biaxial loading system. The dimensional movement is recorded using ceramic probes which pass through each of the loading blocks. This experimental technique can be readily adopted for the study of joints or the alteration in temperature gradient by adding insulation or water cooling to the cold face.

An example of how the thermal expansion of 42% Al_2O_3 firebrick is influenced by a temperature gradient and stress is given in Figure 4. The expansion marked α_N has been recorded by a standard procedure where a small sample has been heated uniformly and α_{TG} is the expansion under the influence of a temperature gradient. Comparison of the two expansions show distinct differences throughout the heating cycle. The most significant changes are, however, produced on the application of a moderate stress of 3.5 MN/m^2, at 600 °C the expansion is fully restrained followed by creep at higher temperatures.

In the case of a refractory concrete, consisting of graded refractory aggregate bonded by a hydraulic cement, the expansion characteristics are very dependent on the experimental conditions as shown in Figure 5. α_N, again determined on a small sample heated uniformly, shows a contraction between 200 °C and 400 °C as the hydraulic bond loses water which is characteristic of these materials. When a larger specimen is heated under a temperature gradient much greater expansions are recorded which depend on heating rate, bond maturity, cast density and free water content. The combination of a temperature gradient and compressive stress of 3.5 MN/m^2 shows the expansion fully restrained at 400 °C followed again by creep at higher temperatures.

From the range of values for thermal expansion illustrated by these two examples it is important to build up a data bank showing the influence of various factors particularly when the refractory ceases to behave elastically.

2.1.2. <u>Modulus of elasticity</u>

There are two methods by which Young's modulus can be measured – dynamic and static. In a dynamic method a rapidly alternating stress is applied and the resulting strain measured. One of the more popular dynamic methods is to vibrate a thin rod in the transverse mode at its resonant frequency, as illustrated in Figure 6, and from the length of the bar, resonant frequency and density then the modulus of elasticity is calculated. Under these conditions the bar will remain close to the origin in the stress-strain curve.

It is not so ideal under service conditions where the refractories are subjected to higher temperatures and stresses and some portion of the lining will be operating above its elastic limit.

A static technique has been developed to overcome some of these deficiencies. The test piece configuration, shown in Figure 7, requires the upper and lower surfaces to be machine ground flat and parallel and a hole drilled vertically through the centre to carry an alumina probe. This hole does not penetrate through the full length of the specimen to avoid bringing two additional surfaces into the extensometer measuring system. In the actual test procedure the specimen is heated uniformly and the height is kept constant by applying a stress through loading columns. The modulus is then taken as the stress required to suppress the expansion divided by the strain which is obtained from the amount of thermal expansion that has been restrained. This technique can be readily adopted for the study of the influence of joints whether they are filled with cement or otherwise.

The specimen and loading columns are then assembled between the supports and cross-head of a universal testing machine and surrounded by a split furnace as shown in Figure 8. A further improvement in the design of the equipment allows similar measurements to be made under the influence of a temperature gradient.

A comparison of the data obtained, by the dynamic and static techniques described, for a 42% Al_2O_3 firebrick are shown in Figure 9. In an unreal joint free situation, as expected, at lower temperatures and stresses the modulus is lower in the static technique but at higher temperatures and higher stresses the positions are reversed. Much lower moduli are recorded on the static technique once the cement joints are introduced and this has been found to be directly related to the ratio of cement to refractory. If the results of modulus with natural surfaces were included then these would fall between the values obtained for 1 and 2 mm. cement joints.

It is often more convenient in theoretical calculations to have the cement modulus separate from the fired refractory. To carry out this separation, additional experiments are required to measure the moduli of the refractory at different temperatures and stress levels. Once this information is known then strain recorded in the refractory σ_C/E_R can be subtracted from the strain in the refractory cement sandwich ε_C giving the strain in the cement. The modulus is then the ratio of the stress to the strain. The moduli of three cement joints, given in Figure 10, imply the thinner joint is temperature sensitive. This is not the case, the higher values at the higher temperatures are directly dependent on the higher stresses developed in this particular suppressed expansion test.

A further comparison of moduli determined by the two methods on a refractory concrete, given in Figure 11, show that apart from those obtained at low temperature the values obtained are very different , the static technique producing much lower values.

2.1.3. Thermal conductivity and Poisson's Ratio

There are several methods of measurement for thermal conductivity and for most applications the values obtained are acceptable. Under service conditions where the refractory is subjected to infiltration from foreign bodies like slag, moisture, gases, or where high or low pressure environments exist, then the thermal conductivity values have to be modified accordingly. This is a separate topic and is not covered in this general paper. A detailed investigation of Poisson's Ratio has not yet been undertaken but from literature and room temperature measurements it ranges from 0.1 to 0.3 and is often assumed to 0.20.

From the evidence presented once a material diverges from an elastic condition then property evaluation must include in part some simulative conditions. In the case of structures containing joints with the possible exception of constructions containing bricks with machine ground surfaces contoured to the required shape then this divergence can be expected over the full temperature range.

3. THEORETICAL MODEL

The second requirement in formulating the design of a refractory structure is a theoretical model that will realistically describe the conditions that exist. One approach is to consider the structures as consisting of simple shapes like cylinders, walls and arches which can be combined to form real structures. Of these a cylinder is the easiest to describe mathematically, it is also the shape of many large industrial structures. It has therefore been chosen as an example of how theoretical modelling is applied.

The model is based on strain axisymmetric finite - element analysis of a hollow disc subjected to a temperature gradient initially developed by Timoshenko[1] and extended by Tang.[2] Modifications to the model take account either directly or indirectly, of the conditions that exist in service including the influence of joints, compressive and tensile moduli, temperature dependence of various properties, relaxation effects and the influence of internal pressures.

The model divides the lining into a series of rings and to each of these rings is then assigned a temperature. Mechanical properties for each of these rings are then selected by interpolation on the basis of temperature and stress level. Difficulties do arise when radial and hoop stresses are of the opposite sign. It is an essential feature of the model that there is a continuous transfer of stress and movement between rings. The equations given in Figure

12 are standard and are included to show how various engineering properties are likely to influence the stress or the displacement. For example, high thermal expansion, α, or high modulus, E, results in higher stresses. In the actual model modifications have been made to these equations to the account of the various conditions.

The successful application of mechanical design, as I have described it, is equally dependent on derived engineering data and theoretical modelling and both have their limitations. Five examples have been selected to show the stress levels and distributions that are predicted from this approach.

The theoretical predictions of stress in a steel shell lined with either refractory concrete or bricks are given in Figure 13. These results show very high stress predictions based on modulus values obtained from the dynamic technique implying that both the concrete and the brick would fail by crushing in service. In the case of the concrete the steel shell would also fail. This example was based on an actual working vessel lined with concrete that had been operating for over three years and no recorded failures in either the refractory or the shell had been observed. Calculations based on a dynamic modulus are obviously deficient and in the past such theoretical predictions have been treated with a certain amount of ridicule. With the static modulus, stress levels are predicted which imply that in the case of the castable both the refractory and the steel shell would be safe. In the example of the brick structure the stresses in shell and refractory are too high to give a reasonable safety margin and an expansion allowance would be required in the lining.

From the values of the engineering data referred to earlier, the types of joints are going to significantly alter the stresses in a given system. A comparison of these stress levels for machine ground joints and cement joints of various thicknesses are given in Figure 14. Very high stresses of 60 MN/m² are predicted for the machine ground dry joint reducing to 30 MN/m² for a 1 mm. cement joint and a further lowering as the cement joint is increased. An expansion allowance would be required for the dry joint and 1 mm. thick cement joint. The chemistry or type of cement joint can also be important and the results given in Figure 15 shows how the distributions change with a heat setting and air setting cement for the same joint thickness.

A number of theoretical calculations[3] have been completed on a typical 9m. blast furnace lined with carbon refractories where different design changes are considered. In the first example given in Figure 16 two variables are considered: the first being a change in lining thickness and the second the influence of shell thickness. In each case the stresses within the carbon lining are only moderately effected within the limits examined but the shell appears to be much more susceptible to the changes. Similar conclusions were drawn in the study of the influence of a ramming material placed between the lining and the shell and these results are reproduced in Figure 17.

4. IMPLICATIONS OF DATA AND CALCULATIONS

The implications of these predictions need to be examined since whatever design is accepted certain safety factors have to be included for both the refractory and the retaining steelwork.

In the case of the steel shell, if a blast furnace is taken as an example, the allowable stress is taken as 80% of the yield stress divided by the stress concentration factor. The concentration factor depends on the individual design and will include holes, quality of welds, supports etc. Often, designs fall outside accepted Codes of Practice and decisions are based predominantly on previous experience. Inevitably this produces conservative safety margins and in case of some blast furnaces this is as low as 20% of the yield stress of steel.

As a starting point ideal behaviour could be assumed in the case of the refractory lining where each brick is subjected to exactly the same thermal and mechanical stress. An 'average brick' could then be selected, subjected to the thermal-mechanical stresses predicted theoretically and the conclusions about service behaviour could then be based on the results. Such conclusions would be subject to some very obvious and valid criticisms. The only acceptable approach is the direct measurement of stress or strains in a real situation and compare these with the theoretical predictions. Unfortunately such an exercise is extremely costly, the opportunity of carrying out measurements on a virgin structure does not happen often and controlled measurements are difficult. A less acceptable alternative is to use scaled down models.

The apparatus shown in Figure 18 is an example of an experimental rig containing a 1 metre diameter tube supported on a movable bogie. Strains on the shell are measured by high temperature strain gauges. The radial and longitudinal displacement is measured by means of linear displacement transducers mounted on a water cooled frame and, as shown in Figure 18, are protected from radiation and convection by heat reflecting shields. The temperature gradient is measured by thermocouples placed in the lining. Six large silicon carbide heating elements located in the centre are connected to suitable ancillary equipment to enable hot face temperatures of 1400°C to be achieved at controlled rates of heating. The frame is so designed that vessel diameter can be doubled so the influence of vessel size can be studied. A number of detailed experiments are currently in progress on various types of structure. These results will then be compared with theoretical predictions. A judgement can then be made by comparing the condition of the refractory with the theoretical predictions and also an appropriate strength value. It is often difficult to judge the condition of the refractory, and additional information is obtained by attaching an acoustic listening device to the refractory lining.

Considerable confidence has already been shown in the results of this study and examples where the analysis has been applied includes blast furnaces, a number of petro-chemical vessels, oxygen converters, electric arc furnace roofs, cyclones, pressure vessels, cement kilns, direct reduction kilns, slagging gasifiers and hydrogenators.

5. FUTURE RESEARCH ACTIVITIES

Poisson's Ratio has only been briefly mentioned. It does have some influence on the predicted stresses and so there is a need to know exactly how it is affected by stress and temperature. The limitations of standard methods for determining thermal expansion and modulus have been highlighted. There are several alternative routes that can be taken to provide engineering data for the theoretical model, and these are summarised in Figure 19. It has not yet been proved which route is the most appropriate and this will depend on the results obtained in scaled down model experiments and also some field work that has been undertaken as part of a private contract. A comparison between predicted and measured stresses is likely to lead to some modifications to the theoretical model.

REFERENCES

1. Timoshenko S.P. and Goodier J.A., Theory of Elasticity, 3rd. Edition. Engineering Societies Monographs 1970, p.448.

2. Tang S. Thermal Stresses in Temperature Dependent Anisotropic Hollow Cylinders. Canadian Aeronautics and Space Institute Transactions 31, 72-76, March 1970.

3. Hodson P.T.A. and Speed K.H. Refractories for Ironmaking. Proceedings of the British Ceramic Society No. 29, October 1980.

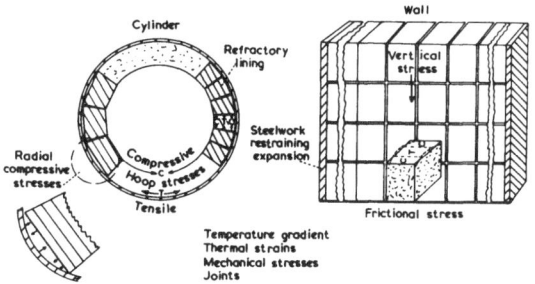

FIGURE 1.   SIMPLIFICATION OF SERVICE CONDITIONS

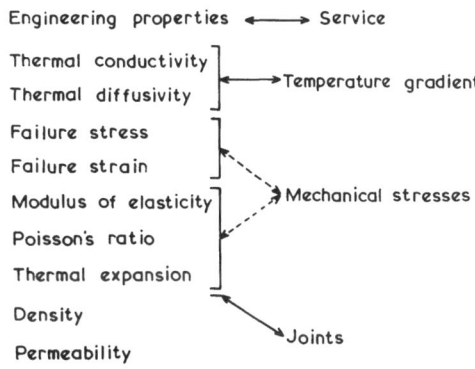

TABLE 1.   ENGINEERING PROPERTIES

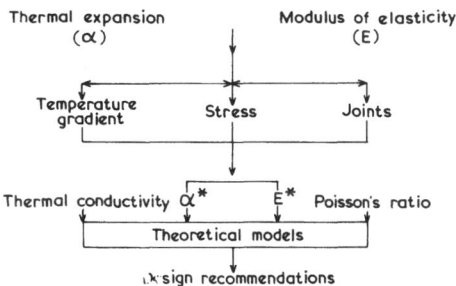

FIGURE 2.   FLOW DIAGRAM OF RESEARCH PROGRAMME

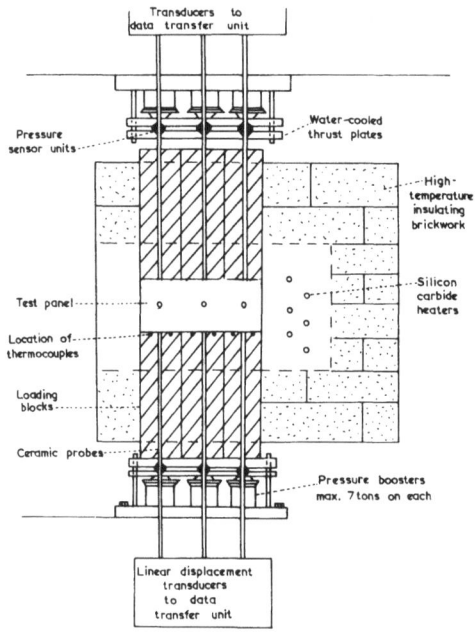

FIGURE 3. CROSS-SECTIONAL VIEW OF BIAXIAL STRESS APPARATUS

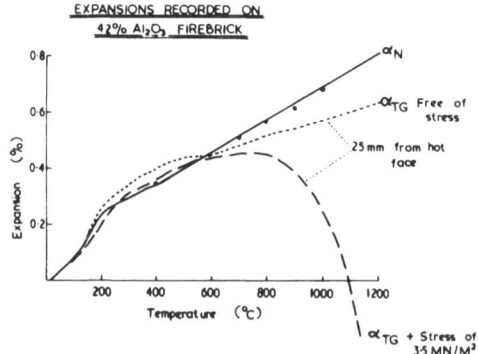

FIGURE 4. THERMAL EXPANSIONS RECORDED ON 42 Al₂O₃ FIREBRICK

130

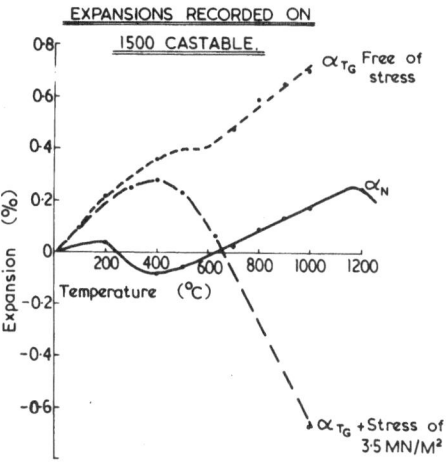

FIGURE 5. THERMAL EXPANSIONS RECORDED ON A REFRACTORY CONCRETE

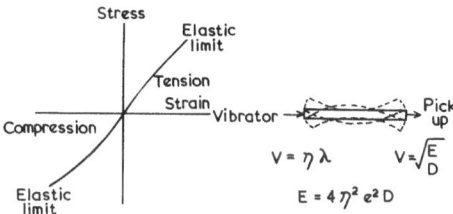

FIGURE 6. DYNAMIC METHOD FOR DETERMINING YOUNG'S MODULUS

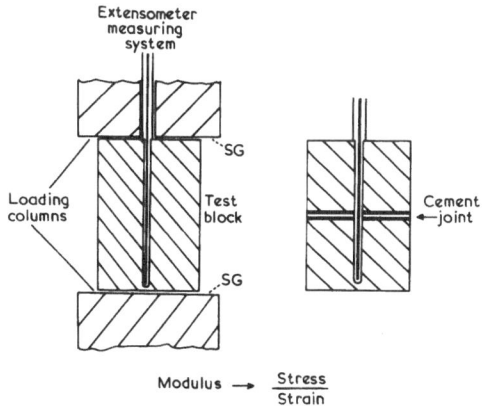

FIGURE 7. TEST PIECE CONFIGURATION FOR SUPPRESSED THERMAL EXPAN-
SION TEST

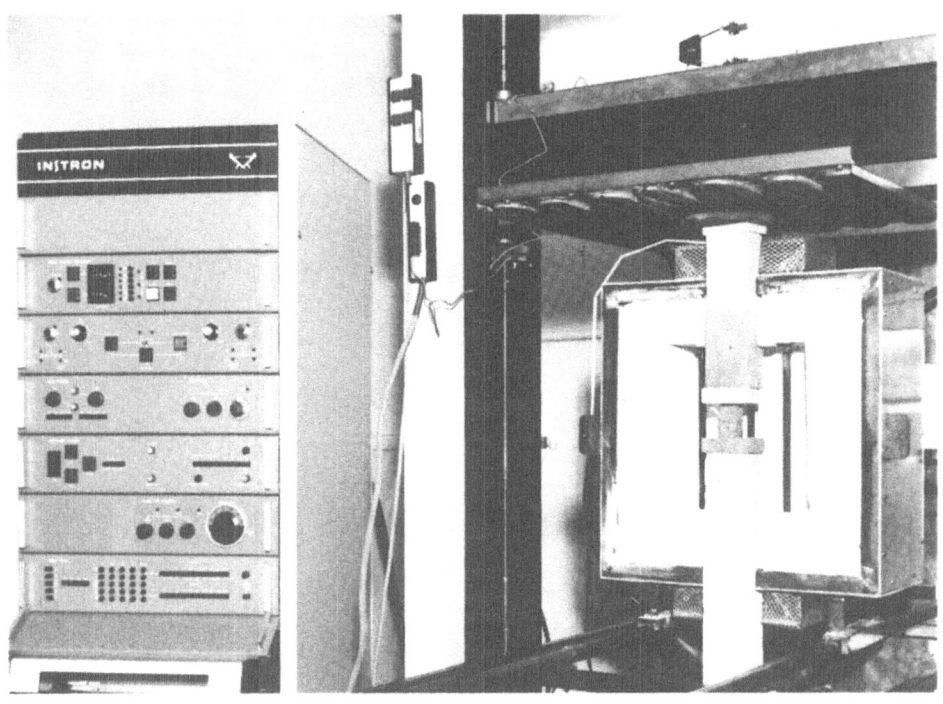

FIGURE 8.   ASSEMBLY FOR SUPPRESSED THERMAL EXPANSION TEST

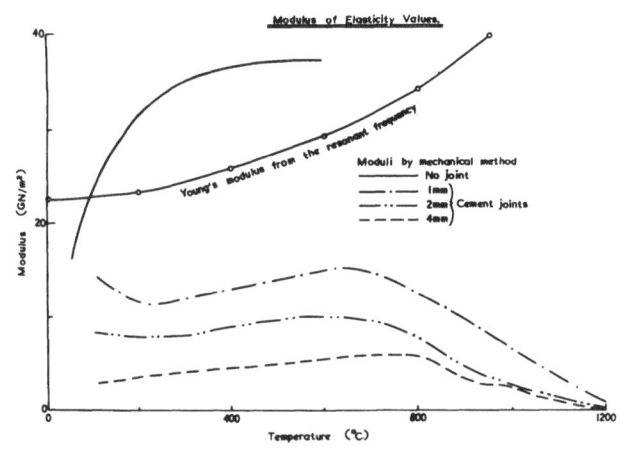

FIGURE 9.   MODULUS VALUES OF REFRACTORY JOINTS

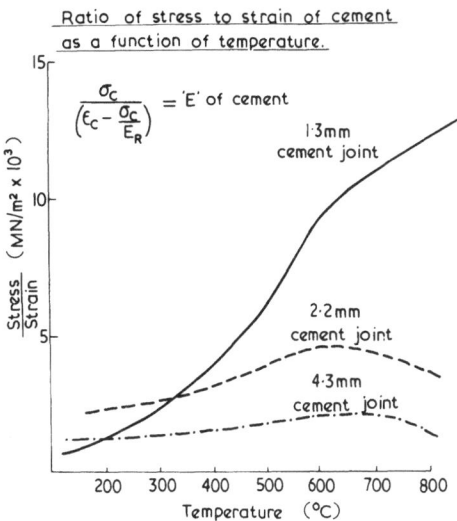

FIGURE 10. MODULUS VALUES OF CEMENT JOINTS

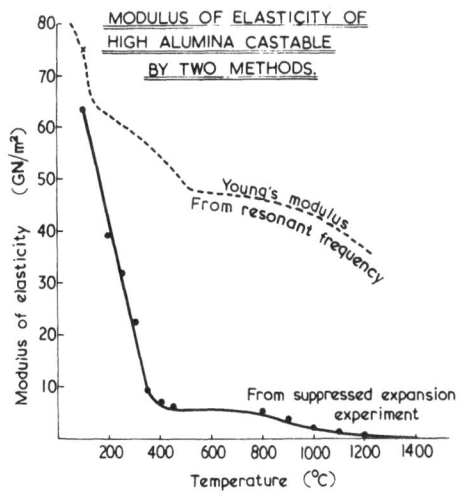

FIGURE 11. MODULUS VALUES OF A REFRACTORY CONCRETE

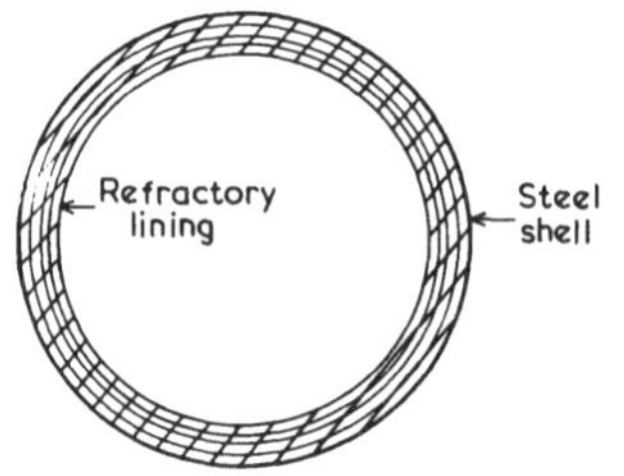

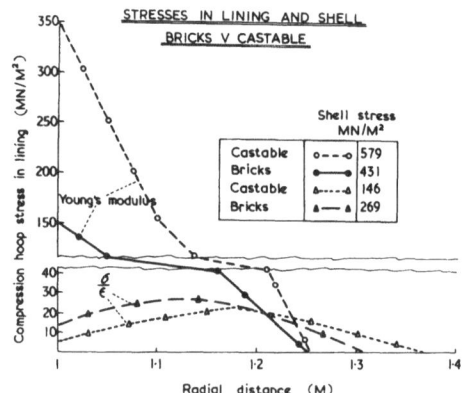

Refractory lining
Steel shell

$$\text{Radial stress} = -\frac{\alpha E}{r^2}\int_a^r Tr\,dr + \frac{EC_1}{1-\nu} - \frac{EC_2}{(1+\nu)r^2}$$

$$\text{Radial displacement} = \frac{(1+\nu)\alpha}{r}\int_a^r Tr\,dr + C_1 r + C_2 r$$

$$\text{Hoop stress} = \frac{\alpha E}{r^2}\int_a^r Tr\,dr - \alpha ET + \frac{EC_1}{1-\nu} + \frac{EC_2}{(1+\nu)r^2}$$

FIGURE 12. THEORETICAL MODEL

STRESSES IN LINING AND SHELL
BRICKS V CASTABLE
Compression hoop stress in lining (MN/M²)
350
300
250
200
150
40
30
20
10
Young's modulus
σ/ε
Shell stress MN/M²
Castable 579
Bricks 431
Castable 146
Bricks 269
1
1·1
1·2
1·3
1·4
Radial distance (M)

FIGURE 13. STRESSES IN A CYLINDRICAL STRUCTURE

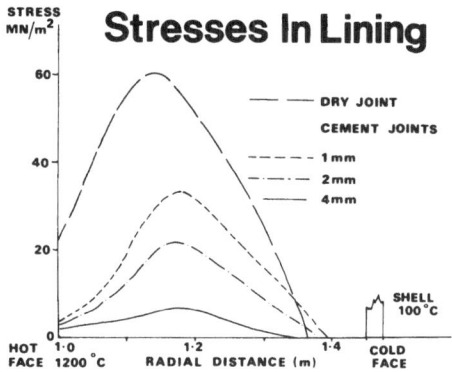

FIGURE 14.   INFLUENCE OF JOINTS ON STRESSES IN A CYLINDRICAL
STRUCTURE

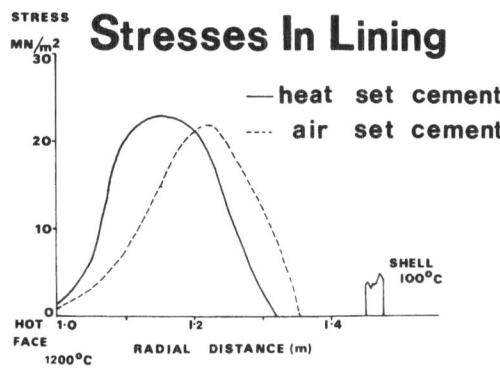

FIGURE 15.   INFLUENCE OF CEMENT TYPE ON STRESS IN A CYLINDRICAL
STRUCTURE

Effect of lining thickness

Lining thickness (mm)	Maximum stress (MN/m²)	
	In lining	In shell
316	17	45
566	18	84
816	18	105

Effect of shell thickness

Shell thickness (mm)	Maximum stress (MN/m²)	
	In lining	In shell
25	16·5	132
50	18	84
75	19	62

FIGURE 16.   STRESSES IN A BLAST FURNACE

Effect of compressibility and thickness of backing
layer on the stresses in the lining and shell of
a blast furnace.

		Maximum stress in lining (MN/m^2)	Maximum stress in shell (MN/m^2)
Compressibility	Halved	19	100
	Standard	18	84
	Doubled	16	65
Thickness (mm)	0	21	124
	38	21	118
	63	18	84
	75	16·5	66

FIGURE 17. INFLUENCE OF RAMMING ON THE STRESSES IN A BLAST
FURNACE

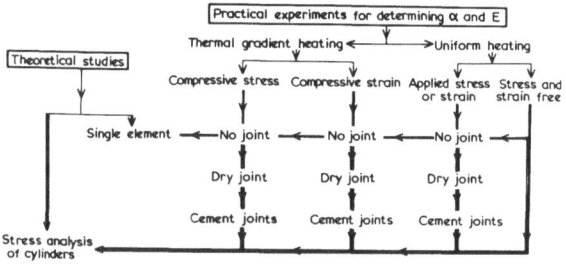

FIGURE 19. SUMMARY METHODS OF DETERMINING α AND E.

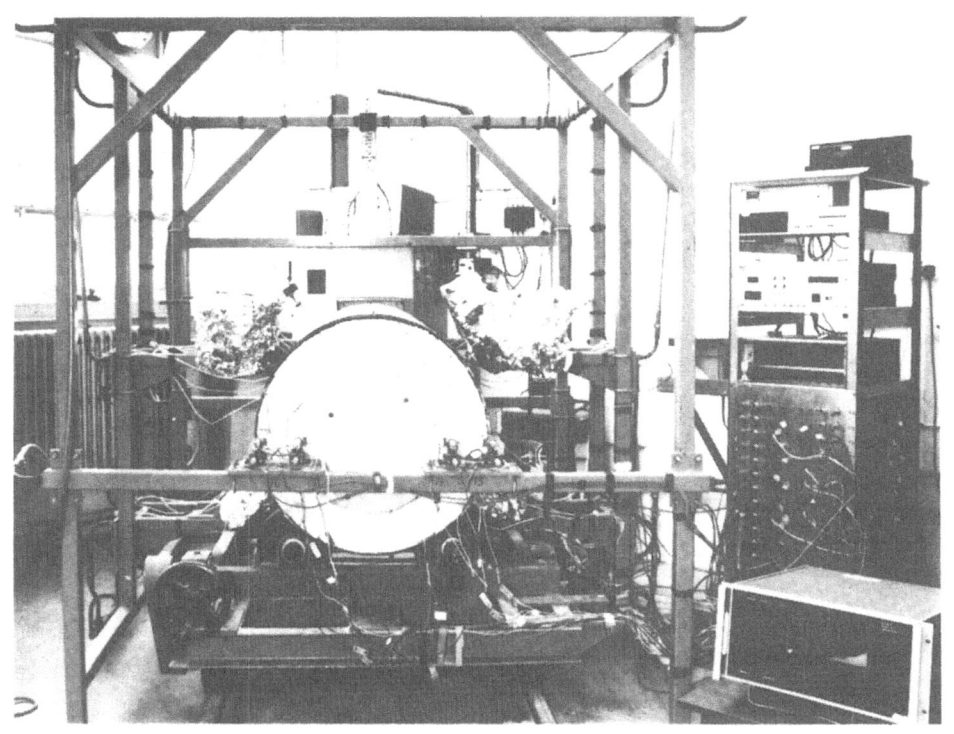

FIGURE 18. CYLINDRICAL RIG FOR MEASURING STRESSES

JOINING OF CERAMICS

FOR HIGH TEMPERATURE APPLICATIONS

M. Erz; H.W. Hennicke

Institut für Steine und Erden

Technische Universität Clausthal

Abstract: Ceramics for engineering applications at elevated temperatures have considerable advantages over metals. Because of limitations in their manufacture it is desirable to restrict their
use to only these zones where the specific properties are required.
For this purpose techniques have been developed to join ceramic
with ceramic or metal components.
In the first part of this review the conditions will be mentioned
which provide a compound, i.e. compatibility, close contact and
surface activity. The adherence is formed by physical adhesion,
chemical reactions and sintering mechanisms.
The second part reports of several joining techniques which are
applicable. They are classified in three groups: mechanical joining, brazing with application of intermediate layers and welding
with or without forming of a liquid phase. Application of these
methods will be shown in the area of ceramics based on alumina,
silicon nitride and silicon carbide.

1. Introduction

Joining means to put two or more components together involved
with forming of a local connection between the parts. Thereby it
must be possible to transfer forces across the interface. In contrast to joining, coating means to cover a component with shape-
less material, for example enameling.

Joining techniques must be classified into mechanical joining
and direct joining involved with chemical reactions.

Mechanical joining is based on mechanical interlocking and mechanical forces. The force transmission results from the shape of
the connected parts, for example screws, rivets, cramps etc..
In the other case a chemical reaction occurs between the parts
to be joined, sometimes involved with application of supplementary intermediate layers. These methods are brazing, welding and
adhesive joining.

The joining of ceramics with ceramics or with metals is important
for several instances. First, size and shape of the components to
be manufactured is limited by physical and economical reasons.
Second, different materials with different required properties
have to be composed.

2. Fundamentals

The first condition of joining is a close contact of the components to be joined. While manufacturing the compound there must be
a close contact between the surfaces to get a joint interface.
There are two different ways to proceed: One of the materials,
generally the metal, has a sufficient ductility to come into contact with the ceramic surface. Such compounds must be produced
with application of high pressure (hot -isostatic- pressing,
pressure welding at room temperature in ultra high vacuum). For
pressureless joining it is mostly necessary to get a liquid phase
at the interface to wet the materials and to provide contact.

Special properties of the materials to be joined are required
to get a fixed connection. They must be compatible at room temperature and at elevated temperature. Especially the latter is
important for some applications of composite materials.
Chemical and mechanical compatibility are required, that means materials should have nearly the same thermal expansion coefficient
and there must not occur any structural transition. Thereby the
internal stresses due to increasing temperature or thermal shock
can be minimized. While at elevated temperatures a reduction of
thermal stress occurs due to plastic deformation or viscous flow,
at lower temperature the component will break as a result of
brittleness in the ceramic.

Many effects are responsible for adherence of materials: physical interactions (adhesion), chemical reactions with forming of
new phases like spinels, nitrides, carbides, diffusion and sintering phenomena across the interface and mechanical interlocking in microscopic scale due to surface roughness. Which kind of
mechanism is dominating depends on joining technique and the parameters of manufacturing, i.e. time, temperature, pressure and
atmosphere.

2.1. Physical Interactions

If two materials are to be joined, two surfaces must get into close contact to form one interface. Thus, first interest should tend to surface characteristics of the materials involved. Generally, atoms at the surface have a higher free energy than those in the volume of a solid or a liquid. This elevated free energy is defined as free surface energy or interfacial energy between solid and vapor phase. In the same manner there are interfacial energies between solid and liquid, solid and solid phases etc..

A system composed of solid, liquid and vapor phase (fig.1) in equilibrium can be calculated with the YOUNG-equation:

$$\gamma_{sv} = \gamma_{sl} + \gamma_{lv} \cos\theta \tag{1}$$

γ_{sv} : surface energy between solid and vapor phase
γ_{lv} :   "   "   "   liquid and vapor phase
γ_{sl} : interfacial energy betw. solid and liquid phase
θ : contact angle

The adhesion energy corresponds to the work to be done to seperate one unit of the interface into two identical surface-units at constant temperature. The specific energy of adhesion between the solid and the liquid phase can be calculated with the equation of YOUNG and DUPRE:

$$W_A = \gamma_{sv} + \gamma_{lv} - \gamma_{sl} \tag{2}$$

Equ.(1) in combination with equ.(2) shows the relation between adhesion energy, surface energy and contact angle of the liquid.

$$W_A = \gamma_{lv} (1 + \cos\theta) \tag{3}$$

However, adhesion energy may be considered only as qualitative measure for adherence. Wetting behaviour improvement of the liquid phase results in increasing adhesion energy. Several effects are related to the contact angle, for example atmosphere, surface roughness and infiltration. Increasing temperature results in decreasing contact angle. Chemical reaction between solid and liquid phase, i.e. solution of the solid in the liquid phase, improves wetting behaviour and adhesion (1,2,3,4).

2.2. Chemical Interactions

The tendency of two materials to form new phases during chemical reactions is related to the type and strength of chemical bonds of the materials involved. A true chemical bond exists, if a continous structure of electrons occurs across the interface, that implies the presence of chemical equilibrium across the interface.

Reactions can be imagined in metal/metal- or ceramic/glass-systems to be caused on a similar type in chemical bonds and structure of electrons. Much more complicate are reactions in metal/glass- and metal/ceramic-systems. Here reactions are imaginable between the oxide material and the oxide layer always existing on metal surfaces. In the same way bonding results of forming nitrides and carbides.

The four types of chemical bonds are metal, ionic, covalent and van-der-Waals-bond (5). The actually in materials existing bonds are always composed of different types. In ceramic materials there is often transition from ionic to covalent bond. This is caused by deformation in electron orbitals of one ion by an oppositely charged one. In case of metals a transition between metal and co-valent bond may be present.

It is possible to get a quantitative measure for the type of bonds.
The theory of field intensity according to DIETZEL (6) describes the transition from ionic to covalent bond. The force of attraction, respectively the field intensity between two oppositely charged ions is proportional to their valencies and inversely proportional to their distance. Increasing field intensity value implies an increasing amount of covalent bond between two ions.

Another method, to estimate the amount of ionic bond has been derived by PAULING (5). This theory is based on a scale of electronegativities, which refer to the behaviour of a single bonded atom within an atomic connection and not to isolated atoms in inert gas condition. An estimation of the bond character shows: the Si-O - bond has 50% and the Al-O - bond has 63% ionic character. At silicon nitride results a small amount of 30% ionic bond, and silicon carbide has with 12% ionic character almost a covalent bond type.

A method to estimate the bond strength of an ion within a crystalline solid is the lattice energy. That means the work to be used to separate the atoms in a crystall,after an imagined splitting,so far,that no more interaction occurs. This is a theoretical measure of the stability of a crystalline structure without respect of defects in lattice.

2.3. Sintering and diffusion

Sintering can be described as strengthening the mechanical connection between single particles without chemical reactions taking place. Sintering is defined as thermal treatment of particles in contact together or pressed powder compacts with or without application of pressure. Thereby the properties of the porous compact will move to those of a poreless solid. Driving force of all sintering mechanisms is the decrease of the particles surface area, that means reducing the free surface energy by minimizing solid/vapor interfaces. Implied with that is the forming of solid/solid interfaces which have less free energy. Although the driving force of sintering is always the same, there are several kinds of material transfer to receive a solid with increased density (7,8).

Evaporation-condensation mechanism is based on different vapor pressures at different parts of the particles caused on different surface curvatures. At surfaces with a positive radius of curvature the vapor pressure of material is somewhat larger than would be observed at flat surfaces or those with a negative radius. Just at the contact point between two particles there is a neck with a small negative curvature. The vapor pressure difference between neck area and particle surface leads to material transfer into the neck area. Here no shrinkage occurs, only variation in the shape of pores, respectively particles, takes place. This results in a variation of properties without affecting density of the solid.

If the vapor pressure of the material is low, material transfer occurs more readily by solid-state sintering processes. Here one must differentiate between self-diffusion and chemical diffusion. Self-diffusion takes place in uniform substances, whereas in materials composed of different substances chemical diffusion dominates. In the latter case driving force is additionally based on compensation of electrochemical potential gradients. Solid state reactions will be done in a temperature range of about 90% of solidus temperature of the lowest melting phase.
Like evaporation-condensation solid state reactions are processes with thermal activation. The material transport is dependend on mobility of atoms in the crystal lattice. With increasing temperature the ability of atoms to change their position in the lattice increases. Defects are necessary to enable the atoms to move. Thus, diffusion at surfaces is larger than at grain boundaries which is larger than in the matrix. Atoms leave their lattice site to occupy vacancies or interstitial face to the next particle.

In metal crystals diffusion is easier than in ionic crystals. In a metal lattice all positions are equivalent, whereas a ionic crystal is composed of two lattice parts, one of anoins and one

142

of cations. Here diffusion is only possible if the condition of electroneutrality is fulfilled.

Another mechanism which is mostly be found in ceramic materials depends on the presence of a liquid phase. An important condition is a sufficient wetting behaviour, which is involved with a certain solubility of the solid in the liquid phase. If liquid wets solid particles, each pore becomes a capillary in which capillary pressure is developed. Densification results from different processes: When the liquid phase is fluid, a rearrangement of the solid particles takes place due to sliding. At contact points, where there are bridges between particles, a high local stress leads to plastic deformation in the solid phase resulting in a further rearrangement. At the same time the solubility of the solid increases at these contact points. This leads to a material transfer away from the contact areas so that particle centers approach one another and shrinkage results. During sintering there is a dissolution of smaller and growth of larger particles by material transfer through the liquid phase.

2.4. Surface preparation

The same mechanisms, which are true for sintering of microscopical small particles of a powder compact, must be applicable also for reactions across an interface between two macroscopic components to be joined. In this case the problem occurs, that these components are fabricated products, they have been fired before. Although, thermodynamical equilibrium usually will not be attained, it must be possible to get further reactions at sufficient high temperature.

Very important is the roughness of the surface. To get solid state reactions, a flat polished surface is required to provide a large contact area. With application of pressure plastic deformation of one material may lead to a better contact. For reactions involving formation of a liquid phase surface, however, roughness is desirable. The fluid will wet the solid and mechanical interlocking results by interaction between surface irregularities.

It is possible to produce an artificial surface roughness with two intentions. First is improvement of mechanical interlocking, second is a intended variation of material properties just at the surface to get an activation which leads to reactions with other components. It was studied, if it is possible to affect the surface of reaction sintered and hot pressed silicon nitride by high energy laser treatment. At the laser beam affected surface, the temperature will raise much higher than dissociation temperature of silicon nitride. Material will be decomposed and then evaporates together with impurities and grain boundary phase. At this point a small crater is excavated (fig.2). When the laser impulse is over, the surface will cool very fast. The phase with the lowest vapor pressure, in this case it is silicon, sublimates at the surface. Around the craters a zone results which is enriched with silicon. In the case of hot pressed silicon nitride a thin silicon layer at the surface is developed (fig.2a), which is not observed on reaction sintered material (fig.2b). The latter material is porous and silicon is able to penetrate into the surface.

With pulsed laser treatment a surface results which is dotted with a lot of small craters. The depth of these craters depends on laser energy, their distance depends on impulse period of the beam and on rate of feed of the samples fixing device. With variation of the laser parameters, energy, impulse period and rate of feed it is possible to produce a required surface roughness, (9). For joining applications small craters are suitable.

3. Joining technologies applied for engineering ceramics

In the second part of this review joining techniques will be presented in the area of engineering ceramics for high temperature applications. Compounds for this use are mainly metal/ alumina joints for vacuum tight components at elevated temperature and, in the area of nonoxide ceramics, silicon nitride and silicon carbide joined with one another or with metals for application in turbine engines, heat exchangers, ceramic motors etc..
Three general methods are applicable to produce such joints: mechanical joining, brazing and welding.

3.1. Mechanical joining

Mechanical joining is possible by using srews, metal clamps etc. This application is limited by comparatively low tensile strength and brittleness of ceramics.
An example of this type of joining in the area of turbine engines is a rotor, where ceramic turbine blades are fixed in the metal rotor with metal shims. The shape of blade basis and shims influences the developing stresses and must therefore be optimized (10).

The most applied technique of mechanical joining is shrinking of metal components onto ceramics. The metal component with the higher thermal expansion coefficient will be hot mounted around the ceramic part. During cooling the metal shrinks onto the cera-

mic and provides a tight connection, where the ceramic is put into compressive stress.
Applications of this technique are found in the wide areas of high voltage ceramics, where conductive metal parts must be joined with
ceramic insulating material in spark plugs, overload cut-outs etc.

3.2. Brazing

The traditional definition of brazing or soldering is related
to joining of metals with application of a liquid intermediate
material. The braze or solder metal has a lower melting point than
the components to be connected. With respect to application temperature it must be distinguished between brazing and soldering.
Soldering will be done below 450°C, brazing at temperatures above.
Based on this definition most ceramic joining processes are made
by brazing. The term soldering is often used for sealing
glasses, even if it is done above 450°C. An important difference
to metal brazing is involved with brazing of ceramics. Whereas
in conventional metal brazing only the filler metal is heated,
brazing of ceramic requires   heating of the total assembly
to prevent it from internal stresses.
Brazing of ceramics may be done in two steps with surface preparation by metallizing or in only one step which is called direct
brazing.

3.2.1. Metallizing

Generally, brazing of oxide ceramics needs a surface preparation
by metallizing. The surface will be coated with a metal layer
which is able to be brazed with conventional brazes to get a vacuum tight compound. Mostly this will be done by silkscreening or
spraying or similar processes. Another way, which is also be used,
is vacuum-deposition, sputtering or reduction of metal oxides by
firing in hydrogen atmosphere.

The adherence results from chemical reactions between metal and
ceramic and from penetration of silicate liquid phases into the
metal to develop mechanical interlocking. The most used process
in the area of high tension ceramics is the Mo-Mn-technique (11,
12,13,14). For application at elevated temperature, where alumina
with a very low content of liquid phase is required, it is feasible to add oxide fluxing agents to the metal powder.

TWENTYMAN (15) reports on metallizing of silicon nitride and
silicon carbide. A thin coating, which consists of tungsten powder
in a vitreous matrix, will be fired in nitrogen/hydrogen atmosphere.

This layer is then coated with nickel oxide, which will be reduced to nickel in a second firing. This layer provides a coating easy to be wetted by the braze. Here the applicable brazes consist of Pd-Cu-Au or Au-Ni.

3.2.2. Direct brazing

New techniques tend to braze ceramics with metals in vacuum or inert gas atmosphere without any metallizing. This could be called "direct" brazing, because a surface preparation like metallizing is not necessary. Alloys which are applicable are based on metals with a high affinity to oxygen: Ti, Cu, Be, Al, Zr. Reactions between metal and oxide ceramic forming new oxide phases provide good adherence. In case of non-oxide ceramics the compound results from the always existing oxide layer respectively the grain boundary phase.

Intermetallic compounds of active metals are very brittle. During cooling this can lead to failure of metal/ceramic compounds with different thermal expansion behaviour. Thus, suitable alloys must be selected.

The temperature program of brazing processes is very important. No fixed rule exists, for each brazing task empirical programs are to develop for each brazing task. The preheating program depends on size of components. With increasing size the time of preheating should also be increased. More steps with constant temperature are necessary to protect the compound from high thermal gradients, At the highest level of the preheating program, nearly 100 °C below solidus temperature of the braze, adhesion may be developed due to solid-state reactions. Usually the brazing temperature is about 50 °C above liquidus temperature of the braze alloy. If the alloy has a high vapor pressure, time at maximal temperature must not be too long, because evaporation of braze results in a compound with pores.

Fig.3 shows an example for direct brazing (10). Reaction bonded silicon nitride is joined with hot pressed silicon nitride with application of a Ti-Cu-Be alloy. Fig.3a shows the optimum compound with a Be-rich diffusion zone at the side of hot pressed silicon nitride and a Cu-rich zone at the reaction sintered side. Fig.3b shows the same compound with too long annealing time, which led to pores in the intermediate layer.

TWENTYMAN (15) reports of a compound of hot pressed silicon nitride with stainless steel using alloys based on Pd-Ni and Pd-Ag. Because of incomplete wetting of the steel by the braze, chromium or manganese were added to improve wetting behaviour.

146

Thereby the steel was dissolved by the braze alloy, which results in good adherence.

ISEKI (16) reports of joining reaction bonded silicon nitride with liquid germanium metal. Here the compound is based on solid solution between germanium and silicon.

An interesting way of brazing is the application of the metal in form of foils instead of powders.
Joining of hot pressed with reaction sintered silicon nitride was successful by the use of aluminium foil. Experiments have been done in vacuum (10) and in nitrogen atmosphere. A reaction occurs mainly with the reaction bonded silicon nitride. Analysis with EMP shows a diffusion zone of aluminium into silicon nitride of about 400 µm. To prevent too much diffusion of aluminium it is feasible to put a foil of copper between aluminium and reaction bonded silicon nitride. This technique leads to good adherence due to forming of Si-Al-O-N phases.

In the case of joining silicon nitride the transition between metal and non metal brazing is the addition of alumina and zirkonia to the brazing alloy to improve wetting behaviour.

The most applied non metallic brazes are vitreous, which exactly should be called solders. Besides that, there are mixtures of vitreous and crystalline phases or eutectic mixtures of crystalline phases. The wetting behaviour mostly is sufficient, because oxides are able to wet the oxide layer on the metal.

An application of ceramic brazes or solders in the area of high temperature engineering ceramics is based on alumina (17). Direct brazing of alumina applies to ceramic fritts, composed of Al_2O_3 -MnO_2 - SiO_2 and Al_2O_3 -CaO-MgO-SiO_2, where the latter has a temperature capability of maximal 1700 °C depending on composition and joining procedure.

Another way of joining is to use ceramic slurries, where a wide variety of material combinations is possible (10). An example are blades of reaction sintered silicon nitride which are placed in openings of a rotor consisting of hot pressed silicon nitride and then fixed with a special developed silicon carbide slurry.
The slurry embedding technique is in the same way applicable for a metal rotor with ceramic blades. Laboratory tests proved the superiority of slurry embedding technique compared with the simple mechanical joining using shims (cp.3.1.)

3.3. Welding

Welding is a process to get a close connection between two components, without application of a supplementary materials, instead of brazing or soldering.
One must distinguish between solid state and reaction welding without or with very low pressure and pressure welding involved with high pressures.

3.3.1. Solid state welding

The purpose of solid state welding is to get a joint based on solid state reactions, that means a process without occurence of a liquid phase.
Miscibility of the materials to be joined has to be presupposed. One of the components often contains small amounts of active alloying constituents, which will assist to reactions at the interface and to formation of chemical adherence. Solid state welding is done in vacuum or inert gas atmosphere at a temperature below melting or softening of the materials involved. Usually, solid state processes require a long time which limits the economic efficiency.

An application of this method is found when alumina is brought into close, adhesive contact with aluminium metal (18). This method was developed to provide a compound in a simple procedure to allow its application for mass-produced articles, like metal film resistors or ceramic capacitors. It makes use of the fact that the always existing oxide layer of aluminium adheres extremely well. This good adhesion also results if pure aluminium is brought into a close contact with foreign aluminium oxide or alumina ceramics. The removal of the oxide layer may be done by rubbing or deformation close to melting temperature of aluminium. Thus, an immediate bond between aluminium and alumina occurs.

DE BRUIN (19) reports of vacuum tight joints of ceramic oxides with metals. According to him solid state reaction with noble metals with application of low pressure results in a surface reaction with a sharp discontiniuity at the interface. By using various transition metals, like Fe, Co, Ni, a macroscopic reaction will take place with diffusion processes across the interface.

Fig. 4 shows a compound of silicon carbide with a Fe-Cr-Ni-alloy which could be sintered (20).

3.3.2. Reaction welding

Such compounds are produced by sintering metal powder compacts onto the ceramic.
TWENTYMAN (16) reports of a technique which uses an annulus formed of a metal powder compact. This ring will be placed over a ceramic rod or tube. During firing the metal sinters and shrinks onto the ceramic. Such a joint is feasible with powders consisting of stainless steel, Ni, Cu or Mo. However,the joint is mainly a mechanical one, because the process is more based on shrinking than on chemical reactions.
Joint strength depends on the sintering program. The idea of this method is to control compressive stress in the ceramic developing during sintering. With a defined heating program it is possible to get a required densification in the porous metal powder and thereby a required shrinkage. It is presumed that elasticity in a sintered metal is proportional to the fraction of theoretical density which is attained. It is therefore feasible to control the stress in the ceramic material by varying temperature and time of the sintering process. Insufficient sintering leads to dispersed metal powders. Firing too long at elevated temperature causes failure in the ceramic due to internal stresses.

A method to get chemical reactions combined with mechanical joining is to glaze the ceramic and then sinter the metal compact onto the glazed surface, so that the metal combines with the glass. Experiments had been done with sintered silicon carbide and silicon nitride glazed with a borosilicate and a manganese aluminosilicate glass (16). Chemical bonding occured with metal components containing nickel. However, the results indicate that at room temperature no higher strength occured with the glazed ceramic and at elevated temperature it was possible to pull off the metal annulus.
Another way of producing a chemical bond between metal powder compact and ceramic component is to use a metal alloy, which forms a liquid phase during sintering. This liquid will wet and bond to the ceramic (16). Alloys of Mo-Mn-Ni are applicable for joints with hot pressed and reaction sintered silicon nitride.

Another way of reaction welding is to join an isostatically pressed silicon powder compact to a ready-made silicon nitride component. During nitridation of the assembly a material will be developed which consists only of silicon nitride.
Samples of silicon nitride, which have been used, were prepared with high energy laser treatment (cp. 2.4.) to get a rough surface with enrichment of silicon. The powder compact was placed on the ceramic applying a silicon slurry, which during sintering provides the contact between the powder compact and the rough ceramic surface. After nitridation, the joint strength is based on mechanical interlocking together with nitrides which have been

formed across the interface.
Fig.5 shows such a joint with hot pressed silicon nitride (fig. 5a) and with reaction bonded silicon nitride (fig.5b). If laser energy is too high, the craters are too large to be filled with the silicon slurry, which results in a porous compound.

3.3.3 Pressure welding

Pressure welding employs hot pressing or hot isostatic pressing of powder compacts consisting of metals or ceramics.
Metal powders, which are applicable in this method, are high mel- ting refractory metals and superalloys. However, lower melting metals are also able to be joined if preformed sintered ceramic components will be applied.
An important condition is chemical and mechanical compatibility of the materials used. To adjust properties like thermal expansion it is possible to use intervening layers consisting of mixtures of the materials in different quantities. In this way it is possible to produce a material which would be on one side pure ceramic and at on the other pure metal with ceramic/metal intervening layers.
Parameters of pressure welding are temperature, time, pressure, atmosphere, heating and cooling rate. They have to be empirically fitted to different tasks of joining.

An example for pressure welding is shown in fig.6. Silicon carbide was hot pressed with molybdenum. Whereas silicon carbide is applicable for pressure welding with direct contact to the metal component, problems occur with usage of silicon nitride due to thermal expansion mismatch. Adherence may be developed by application of intervening layers with increasing metal pow- der content, which leads to a transition in thermal expansion behaviour (20).

3.3.4. Fusion welding

At the end, this review of joining techniques should be comple- ted with mention of fusion welding.
It is achieved by filling the joint zone between the components with a liquid phase obtained by melting the materials surfaces.
The liquid phases must be compatible and miscible.
This technique is applicable only for material with low vapor pressure and low tendency to sublime. For example, silicon nitride and silicon carbide sublime without melting and therefore are not able to be fusion welded at normal pressure.

The problem of thermal stresses occurs again. The components must be heated up to melting only in the connection zone. Thus, temperature gradients are always present, which have to be decreased with supplemental heating of the total assembly.
Fusion welding technique is based on electromagnetic radiation or electric conductivity of the materials to be joined. Here limitations occur in the applicability of ceramics. However, temperature would mostly be high enough to provide sufficient conductivity.

4. Conclusions and suggestions for further development

A review of techniques was given which are applicable for ceramic/ceramic or ceramic/metal joints in the area of engineering ceramics for high temperature application.
Ceramics used in this area consist of oxides, alumina, magnesia or zirconia, or of non-oxides like silicon nitride and silicon carbide. These materials have considerable advantages over metals, especially at elevated temperatures. However, because of their relatively low tensile strength and limitations in their preparation - particularly of ceramic materials with nearly theoretical density - it is often required to used these materials only in the hottest zones. In the other parts metals may be employed or ceramics with lower density, which are easier to fabricate.

Applicable joining techniques are brazing, welding and mechanical joining. Because of economic limitations techniques are promising which are easy to manufacture. In the main, these techniques will be brazing, soldering and reaction welding, since these methods do not require the application of pressure.
Brazing is a widely used method, especially in the area of oxide ceramics. The joints are quite versatile, they provide good strength and vacuum-tightness and a sufficient temperature capability. There is a development trend towards direct brazing without prior metallizing.
Reaction welding is another promising method, applicable if a sintered component must be joined to a metal powder compact. The following sintering leads to good joints with sufficient strength. Simple shaped hot pressed components can be built up to complex configurations by joining with preformed powder compacts, which are easy to produce. In the case of silicon, reaction bonded silicon nitride is formed during firing in nitrogen atmosphere and a complete ceramic compound results.

Methods involving application of pressure at elevated temperature, i.e. solid state welding, are limited to components with relatively simple shapes. This technique will be of more importance if hot isostatic pressing will be improved.

In addition to the aspect of adhesion and bonding, another important effect in forming joints is stress-induced adherence. These stresses are caused by different thermal expansion coefficients and different structural transitions. To get an optimum joint, expansion behaviour of ceramic, metal and intermediate layer must be the same. However, this is not always true, since metals expand or shrink more than ceramics in most cases.

It is a common practice to design joints which put the ceramic into compression. An example for this kind of design is the tubular-overlap joint with a metal annulus around a ceramic tube or rod. Here it is possible to combine mechanical with chemical joining mechanisms. Butt-joints, are more difficult, where the adherence results only from chemical bonding.

Thus it follows, that joining techniques are successfull if the materials to be joined are compatible and if the joint design prevents the compound from internal stresses.

5. References

(1)   Humenik, M.Jr.; Kingery, W.D.: "Metal-ceramic interactions: III, Surface tension and wettability of metal-ceramic systems", Journ.Am.Ceram.Soc., 37(1954), No.1, 18-23

(2)   Kingery, W.D.: "Role of surface energies and wetting in metal-ceramic sealing", Bull.Am.Ceram.Soc., 35(1956), No.3, 108-112

(3)   Rhee, S.K.: "Wetting of ceramics by liquid metals", Journ.Am.Ceram.Soc., 54(1971), No.7, 332-334

(4)   Geirnaert, G.: "Ceramiques-metaux liquides. Compatibilites et angles de mouillages", Bull.Soc.Francaise Ceram., No.104(1977), 7-50

(5)   Christen, H.R.: "Grundlagen der allgemeinen und anorganischen Chemie", Verl. Sauerländer, Frankfurt a.M. 1968, 78-146

(6)   Scholze, H.: "Glas - Natur, Struktur, Eigenschaften", Springer Verl., Berlin 1977, 89-94

(7)   Kingery, W.D.; Bowen, H.K.; Uhlmann, D.R.: "Introduction to Ceramics", John Wiley a. Sons, Inc., New York 1976, 469-503

(8)   Salmang, H.; Scholze, H.: "Keramik, Teil 1: Allg. Grundlagen und wichtige Eigenschaften", Springer, Berlin 1982,154-185

(9) Copley, S.M.; Bass, M.; Wallace, R.G.: "Shaping of silicon compound ceramics with a continous wave carbon dioxide LASER", Proc. 2nd Symp."The Science of Ceramic Machining and Surface Finishing", spec. publ. Nat. Bureau of Standards, No.562, Washington 1979, 283-293

(10) Hennicke, H.W.; Müller-Zell, A.; Siebels, J.: "Fügetechnik von Siliciumnitrid und Siliciumcarbid untereinander und mit Metallen", 2.BMFT-Status-Seminar-1980: "Keramische Bauteile für Fahrzeug-Gasturbinen, Bad Neuenahr.

(11) Rice, R.W.: "Joining of Ceramics", Advances in Joining Technology, ed. by Burke,J.J.; Gorum,A.E.; Tarpinian,A.; Brook Hill Publishing Comp., Chestnut Hill, Massechusetts 1975, 69-111

(12) Twentyman, M.E.: "High-Temperature Metallizing, Part 1: The mechanism of glass migration in the production of metal-ceramic seals", Journ.Mat.Science, 10(1975), 765-776

(13) Pattee, H.E.: "Joining Ceramics to Metals and Other Materials", WRC Bulletin 178, 1-41

(14) Heimke, G.: "Keramik-Metall-Verbindungen", Sonderdruck aus Konstruktion, Elemente, Methoden, Nr.6,7 (1972), 2-7

(15) Twentyman, M.E.: "The Joining of Engineering Ceramics", British Ceram. Res. Ass., Final report, 1976-1978

(16) Iseki, T.; Yamashita, K.; Suzuki, H.: "Joining of self-bonded SiC by Ge metal", 7th Symp. on Special Ceramics, London, Dec. 1980

(17) Klomp, J.T.; Botden, Th.P.J.: "Sealing pure alumina ceramics to metals", Bull.Am.Ceram.Soc., 49(1970), No.2, 204-211

(18) Schmidt-Brücken, H.; Schlapp, W.: "Verbinden von Aluminium-metall mit Aluminiumoxidkeramik", Ker.Zeitschr. 25(1973), 141-143

(19) De Bruin, H.J.; Moodie, A.F.; Warble, C.E.: "Ceramic-metal reaction welding", Journ.Mat.Science, 7(1972), 908-918

(20) Müller-Zell, A.: "Fügetechnik von Siliciumnitrid und Siliciumcarbid in Mischungen und/oder in Kontakt mit hochschmelzenden Metallen und Superlegierungen", Dissertation, TU Clausthal, Juli 1980

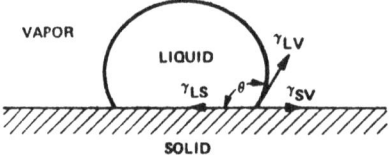

Fig. 1: Sessile drop at equilibrium with contact angle θ

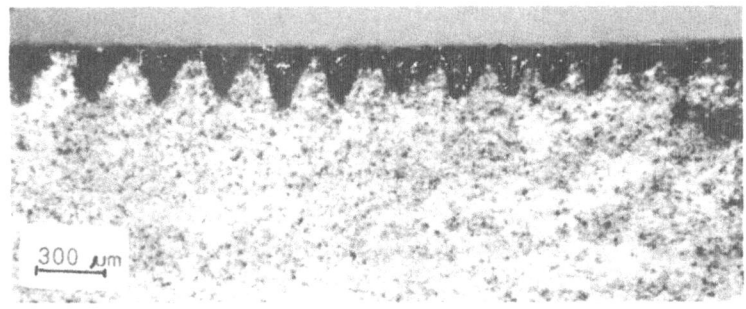

b)

Fig. 2: Surface of silicon nitride after high energy laser
treatment: a) RBSN, energy density 1,1 J/mm²
            b) HPSN,   ″   ″   10,1 J/mm²

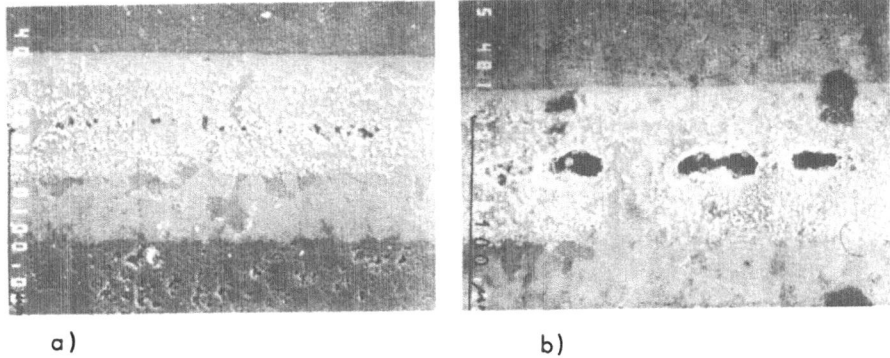

a)   b)

Fig. 3: HPSN/RBSN - joint with Ti-Cu-Be - braze
         a) optimum brazing
         b) pores in braze-alloy due to too long annealing time

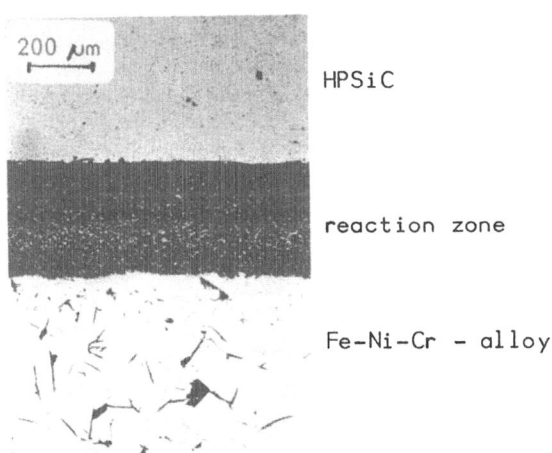

HPSiC

reaction zone

Fe-Ni-Cr - alloy

Fig. 4: Solid state welding of hot pressed silicon carbide
        with a Fe-Ni-Cr - alloy

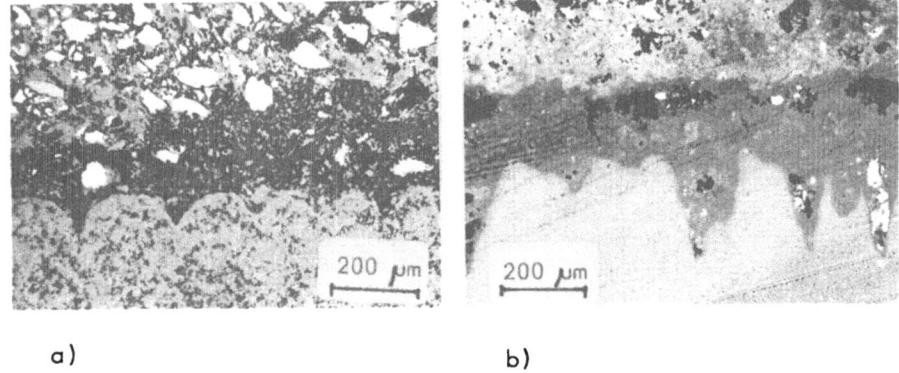

a)   b)

Fig. 5: Silicon nitride with Laser-treated surfaces
reaction welded with a silicon powder compact
in nitrogen atmosphere
a) reaction bonded silicon nitride
b) hot pressed silicon nitride

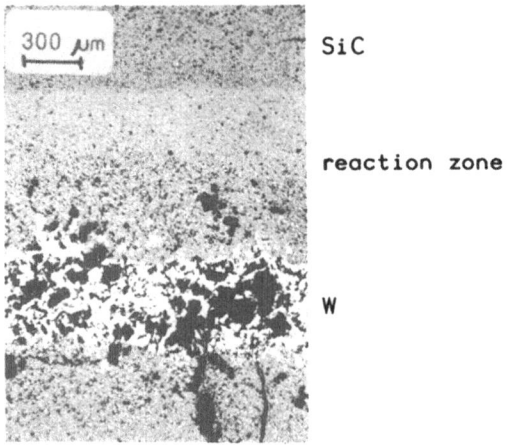

SiC

reaction zone

W

Fig. 6: Pressure welding of silicon carbide in close contact
to tungsten

156

NONDESTRUCTIVE TESTING OF CERAMICS

K. Goebbels and H. Reiter

Fraunhofer-Institut für zerstörungsfreie Prüfverfahren,

Bldg. 37, University, D-6600 Saarbrücken 11, FRG

SYNOPSIS: Quality assurance of ceramic components needs special efforts. The material's brittleness is prohibitive against the release of stresses by plastic deformation at defects under load. Therefore, critical defect dimensions, e.g. following fracture mechanics evaluations, are some microns until several hundred microns depending upon the kind and the position of the defect. The state of the art in nondestructive evaluation (NDE) is described, covering especially X-ray methods and ultrasonic techniques. The variety of other NDE-methods seems to have only a limited chance for practical applications. Unsolved problems are defined and recommendations are given for further research and development activities.

I) Introduction

Ceramic materials are the appropriate solution for high temperature, high strength and high temperature - high strength components. While oxide ceramics show an excellent resistivity against oxidation but a high sensitivity to thermal shocks, non-oxide ceramics seem to withstand thermal shocks while suffering under oxidising atmospheres. Both are brittle materials due to the covalent nature of bonding forces. This brittleness is responsible for the fact that stresses by mechanical or thermal loading cannot be released by plastic flow. Especially peak stresses at defects are therefore responsible for the failure of ceramic parts. Under this aspect the smallest defect can be the single grain of a polycrystalline material as well as any foreign inclusion or a pore because of the elastic and thermal anisotropy. So nondestructive testing (NDT) methods have to detect and to evaluate defects, to describe the materials structure and to determine stresses. The distributions of temperatures and stresses across a loaded structure are resulting in the fact that at different locations different types and sizes of defects will lead to failure. Fracture mechanics can help to find out the critical defect dimensions which are some microns for a crack at the surface until several hundred microns for an inclusion. This defines the detection requirements for NDT techniques and the necessary interpretation of NDT results (i.e. nondestructive evaluation - NDE). From the safety point of view a high reliability is required for e.g. bioceramics and ceramic gas turbine components. From the more economical and costs point of view the same criteria are valid for other parts like ceramic tools, ball bearings, piston caps and a lot of specific applications. There are worldwide activities to develop the appropriate NDT techniques including any physical interaction process with matter and defect, e.g. X-rays, ultrasonics, electrical and thermal methods. The state of the art will be given as well as a discussion of outstanding development areas.

The important role of quality assurance by NDT methods for ceramic materials can be emphasized by comparison with the basic safety concept for nuclear power plant components:

- The mean grain size of reactor pressure vessel steels is about 10 µm. The size of defects which have to be detected is about 3 mm, i.e. two or three orders of magnitude higher. The defect size which will be responsible for the pressure vessel failure is about 100 mm, again more than one order of magnitude above [1];

- the mean grain size of gas turbine ceramics is about 1 µm and the size of defects which have to be detected is only one order of magnitude higher, i.e. $\approx$ 10 µm. Finally, the defect size which will cause failure is of the same order of magnitude or just slightly above [2].

This comparison gives an impression for the narrow safety margin referring to ceramic components. Without detailed knowledge about a defect detected, only a small credit can be taken from NDT results. But with additional information about the kind of defects, their size and their orientation this margin can be improved and less material has to be rejected.

II) Components

With some time shift the development of high temperature, high strength ceramics is accompanied by the development of NDT methods to find and to evaluate defects with sizes from some microns to several hundred microns. The main steps forward were programs running in the USA and the FRG for ceramic vehicular gas turbines based especially on silicon nitride and silicon carbide. For hot-pressed and for reaction-bonded material quality inspection techniques have to be developed for single turbine blades, for rotors and stators as well as for flame tubes and inlet cones. The development of these non-oxide ceramics allowed their application in other industrial areas, too. So today, ball bearings and piston caps (Si_3N_4), heat exchanger tubes (SiSiC) and bioceramics (CSiC) as well as a series of other parts with simple shape or complex geometry are manufactured and correspondent NDT techniques are asked for that.

The input into NDT for small defects' detection includes now better and new inspection methods for dense oxide ceramics like used in bioceramics, for the Na-S-battery or for cutting tools (Al_2O_3) or components like wire drawing tools (ZrO_2) and several other industrial ceramic components. A listing of the most relevant ceramic parts and their inspection problems together with the hitherto existing NDT experiences is given in Table 1.

III) Problems

Structure

The strength of ceramics strongly depends upon the material microstructure including the grain size (distribution), the porosity (pore size distribution) and the overall homogeneity described integrally by the density ρ. For reaction-bonded materials as well as for hot-pressed and hot isostatically pressed qualities the density could be increased considerably for carbide and oxide ceramics [3]. For a single component density and homogeneity have to be checked before a detailed analysis will look for isolated heterogeneities like coarse grained regions or foreign inclusions or cracks.

Stress

Components are designed and manufactured to withstand mechanical and/or thermal loads. The ability for this is a function of the local stress situation, the sum of residual and applied stresses. Besides the information about the microstructure and the defects the characterization of the quantitative stress situation, especially the residual stresses, is an important demand. With increasing knowledge and experience using nondestructive methods for stress determination, this up to now unsolved problem reaches appropriate solutions, for the surface as well as for the interior. For example, surface finishing is responsible for the occurrence of residual stresses which can reduce the strength by up to 40 % [4].

Defect

A long time philosophy was to manufacture materials free from defects. But this caused unjustifiable costs and finally it is impossible. So one has to live with defects and only the information about their kind, size, shape and orientation can help for the decision whether a part has to be accepted or rejected. Within the above mentioned small safety margin for ceramics the developments of NDT methods have to include such interpretations of the results for a finally NDT based fracture mechanical judgement (accept/reject criteria). Especially for hot-pressed silicon nitride (HPSN) criteria were developed by Evans [2] and the result of theoretical and experimental studies is shown in Fig. 1. It can be seen from this figure that the most important defect is a surface crack while an internal void or a spherical Fe-inclusion will cause failure at the same load only if it has twice,respectively,four times the size of the crack.

Not only the single crack has to be considered. During the last years the important role of surface finishing induced cracks was extensively studied. This process leads to cracks as well as to strength reducing residual stresses [4].

It belongs to the problem of defect evaluation that defects which will cause failure at room temperature will not in any case be the relevant defects at elevated temperatures. First results are known about this fact [2,5] but they have still to be discussed finally for the different ceramic materials.

There are several ways to join ceramics via metal-to-ceramic or ceramic-to-ceramic bonding procedures. They are responsible for typical defects (lack of bonding, hard inclusions, porosity, cracks) in the small bonding layer and these defects are not easy to detect [6].

Table 2 is a trial to map the relevant types of defects in ceramics.

IV) Solutions and State-of-the-Art

The interaction between a defect and analysing radiation can be reduced to a few parameters. The essentials are:

- the difference in the physical properties of surrounding matrix and defect, e.g. X-ray absorption coefficient μ, sound impedance density ρ times velocity v, electrical conductivity σ_{el}, magnetical permeability μ_{magn}, dielectric constant ε and thermal conductivity λ_w. Without any of these differences there will be no measurable effect;

- the extension or size d of the defect. With increasing size there will be an increasing signal;

- the relation of d to analysing beam parameters, e.g. d/D with D = beam diameter in ultrasonics or D = focal spot size for X-rays, d/λ with λ = ultrasonic wavelength or microwave wave-

length, d/h with h = material's thickness for X-rays and a/b
for X-rays too with a = distance focus - object and b = distance
object - film or image intensifier;

- the <u>orientation</u> of non-spherical defects like cracks; e.g. for
  X-rays an orientation parallel to the beam is optimal while an
  orientation perpendicular to the beam is optimal for ultrasonics.

This consideration stands for the fact that not all of the possibly occurring defect types can be found best by one NDT technique
only. And in this context the following discussions are based on
the philosophy that the optimal way for the quality assurance of
ceramic components consists of the overlapping use of two or more
NDT techniques. The amount of work which is necessary to assure
that the relevant defects with a given threshold will be detected
depends therefore strongly upon the matrix material and the defects which have to be considered.

<u>Philosophy</u>

For safety relevant ceramic components the philosophy of industrial quality assurance consists of visual inspection including
dye penetrants plus proof testing. Partly destructive tests for
given percentages of a series deliver information about the structure (ceramography, density measurement) and the strength. But
this type of quality assurance cannot deliver safety about any
type of relevant defects, it is expensive and the random tests
give uncertainty for the single part. In special cases ultrasonic
C-scans are used to detect greater internal defects. The philosophy
developed using NDT methods covers the visual inspection with dye
penetrants (some open questions, see below), the X-ray analysis
with high-resolution and filmless image evaluation plus ultrasonic
testing. For large series of equally fabricated parts, e.g. bioceramic prostheses or gas turbine blades, the integral characterization by vibration analysis is recommended. The following sections
describe these procedures in detail.

<u>Dye Penetrants</u>

The use of dye penetrants especially with fluorescent matter for
ceramics is quite easy and widely distributed. But the penetrants
developed for metallic materials show some uncertainties about
their application to ceramics referring to the capillary effect
and the limits of resolution. This was stated by industrial users
but, up to now, no research was done to analyse the combination
penetrant-ceramic-defect and to optimize it. The method therefore
seems to be valid only for a qualitative check of restricted value.

X-Rays

As well single defects as extended structure heterogeneities (density variations, anisotropy of the structure, phase boundaries and bonding) can be imaged and analysed with X-rays. The main question of the resolution and detectability (signal-to-noise ratio, contrast) has to be answered by the focal spot size of the X-ray tube, the imaging technique and the receiver (film, fluorescent screen, image intensifier). The geometrical unsharpness U_g, defined by [7,8]

$$U_g = f \cdot b/a$$

is proportional to the focal spot diameter f inside the tube (a = distance X-ray source to specimen, b = distance specimen to receiver). It follows from this relation that usual X-ray analysis with $f \approx$ mm cannot be used for the detection of defects ≥ 10 µm. Even the compensation of f by a small b/a value needs - besides other disadvantages - an expensive secondary enlargement, to make visible the microscopic defects.

For the high definition X-ray analysis with fine focus tubes ($f < 100$ µm) and projection technique ("microradiography"), see Fig. 2, a secondary enlargement is unnecessary. The imaging technique reproduces on film or image intensifier specimen details with a primary magnification m = (a+b)/a.

With usual values b = 2000 mm, a = 100 mm this corresponds to a magnification x20. Focal spot sizes of 15 µm are commercially available so that the geometrical unsharpness does not influence too much the resolution of defect details: a defect of 50 µm diameter will be imaged on the film as a 1 mm object with 0.3 mm penumbral unsharpness.

The second important feature is the contrast K defined by [7,8]

$$K = \text{const } \Delta\mu \cdot d/(1+I_S/I_D)$$

where the constant includes the receiver parameters (film, image intensifier), d = defect size in X-ray beam direction, $\Delta\mu$ = difference in X-ray absorption for matrix and defect, I_S = scatter intensity, I_D = image intensity. With the projection technique, shown in Fig. 2, $I_S/I_D \rightarrow 0$.

Up to now, the film is the most sensitive imaging medium resolving defects with $d \geq 20$ µm. It belongs to the state of the art to use it for any type of ceramics, especially with complex shape, e.g. turbine blades, bioceramics, ball bearings and others [9]. Since the reproduction of X-ray films in a paper is difficult and not satisfying, no figures are given here.

The X-ray film is an expensive receiving medium. To simplify the technique image intensifiers and X-ray sensitive cameras with image processing techniques are under development. Up to now, the filmless analysis allows only to detect defects ≥ 100 µm but, as shown

in Fig. 3, averaging techniques are improving the detectability
and they can be realized as a quasi- on-line technique (≈ 1000
averages in less than half a second).

The filmless methods include finally radiometry and tomography.
Figure 4 shows a reaction-bonded silicon nitride (RBSN) stator;
several tomograms using a medical unit are given in Fig. 5. With
today hardware and software data processing it seems possible to
develop appropriate instrumentation for materials' testing problems. The basic question of resolution in the region of 10 μm to
100 μm has to be answered by microfocus X-ray tubes.

The imaging quality can be improved for the surface opening defects using dye-enhanced radiography as demonstrated elsewhere [10].

Ultrasonics

For the detection of defects with ultrasonic waves the change of
sound impedance $\rho \cdot v$ at the phase boundary and the relation d/λ
are the essential parameters. The high velocity of ultrasound in
ceramics, e.g. 11 mm/μs for hot-pressed silicon nitride, makes it
necessary to use high frequencies f ($\lambda = v/f$). But besides values
of f $\gtrsim$ 100 MHz low frequency ultrasonics have their own possibilities:

- The determination of elastic properties like E-, G-modulus is
  possible too at low frequencies by velocity measurements.
  Additionally, the velocity is correlated to the density, so
  measurements at different positions allow to make judgements
  about the structural homogeneity [11].

- The possibility to polarize shear waves enables a nondestructive measurement of residual and/or applied stresses in analogy
  to the optical birefringence [9]. The sign of stresses and the
  stress distributions qualitatively can be mapped by the wave
  velocities while a quantitative determination needs knowledge
  about the third-order elastic constants.

- For small d/λ-values scattering effects are prevailing reflection properties. With high dynamic ranges ($\gtrsim$ 80 dB) the scattering signals from defects with small amplitude can be measured,
  and they show the advantage to be less influenced by the defect
  orientation [11].

- The detection of defects with free longitudinal and shear waves
  is optimal with $d/\lambda \approx 1$ (see below). Guided waves are reflected
  strongly already for $d/\lambda \gtrsim 0.01$, so the same defect can be
  found with wavelengths about two orders of magnitude smaller.
  This implies the use of low frequency and low attenuation ultrasonic waves. Tube waves in SiSiC heat exchanger tubes have

detected with this technique defects with dimensions less than 4 % of the wall thickness (5 mm) [9].

The goal to use high frequency ultrasonics is the characterization of defects referring to their kind and size. The frequency response of a defect depends strongly upon these two parameters as shown in Fig. 6 where $ka = \pi \cdot d/\lambda$ is the abscissa and the backscattered or reflected amplitude $f_N(\pi)$ the ordinate value [9]. To determine these signatures either small bandwidth measurements have to be made over a given range of frequencies or measurements with broadband probes and systems where the Fourier transformation corrected for the probe characteristics delivers the spectral distribution. For this a series of techniques were developed to excite and to detect high-frequency ultrasonic waves [9]:

- sapphire-single crystals with evaporated piezoelectric films,
- piezoelectric layers glued on HPSN-delay lines,
- piezoelectric films (PVF_2) or single crystals ($LiTaO_3$),
- opto-acoustic transmitting and/or receiving techniques.

Schematically some of these high-frequency ultrasonic "probes" are shown in Fig. 7. Up to now, the equipment for the piezoelectric realization allows the detection of single defects $\gtrsim 20$ μm. Referring to the interpretation of ultrasonic data, systems with enough dynamic range are still under development.

Additionally to the usual detection techniques imaging procedures and systems were developed. Up to now commercially available is the Scanning Laser Acoustic Microscope (SLAM) with frequencies of 30, 100 and 500 MHz [12]. Scanning Acoustic Microscopes (SAM) are still under development, but they seem to be available in about 1983 [13]. Figure 8 shows some examples of defect detection with the SLAM. Following recent experiences, the two-dimensional image including the diffraction pattern of the defect can be evaluated to the defect size and kind [14].

Photoacoustic Microscopy and Photothermic Microscopy

With a chopped laser beam it is possible to periodically heat up object surfaces. The focused beam allows a lateral resolution of some microns while the analysed depth covers ≤ 1 mm. This is a sensitive and contactless working NDT technique for the surface and the surface near regions using as receiver microphones or piezoelectric probes depending on the chopping frequency (kHz to MHz). Positive results are known with defects in turbine blades [15] and ball bearings (both materials Si_3N_4).
The same technique using IR-detectors relates the measured signal not to the acoustic but to the thermal properties of the surface region. So additional information can be gathered.

Acoustic Emission

Generally, the potential of acoustic emission analysis is very high for brittle materials. The elastic energy stored during a loading process cannot be released by plastic deformation but will be used almost totally for the generation of cracks. Especially for oxide ceramics (with grain sizes in the ten μm region and above) this was proved in the past [16]. In correspondence with the known literature experiments have shown [9] that for RBSN specimens with and without Knoop identations no correlation could be found with the strength or the damage. Until now the reason for this is assumed to be related with the fine grained structure (grain sizes smaller than 1 μm). It seems that too many bridges are broken instead of storing the elastic energy until the level of an "audible" noise is reached. But a complete judgement for this behaviour needs still more detailed analysis of acoustic emission experiments. First promising results were obtained for HPSN during double torsion experiments [17]. The acoustic emission could be positively related to the room temperature crack growing process.

Vibration Analysis

The excitation of mechanical long wavelength vibrations - where the wavelength corresponds to the dimensions of the specimen under test - leads to resonance vibrations which can be measured by piezoelectric transducers or just microphones. The simple solution uses an impact excitation (broadband spectrum) and the analysis of the specimen-related main vibration). For simple geometries (bar, disc, sphere) the values are related to the elastic modulus, for complex shaped components like turbine blades or whole rotors they are characteristic parameters of the parts. More detailed information can be obtained exciting and analysing with small bandwidth the spectrum of ground modes and higher harmonic vibration modes, the so-called modal analysis. Caused by the overall excitation the usual vibration analysis allows only an integral component characterization. But for a series of equally fabricated parts this seems to be a simple and reliable judgement about the homogeneity of the fabrication and manufacturing process [18].

Electrical and Electromagnetic Methods

SiC as well as SiSiC or CSiC show a limited electrical conductivity. The proper choice of the relevant eddy current frequency makes it possible to detect defects in such components by eddy current probes [19].
In analogy to the eddy current testing of "metallic" materials the microwave testing can be seen for materials without electrical conductivity. Especially then the semiconducting Si as one of the most important types of defect in Si_3N_4 can easily be detected with microwaves. Detailed analysis has shown their capability to

detect other defects too like WC, Fe, BN, pores and C [20]. The
necessary frequency range of more than 100 GHz at the moment unfortunately includes disturbing signals from the component geometry
(curvature, edges etc.) which cannot be suppressed. So up to now,
it seems not to be possible to develop this technique without
extensive work for application in practice.

Thermal Methods

The detection of defects by photothermal microscopy was mentioned
above. With less resolution the thermographic method too can be
used. It shows especially the integral quality of series of products, e.g. heat exchanger tubes made from SiSiC, referring to
their heat conducting properties [10,21]. Photothermal microscopy
and thermography seem to have still more potential for NDT of
ceramics.

Optical Methods

With the optical-holographical interferometry it is possible to
image defects in specimens under load (mechanically, thermally).
The ability was shown by several experiments [10,11,21]. But compared to other techniques like ultrasonic testing and X-rays the
lack of resolution especially for internal defects does not re-
commend this technique up to now.
From another point of view the scattering of focused laser light
at a reflecting surface can be evaluated for the surface characterization. Especially for polished surfaces, e.g. ball bearings
or bioceramic parts, the further analysis of this technique (small
and wide angle scattering) can be recommended. Experimental results are published for surface roughness characterization of
Si_3N_4 ball bearings [22].

V) Discussion

State-of-the-Art

Defects with linear dimensions from some microns to several hundred microns can cause failure of ceramic components under load.
At the moment there exists only a small safety margin between the
defect size which has to be detected and the defect size which
will lead to fracture. This is the reason why NDT methods are required delivering information about defect type, size, shape, and
orientation. Nondestructive quality assurance methods fulfilling
this demand do not belong to the state of the art. They are still
under development.
Up to now, defects can be detected quite surely if their dimensions are above 50 µm, under special circumstances below, too.
But today's methods do not allow enough interpretation. The basic

philosophy developed during the last years and considering high
temperature, high strength ceramic components with complicated
geometry consists of

- visual inspection with dye penetrants for qualitative surface
  inspection,
- high resolution X-ray testing with film evaluation for structure analysis and defect analysis,
- low and high frequency ultrasonic testing for defect detection,
  structure and stress analysis,
- vibration analysis for the integral characterization of lots
  of equally fabricated components.

Other NDT methods have only a limited value or they are still far
from application.

Short Time Developments

Within running projects developments are in sight which will improve the detection and interpretation characteristics of NDT
methods. This implies especially

- filmless radiography with on-line signal processing, enabling
  quantitative evaluation of structural homogeneity and multi-
  angle defect analysis with the direction of tomography-like
  X-ray testing,

- high frequency ultrasonics ($f \lesssim 350$ MHz) for volume inspection
  and

- ultrasonic microscopy ($f \lesssim 3$ GHz) for surface inspection.

Future Aspects

It belongs to the important points to further develop NDT based
accept/ reject criteria for ceramics. Up to now realized for hot-
pressed silicon nitride they have to be generated too for reaction-
bonded material with a high degree of porosity as well as for
other ceramic materials like SiC, Al_2O_3, ZrO_2 referring to their
use as safety relevant components. These questions can only be
answered by a mapping of the defects which can occur in the material, in the real component and which will cause failure under
mechanical and/or thermal load.
Outstanding problems which rely to any NDT technique cover the
long time behaviour (e.g. oxidation and thermal shock), the on-
site testing as well as the high temperature NDT with favour for
the contactless working methods and finally - if at all possible -
the nondestructive strength evaluation. For the single nondestructive testing method promising directions will be:

- high frequency ultrasonic phased arrays,
- ultrasonic stress measurement quantitatively, using third-order elastic constants,
- photoacoustic and photothermic microscopy as surface inspection methods with high resolution,
- development of acoustic emission analysis as a reliable technique to find defects and to survey systems,
- further development of vibration analysis into the direction of modal analysis,
- studies referring to electrical, thermal and optical methods for special applications.

Recommendation

The ceramic materials as well as the small defect sizes responsible for failure are setting special NDT problems. After less than one decade of development activities these problems are generally defined but not yet solved. The headlines
- design for inspection and
- inspection for quality assurance
show that NDT and NDE are playing the key role for the application of ceramic parts in practice.
Under the aspects of interpretation of NDT results and NDT based accept/reject criteria there are a lot of outstanding unsolved problems but also a lot of promising better or new NDT techniques. The only recommendation which can be made is to continue this way corresponding to the importance of ceramic components.

Acknowledgements

This contribution is based on work performed with the support of the German Ministry for Research and Technology and within projects for manufacturers of ceramic materials and components.

References

1 G. Deuster, and L. Issler: Proc. 4th Intern. Conf. NDE in Nuclear Industry, Lindau (FRG) 1981; pp. 655-682; 1981, Berlin, Deutsche Gesellschaft für Zerstörungsfreie Prüfung.

2 A.G. Evans: "Nondestructive failure prediction in ceramics." Proceedings NATO-ASI "Nitrogen Ceramics", July/August 1981, Brighton (UK).

3 W. Bunk, and M. Böhmer: Keramische Komponenten für Fahrzeug-Gasturbinen II. p. 5; 1981, Berlin, Springer-Verlag

4 D.B. Marshall: "Surface damage in ceramics." Proceedings
   NATO-ASI "Nitrogen Ceramics", July/August 1981, Brighton
   (UK)

5 G.D. Quinn: Characterization of turbine ceramics after long-
   term environmental exposure. Report AMMRC-TR-80-15, April 1980,
   Army Materials and Mechanics Research Center, Watertown, Mass.
   (USA)

6 H.W. Hennicke, A. Müller-Zell, and J. Siebels: Keramische
   Komponenten für Fahrzeug-Gasturbinen II. p. 535; 1981, Berlin,
   Springer-Verlag

7 W.N. Reynolds, and R.L. Smith: British J. NDT 24 (1982) 145

8 H. Reiter, and K. Goebbels: Mikroradiographie. IzfP-Report
   780106-TW, 1978, Saarbrücken (FRG)

9 K. Goebbels, and H. Reiter: "ZfP keramischer Bauteile."
   Science of Ceramics, Vol. 11 (1981), pp. 361-374

10 D.S. Kupperman et al.: Preliminary evaluation of NDE techniques
   for structural ceramics. Report AFML-TR-78-209, p. 214; 1979,
   Dayton, Ohio (USA)

11 K. Goebbels, and H. Reiter: Keramische Komponenten für Fahrzeug-Gasturbinen II. pp. 163-193; 1981, Berlin, Springer-Verlag

12 L.W. Kessler, and D.E. Yuhas: Proc. IEEE 67 (1979), 526

13 R.A. Lemons, and C.F. Quate: Physical Acoustics, Vol. 14, p. 1;
   W.P. Mason and R.N. Thurston, Eds.; 1979, New York, Academic
   Press

14 D.E. Yuhas, M.G. Oravecz, and L.W. Kessler: 1981 IEEE Ultrasonics Symposium Proceedings, p. 541

15 Y.H. Wong: Scanned Image Microscopy (E.A. Ash, Ed.), p. 247;
   1980, London, Academic Press

16 L.J. Graham, and G.A. Alers: Rockwell Science Center Reports
   AD-745000 (1972), AD-754839 (1972), AD-778015 (1974). Thousand
   Oaks, Ca.

17 H. Iwasaki, and M. Izumi: Annual Meeting, Japanese Soc. NDI,
   May 1981

18 K. Goebbels, and H. Reiter: "NDE of ceramic gas turbine components." Proceedings NATO-ASI "Nitrogen Ceramics", July/August
   1981, Brighton (UK)

19 H. Reiter, R. Becker, and G. Coen: IzfP-Report 800121-TW, June 1980, Saarbrücken (FRG)

20 A.J. Bahr: Microwaves NDE of ceramics. Report AD/A-048582, AMMRC-CTR-77; 1977, Stanford, Ca (USA), Stanford Research Institute

21 D.S. Kupperman et al.: "NDE techniques for SiC heat exchanger tubing." DARPA-Proceedings Review of Progress in Quantitative NDE, AFWAL-TR-80/ 4078, July 1980

22 T.V. Roszhart: Holographic characterization of ceramics. IV TRW-Report AD-776454, 1973, TRW Systems Group, Redondo Beach, CA (USA)

Table 1: Experiences with NDT of ceramic components

component (material)		defects	structure (homogeneity)	residual stresses (qualitatively)
ceramic gas turbine compo- nents	flame tube (SiC, RBSN)	X-rays	X-rays ultrasonics (v)	
	inlet cone (RBSN)	X-rays		
	inlet cone bonding (Si)	X-rays		
	rotors (HPSN, HPSiC), blades (RBSN)	X-rays ultrasonics vibration analysis dye penetrant SLAM holographic interferometry	X-rays ultrasonics (v)	ultrasonics (P(T))
	stators (RBSN)	X-rays		
	single blades (RBSN)	X-rays photoacoustic microscopy SLAM dye penetrant	X-rays vibration analysis	
diverse compo- nents	bioceramics endoprostheses (Al$_2$O$_3$)	X-rays dye penetrant	X-rays	
	heat exchanger tubes (SiSiC)	X-rays ultrasonics (esp. tube waves) eddy current holographic interferometry thermography	X-rays	
	ball bearings (HPSN)	X-rays ultrasonics photoacoustic microscopy	X-rays ultrasonics (v) vibration analysis	
	wire drawing tool (ZrO$_2$)			ultrasonics (P(T))
	cutting tools (Al$_2$O$_3$, ZrO$_2$)	X-rays		

legend: X-rays: microfocus X-ray unit ($\geq$ 15 μm focal spot size, projection technique), partly dye enhanced radiography and tomography
ultrasonics: v = velocity measurement
P(T) = polarized shear waves birefringence measurements

Table 2: Strength Controlling Defects in Ceramics

		production	machining	in-service
structure		inhomogeneities anisotropy		corrosion (oxidation) erosion
stress		residual stresses	residual stresses (e.g. by surface machining, bonding)	
defect	surface	cracks surface connected pores and inclusions	cracks corner chips	cracks (incl. crack growing)
defect	bulk	cracks pores inclusions shrinkage cavity delamination	cracks pores inclusions delamination (e.g. induced by bonding layers)	cracks reactions of inclusions with matrix material

Illustrations

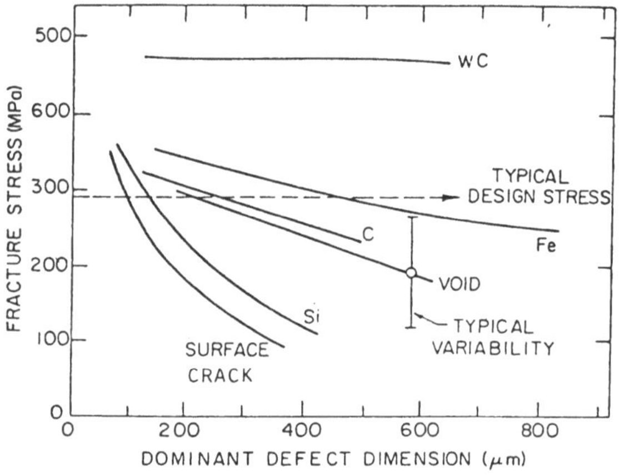

Fig. 1. Defect criticality for hot-pressed silicon nitride /2/

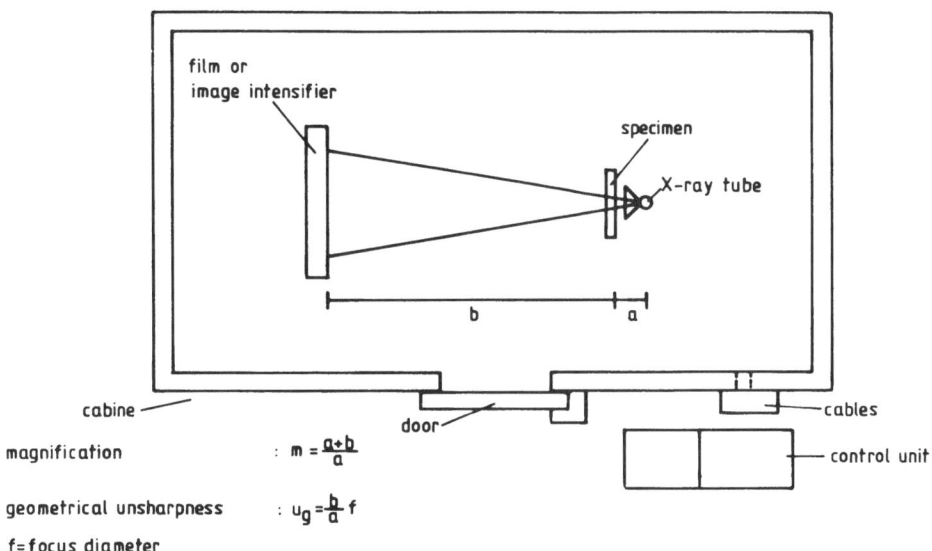

film or
image intensifier

specimen

X-ray tube

b   a

cabine

door

cables

control unit

magnification $: m = \frac{a+b}{a}$

geometrical unsharpness $: u_g = \frac{b}{a} f$

f=focus diameter

Fig. 2. Microfocus X-ray unit with projection technique

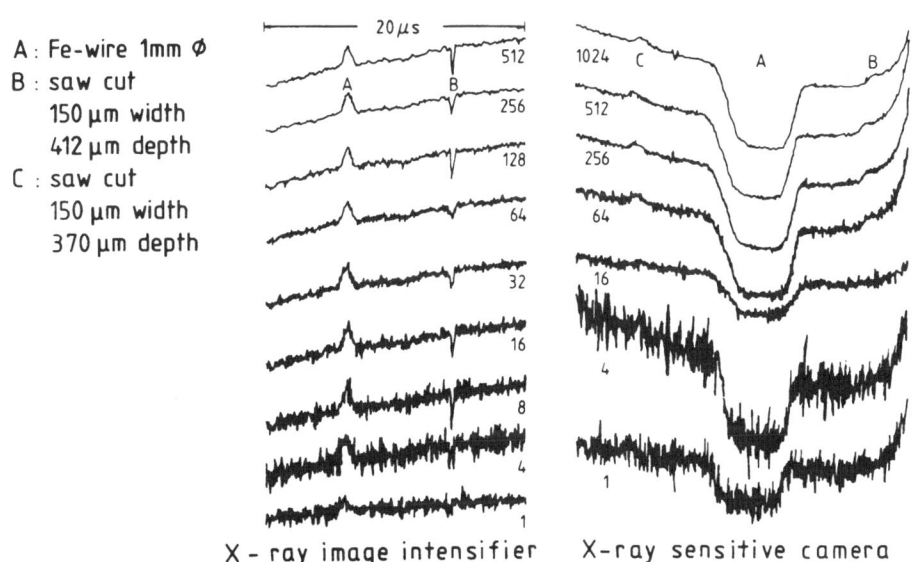

A : Fe-wire 1mm ⌀

B : saw cut
    150 μm width
    412 μm depth

C : saw cut
    150 μm width
    370 μm depth

20 μs

512
256
128
64
32
16
8
4
1

1024 C   A   B
512
256
64
16
4
1

X - ray image intensifier   X-ray sensitive camera

Fig. 3. Signal enhancement by averaging of digitized
        video signals

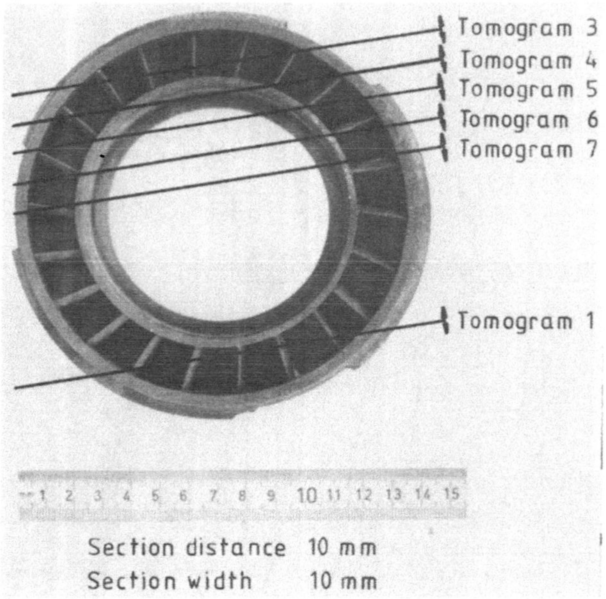

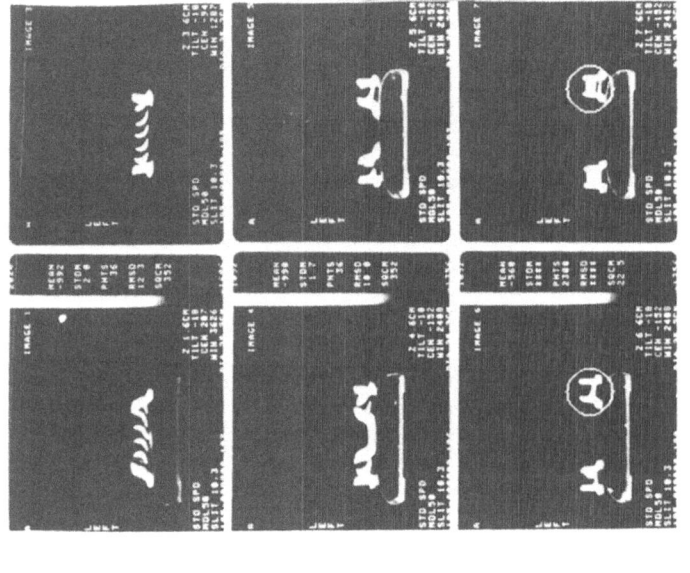

Fig. 5. X-ray tomography of different cross sections through the component shown in Fig. 4

Fig. 4. RBSN-stator, positions of the tomograms shown in Fig. 5

Tomogram 3
Tomogram 4
Tomogram 5
Tomogram 6
Tomogram 7

Tomogram 1

Section distance   10 mm
Section width   10 mm

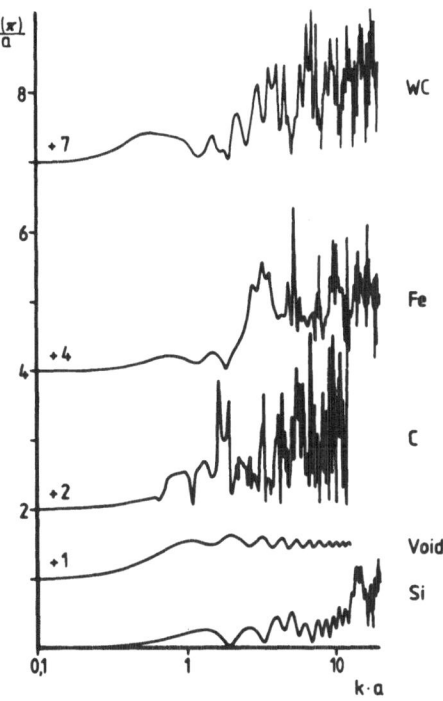

Fig. 6. Backscattering amplitudes for different
types of defects in HPSN

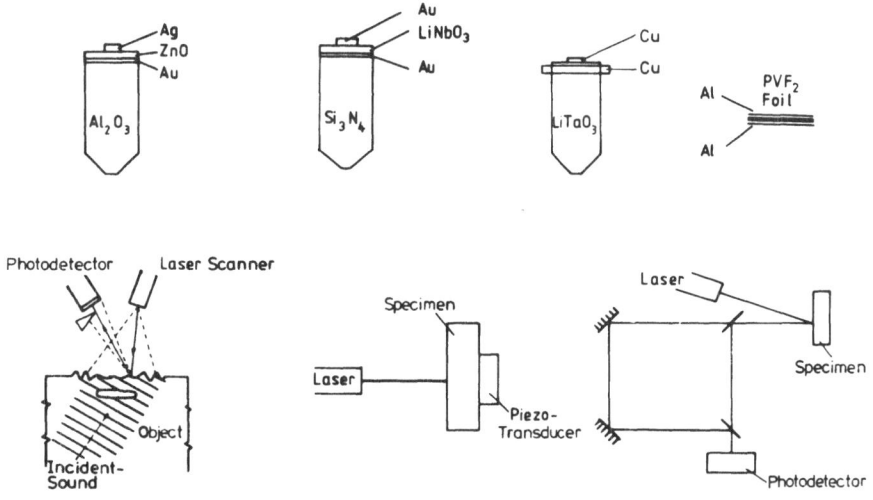

Fig. 7. Excitation and detection of high frequency
ultrasonic waves

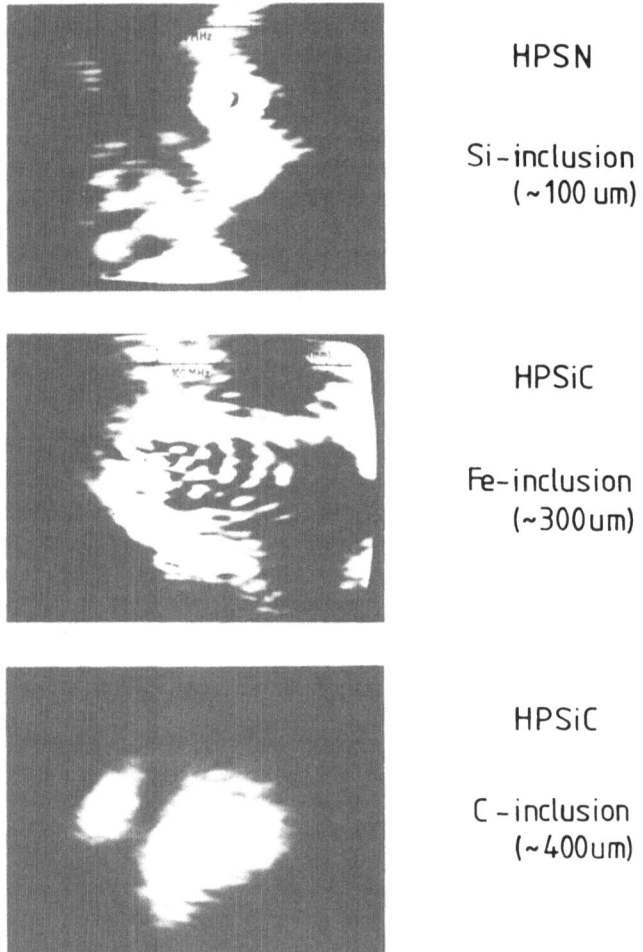

HPSN

Si-inclusion
(~100 um)

HPSiC

Fe-inclusion
(~300um)

HPSiC

C-inclusion
(~400um)

Fig. 8. Defect detection with the SLAM (100 MHz)

DISCUSSION

W.E. Buresch: In most ceramics the macroscopic deterioration
is governed by the accumulation and coagulation of coincident
colinear microcracks in the high stressed region of the process.
In specific cases the thermal shock resistance of materials
increases with increasing microcrack density, say in the range
of 5 to 10% by volume. To assess the lifetime of a thermally
stressed component it is important to know the state of
deterioration that is the microcrack density by nondestructive

testing. The question is: how can you measure microcrack densities in the range of 5 to 10%?

K. Goebbels: Three ways seem to be possible for this:

a)   The measurement of the velocity (e.g. with an ultrasonic goniometer) refers to the microcrack density: a density of 10 Vol. % will result in a decrease of the velocity by 5%. The relation is here about a linear one.

b)   The fine grained material usually (frequencies < 100 MHz) does not show any significant ultrasonic scattering. The distribution of cracks will lead to scattering signals proportional to crack size and crack density. Unfortunately, no quantitative figure can be given for that.

c)   Highly sensitive acoustic emission measurements should be one of the most promising ways for the quantitative evaluation of crack formation and crack density. For coarse grained ceramics (grain size ~ 50 μm) this seems to belong to the state of the art (qualitative information), for fine grained ceramics (< 10 μm) this seems to be still difficult to realise.

ON THE USE OF CERAMICS IN DIESEL ENGINES

Bengt Palm

Saab-Scania AB

S 151 87 Södertälje, Sweden

SYNOPSIS

In the future diesel engine, ceramics may be used in a number of functions: as turbocharger material, for wear parts and as combustion chamber insulation. Mainly the latter function is discussed here.

The idea of insulation is not new, but until recently few practical results have been reported. One reason is the lack of suitable high temperature materials. By combustion chamber insulation a major part of that heat, which is normally cooled away by the engine water cooling system, is transferred to the exhaust gas. Thus more of the rejected heat leaves the engine at a temperature sufficiently high for efficient energy recovery systems. Less cooling may also improve the internal engine efficiency somewhat.

Different ceramic gas turbine programs have focused on the development of high strength, high temperature structural ceramics, such as Si_3N_4 and SiC. Because of their strength, these materials have been used in a first generation of ceramic diesel engine components. However, said materials have fairly high thermal conductivity and thus some designs have incorporated additional insulation, metallic or ceramic. Furthermore, the high temperature capability of the gas turbine materials are hardly needed in a diesel engine. The ideal ceramic diesel engine insulation material is light, as well as strong and tough up to about 1000°C, and has the lowest possible thermal conductivity and heat capacity. To this should be added a thermal expansion coefficient making it compatible with a surrounding metal structure. Different ceramic materials may be used for different parts. Complex ones, like inlet- and exhaust ports, ought to have such high a thermal-shock resistance that they can be included in a casting procedure, while others require the strength to take the stresses of an interference fit or some other kind of mechanical attachment. The aim should be to use as few parts as possible, made from materials having as many of the above properties as possible. One such material favored lately is ZrO_2, but the ideal material is still missing.

At present the running time with ceramics in diesel engines is limited, and long-term strength of the materials used in this environment largely unproven. Thus one of the most important tasks is to increase the number of test hours under realistic conditions and to study such phenomena as slow crack growth, oxidation, creep and material transformation. Another important area is the machining and grinding of ceramics. This must - if it can not be avoided altogether - be done in an economical way and without reducing the strength of a good material.

BACKGROUND

For a long time the turbocharged (TC), direct injection (DI) diesel engine has been the main power plant for trucks and busses because of its low fuel consumption. After the recent fuel crisis, more and more diesel engines are also to be found in passenger cars.

Since the rapid introduction and extraordinary success of the gas turbine in the aviation field after the second world war, a number of attempts have been made to introduce an automotive gas turbine as an alternative, or a complement, to the diesel engine. The major advantages would be compactness, light weight, potential of very low emissions, long life and low maintenance cost. However, so far the higher fuel consumption - especially at part load and of small engine sizes - has been a draw back. About ten years ago, certain ceramic materials (silicon nitride, silicon carbide, sialon and others) had reached such a level of development that they could be considered for the hot gas path and the turbine rotor of gas turbine engines, thus making higher cycle temperatures and lower fuel consumption possible.

During all*the gas turbine development, from the first Rover engine (1)* to the ceramic designs of to-day (2), the diesel engine has been a moving target reaching lower and lower fuel consumption figures. When ceramic high temperature materials became available for the gas turbine, the diesel engine designer asked himself if he too could make use of these, or similar, materials (3).

*
Numbers in parentheses designate references at end of paper.

In a gas turbine the fuel consumption is directly depending on the allowable turbine inlet temperature (TIT). Thus the place for ceramics is as a high temperature structural material when high temperature metal alloys can not be used any more.

In a diesel engine, however, the cycle top temperature is already high and the components making up the combustion chamber - cylinder head, cylinder liner and piston - can easily be cooled by water, oil or air. By this cooling a substantial part of the energy supplied with the fuel leaves the engine at a low temperature. If the combustion chamber could be insulated a large part of this energy would remain with the combustion gas, to be turned into useful work on the piston or leaving the cylinder at a higher temperature level. An efficient exhaust energy recovery system may then be used to extract further work. With an insulated combustion chamber the cooling need will be substantially reduced and so cooling fan and waterpump power consumption. It may eventually become possible to design an engine without the conventional water, or air cooling circuit.

Thus, one use of ceramics in the future diesel engine will probably be as insulation material. Other areas of interest for ceramics in this engine are as wear parts and as turbocharger material. The remainder of this paper will mainly discuss the insulated diesel engine and the materials that will be needed.

The idea of insulating the internal combustion engine is not new as is illustrated in figure 1 from a patent filed in 1921 (4). Typical of that period the inventor is mainly concerned about the functioning and durability of his engine, rather than the possibilities of improving its efficiency. This in contrast to a much later paper by Kamo (5) where the potential for very low fuel consumption of an adiabatic turbocompound engine is shown. Some years ago it was reported (6) that an engine of this kind had been demonstrated.

THERMODYNAMIC ASPECTS

Energy balance

Part of the energy supplied to a diesel engine with the fuel is converted to shaft-work, while the rest leaves as heat, either by different coolers or with the exhaust gas. The different energy flows to and from a typical turbocharged (TC), charge cooled truck diesel engine are shown on the schematic in figure 2. In the radiator, heat from the cooling of the combustion chamber and heat caused by friction leaves the system. Coolant flow circulates through the cylinder head and around the liner while the piston may be cooled by oil jets. Part of the energy in the exhaust is used to drive the turbocharger and is returned to the system as high temperature, high pressure air by the turbo-charger compressor. More or less of this heat will then be cooled away in the charge cooler and leave the system.

The relative magnitudes of the energy flows in figure 2 are illustrated in figure 3 for a typical full load case. The part labelled "internal cooling" is what can be influenced by insulating the engine combustion chamber (see figure 4). If the combustion chamber could be completely insulated, this part would be redistributed. Most of it to turn up as increased exhaust energy and a small part as increased shaft-work (increased indicated efficiency). With insulation materials available to-day 60-70 % reduction of the internal cooling seems feasible (7). In other words, the cooling demand on the radiator - also including friction - would be about halved.

Besides the increase of the exhaust energy, insulation will also influence the diesel engine in other ways. As mentioned, the indicated efficiency may be improved somewhat, while the amount of fresh air supplied to the cylinders will be reduced due to more heating up during the intake stroke. In the case of the TC engine this lowering of the volumetric efficiency can be compensated for by increasing the boost to restore maximum torque, while if a naturally aspirated (NA) engine is insulated, torque as well as rated power will suffer (8). In the case of the turbocharged engine the greatest efficiency improvement potential seems to be by exhaust energy recovery.

Exhaust energy recovery

By insulation a large part of the low temperature internal cooling energy will be converted to high temperature exhaust energy. Among the ways to recover as much as possible of this energy two main routes can be recognized:

1) expansion in one or more turbines

2) the use of some kind of heat engine (i.e. Rankine or Stirling) after the diesel proper - a bottoming cycle.

In figure 5 three typical systems are shown. The first two belong to cathegory 1 and the third is an exemple of a bottoming cycle, using a steam (rankine) engine (9).

The first exemple is a straight TC diesel engine. Exhaust gas is used to drive the turboturbine, which in turn drives the turbo-compressor. The turbo (charger) unit is free-floating, i.e. with no mechanical contact to the engine. By turbocharging, more air will be available, more fuel can be injected and the power increased. As engine friction does not increase at the same rate and as turbocharging can have a positive influence on the pressure drop across the diesel - resulting in less pumping work - the efficiency of a TC engine is usually somewhat higher than that of a NA engine.

If the overall efficiency of the turbocharger is high, more exhaust energy than needed will be available. Insulation of the combustion chamber will make this condition more pronounced. In order to make use of as much of the exhaust energy as possible, a second turbine, coupled to the engine output shaft via a transmission, may be placed after the turboturbine. The second system in figure 5 illustrates this principle, called the turbocompound engine (10).

Another form of compound engine is the bottoming cycle (third system in figure 5). As in the case of the turbocompound system, power is supplied to the output shaft via a transmission. One drawback of the bottoming cycle alternative is the large number of heat exchangers. The major part of the heat taken out of the exhaust is cooled away in a condensor, thus counteracting one of the advantages of insulation, that of reduced cooling. Another problem area is the steam generator which may need frequent cleaning (10).

As the exhaust of the turbocompound engine is still of quite high a temperature the addition of a bottoming cycle is possible - at the cost of increased complexity and cooling need (5).

In the turbocompound engine the amount of power to be extracted from the exhaust depends on the expansion ratio across the turbines. With a reasonable expansion ratio in the range 2-4 and with a turbine efficiency of about 80 %, 10-20 % of the additional exhaust energy flow will be converted to shaft power. This is roughly equivalent to the efficiency to be expected from a good steam bottoming cycle (9).

At full power the turbocompressor compression ratio of a typical straight TC truck engine of to-day is about 2.5. With a turbocharger of good efficiency a turboturbine expansion ratio of less than two is required, especially so in the case of an insulated engine. In order to get the most out of the exhaust a higher expansion ratio is needed. Figure 6 - from a simplified study of the turbocompound concept - illustrates this. Total turbine expansion ratio has been varied, while holding the boost level constant.

With increasing back pressure - or overall turbine expansion ratio - the pumping losses of the diesel engine will increase too, and, depending on TIT and turbomachinery efficiencies, an optimum can be reached.

In figure 6 the importance of very efficient turbomachinery and the moderate effect of insulating a straight TC engine are quite clear. An exhaust gas energy increase corresponding to about 70 % of the internal cooling of a charge-cooled turbocompound diesel engine may increase the overall efficiency by about 4 %, roughly equivalent to a BSFC decrease of 8 g/kWh. To this should be added any increase of the indicated efficiency due to insulation.

Influence of insulation on indicated efficiency

So far different ways of utilizing the additional exhaust energy of an insulated engine have been discussed. However, also the internal diesel cycle is influenced, particularly because of the change in heat-flux from the combustion gas in the cylinders to the surrounding walls. Also the combustion may be influenced and an improved ability to burn low-cetane fuels have been demonstrated (7). However, to find the optimum combustion system of an insulated engine is mainly an experimental task, outside the scope of this paper.

Figure 7 shows calculated temperature fluctuations in a diesel engine combustion chamber wall. Gas temperature and gas-to-wall heat transfer fluctuations as well as the conditions on the outside of the wall have been held constant while the material properties of the wall itself have been varied. This is a simplification, but will hardly influence the below discussion. Data for the four materials in the calculation are given in fig 7. Materials A and B (cast iron and zirconia, ZrO_2) are available to-day, while C and D are non existing. C has the same properties as ZrO_2 but for the thermal conductivity which is assumed 10 times lower. Material D corresponds to a ZrO_2 wall with a thin additional insulation layer with the properties of air. Of course this combination is totally unrealistic but has been included to show the difference between a fully adiabatic engine and one which is only uncooled (11).

The diagram in figure 7 has been divided into four periods, one for each of the four strokes of a four-stroke engine. The main effect of the insulation is a marked increase of the mean wall temperature, while material temperature fluctuations are comparatively small, even in the case of material C. Only when density and thermal conductivity have been radically reduced, as in case D, the wall temperature follows that of the gas. In all the other, realistic, cases there is marked cooling of the gas when its temperature is higher than that of the wall and vice versa. As the gas is alternatively heated and cooled the process is far from adiabatic, even though no heat leaves the combustion chamber. The gas is cooled during the expansion stroke also in a very well insulated engine. Thus only a portion of the efficiency gain theoretically possible with a fully adiabatic engine can be realized in practice. This is made more clear in figure 8 where integrated heatflux, total as well as for each of the four strokes separately, is shown. When comparing materials C and D it can be seen that although there is little difference in total heat-flux, there is a marked difference for the expansion stroke, and less so for the other periods. Even though material C must be considered extremely good thermal insulator, the heat-flux during the expansion stroke has only been about halved, compared to A.

With an insulation material giving about 70 % total heat-flux reduction there may be a reduction in the indicated fuel consumption with, say, 5 g/kWh, mainly due to reduced cooling during the expansion stroke. This is to be compared with perhaps 15 g/kWh in the ideal, adiabatic case. For calculations like those behind figures 7 and 8 the heat transfer model assumed is important. Here the model proposed in (12) has been used.

Thus, from an efficiency point of view, the ideal wall material for an insulated diesel engine combustion chamber is one which, besides having the lowest possible thermal conductivity, is light and has low heat capacity. In other words a material which allows the wall temperature to follow that of the gas as close as possible.

EFFICIENCY POTENTIAL

Figure 9 is an attempt to illustrate the potential brake specific fuel consumption (BSFC) reduction, when the possibilities discussed above are utilized. Baseline is to-days TC, charge-cooled, conventionally cooled, DI diesel engine at full load (A). With higher turbocharger efficiency a certain improvement in BSFC can be achieved (B) and by insulation of the combustion chamber a further 6-7 g/kWh(C), most of it from an assumed internal cycle improvement of 5 g/kWh (see the above discussion). The next bar (D) shows the potential of an ordinary cooled turbocompound engine and bar E, finally, how such an engine may be improved further by insulation. As temperature controlled cooling fans are used by most manufacturers nowadays, possible reduction of fan power due to less cooling has not been included in figure 9. The same holds for effects on the coolant pump power consumption. In all, one gets a potential improvement in BSFC of about 28 g/kWh, or 14 % over the baseline engine. Of this 11-12 g/kWh is due to the insulation and the rest because of turbocompounding and assuming improved turbomachinery. If all these improvements could be realized a truck engine with a maximum efficiency close to 50 % would be within reach.

INSULATION MATERIAL ASPECTS

Above the properties wanted, from an efficiency point of view, in
an insulation material suitable for a diesel engine combustion
chamber (figure 4) were discussed. To this should be added a
capability to stand the mechanical and thermal loads in the
engine and to be able to work together with surrounding metal
parts. An additional problem is the design of a suitable attachment for the ceramic parts. It must not be forgotten that ceramics are non-ductile and ought to be kept under compressive
stress.

Even though non-ceramic insulation materials (14) as well as a
combination of metal and ceramics (15) have been proposed, the
ideal combustion chamber insulation seems to be a simple, inexpensive ceramic material, able to stand a temperature of about
1000°C and the environment present in a diesel engine. Futhermore
it should be non-porous or have closed pores. Besides low thermal
conductivity, low heat capacity and low density, good thermal
shock resistance, high toughness, sufficient strength and suitable
tribological characteristics - especially important for liner,
valve seats etc - are asked for. Probably it will prove difficult to combine all these qualities in one and the same material.

Even though it is possible to machine and grind most ceramic
materials, this should be avoided whenever possible, because of
the high cost involved. Ideal is when an as-sintered ceramic
component can be included in the normal casting procedure of the
part to be insulated. This is especially important in the case of
complex shapes, like those of exhaust and inlet port liners.
In some cases where close tolerances are required some grinding
can not be avoided. For instance in designs where the attachment
is in the form of an interference fit. However, as much as
possible of the forming should always be done before sintering,
with the ceramics in the green state.

An alternative to solid insulation parts is to plasma spray a
ceramic layer - in most cases yttria stabilized zirconia - on
to the metal component to be insulated. Up to now it has proved
difficult to achieve layers of sufficient thickness and strength,
particularly on large parts like cylinder heads and pistons (9).
Greater success has been achieved with smaller components, like
inlet and exhaust valves. The main drawback with sprayed-on
layers is the insufficient level of insulation due to the limited
thickness possible.

Table 1 shows typical properties of some ceramic materials which
have been used in, or considered for, insulated diesel engines.
Silicon nitride (reaction bonded, RBSN, hot pressed, HPSN, hot
isostatically pressed, HIPSN and sintered SSN) and silicon

carbide (SiC) are strong high temperature materials with good thermal shock resistance and have primarily been developed for the use in gas turbine engines. However, they are bad thermal insulators and need some kind of additional insulation in order to be of much use in an insulated diesel engine. This holds for the fully dense qualities (HPSN, HIPSN, etc.) in particular. Strictly speaking, these ceramics are in most cases too good and too expensive to be used in a diesel engine, where little of their high temperature capability is needed, especially so in a charge cooled engine. As additional insulation, layers of roughened metal shims (13) and airgaps as well as other, weaker ceramics have been proposed. Aluminumtitanate, Al_2TiO_5, (14) is such a fairly weak material with low thermal conductivity and excellent thermal shock resistance due to very low thermal expansion and modulus of elasticity. These properties also make it very suitable to be included in a casting process.

A material having found great interest recently is partially stabilized zirconia (PSZ), in which a high strength and a thermal conductivity as low as that of Al_2TiO_5 are combined. High density and a comparatively low thermal shock resistance are drawbacks, partly compensated for by high toughness. In (15) the application of PSZ in a diesel engine is described. The fact that the thermal expansion of PSZ is close to that of cast-iron is pointed out as an advantage. This may not be completly true in the case of a well insulated combustion chamber wall, where high thermal stresses will appear (see below).

The reason for the high strength of PSZ is that the basic compound contains particles of tetragonal phase. A crack propagating through the material will transform these particles, under an increase of volume, to monoclinic phase. Simultaneously stresses and/or microcracks are introduced, rendering the growth of the original crack more difficult. This phenomenon may also be used to strengthen other ceramic materials suitable for use in a diesel engine. All in order to get an inexpensive and easy to produce material. In (16) experiments with alumina (Al_2O_3) containing different amounts of ZrO_2 are reported. The strength was drastically increased by the use of about 9 % ZrO_2.

Besides sufficient strength ceramics to be used in engines should have as high a fracture toughness as possible. An example will illustrate this. Assume two identical components made from two different materials, A and B, with the fracture toughness of B being half that of A. In all other respects the materials are assumed identical and both are subject to the same stress.

Made from material A the component will be able to stand the stress, having a four times larger defect, than if made from B. 4-5 is a typical fracture toughness value for SiC and Si_3N_4 as compared with 8-10 for ZrO_2. By adding ZrO_2 to a material its fracture toughness can be considerably improved upon, as is illustrated in figure 10 from (17).

In order to be considered for future diesel engines, ceramics must be able to stand more than 10000 hours without losing too much strength or changing in other ways. Up to now only a few hundred hours have been reported for ceramic combustion chamber parts in real engines (7). Thus it is important to try to run candidate materials under realistic conditions for thousands of hours as soon as possible in order to study such problems as oxidation, phase reactions, creep, slow crack growth (SCG) etc.

Temperatures below $1000^{o}C$ are usually considered as safe for most ceramics. However, there are examples contradicting this opinion. For example additives can cause degradation of a material due to internal phase reactions. Different oxidation kinetics will cause non-protective oxide layers at low temperatures (18). It may also happen that the long-time characteristics of a material is quite acceptable at a high temperature, but far from that at a lower (19).

The meaning of SCG is that a crack present in the material will continue to grow, although the stress level may be very low. The crack growth rate is a function of material type, temperature, environment etc. Al_2O_3 and ZrO_2 show SCG already at room temperature (20), while for Si_3N_4 and SiC this phenomenon usually does not appear below $1200^{o}C$. At low temperatures the crack growth rate usually is very low and it may take perhaps a thousand hours before a crack has reached critical size and is observed, perhaps in the shape of a complete break-down. As the presence of SCG may determine the stress level at which a certain - perhaps in static tests very strong - material can be safely used, it is important to study the long-time behaviour at an early state of any development program.

Even though machining of ceramics should be avoided, a final grinding operation may be necessary. In the case of a very strong material, like HPSN, careless grinding can reduce the strength considerably, especially if done transversal to the main stress direction. Alternatively, too careful grinding may prove economically unrealistic. Thus, grinding parameters and material type must be properly matched. Figure 11, from (21), shows what can happen to the strengh when two of the parameters, the direction and the grinding mesh, are varied.

THERMAL STRESSES

Most parts of an insulated diesel engine combustion chamber are of complex shape and a complete stress analysis is a formidable task. However, for a general discussion of how material properties will influence the stresses in a highly insulated component, a simple model in the form of a tube may be sufficient (see figure 12). In the real engine this could be, for instance, part of the cylinder liner. The model consists of an inner, insulating tube and an outer, loadcarrying one. Mean values for gas temperature, coolant temperature and the corresponding heat transfer numbers are assumed.

Without any external load maximum thermal tensile stresses in the tangential (φ) and axial (Z) directions due to the temperature difference T between gas and coolant, will be found on the outside rim of the insulation tube. A state of plain stress is assumed so that the axial component is fully developed at a point some distance from the tube end.

$$\sigma_{\varphi_{MAX}} = \sigma_{Z_{MAX}} = \frac{1}{1 + A\lambda_I} \cdot \frac{\beta_I E_I}{1 - \nu_I} B \cdot \Delta T$$

where

σ	=	stress (tensile)
λ	=	thermal conductivity
β	=	thermal expansion
E	=	modulus of elasticity
ν	=	Poison´s number
A, B	=	"constants" containing geometrical and heat transfer data only

Index I refers to the insulation tube.

The tensile stresses in the tangential direction can be counteracted by a pressure load on the outer side, caused by the container having an interference fit. The axial stress component is more difficult to counteract as this must be done by a combination of friction and axial loading.

According to table 1 the two ceramic materials PSZ and Al_2TiO_5 have about the same thermal conductivity, but otherwise quite different properties. They have both been proposed as suitable insulation materials for diesel engines. PSZ is a much stronger and tougher material, more suitable to handle the mechanical loads in an engine. If material data from table 1, together with suitable dimensions, temperatures and heat transfer numbers are put into the above equation, one gets the following maximum tensile stresses:

PSZ: 510 Mpa

Al_2TiO_5:   5   "

Thus, although PSZ is more than ten times stronger than Al_2TiO_5, all this extra strength is needed to cope with the higher thermal stresses caused by higher modulus of elasticity and thermal expansion.

In the insulated diesel engine the stresses in the high temperature parts due to mechanical loads may not be particularly high. The highest stresses may instead be thermal stresses caused by the material itself - or rather its properties - when full- filling its role as an efficient insulation. Thus it is important to find a material with properties minimizing the thermal stresses but at the same time having sufficient strength and toughness to take the mechanical loads of a diesel engine com- bustion chamber.

DESIGN EXAMPLES

Comparatively few examples of ceramic diesel engine combustion chamber components have been published and most of them originate from the Cummins/TARADCOM (US Army Tank - Automotive Research and Development Command) program aimed at the development of an "adiabatic" turbocompound diesel engine. (5) is one of the first papers published on this project while in (6) it is reported that a fuel consumption of .285 lb/hph (174 g/kWh) - the original goal was .28 - has been achieved with a multicylinder demon- stration engine.

Early experiments with ceramic pistons were made in England and are reported by Godfrey in (22). A monolithic RBSN piston with piston rings and gudgeon pin of the same material was run successfully in a small petrol engine and a bigger piston of similar design in a Gardner single cylinder diesel engine. Paper (23) tells about further English tests with ceramic pistons of monolithic as well as composite design. RBSN without any addi- tional insulation was used. The main objectives of the experiments in (22) and (23) seem to have been to demonstrate that ceramics can function in a piston engine, rather than to develop a highly insulated combustion chamber.

Probably the first piston with separate insulation was the one used by Cummins in the demonstration engine (figure 13). It consists of a HPSN piston cap, separated from the rest of the piston by an insulation pack of roughened metal shims, held in place by a high temperature alloy bolt. A section of one version of the engine of which this piston is a part is shown in figure 14 (6). Here the cylinder head insulation can be seen as well. It too is made up of metal shims behind a hot-plate of HPSN (13). An insulated exhaust port liner is also shown. It is said to be made of solid ceramics and included in the cylinder-head casting process. An alternative design discussed is a port liner of sheet metal with an insulation material sprayed on to the outside before casting. One of the cylinder liners described in (6) is of composite design with layers of zirconia and chrome oxide in an ordinary cast-iron liner, another with a Si_3N_4 insert. In parallel with these concepts Cummins has been working on other solutions like plasma sprayed insulation layers of zirconia (8) and, lately, solid insulation parts of PSZ (15). This latter material has been used in pistons as well as in cylinder heads and liners. One of the main advantages with this new design is that the number of parts have been drastically reduced - from 35 to 1 or 2 in the case of the piston according to (15). The single PSZ cap fullfills the function of high temperature part as well as insulation.

Lately also piston manufacturers - most engine makers do not produce or develop their own pistons - have begun to show interest in insulated pistons. In (24) Wacker and Sander from Karl Schmidt GmbH (KS) in Germany discuss a number of design concepts, from all metal, high temperature alloy piston caps with air gaps for insulation to a piston having a solid PSZ cap (figure 15).

Also valve manufacturers have become active in the field of insulated components and in (25) different designs, developed and tested by the Eaton corporation, are discussed. Figure 16 shows valves from this paper.

In addition to Cummins, Komatsu Ltd of Japan have published results from tests with a completely insulated multicylinder DI diesel engine (7). When the engine was first tested with insulated components a marked increase in fuel consumption was observed. By optimizing the combustion system, including the shape of the piston bowl, the fuel consumption was brought down to about the same level as the non-insulated baseline engine. However, very few design details are given in (7), but that the piston has a ceramic cap bolted to the rest of the piston.

Paper (26) describes another japanese ceramic diesel engine program. The capability of sintered Si_3N_4 from Kyoto Ceramics Co Ltd (Kyocera) to function - without additional insulation - in a small NA diesel engine was demonstrated. But for some increase in exhaust temperature, no performance results are given. Total running hours between 290 and 350 are reported for different combustion chamber components.

A joint effort by the Carborundum Co of Niagara Falls, N.Y. USA and prof. S Timoney at the Dublin University, Ireland, to use SiC components in a two-stroke, opposed piston diesel, is mentioned in (27). This kind of engine has no inlet- and exhaust valves and no cylinder head. This simplicity makes it particularly suited for the SiC fabrication techniques. However, SiC is a bad insulator.

In the future high temperature combustion chamber parts may not only be found in car and truck power plants. In (28) an "adiabatic" cruise-missile diesel/gas turbine engine is outlined and (29) contains the concept of an insulated helicopter engine.

CONCLUSIONS

By insulation of the combustion chamber and the introduction of an exhaust energy recovery system in the form of turbocompounding, the specific fuel consumption of a diesel engine may be reduced by about 14 %, making the overall efficiency approaching 50 %. This, however, will require a good insulation material and very high turbomachinery efficiency.

The ideal insulation material for this kind of engine seems to be an inexpensive ceramic material able to stand a temperature of about 1000°C and the environment of a diesel engine. Besides low thermal conductivity, the material also ought to have low thermal expansion, heat capacity, density and modulus of elasticity together with high fracture toughness and sufficient strength. In other words, quite a different material to what is being developed for gas turbines.

To-day there is no known material fulfilling all the wanted properties listed above. Partially stabilized zirconia (PSZ) is a good insulation material with high strength and toughness and has been proposed for the diesel engine. However, high thermal expansion and modulus of elasticity also cause high thermal stresses. In aluminumtitanate, which has about the same thermal conductivity as PSZ, the same stresses will be much lower, but the strength will probably prove too low as well. A combination of the most suitable properties of these two materials would make a good diesel engine combustion chamber insulation material. If this can be achieved is another question. Perhaps it is not possible to get all the required properties in one single material, but in a combination of different materials.

The long-term behavior (such as SCG, oxidation, creep etc.) of a candidate material must be studied in a representative diesel engine environment as early as possible in any development program. To-day few ceramic components in diesel engines have run for more than a few hundred hours. Little detailed information from tests has been published. In several cases the main purpose seems to have been to demonstrate that ceramics can survive, rather than to develop an efficient insulated combustion chamber.

Even if a suitable material can be found, other problems in connection with an insulated combustion chamber will have to be solved. High cylinder wall temperatures will make lubrication of the piston more difficult than to-day and insulation will also influence the combustion process, requiring redesign.

REFERENCES

1. Penny, N: Rover Case History of Small Gas Turbines (SAE 634A, 1963)

2. Helms, H.E. and Byrd, J.A.: Ceramic Components for Automotive and Heavy Duty Engines - CATE and AGT 100 (ASME 82-GT-253)

3. Katz, R.N. and Lenoe, E.M.: Ceramics for Diesel Engines: Preliminary Results of a Technology Assessment (Automotive Development Contractors Meeting, 28 October 1981, Dearborn, Mich. USA)

4. US Patent 1, 462, 654, July 24, 1923 (Internal Combustion Engine and Parts thereof)

5. Kamo, R.: Cycles and Performance Studies for Advanced Diesel Engines (Ceramics for High Performance Applications. Proceedings of the Fifth Army Materials Technology Conference, Brook Hill Publishing Co, Chesnut Hill, Mass., USA 1978)

6. Bryzik, W.: Adiabatic Engine Program (Automotive Technology Development Contractor's Coordination Meeting, 11-14 Nov. 1980, Dearborn, Mich., USA)

7. Yoshimitsu, T et al: Capabilities of Heat Insulated Diesel Engine (SAE 820431)

8. Kamo, R. et al: Thermal Barrier for Diesel Engine Piston (ASME 80-DGP-14)

9. Koplow, M.D. and DiNanno, L.: Status Report on Diesel Organic Rankine Compound Engine for Long-Haul Trucks (same source as ref 3)

10. Brands, M. et al: Vehicle Testing of Cummins Turbocompound Diesel Engine (SAE 810073)

11. Zapf, H.: Grenzen und Möglichkeiten eines wärmedichten Brennraumes bei Dieselmotoren (VDI Berichte nr 238, 1975)

12. Hohenberg, G.F.: Advanced Approaches for Heat Transfer Calculations (SAE 790825)

13. Kamo, R. et al: Ceramics for Adiabatic Diesel Engine (Energy and Ceramics, Prodceedings of the 4:th International Meeting on Modern Ceramics Technologies, Saint-Vincent, Italy, 28-31 May 1979)

14. Pohlman, H.J. et al: Untersuchungen an Werkstoffen im System Al_2O_3-TiO_2-SiO_2 (Bericht der Keramischen Gesellschaft, Band 52 (1975) Nr 6, s. 179-183)

15. Woods, M.E. and Oda, I: PSZ Ceramics for Adiabatic Engine Components (SAE 820429)

16. Becker, P.F.: Transient Thermal Stress Behaviour in ZrO_2 - Tonghened Al_2O_3 (Journal of the American Society 64: (1981): 1 pp 37-39)

17. Klaussen, N. and Petzow, G.: Strengthening and Toughening Models in Ceramics Based on ZrO_2 Inclusions (same source as ref 13)

18. Quackenbush, C.L.: Oxidation in the Si_3N_4 - Y_2O_3 - SiO_2 System (same source as ref 6)

19. Qvinn, G.D.: Characterization of Turbine Ceramics after Long-Term Environmental Exposure (Technical Report AMMRC TR 80-15, April 1980)

20. Lishing Li: Bestimmung kritischer und unterkritischer Rissausbreitung keramischer Werkstoffe bei Raum- und Hochtemperatur (Dissertation, Universität Stuttgart, 1980)

21. Andersson, C.A. et al: Progress on EPRI Ceramic Rotor Blade Program (same source as ref 5)

22. Godfrey, D.J.: Ceramics for High Temperature Engineering (Proceedings 22, British Ceramics Society, Stoke-on-Trent, July 1973)

23. Parker, D.A. and Smart, R.F.: An evaluation of Silicon Nitride Diesel Pistons (from Associated Engineering Developments Ltd, Cawston House, Cawston, Rugby, Warwickshire,

England, 1980)

24. Wacker, E. and Sander, W.: Piston Design for High Temperature Combustion Pressures and Reduced Heat Rejection to Coolant (SAE 820505)

25. Worthen, R.P. and Tunnecliffe T.N.: Temperature Controlled Engine Valves (SAE 820501)

26. Hamano, Y. and Nakahara, Y.: Ceramic Parts for a Diesel Engine (the same source as ref 6)

27. Long, W.D. et al: High Performance Ceramic Engine Applications (the Carborundum Co, Niagara Falls, N.Y. USA, 1982)

28. Robinson, C.Jr.: USAF Readies Advanced Cruise Missiles (Aviation Week and Space Technology, March 10, 1980)

29. Wilstead, H.D.: Preliminary Survey of Possible Use of the Compound Adiabatic Diesel Engine for Helicopters (SAE 820432)

Table 1: Typical properties of ceramics considered for the use in an insulated diesel engine

		λ	$E.10^{-3}$	$\beta.10^{+6}$	ϱ	c	$\sigma_{B_{RT}}$	K_{IC}
RBSN		18	170	3	2600	850	200	2,5
HPSN/ HIPSN	SILICON NITRIDE	25	310	3,4	3200	850	700	5
SSN		22	300	3,3	3100	850	500	4,5
SSiC	SILICON CARBIDE	75	400	4,6	3100	1000	500	4,5
HPSiC		90	450	4,7	3200	1000	700	5
Al_2O_3	ALUMINA	33	380	8	4000	800	300	3,5
PSZ	ZIRCONIA	2,5	200	9,8	5700	400	500	10
Al_2TiO_5	ALUMINUM-TITANATE	2	13	1,5	3200	880	35	-

λ = THERMAL CONDUCTIVITY (W/M, K)

E = MODUL OF ELASTICITY (MPA)

β = THERMAL EXPANSION (1/K)

ϱ = DENSITY (KG/M^3)

c = HEAT CAPACITY (J/KG, K)

$\sigma_{B_{RT}}$ = FLEXUAL STRENGTH AT ROOM TEMPERATURE (MPA)

K_{IC} = FRACTURE TOUGHNESS (MPA . M$^{\frac{1}{2}}$)

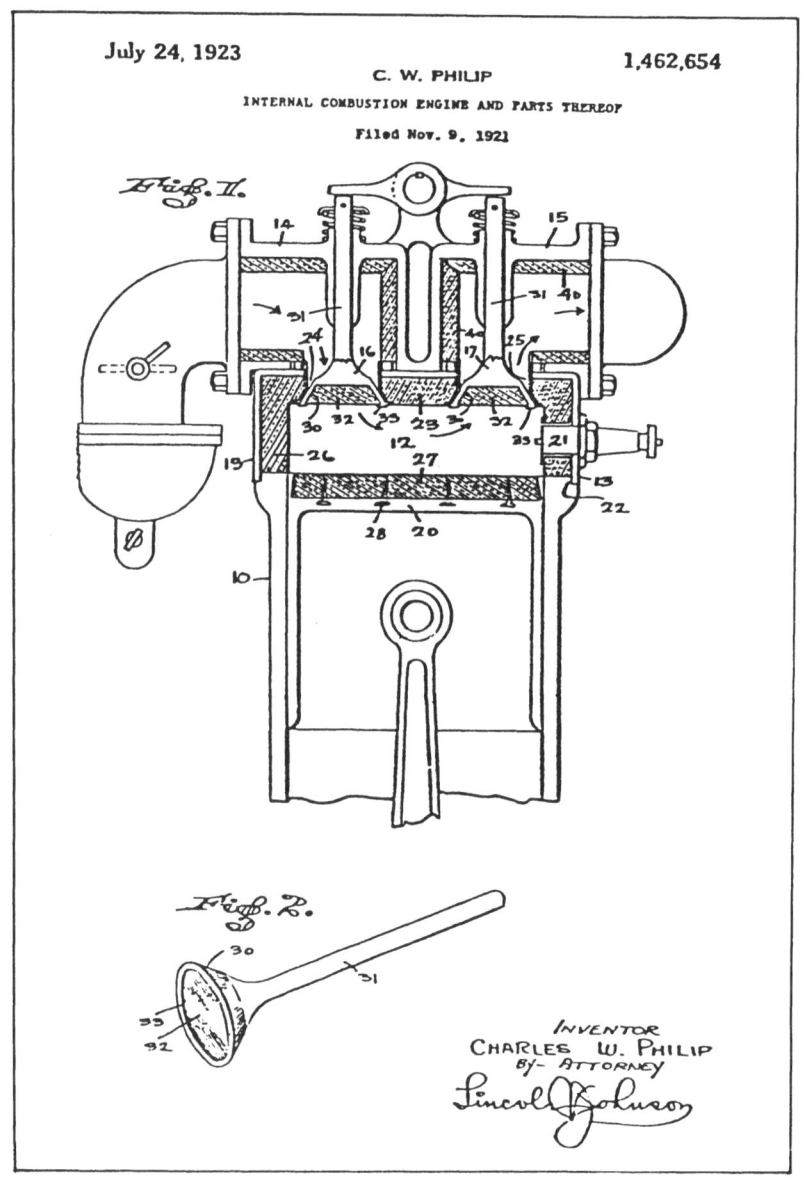

Figure 1. Insulated engine concept from 1921.

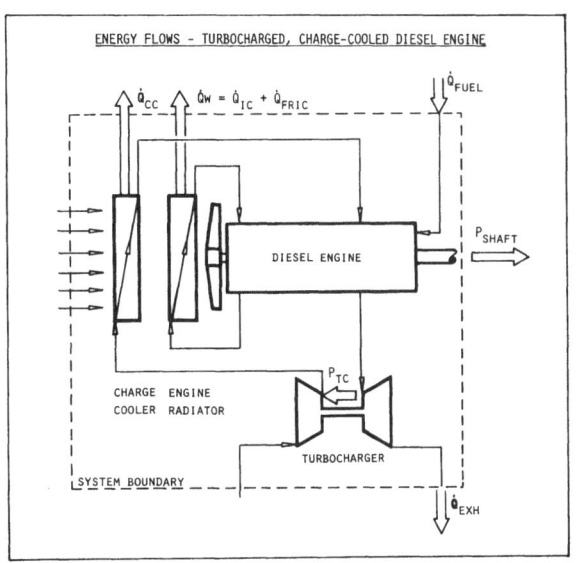

Figure 2. Energy flows. Turbocharged, charge cooled diesel engine.

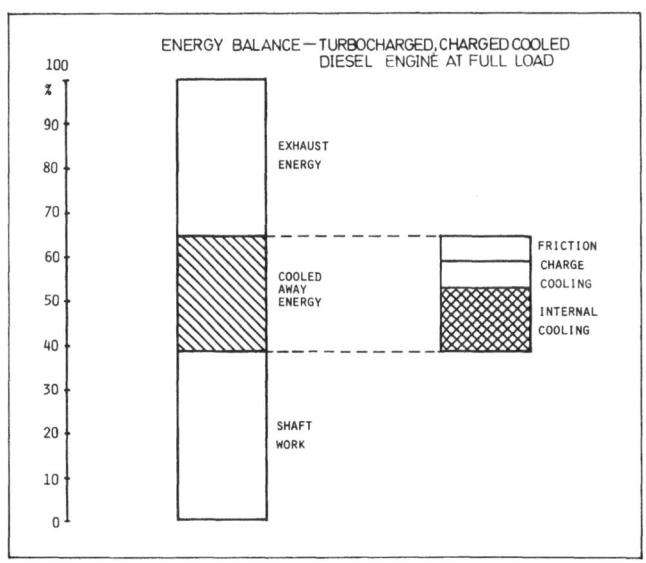

Figure 3. Energy balance. Turbocharged, charge cooled diesel engine.

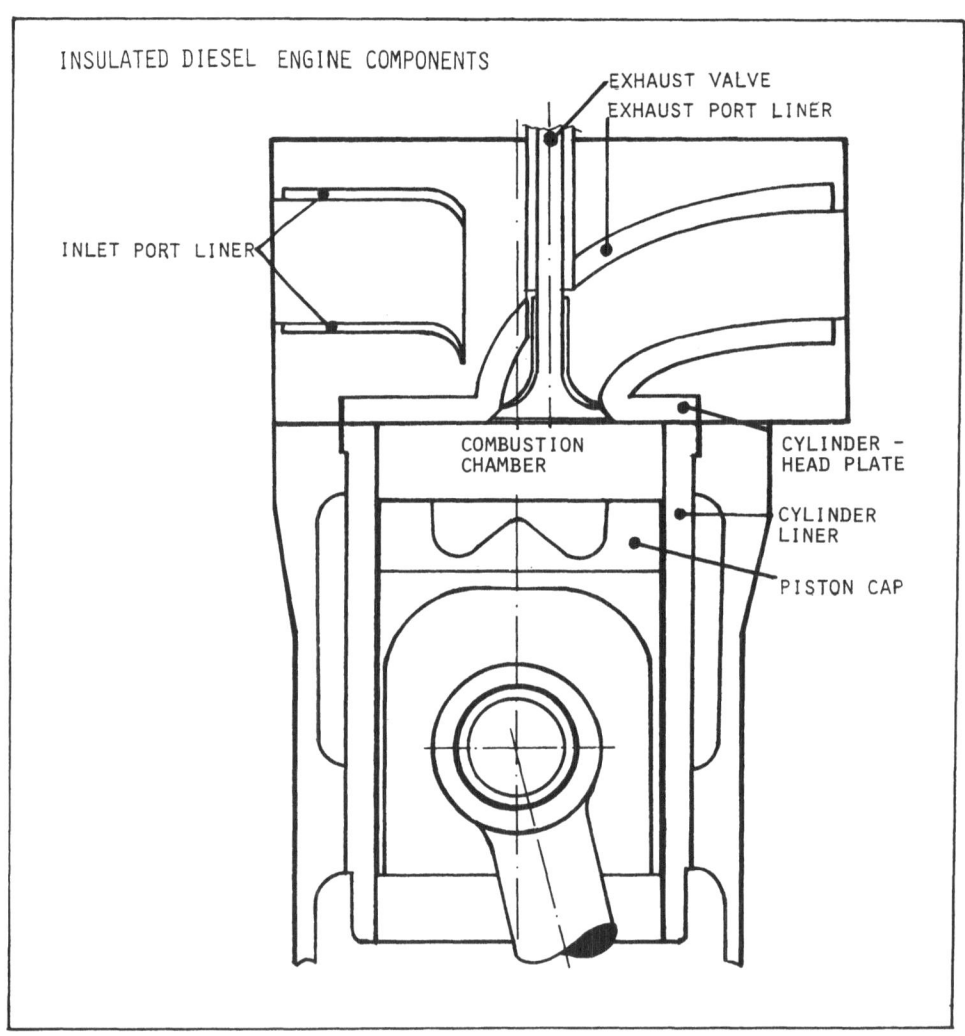

Figure 4. Insulated diesel engine components.

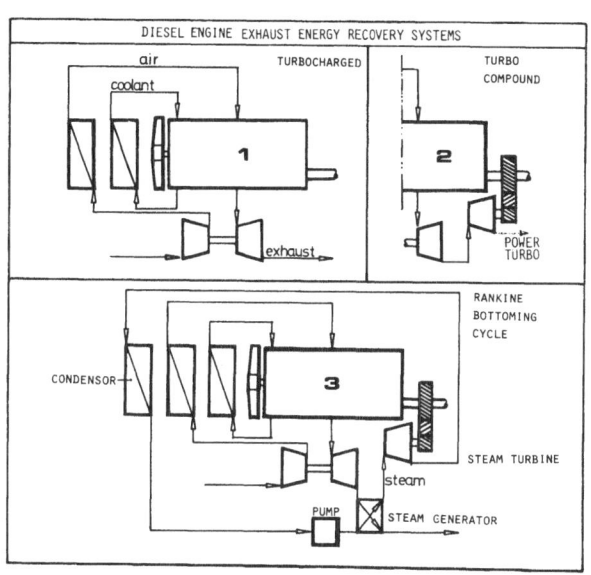

Figure 5. Diesel engine exhaust energy recovery
systems.

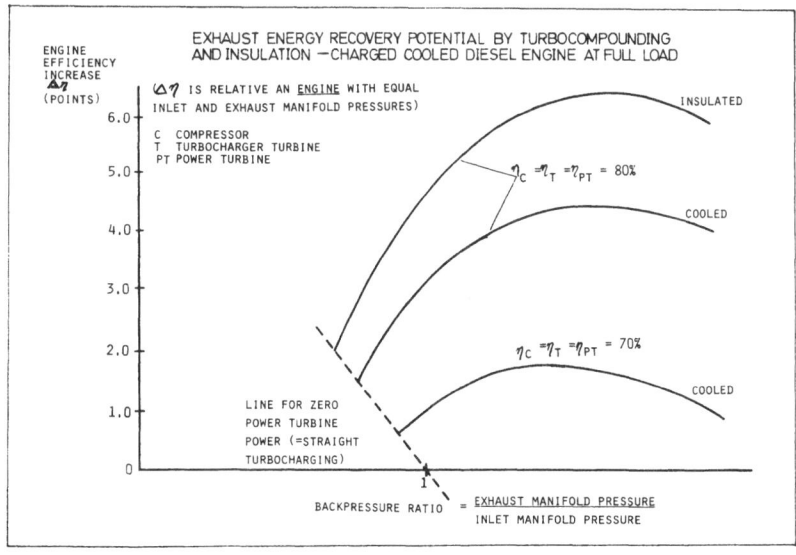

Figure 6. Exhaust energy recovery potential by
turbocompounding and insulation.
Charge cooled diesel engine at full road.

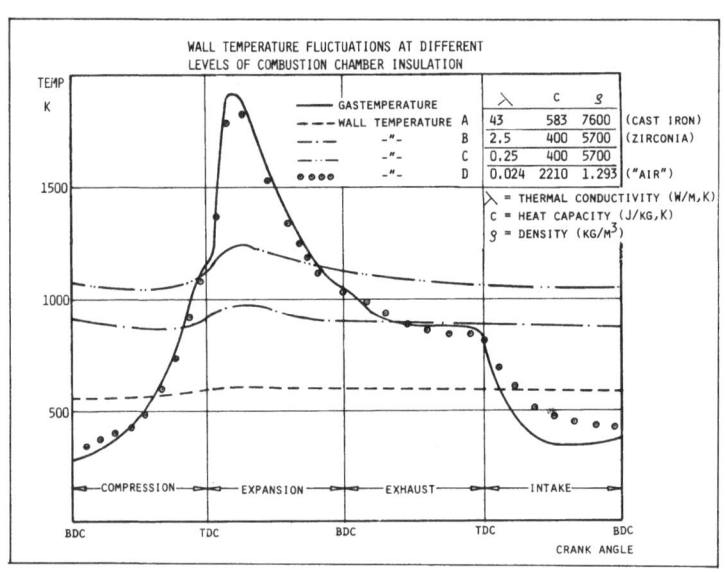

Figure 7. Wall surface temperature fluctuations at different levels of combustion chamber insulation.

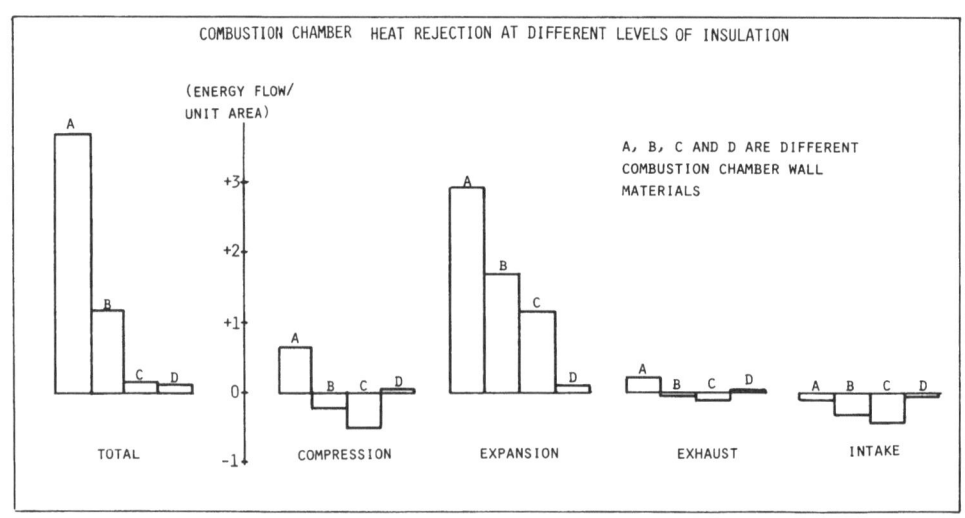

Figure 8. Combustion chamber heat rejection at different levels of insulation.

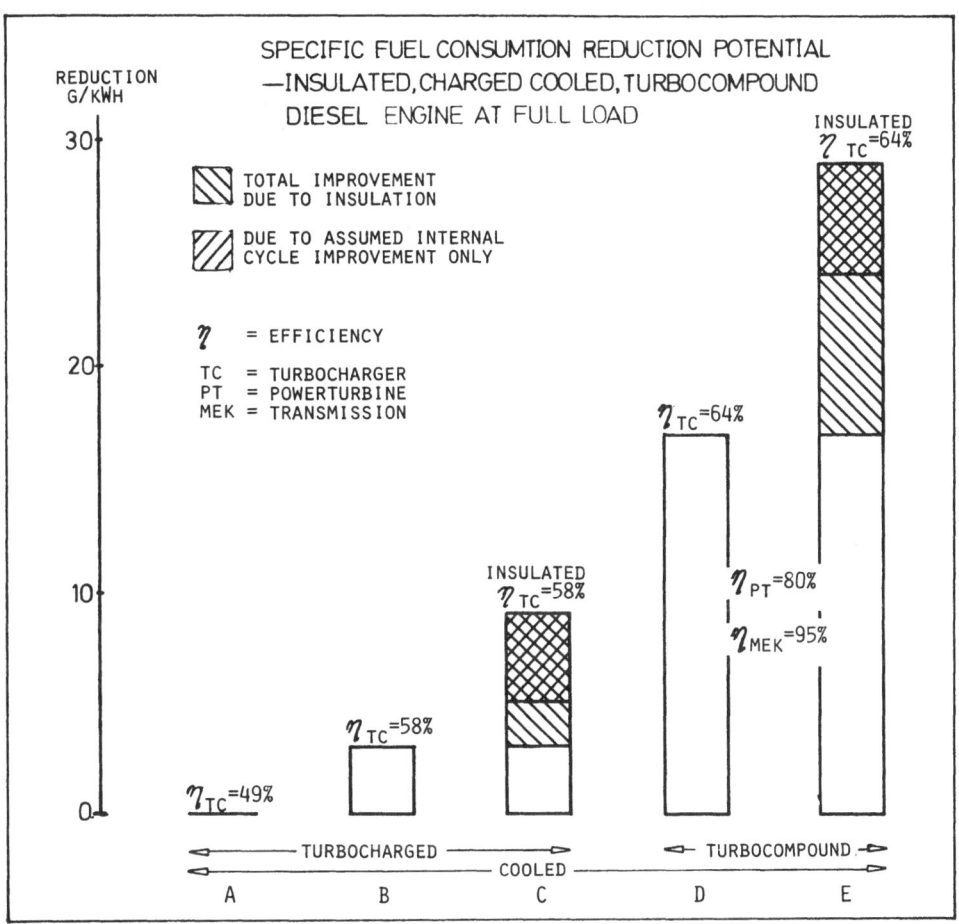

Figure 9. Brake specific fuel consumption reduction potential. Insulated charge cooled turbocompound diesel engine at full load.

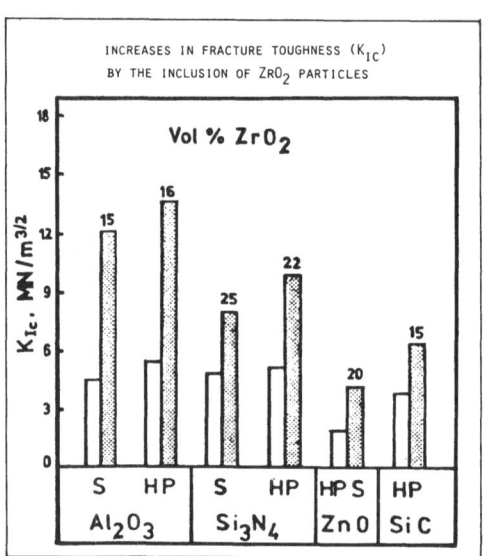

Figure 10. Increase in fracture toughness (K_{IC}) by
the inclusion of ZrO_2 particles (vol. %
on top of dotted bars); white bars: K_{IC}
of matrix material alone.
From (17)

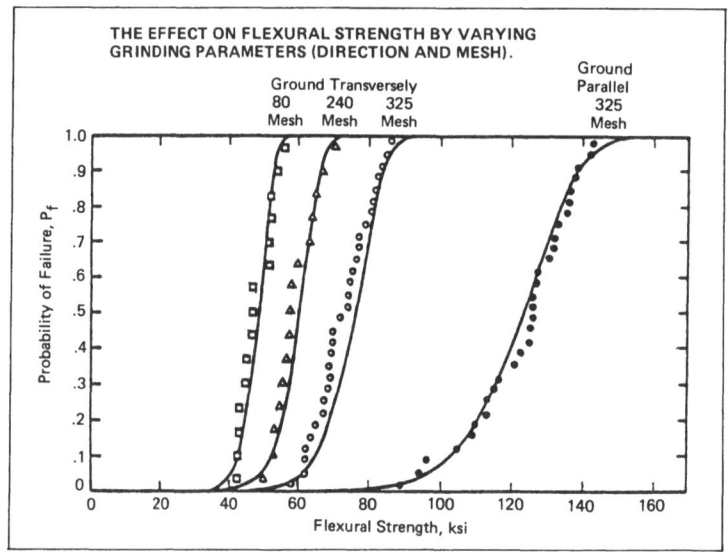

Figure 11. The effect on flexural strength by varying
grinding parameters (direction and mesh).
From (21)

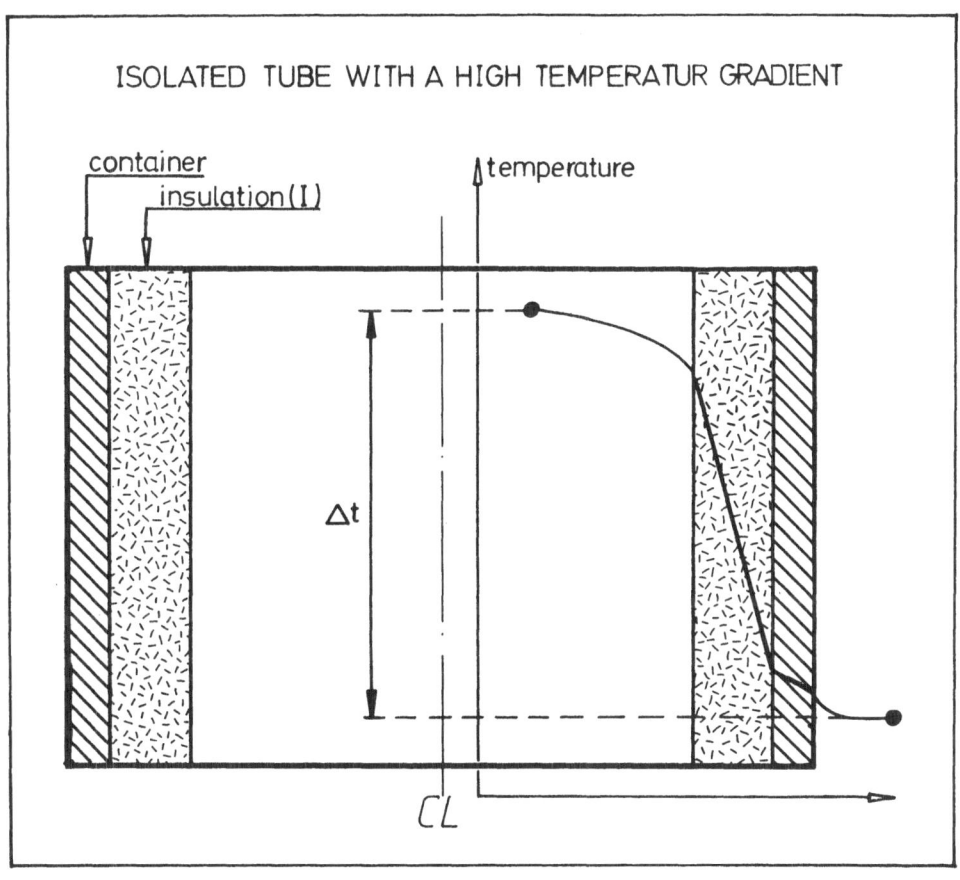

Figure 12. Insulated tube with a high temperature
gradient.

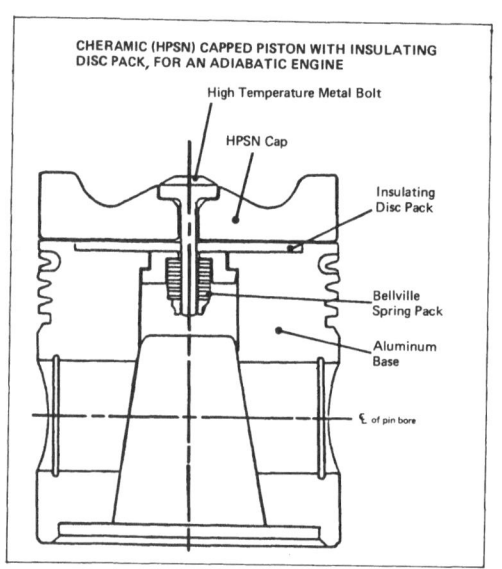

Figure 13. Ceramic (HPSN) capped piston, with insulating disc, from an adiabatic engine.
From (8)

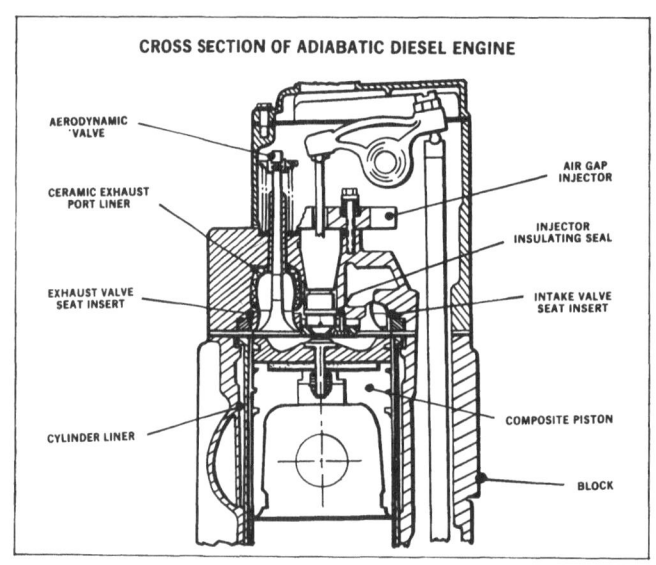

Figure 14. Cross section of adiabatic diesel engine.
From (6)

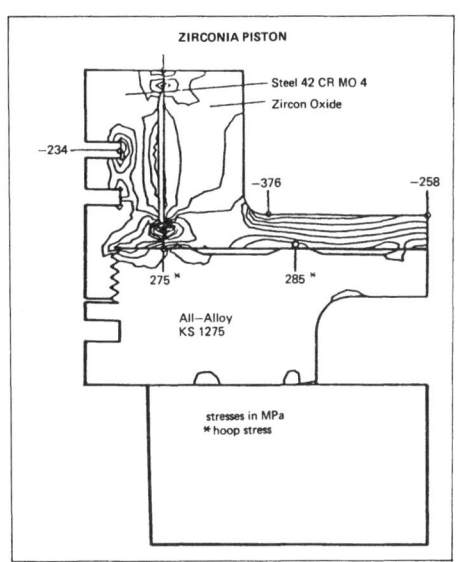

Figure 15. Zirconia piston - Thermal stress plot of
ceramic insulated piston.
From (24)

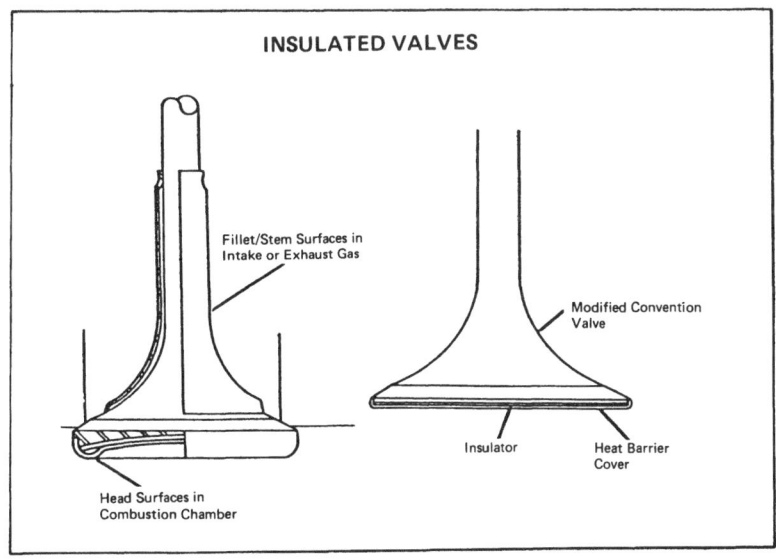

Figure 16. Insulated valves.
From (25)

DISCUSSION

G. Grathwohl: Have you any idea whether these insulated
components can withstand cyclic fatigue on the expected
stress levels (thermal and mechanical stresses)?

B. Palm: This of course depends on design as well as on the
ceramic materials used. In the different designs reported so
far the difference is great. Certain high temperature, high
strength gas turbine materials have very good thermal fatigue
properties. However, reported experience of high temperature
durability and thermal fatigue properties of ceramics also
being good insulators, like ZrO_2, is very limited. This is
certainly a problem area which should be addressed as soon as
possible. (See a presentation by R.N. Katz and E.M. Lenoe at
DoE Automotive Technology Development Contractor's
Coordination Meeting 26 October 1982 in Dearborn, Mich., USA:
"Ceramic Technology Progress Report - DoE/AMMRC Ceramic
Materials Program").

R.J. Brook: Is there agreement about the amount of porosity
that can be tolerated in insulating liners for use in
cylinders?

B. Palm: If there have to be pores, for insulating reasons,
in a combustion chamber wall material, they ought to be
closed, otherwise engine performance will suffer. In the case
of the liner surface, tribological aspects have to be
considered as well.

F. Starr: What are the peak and average heat transfer rates
and coefficients into diesel engines? What temperature
gradients are induced in ceramic liners?

B. Palm: Peak heat transfer coefficients have values of
$3000-4000$ W/m^2,K and corresponding mean values are $700-800$
W/m^2,K, at full load. The temperature gradient in the wall of
an insulated liner will depend on the insulation material,
geometrical dimensions, the amount of cooling on the outside
as well as the allowable temperature of a support structure,
if any. As several concepts have been proposed, with very
different degrees of insulation, it is not possible to give a
straight answer. Some information may be found in SAE paper
780069 (Stang: Designing Adiabatic Engine Components).

J. Kruithof: Please give an assessment of the distribution of
the heat flow over the various engine components. In an
adiabatic turbocompound diesel engine, would you agree that

isolation of exhaust system and turbocharger housing would be needed?

B. Palm: From our experience a typical combustion chamber heat flow distribution is:

to piston	36%
to liner	26%
to cyl.head	25%
to exhaust port	13%

The heat flow to the cooling water is about equally divided between liner and cylinder head, the liner in this case receiving heat from piston heat transfer as well as friction. Insulation of the exhaust system and turbine housing should be beneficial as is reported in a recent Cummins presentation (Hoehne J.L. and Werner J.R.: "Report on the Cummins Advanced Turbocompound Diesel Engine Evaluation". Same source as in the answer to Mr. Grathwohl.)

O. de Pous: You increase the temperature of the engine. How did you solve the problem of lubricant?

B. Palm: The Cummins Engine Company have reported that they use a certain high temperature synthetic oil in combination with piston rings having a coating consisting of alumina and titania (ref. 6 in my paper).

F.K. van Dijen: Did you consider Si_2N_2O as an insulating material?

B. Palm: No, I have not and I do not know of anybody who has.

G. Martinez: In the actual concept of engines using ceramic insulation, one has to maintain ceramic parts with the help of other parts of superalloys, which complicates the design and the execution of the system. Can one envisage direct joinings of metal/ceramic and ceramic/ceramic with a type of cement or glue? The first kind of joining would simplify the design, the second one would permit to associate ceramic materials in order to realise a combination of insulating properties and mechanical strength.

B. Palm: I do not know of such a cement or glue, but would like to. Some experiments with brazing of ceramics have been reported. One of the great problems with this kind of a joint is usually that the materials to be joined together have different thermal expansion coefficients. One has the same problem with sprayed-on coatings where compliant layers have to be used and only thin insulation layers are possible.

RECUPERATIVE CERAMIC HEAT EXCHANGER

St.Schindler, A.Krauth

ROSENTHAL TECHNIK AG

D-8672 Selb

Abstract

The development of silicon nitride recuperative heat exchangers (HX) for gas turbine application, heat technology, process engineering and automobile engines as well is described. The technology used includes tape casting, punching, lamination and embossing as well. Heat exchanger (HX) areas up to 2500 m^2/m^3 can be fabricated. Materials and heat exchanger (HX) data will be given and selected applications will be discussed.

1. Introduction

The use of ceramic material for static (recuperative) HX opens a wide field of applications in heat technology, process engineering and for automobile engines due to following facts (see also fig. 1):

- high corrosion resistance to aggressive media (gases and liquids)
- range of operational temperature up to above 1000 °C

– thermal shock resistance and

– world-wide availability of raw materials without restriction.

Suitable HX materials are cordierite-type material and silicon nitride. We prefer the latter one: it does not shrink when being fired and its permeability can be reduced by a sealing media down to below 10^{-9} kg/s cm² (test specimen $\emptyset$ 30 x 0,8 mm, Δp = 5 bar), which is even enough for a recuperator to be used in vehicular gas turbines. The paper presented reviews the development works concerning ceramic HX for vehicular gas turbines (Rosenthal), heat technology, process engineering and automobile engines (cooperation Rosenthal-KFA, Nuclear Research Center at Jülich, FRG).

2. HX for gas turbine application

In 1974 the HX development has been started as part of the R&D-project "Ceramic Materials for Automobile Gas Turbines", sponsored by the Ministry of Research and Technology of FRG. It aimed at a ceramic counterflow matrix having a wall thickness comparable to that of a regenerator disc, which was part of the USA-Gas-Turbine-Development-Program started some years earlier.

Fig. 2 shows a scheme of a gas turbine which is the goal of the USA R&D-project, fig. 3 a section out of a regenerator disc matrix. In the course of that project, which is still carried on, testing in simulated application conditions, progress in material development and production technique made it possible to reduce the wall thickness of such re-generator discs to less than 0,2 mm.

So with respect to that wall thickness the development of a comparable recuperator matrix was the most challenging target for Rosenthal within the FRG-Automobile-Gas-Turbine-Project.

Some ceramic components manufactured within this project can be seen in fig. 4. Fig. 5 shows a special KFA-Gas-turbine-concept with an annular arrangement of recuperator modules made from silicon nitride. The design principles of such a recuperative HX matrix are given in fig. 6. Due to the openings at the top and bottom side a split counterflow for both the high pressure and low pressure air is obtained.

Another possible arrangement of recuperators – comparable to that of the rotating regenerator discs – is simply on both sides of a gas turbine. Fig. 7 shows a scheme of such a recuperative HX module with integrated air ducts, being the objective of the R&D-project.

2.1 Fabrication techniques for ceramic HX

The fabrication technique is strongly influenced by the design of the HX . A most flexible technique, especially regarding mass-production, is based on a ceramic tape produced by the doctor-blade-process. Tape thickness down to 0,2 mm can easily be produced. Shaping is accomplished by corrugation or more simply by a combined foil, punch and emboss technique. Subsequently the shaped parts are laminated to cross-or counterflow HX, as shown in fig. 8.

The development work so far was to find out the limits of this technique with regard to the specifications of the HX matrix. Table 1 compares the specifications required with those reached.

Fig. 9 to 12 show
- the cores of counterflow HX modules produced by laminating embossed single layers (Fig. 9)
- an enlargement of the HX matrix (Fig. 10)
- a polished section of embossed and laminated single layers (Fig. 11) and
- a polished section of laminated single layers (Fig. 12).

The development of this type of HX will be carried on. Test results of matrices with 0,2 mm wall thickness will not be available before next year.
Apart from vehicular gas turbines Rosenthal and KFA found some other potential markets for HX applications in the field of heat technology and process engineering and automobile engines . Selected examples are described below.

3. Ceramic HX for home heating sector
The newly developed ceramic heater module is an example of designing

a new system. It is an approach to get the heat loss sources (burner, boiler, flue gas) of the home heating sector under control with the aid of a single, totally new aggregate. The heart of this new system is a combination of HX and burner made of silicon nitride. The development of the ceramic heater module has reached an advanced stage. Not only have the ceramic components proved to be functionally reliable, they have also demonstrated their worth in the form of a complete prototype unit that showed off an efficiency increase of up to 20 %.

3.1 Design and function

The burner and the HX are in-line and interconnected by the combustion chamber. They are blockshaped, have a ceramic walled combustion chamber and a ceramic heating section, the HX (Fig. 13).
The burner and the HX are filled with numerous, axially oriented, slender ducts or slots. These ducts serve to conduct the media in co-current flow through the burner and in countercurrent flow through the heating section. The partitions have the largest area possible, so as to provide for a rapid transfer of heat in the HX. The media enter the unit through simple connections on opposite ends of the HX and on the end face and one side of the burner.
Fig. 14 can be of assistance in describing the function of the ceramic heater module.
Fuel and combustion air are mixed at the burner mouth, ignited and burn. The hot combustion gases enter the heating section and re-

linquish practically all of their heat to the space heating water, thus cooling down almost to the dew point. The cold potable water in the second chamber of the heating section serves as a heat sink, causing the combustion gas to drop below the dew point and the water vapor to precipitate. The thereby released heat of condensation serves to heat up the potable water. With the transgression of the dew point not only the water vapor of the combustion gas condenses but also the pollutants (SO_2/SO_3) contained in the gas as a result of combustion precipitate in the form of aggressive sulfuric and/or sulfurous acids. Thus a flue gas is produced which practically has no pollutive character. Propelled by an induced draft fan, the combustion air /flue gas runs rectilinearly through the system. The planar-flame burner ensures an extremely short flame, which in turn allows an extremely compact combustion chamber design.

3.2 Materials and shaping technique

The ceramic material must meet the following functional requirements:

    It must have sufficient high-temperature stability.

    It must be resistant to thermal shock.

    It must be resistant to aggressive media.

Furthermore, with a view to introducing the new heating system commercially, the production technique used for making the ceramic components must favor a high piece rate at relatively low cost. Silicon nitride is a material that not only satisfies the above requirements, but also allows the use of a low cost production technique.

Shaping is accomplished by a combined foil, punch and lamination technique that gives the designer sufficient latitude for an optimum burner/HX layout.

3.3 Prototype system – characteristic dimensions, initial data from preliminary tests

Continuing onward past the principles of design and function, figure 15 shows the burner and HX of the prototype system. Both are elongated so that each provide half the lenght of the combustion chamber. The ceramic components (burner combustion, chamber, HX) have the following characterstic dimensions:

length	30...60 cm
width	5...15 cm
height	4...10 cm
duct width	0,8...3 mm
duct partition thickness	0,8...2 mm
specific area of HX	250...450 m² /m³
capacity per element	20...30 kW (gas or oil).

A different or modified burner head must be employed for oil firing, because the oil has to be prepared (vaporized) for combustion. Initial testing of the ceramic heater module yielded the following results:

- The combustion gases can be cooled back down to within a few degrees centigrade of the lowest cold water intake temperature
- Condensate yields of 50..90 % are attainable (water intake tempe-

rature 10-13 °C).

- A thermal efficiency of 98 % (for gas-fired aggregates) and 97 %
  (for oil-fired aggregates), respectively, is attainable. That amounts
  to a 20 % increase in efficiency or, as you will, 20 % savings on
  fuel as compared to state-of-the-art equipment.
- The unit is suitable for continuous low-load service, i.e. heat losses
  due to an incessant ON/OFF sequence can be completely eliminated.

Cooling of the hot combustion gases in the water-cooled HX was also
investigated (see fig. 16). Within a length of about 5 cm, theses gases
are cooled from a combustion temperature upwards of 1600 °C down
to the near vicinity of the dew point (70 °C)! The rest of the HX
length is more than adequate for cooling the entrained water vapor
to the point of condensation.

The compact construction of the ceramic heater module makes it
possible to install the unit in kitchens (fig. 17). The flue gas, at
temperatures of 20...30 °C, can make do with much simpler, cheaper
flue ducts.

4. HX/burner assembly for process engineering

Slight modification of the ceramic components employed in the ceramic
heater module will produce a system suitable for use as a pollutant
combustion unit. The burning-off of pollution gases via thermal after-
burning is a widely used method of purifiying flue gases. The pollutive
flue gas is boosted to 600...800 °C and held at that temperature long
enough to ensure thorough burning. The off-heat from the hot flue

gas is recovered and used to preheat the unburned pollutive air or for generating steam, hot air or hot water. However the economical and technical value of present-day systems is limited by the fact that certain acidic ingredients such as HCl, SO_2 and SO_3 - often enough appearing only as a direct result of post-combustion - precipitate on the boiler and chimney surfaces when the gas cools to below the dew point. Serious corrosion can result.

Such problems do not occur in the silicon nitride HX/burner assembly. On the contrary, with its resistance to high temperatures, thermal shock and aggressive media, the ceramic material allows intentional transgression of the dew point, thus recovering the heat of condensation.

4.1 Pilot plant: technical disposition, function and first findings

The pollutant combustion unit shown in figure 18 comprises a ceramic HX with an integrated burner, a second ceramic HX for cooling the gas and inducing condensation, and a reaction chamber. The HX/burner assembly is of the same design and execution as the ceramic heater module: block-shape, multislot system, media conducted side by side in counterflow.

The pollutive flue gases enter the HX from the side, are heated up, and continue upwards through the planar-flame burner into the combustion chamber. The alternate pollutive gas and flue gas ducts in the burner ensure intensive mixing of the gases within a short distance and consequently, complete combustion. This also makes it possible to reduce the combustion gas time in the combustion chamber, so that the combustion chamber can be made significantly smaller than in conventional systems. The gases are then directed back to the HX, where they relinquish most of their heat to the incoming pollutive gases. Further downstream, in the second ceramic HX, the flue gases are cooled down to within a few degrees centigrade of the water intake temperature. Since this entails transgressing the dew point, most of the combustion water vapor precipitates, thereby adding its heat of condensation to the residual heat of the flue gases for use in heating water.

The unburnable acidic ingredients precipitate along with the water vapor, producing concentrated acids that can be eliminiated by way of filtration or washing out. The off-gas is therefore practically free of pollutants when it is released to the atmosphere.

A first comparison of ceramic and conventional systems was conducted using a flue gas containing 554 mg/kg methanol, 531 mg/kg xylene and 133 mg/kg dichlormethane as pollutants. The tests were run on a pilot plant with a gas throughput of 30 standard m^3/h. The principle of the plant is a ceramic HX with an integrated burner (fig.19).

The test results are represented in the form of an energy balance diagram comparing the two systems (figure 20). The diagram, the detailed discussion of which is unnecessary at this point, yields the following information:

1. The amount of fuel needed for heating the pollutive gases up to reaction temperature is reduced by up to 50 % in the ceramic system.
2. The efficiency of the ceramic system is enhanced by the recovered energy amounting to 371000 kJ/h as compared to the total fuel input of 180000 kJ/h.

The 50 % fuel savings are attributable to the large specific area of the HX and to the favorable heat transfer in the ceramic HX. While the pollutive gases attain a temperature of around 550 °C in conventional systems employing metal tube bundle HX, the new system heats them up to 670 °C. Such efficient energy recovery is essentially due to the fact that the off-gas exits at a temperature of as low as 30 °C, while conventional systems can only cool it to about 250 °C, otherwise the chimney would quickly soot up.

Field testing of the HX/burner for pollutant combustion will entail combining several modules to form a larger integral unit. Presently available modules allow the installation of pilot plants with gas through-puts ranging from 500...1000 standard m³/h. Larger modules are under development.

5. Ceramic heat exchanger for automobile engine

To close with, there is an outlook for a possible application of ceramic HX in automobile engines.

Fig. 21 shows the ceramic components which are under development for gasoline as well as for diesel engines. The idea of this development is to insulate the "hot path" of the engine. Apart from other advantages, which need not to be discussed here, this insulation results in a higher heat energy of the exhaust gas. This energy can be used in different ways:

. by a turbocharger to improve the efficiency of the engine

. for heating purposes (passenger cabin) by applying for example a crossflow HX to the exhaust pipe or

. for evaporation of liquid fuels by means of a ceramic evaporator,

a schematic graph of which is shown in fig. 22. The principle of this evaporator is as follows: the liquid fuel penetrates the porous ceramic wall, is vaporized and mixed with hot combustion air and than fed to the engine. Both liquid fuel and combustion air must be preheated in case of a cold start. Up to now this development is in the first experimental stage.

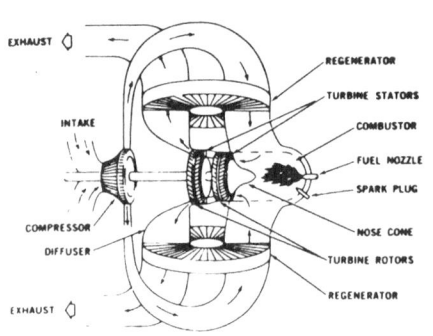

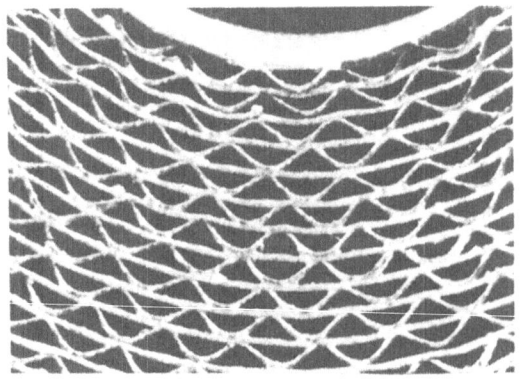

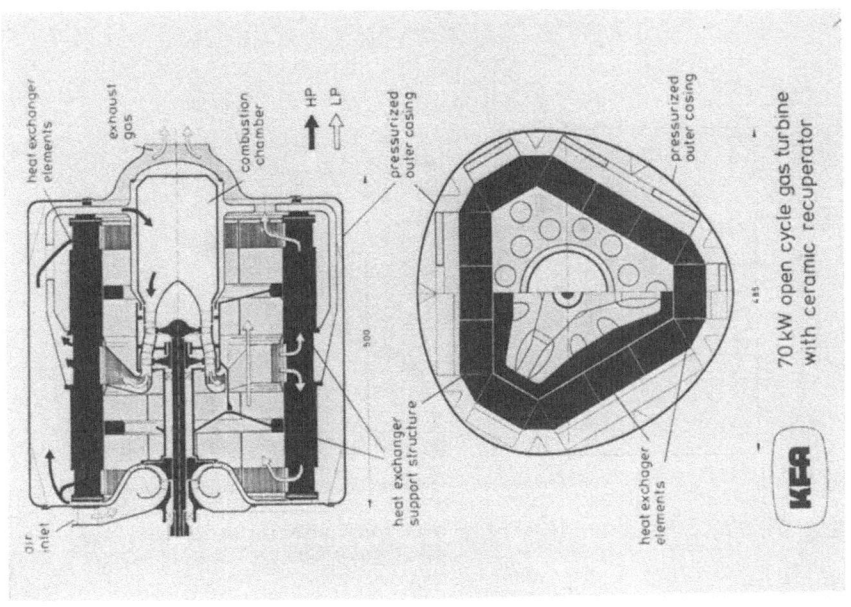

Fig.5: 70 KW open cycle gas turbine
application

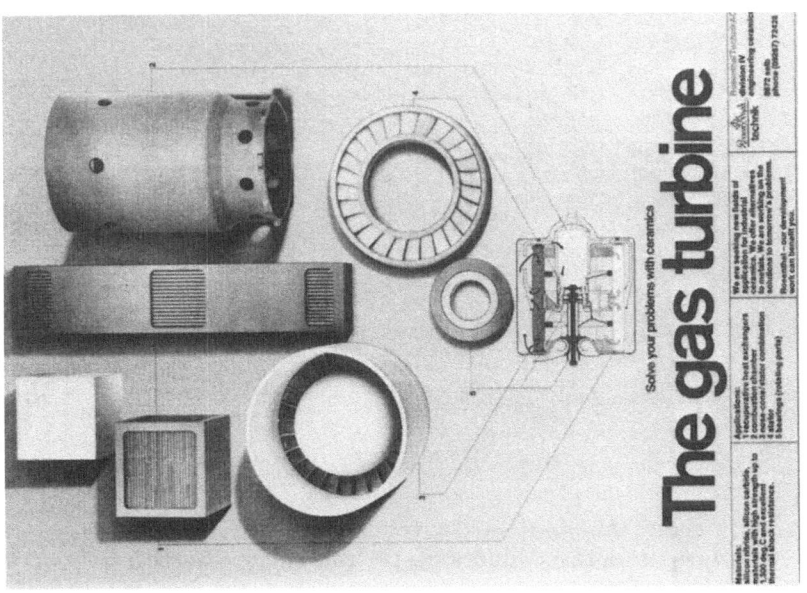

Fig. 4: Ceramic components for gas turbine
application

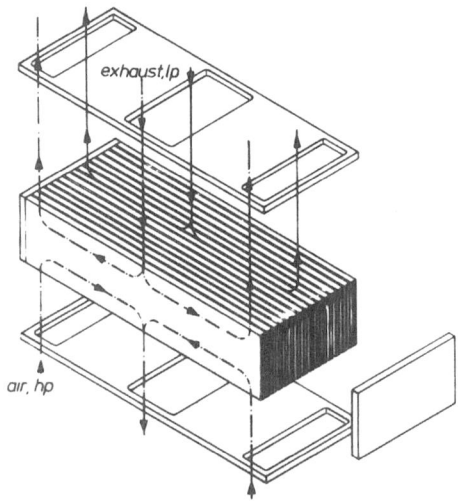

Fig. 6: Design schematic of a recuperator module (Si_3N_4)

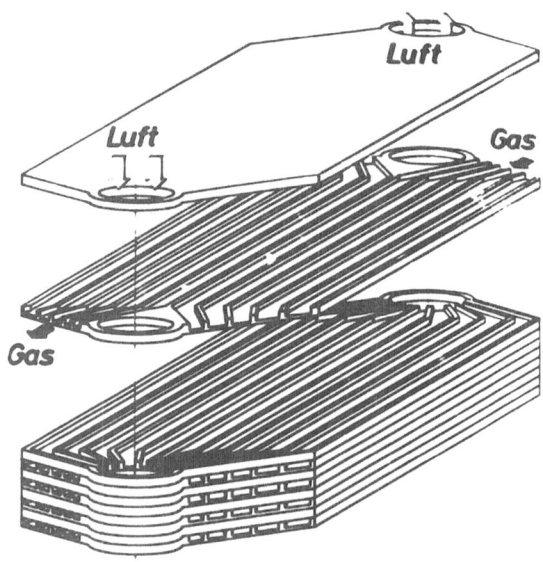

Fig. 7: Design schematic of a recuperative HX module
with integrated ducts, tape technique (Si_3N_4)

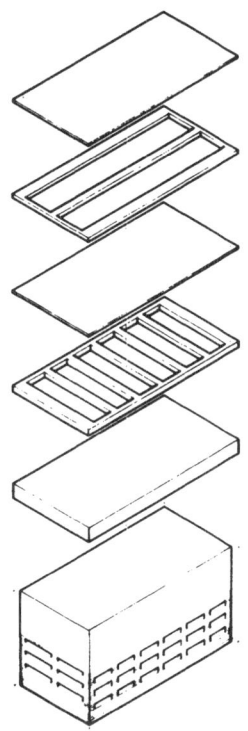

Fig. 8: HX, tape technique

		Required	Reached
Wall Thickness	[mm]	0,2	0,2
Fin width	[mm]	0,2	0,2 - 0,4
Fin hight	[mm]	0,7	0,7
Specific area of HX	[m²/m³]	2500	2500
Material density	[g/cm³]	max.	2,35
Permeability to gas [Δp = 5 bar]	[kg/scm²]	$1,25 \cdot 10^{-8}$	$1,8 \cdot 10^{-9}$
Burst pressure	[bar]	50	8 - 10

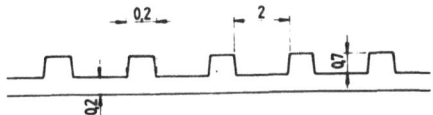

Table 1: Specifications of HX matrix, tape technique

Fig. 9: Counterflow HX, cores produced by laminating embossed single layers

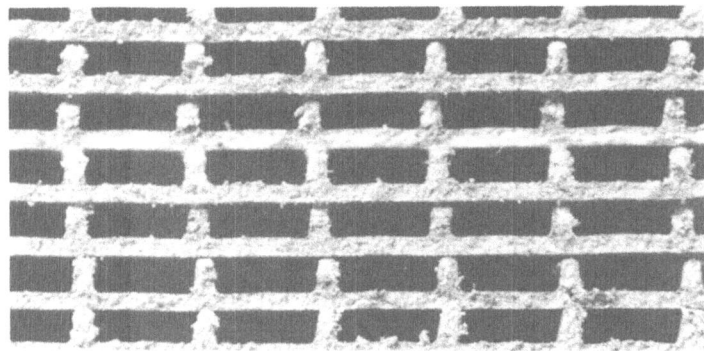

Fig. 10: Enlargement of HX matrix of fig. 9

Fig. 11: Polished section of embossed and laminated single layer of fig. 9

224

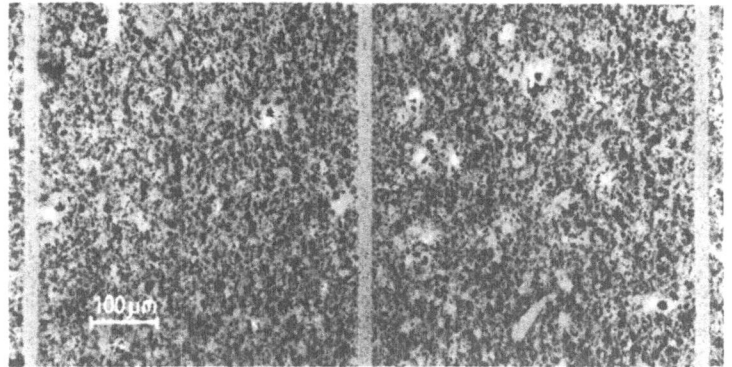

Fig. 12: Polished section of plane laminated single layers

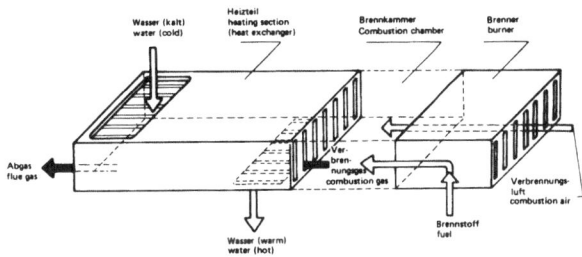

Fig. 13: Design schematic of a ceramic heater module

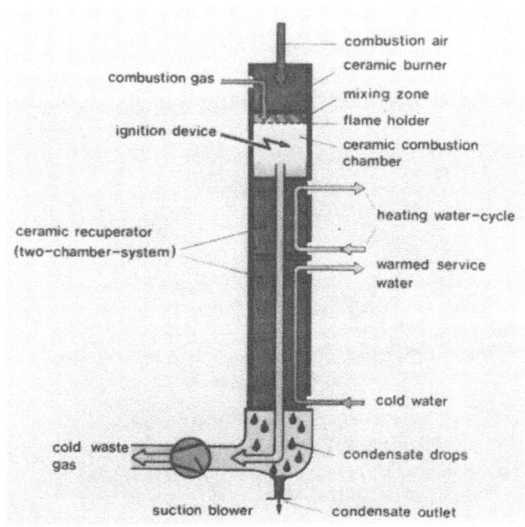

Fig. 14: Diagrammatic section of a ceramic heater module

Fig. 15: Ceramic heater module, ceramic components (Si_3N_4)

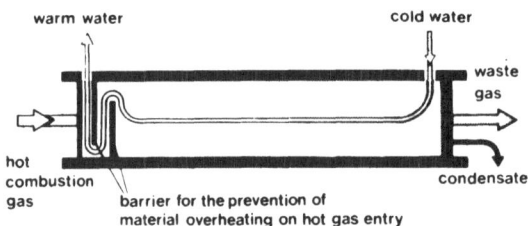

Fig. 16: Water-cooled HX with hot combustion gas cooling curve

Fig. 17: Ceramic heater module installed in the kitchen (assembly photo)

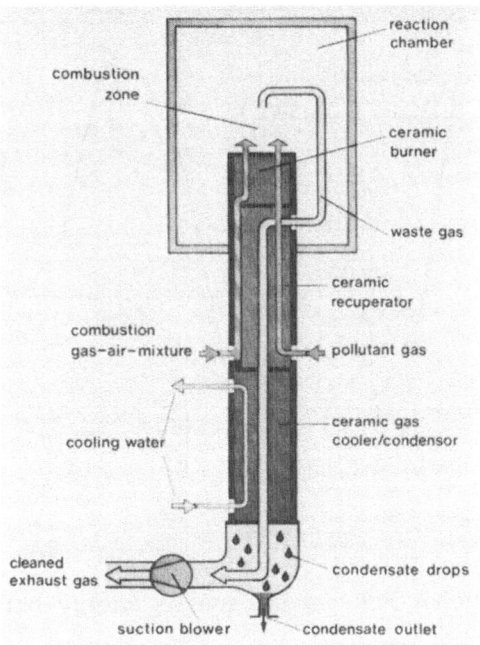

Fig. 18: Diagrammatic section of a pollutant combustion unit

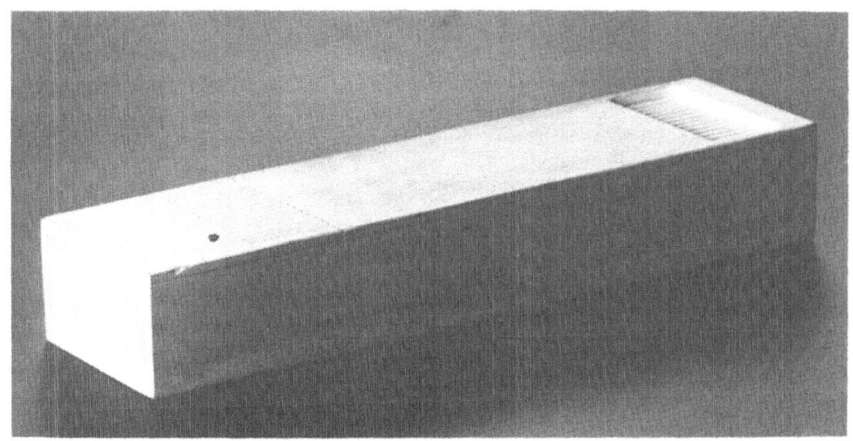

Fig. 19: Integral HX/burner assembly (pollutant combustion)

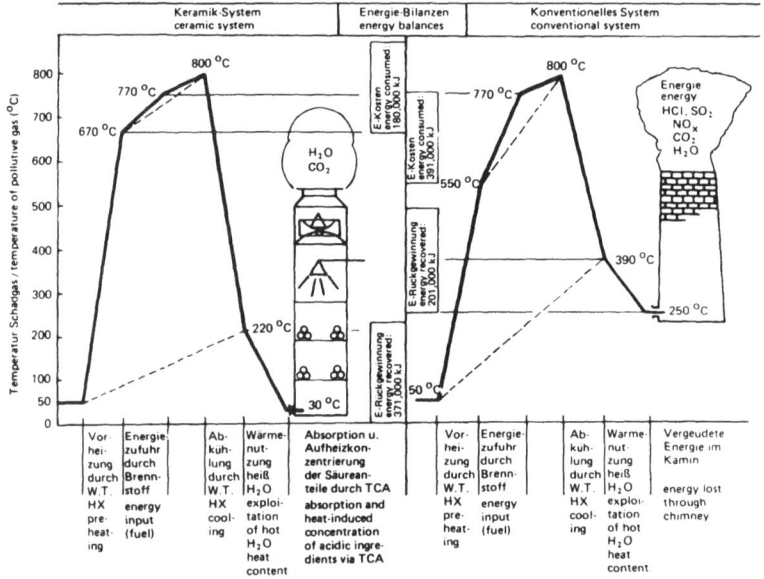

Fig. 20: Pollutant combustion: comparison of energy balances

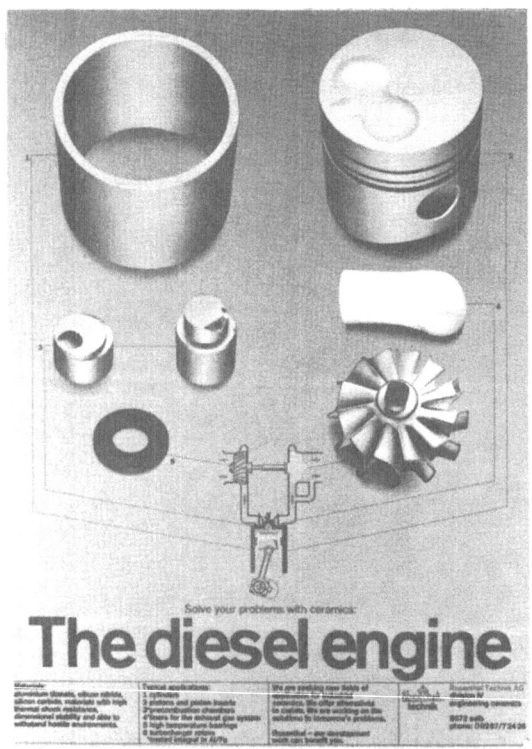

Fig. 21: Ceramic components for automobile engines (diesel engine)

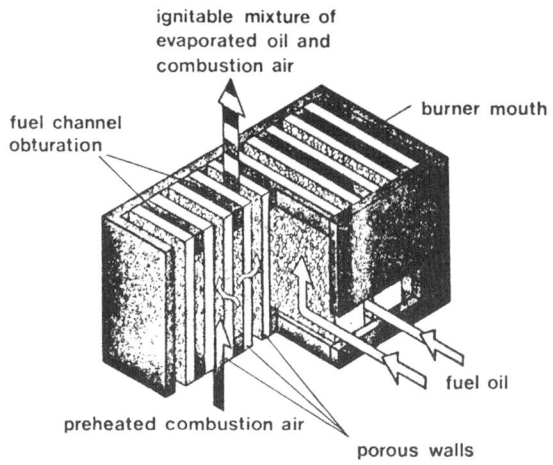

Fig. 22: Design schematic of a ceramic liquid fuel-evaporator

DISCUSSION

G. Grathwohl: When will this combined system, oil or gas burner with a ceramic heat exchanger, be ready for commercialisation?

A. Krauth: From 1984 onwards.

W. Dönitz: What is the overall heat exchange surface of your 15-20 KW system and what will be the approximate cost of this system?

A. Krauth: The heat exchanging surface is about 1,3 m^2 or in terms of specific heat exchanger area 250 m^2/m^3 max.

R.W. Davidge: Could you give any indications of the lifetimes of your components under realistic operating conditions?

A. Krauth: Intensive field tests have been started some time ago and are still under way.

P. Pilavachi: How do you clean your heat exchanger?

A. Krauth: Up to now there is no need for cleaning.

F. Starr: Can you tell us the point of the gas turbine heat exchanger which caused failure at 8-10 bar. Did it leak at the sides or did one of the plates crack? Did the failure take place at high or low temperature?

A. Krauth: The cracks arose at about 900°C wall temperature and went right through the matrix.

K. Goebbels: One of the tables shows the leak rate realised as 10^{-3} against 10^{-1} as planned. Is that right?

A. Krauth: Yes, that is right.

P. Popper: I noticed from your slides that there were areas of widely different densities; some very porous, some very dense. How was this achieved?

A. Krauth: The dense areas are necessary to reduce the leak rate. They are achieved by a special treatment during the build-up of the heat exchanger.

SiSiC - A MATERIAL FOR HIGH TEMPERATURE CERAMIC HEAT
EXCHANGERS

R. Röttenbacher, G. Willmann

Dornier System BmbH, Friedrichshafen, BRD

ABSTRACT

SiSiC (reaction-bonded silicon carbide infiltrated with silicon)
is an advanced material suitable for engineering applications,
especially for heat exchangers. In the R&D-project GAST, SiSiC-
tubes are used to construct a receiver (= heat exchanger) to heat
air by solar radiation. The state of the art and the objectives
for further developments are presented in a discussion of SiSiC-
components, joining techniques, mechanical fatigue behaviour of
SiSiC, resistance to oxidation, and non-destructive evaluation.

1. INTRODUCTION

The efficient use of all energy resources is one of the most ur-
gent tasks for the next few years. One "source of energy" not yet
optimally used is industrial waste heat.
Industrial waste heat with temperatures of roughly 800°C is often
associated with gases which are very aggressive chemically and
possibly erosive because of their dust content, so that demands
on materials used for large-scale mechanical engineering are very
complex. Very often, concepts for using waste heat cannot be implemented because of a lack of mechanical strength of metallic
materials at high temperatures (Fig. 1). The alternative is the
use of ceramic materials, the potential applications of which
have not yet been fully considered. This applies in particular
to materials based on silicon carbide.

The potential applications of silicon carbide materials (SiC), in
particular those made of silicon infiltrated silicon carbide

(SiSiC), are extremely promising because of their high thermal stability, excellent resistance to corrosion, in particular their resistance to oxidation far beyond 1300°C, high thermal conductivity, good resistance to thermal shocks and their gas tightness (Tab. 1). Components made of SiSiC are available commercially in the most diverse forms.

Possible applications for SiSiC-components are their use in high-temperature heat exchangers for preheating the combustion air and in indirect-fired gas turbine cycles, or for use as heat pipes in a desulphurisation process. One example for the use of such components in projects utilizing alternative energy sources is the GAST project (Gas Cooled Solar Tower Power Plant with 20 MW_e). Here the design and testing of a ceramic receiver is being performed /1, 2/.

Reaction-bonded silicon carbide infiltrated with silicon (SiSiC) is, compared to other engineering ceramics, the material best suited for GAST /3, 4/. SiSiC also meets the requirements of many other possible high-temperature applications /5, 6/. The manufacturing process has reached a state of development where the components required for the above-mentioned applications can be produced at acceptable prices. Fig. 2 shows a model of a tubular SiSiC-receiver as is being planned for use in the GAST project (where a tubular heat exchanger was designed).

The "state of the art" technology in the GAST project is presented, discussing the availability of SiSiC-components, joining techniques of SiSiC-components, the mechanical properties of SiSiC, the resistance to oxidation, and the non-destructive evaluation.

2.   STATE OF THE ART

The GAST ceramic receiver will serve as an example for the typical requirements which high-temperature heat exchangers must fulfil /1-5, 7/:

Up to temperatures of 1200°C, the strength of the material must be sufficiently high. The material has to withstand situations when the temperatures will rise above 1200°C for short periods of time. The air which is heated in the receiver is pressurized (about 10 bar). Therefore, the material has to be gastight.
The material for the receiver has to be stable in air in the temperature range from 25°C - 1200°C. Because of the periodic sun rise and sunset the receiver is heated up and cooled down daily. In case of systems failure and variation of radiation (clouds), the surface temperature of the tubes in the receiver will change rapidly and often. Therefore, the material used has to have a good resistance to thermal shock.

The material also has to have a good thermal conductivity to minimize the existing thermal tensions in the tubes of the receiver.

The application of silicon carbide in the production of heating
elements, refractories and abrasives is well-known and established /8, 9/. Because of its covalent bonding and small coefficient of bulk diffusion, pure silicon carbide can only be sintered to a dense body at very high temperatures using additives. For
the latter applications, the silicon carbide pieces are porous
and therefore cannot be used for advanced heat exchangers. By in-
filtrating reaction-bonded silicon carbide with silicon (= SiSiC)
the open porosity is eliminated and a gastight material with high
strength is produced. Depending on the manufacturing process the
content of free silicon in SiSiC is between 8 wt.% and 12 wt.%.
Typical properties of SiSiC are shown in Tab. 1.

For GAST, the components needed are: Straight, cylindrical tubes
with a length of a few meters and a diameter up to 5 cm. The
exact specifications depend on a design that considers the needs
of the application, the technical feasibility and the cost of the
SiSiC-components. Besides tubes, SiSiC-flanges also have to be
designed and manufactured. The flanges join the ceramic tubes to
the metallic gas ducts.

The tubes have an inner diameter of 32 mm and an outer diameter
of 42 mm. All tubes are manufactured by an extrusion process.
The maximum length is 2.5 m at present. It should be possible to
manufacture tubes up to 3.5 m. The flanges are manufactured by
a slip casting process /5, 6, 10/. Fig. 3 shows various SiSiC-
components.

Tubes and flanges are manufactured by different methods and the
maximum length of the tubes is limited to 2.5 - 3.5 m respectively. Therefore, a joining technique has to be developed to join
flanges to tubes and tubes to tubes.

There are two cases: Joining in the green state and joining in
the reaction sintered state.
Green components can be joined using a cement with a composition
similar to that of the basic material. After joining, the whole
piece is reaction-sintered in a tunnel kiln. Only the manufacturer can perform this technique /11/. For joining reaction-
bonded SiSiC-components, two procedures exist: Diffusion bonding
in the absence of any liquid phase or brazing at temperatures
above the service temperature.

In SiSiC the silicon is the principal basis for the two joining
techniques. Silicon has a melting point of 1410°C, silicon carbide melts incongruently at about 2800°C. Thus, silicon is the
principal basis for the joining procedure.

The parameters that influence the physical properties of the joint are as follows /13/:

- Smoothness of the surfaces that are joined
- Time necessary for joining
- Temperature necessary for joining
- Pressure
- Atmosphere for the reaction.

For diffusion bonding the most important parameter is the smoothness of the surfaces because diffusion bonding is performed in the absence of a liquid phase /14/. The temperature chosen must be below the melting point of silicon, yet as high as possible in order to enhance the diffusion. If the surfaces are pressed together, the bonding process is accelerated, probably because of the plastic deformation of silicon. To prevent oxidation, the diffusion bonding is performed in an inert atmosphere or vacuum. In Fig. 4, the microstructure of such a SiSiC-joint is shown.

For brazing, similar conditions are necessary, but additionally the solder has to wet the silicon and silicon carbide. The coefficient of thermal expansion of bulk material (SiSiC) has to be nearly the same as that of the soldering material in order to avoid thermal tensions. The formation of carbides should be prevented because carbides usually have different coefficients of thermal expansion and are brittle. If these conditions can be attained, the mechanical strength of a brazed joint will be sufficient.

For the soldering material silicon with additives was used. The additives lower the melting point and the reaction to the SiSiC-bulk material is accelerated. The temperature for brazing was 1380°C. In Fig. 5 the microstructure of a brazed joint is shown.

Using both joining techniques, the joints produced were gastight and the strength was between 30 % and 100 % of that of the bulk material. The strength of the joint is extremely dependent upon procedural parameters and conditions. Both joining techniques have certain advantages.

Using diffusion bonding, no liquid phase and no material other than Si and SiC is introduced to the joint. The disadvantages are the long times necessary for the bonding process (up to a few hours) and the restriction to simple geometries.

The brazing process does not have the last two mentioned disadvantages. The time for brazing is below 20 minutes. But for certain soldering compositions, intermetallic phases which reduce the strength of the brazed joint can be formed.

In designing a heat exchanger, the following steps have to be performed: Firstly, all of the functions and operating conditions of the plant must be defined. This then allows the dimensions of the plant to be specified. Knowing this, the material for the apparatus can be selected.

After a stress analysis of the plant design, the strength, service time or the probability of failure and the cost effectiveness of the design can be determined. If the results are not satisfactory, the operation conditions or the geometric design must be changed. Such an iteration was performed for the SiSiC-flange for the hot end of the gas duct (compare No. 1 to No. 2 in Fig. 3).

The methods used in stress analysis are basically the same for metals and ceramics /15/; they are well known and proven. The method of finite element analysis has to be used in designing with engineering ceramics.

In order to performe a stress analysis and to calculate the risk of failure, it is necessary to know all mechanical properties of SiSiC involved, mainly the mechanical fatigue for temperatures up to service temperature. The same is valid for the joint.

The ceramic engineer usually measures the 3-point or 4-point bending strength. These results are not sufficient for design work because the fatigue due to sub-critical crack growth is not described. Therefore, the relevant parameters for fatigue of SiSiC had to be measured in order to provide a basis for our design work.

The sub-critical crack growth and the Weibull-parameters were measured for slip-casted and extruded SiSiC at room temperature and $1200^{\circ}C$ in air. Double-torsion-samples and bars for testing bending strength were used, respectively (methods see /16, 17/).

Using these results, SPT-diagrams were calculated to show the dependence of strength, time and probability of failure (see Fig. 6[1]) /19/. These results and the ones measured by other researchers /20, 21/ show that SiSiC has the highest resistance to sub-critical crack growth, when compared with other SiC-materials.

In Tab. 2 the results are compiled. n and A are the parameters describing the sub-critical crack growth, where:

$$v = A \, K_I^n \qquad\qquad \text{(Equ. 1)}$$

v = crack velocity, K_I = fracture toughness.

---

[1] The SPT-diagram for SiSiC at $1200^{\circ}C$ will be available in December 1982.

For the SiSiC material used the following results were obtained
/19, 22/. The parameter n is greater than 100, i.e. for SiSiC,
sub-critical crack growth can be neglected. The bending strength
does not decrease for temperatures up to 1200°C. The properties
of extruded SiSiC material are better than those of slip-casted
material.

For design purposes, the relevant physical and mechanical properties are known for the bulk material, but not as yet for the
joint.

The oxidation resistance of SiC-materials is dependent on the
formation of a thin, self-healing, passive SiO_2-layer which adheres to the SiSiC /23-26/. In certain cases (depending on temperature and partial pressure of oxygen) the oxidation is "active", i.e. because of the evaporation of SiO no protective SiO_2-
layer is built up /25, 26/.

The heat exchanger for GAST operates at air temperatures up to
1200°C. It can be expected that a self-healing, adhering coating
of SiO_2 is formed.

Compared to static oxidation, the kinetics of the oxidation process should be different in GAST because of the daily thermal
cycling due to sunrise and sunset. The different coefficients of
thermal expansion of the crystallized SiO_2 (cristobalite) and
SiSiC should crack the coating. Therefore, the oxidation process
will be accelerated and a strength degradation is expected because of the cracks. Experiments simulating the thermal cycling
were performed and the thickness of the SiO_2-layer and the
strength were determined as a function of the number of thermal
cycles /27/.

The experimental results are: The oxidation of SiSiC is accelerated when tests are performed under thermal cycling. After simulating 365 cycles (i.e. one year), the thickness of the SiO_2-layer
was about 5 μm. Extrapolating to 30 years the thickness would be
30 μm. The SiO_2-layer cracks, (see Fig. 7), but the coating does
not spall off, although cristobalite is formed.

The strength of polished and oxidized samples decreases at first,
and later increases to a value above the strength before oxidizing (see Fig. 8). Sandblasted SiSiC is used for constructing the
heat exchanger. The strength of sandblasted SiSiC is lower than
that of polished SiSiC. After oxidation, the strength of sandblasted SiSiC increases (see Fig. 8).

Because SiSiC is a brittle material, defects (voids, flaws, etc.)
govern its strength. Flaws with sizes between 10 μm and 100 μm
may be critical and cause failure. To detect defects of that

size in bulk material and joints, techniques for non-destructive
evaluation (NDE) must form the basis of a quality control.
NDE of engineering ceramics is under development /28/. The following techniques are suitable to detect flaws of the critical
size:
Ultrasonic and radiographic (X-ray) testing, especially micro-
radiographic testing. For both techniques, development is still
needed. For components used in GAST NDE using radiography was
performed.

It was possible to detect voids, variations in wall thickness,
inhomogenities, and areas not infiltrated by Si. It was not possible to detect small flaws and to distinguish between free silicon and the SiSiC-matrix. Fig. 9 shows the radiography of
SiSiC-tubes, one with and one without defects; Fig. 10 shows the
radiography of a joint.

3.   PROBLEM AREAS

Because of the results of the strength and fatigue measurements
it can be concluded that a modulus of rupture of 300 MN/m^2 for
SiSiC is adequate for the applications in the GAST project.

Small improvements in the M.O.R. will not influence the risk of
failure of SiSiC-components. To guarantee a small risk of failure it is very important that the homogeneity of large SiSiC-components is equal to that of small pieces, i.e. the Weibull-modulus m should be > 10.
For the manufacturer, this is an important aim.

For the SiSiC-joint more information about mechanical properties,
especially SPT-diagrams are needed for design work.

If SiSiC-components are used in heat exchangers utilizing waste
heat of industrial processes more information about the corrosion behaviour in such severe environments, is necessary /30/.

Another problem area for design work is the variation in the
dimensions of SiSiC-components. The tolerances of the tubes are
in accordance to DIN 40680 "Genauigkeitsgrad mittel". Since gastight joints and flanges are required, the achievement of the necessary geometric specifications is problematical. The possibility of achieving such specifications by grinding is very costly.

Techniques for joining SiSiC-components are known. However, there
is little information available about fatigue and corrosion resistance of joins.

The NDE of SiSiC-components is possible, but not all defects can be detected. Because of this situation it seems preferable to use a proof-test for evaluating the components instead of NDE.

4.   FUTURE DEVELOPMENTS

SiSiC-components are manufactured and sold by various German, British and American companies. The SiSiC-components produced by different companies vary in their properties and sizes available. But there are no technical constraints to the production of large components. Therefore, it can be concluded that the development of SiSiC has reached a stage where it is suitable for many applications. The technical limits and the potential of SiSiC are well known.

The objectives of further research and development are as follows: The properties achieved for small SiSiC-components have to be reproducible for large components too. Here the most important properties are the mechanical ones. The main problem is the attainment of a Weibull-modulus of 10, or even better, of 20.

The manufacturers have to deliver better descriptions of the properties of SiSiC. In particular, more information is needed about creep and the static, dynamic, and cyclic fatigue for the different SiSiC-materials at temperatures up to 1300°C. These data are the basis for design work. The same information is needed for joined components.

If SiSiC-components are used in industrial heat exchangers, their resistance against corrosion and/or erosion has to be determined in order to calculate the lifetime of the SiSiC-components.

Techniques of NDE have to be developed and defined for SiSiC-components and a list of defects has to be collected to establish a basis for a reject/accept-decision in quality control.

REFERENCES

/1/   S. Kostrzewa, P. Wehowsky
      BMFT-FB-T 81-097 (1981)

/2/   H.-J. Henseler, in G. Adomeit, H.-J. Frieske (ed.)
      Neue Wege in der Mechanik, VDI-Verlag (1981) p. 135-141

/3/   G. Willmann
      cfi/Ber. Dt. Keram. Ges. 58 (1981) 153

/4/ G. Willmann
Sprechsaal 113 (1980) 915

/5/ G. Willmann, W. Heider
Sprechsaal 114 (1981) 758

/6/ G. Willmann, W. Heider
Z. f. Werkstofftechnik 13 (1982) im Druck

/7/ H. Gehrke, M. Kuczera, G. Willmann
Science of Ceramics 11 (1981) 71

/8/ E.H.P. Wecht
Feuerfest-Siliziumcarbid, Springer-Verlag (1977)

/9/ J. Schlichting
Sprechsaal 113 (1980) 601

/10/ W. Heider, H. Böder
VDI-Nachrichten Nr. 26, 26. Juni 1981

/11/ W. Heider, H. Böder
aus /12/ 457

/12/ W. Bunk, M. Böhmer (ed.)
Keramische Komponenten für Fahrzeug-Gasturbinen, Bd. II,
Springer-Verlag (1981)

/13/ R. Röttenbacher, G. Willmann
Z. f. Werkstofftechnik 12 (1981) 227

/14/ R. Röttenbacher, G. Willmann
Industrie Diamanten Rundschau 15 (1981) 140

/15/ P.M. Braiden
Proc. Brit. Ceram. Soc. 32 (1982) 315

/16/ D. Munz
aus /18/ 1

/17/ H. Richter
aus /18/ 64

/18/ D. Munz
Fortschr.-Ber. VDI-Z, Reihe 18, Nr. 11, VDI-Verlag (1981)

/19/ H. Richter, G. Willmann, W. Heider
Z. f. Werkstofftechnik 13 (1982) im Druck

/20/ M.A. Walton, R.C. Bradt
Proc. Brit. Ceram. Soc. 32 (1982) 249

/21/ G. Popp
Untersuchungen zum Rißausbreitungsverhalten von Silizium-
carbid und Aluminiumoxid an Luft und im Vakuum bei hohen
Temperaturen, Dissertation, Universität Stuttgart (1981)

/22/ G. Kleer, H. Richter
private Mitteilung (1982)

/23/ J. Schlichting
Ber. Dt. Keram. Ges. 56 (1979) 196 und 256

/24/ P.J. Jorgensen et al.
J. Amer. Ceram. Soc. 43 (1960) 209

/25/ D.W. McKee, D. Chatterje
J. Amer. Ceram. Soc. 59 (1976) 44

/26/ S.C. Singhal
Ceramurgia Int. 2 (1976) 123

/27/ G. Willmann, R. Röttenbacher
Science of Ceramics 11 (1981) 341

/28/ K. Goebbels, H. Reiter
Science of Ceramics 11 (1981) 361

/29/ G. Willmann, R. Hundhausen, W. Ludwig
Sprechsaal 115 (1982) 126

/30/ W.T. Bakker, D. Kotchik
ASME 82-GT-182

/31/ Sigri Elektrographit GmbH in Meitingen bei Augsburg

Tab. 1:   Typical Properties of SiSiC

Specific gravity	3 g/cm^3
no open porosity	
Si content (residual SiC)	8 - 12 %
Flexural strength[1] not less than	300 MN/m^2
Modulus of elasticity[1]	350 GN/m^2
Thermal expansion (0°-1000°C)	4.5 x 10^{-6} K^{-1}
Thermal conductivity at 25°C	200 W/mK
at 1000°C	40 W/mK

[1] In a first approximation, these two characteristics are independent of temperature up to roughly 1350°C.

Tab. 2:   Slow crack growth data for SiSiC, measured by using the
double-torsion technique[1]

material	temperature $^\circ$C	n	log A
extruded	23	200	- 410
	1200	115	- 195
slip casted	23	120	- 240
	1200	125	- 220

[1] The dimension used for the crack velocity v are mm/sec, for
the stress intensity factor K_I N/mm$^{3/2}$

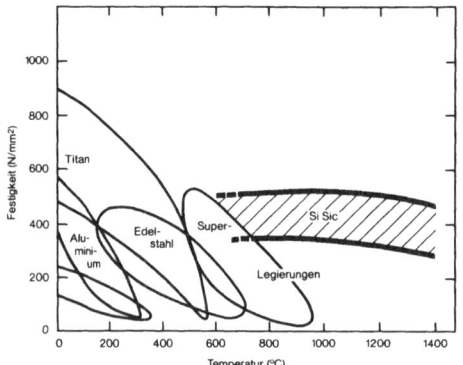

Fig. 1:   Strength of various materials versus temperature

Fig. 2:   Model of the SiSiC receiver for the GAST solar tower
           power plant (scale 1:1 except for the lengths)

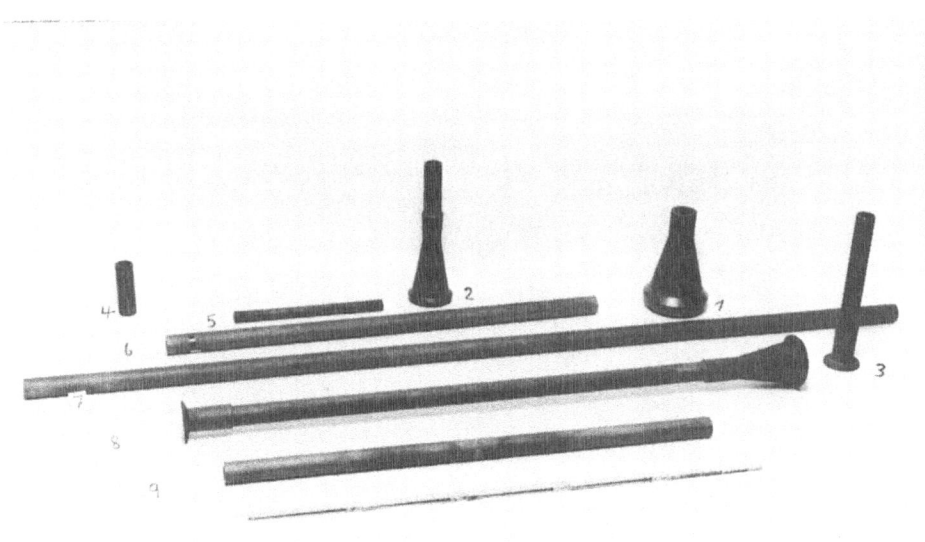

Fig. 3:   Components made of SiSiC, manufactured by Sigri /31/

1   hot end flange, manufactured by slip casting (old design)

2   hot end flange (new design), manufactured by slip casting
     and joined to a tube by cementing in the green state

3   cold end flange, manufactured by slip casting and joined
     to a tube by cementing in the green state

4   tubes joined by diffusion bonding

5   extruded tube, 40 cm long (state of the art in 1979)

6   extruded tube, 1 m long (state of the art in 1980)

7   extruded tube, 2 m long (state of the art in 1981)

8   extruded tube, joined to both flanges in the green state
     and reaction sintered in a tunnel kiln

9   tubes joined by brazing

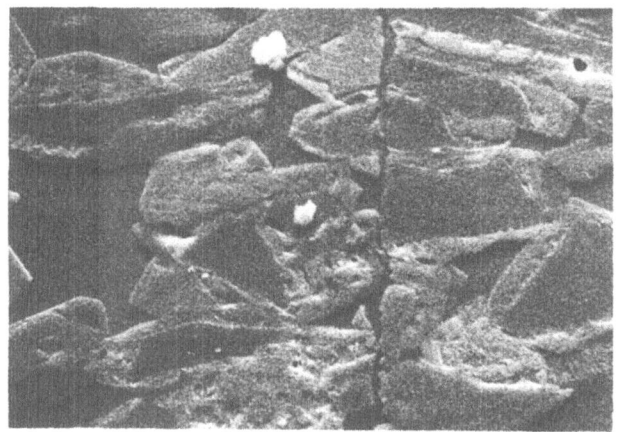

Fig. 4:   Scanning electron microscope picture of an polished
           and etched section of an SiSiC-component joined by
           diffusion welding

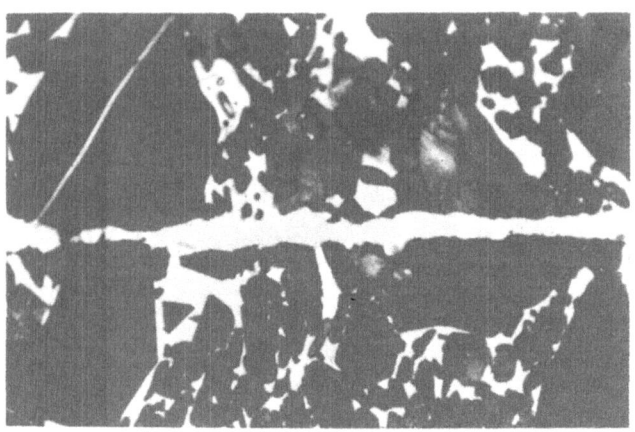

Fig. 5:   Microstructure of a SiSiC-tube joined by brazing

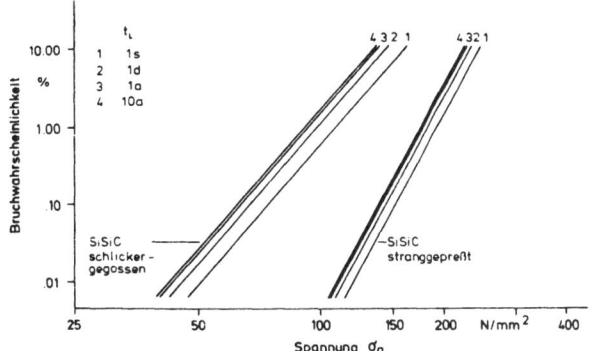

Fig. 6: SPT-diagram of slip casted SiSiC (left) and extruded
SiSiC (right) for room temperature

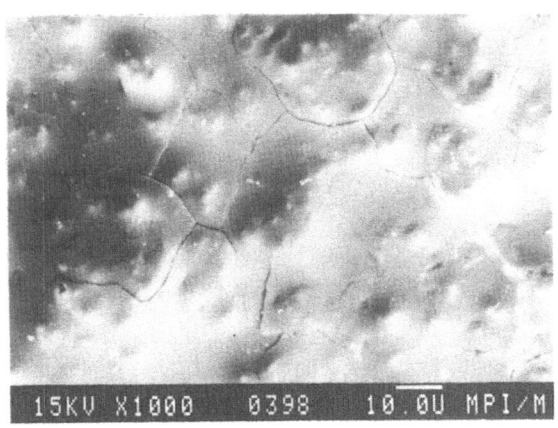

Fig. 7: Surface of a SiSiC-sample oxidized by thermal cycling
(REM)

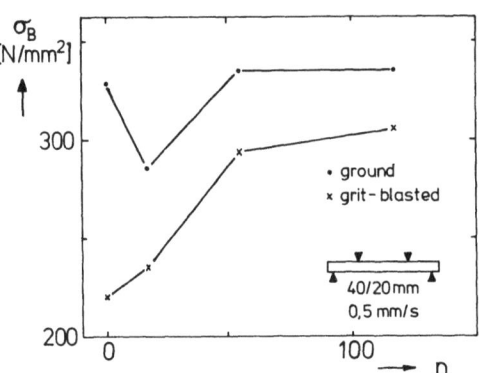

Fig. 8: Bending strength of SiSiC versus numbers of thermal
cycles

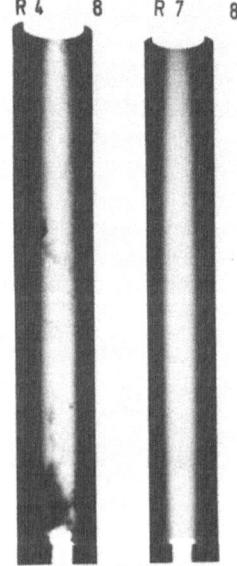

Fig. 9: SiSiC-tubes examined by X-ray, left: with defects,
right: without

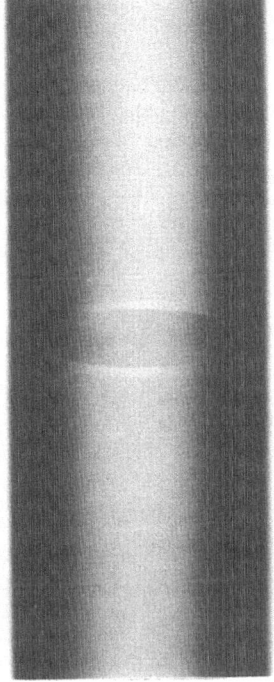

Fig. 10: SiSiC-tubes joined by diffusion bonding and examined by X-ray

DISCUSSION

K. Goebbels: How do you explain the different and well separated film intensity for the tube, the diffusion bonded zone and the elliptical region in between? And if the thickness of the bonding layer is only 0.5 μm, the film shows a broader one; how can you interprete this?

G. Willmann/R. Röttenbacher: The diameters and the ovality of the joined tubes were not exactly the same. Because of

this mismatch you can detect the edges of the inner and outer surfaces.
A measurement of the film intensity shows the same values for the tube and the diffusion bonded zone. Because of the angle between x-rays and tube you see the joint of the circular tubes as an ellipse and nothing can be said about the thickness of the bonding layer.

A. Auriol: Could you comment on the reasons of your choice of Si-infiltrated SiC instead of dense sintered SiC. Apparently large pieces of sintered SiC can now be manufactured by several companies. You mentioned that SiSiC was the best material for resistance to subcritical crack growth; what is the reason?

R. Röttenbacher: In 1979 at the beginning of the project no company was able to manufacture 2 m long tubes out of SiSiC nor out of sintered SiC (SSiC). At this time only the manufacturing technology of SiSiC was so far developed that a production of the desired large shapes could be expected within a short time. The second reason for the decision to use SiSiC were the relatively well known material properties.

The microstructure of SiSiC gives the material a good resistance to subcritical crack growth. The change between SiC-grains and Si-infiltrated regions causes a branching of cracks and lowers the stress energy.

H. Knoch commenting on evaluation and availability of large components of Si-infiltrated SiC in contrast to sintered SiC: In production of sintered parts it has to be noticed that sintering is accompanied by a shrinkage of about 15%, therefore problems arise if the parts are complicated and large. Infiltrated material shows a negligible shrinkage.

S.P. Hansrani: The design and operation of ceramic recuperators has been successfully demonstrated by British Steel Corporation. However, the push of the technology has been slow due to the high costs of ceramic tubes. How do the authors evaluate their material in terms of costs especially when compared to the existing metallic recuperators and ceramic recuperators.

G. Willmann: The costs for SiSiC-components can be calculated. For the production of a large number of tubes (> 1000) the cost will be about DM 50/kg and includes the

manufacturing process. This is comparable to the price of
superalloys.

A. Auriol: Concerning the price of Si infiltrated SiC
compared to dense sintered SiC, the comparison is difficult
for large pieces because of the lack of experience. But for
small pieces such as seal rings, the market price of the two
materials is similar.

R.W. Davidge: A comment on high temperature design data.
For REFEL silicon carbide we found that the material increased
in strength with decreasing strain rate at 1000°C (the opposite
effect to that found at 20°C) due to crack tip rounding
associated with silicon plasticity - R.W. Davidge et al. Powder
Met. Int. 8 (1976) 110.
It is thus not appropriate to consider SPT diagrams (R.W.Davidge
et al J.Mat.Sci. 8 (1973/1699) for Si/SiC composites at 1200°C
because the failure mechanism is too complex for the simple
SPT model.

R. Röttenbacher: Surely it is not appropriate to consider SPT
diagrams for SiSiC at high temperatures because the failure
mechanism is complex. But today it's the only way to make
statements about the failure probability because nobody can
offer a model describing the exact failure mechanisms.

MHD SYSTEMS

L.H.Th. Rietjens

Department of Electrical Engineering,
Eindhoven University of Technology,
Eindhoven, The Netherlands

SYNOPSIS

High efficiencies can be achieved in the conversion of heat into electrical energy by the introduction of a combined magneto-hydrodynamic (MHD) steam cycle. The potential offered by MHD has been demonstrated by studies showing coal-to-busbar conversion efficiencies approaching 50% with an electricity cost lower than that of competitive systems. Low NO_x and SO_2 emission levels are obtained. The development program in the world will be described briefly, problem areas identified and recommendations for new development given.

INTRODUCTION

The last decade has witnessed dramatic progress in the status of MHD power generation[1]. MHD technology has advanced from exploratory investigations and is now firmly established in the area of engineering development and design of major facilities. Both the United States and the U.S.S.R. are in the advanced planning stage of power generating plants. The construction of the first 'commercial' MHD steam power plant in Ryazan, 200 km outside Moscow, has been initiated in the U.S.S.R. Several countries are also exploring the potential of MHD power generation. Much of this progress has been achieved through a broad activity of international information exchange, which has led to formal international cooperation in several large-scale projects.

MHD power generation is based on the direct conversion of heat into electrical energy by expanding a heated, electrically conducting fluid through a magnetic field. The essential difference between an MHD system and a conventional turbine-generator system is that in the MHD system the mechanical rotor has been eliminated. Instead, a partially ionized electrically conducting working fluid is utilized. The accelerated, conducting fluid (see figure 1) interacts with an intense transverse magnetic field inducing an electric field within the fluid. If electrodes are present to collect the current, then electrical power can be supplied to an external load.

MHD power systems are of two general types: open cycles and closed cycles. In an open-cycle system , the working fluid is utilized on a once-through basis and consists of gaseous fossil fuel combustion products which are eventually exhausted into the atmosphere. In a closed-cycle system, thermal energy is transferred from a heat source (fossil fuel, nuclear, etc.) to the working fluid (inert gas or liquid metal/vapor mixture) via a heat exchanger. After supplying energy to the MHD generator and other downstream components, the working fluid is fed back to the heat source. The bulk of the activities to date is focused on open-cycle systems which have also attained the most advanced level of engineering development.

OPEN CYCLE MHD

In the open-cycle systems the gaseous working fluid must be heated to about 2700 K to produce free electrons by ionization, thereby making the gas electrically conducting. In the United States the major effort in MHD is directed towards an open-cycle system using ionized gas produced by the combustion of coal. The U.S.S.R. programme has to date focused on natural gas as fuel, but recent statements from the U.S.S.R. indicate that coal will be emphasized in future years.

The potential offered by open-cycle MHD is supported by data obtained from the U.S. Energy Conversion Alternatives Study (ECAS) as well as other studies. The ECAS study[2] showed that coal-fired MHD systems promise the highest efficiency of all the base-load power systems considered, with low electricity cost and low environmental intrusion. These studies predict that a combined-cycle MHD steam power plant could achieve a coal-to-busbar efficiency approaching 50%. The ECAS determined that the costs of electricity for an MHD steam power plant will be approximately 33 $mills/kWh. This compares with an efficiency of approximately 35% and a cost of approximately 40 $mills/kWh for a conventional coal fired power plant[3]. The study also predicts that SO_2 and NO_x predicted emission levels will be within the U.S. New Source Performance Standards of the Environmental Protection Agency (EPA) without requiring significant, if any, coal preprocessing or stack gas clean-up systems. Compare figures 2 and 3.

The U.S.S.R. National Programme on MHD Electrical Power Gene-

ration pursues a very active path which can, at present, be roughly divided into the following two stages:
1. Construction by 1985 of a commercial MHD combined-cycle natural
   gas plant, followed by the construction of several similar
   plants operating on gas-black oil in metropolitan areas, and
2. Construction of MHD power plants operating on coal, with
   start-up of the first commercial plant in the early nineties.
The two-stage approach is being pursued; however, the predominant
fuel in MHD plants in the U.S.S.R. is expected to be coal. Adop-
tion of the two-stage approach is expected to speed the commercialization of MHD in the U.S.S.R. Stage 1 of the present U.S.S.R.
programme, to be completed by the mid-1980s, has already started.
The first 'commericial' plant will be the U-500 Ryazan Power
Station (250 MWe topping, 250 MWe bottoming cycle). Construction
on site has been initiated, the design of the plant components
has been, or is being, completed and consultation by U.S.S.R.
industries on the manufacturing of the components is taking place.

The U.S.S.R. design philosophy for the first power plant leans
heavily towards reliable operation and utilization of existing
components rather than achieving the maximum possible efficiency.
Still, the cost of power produced will be comparable to that of
conventional power plants. Data and operating experience obtained
at the U-02 facility, the 250 MWt U-25 plant and the U-25B have
provided sufficient technical basis to proceed directly with the
design and manufacture of all the non-standard MHD power plant
components of the commercial plant generator: combustion chamber,
steam generator, seeding and inverter systems, except the superconducting magnet system (SCMS). In regard to the SCMS, an intermediate size magnet (warm bore diameter 2 m, length 12 m, peak
magnet field 4 T) is being designed and constructed.

The United States MHD development programme considers a two-
step approach to commercialization. The first step, Phase I, lays
the engineering groundwork for the design of an MHD commercial
prototype plant termed the Engineering Test Facility (ETF). Commercial readiness is then demonstrated in Phase II by the construction and operation of the ETF.

Phase I focuses mainly on the engineering development of MHD
plant sub-systems, generally of up to 20 MWt size. Engineering
development progress in Phase I is monitored through a succession
of critical 50 to 100 MWt tests in the Component Development and
Integration Facility (CDIF), now in progress in the state of
Montana. During Phase II construction and activation of the ETF
is proposed to be undertaken by an industrial contractor team at
a scale of approximately 200 MWe.

The U.S. Phase I utilizes numerous facilities at various locations in 15 states around the country. Two major government test
facilities accommodate key development and integrated engineering
tests at larger scales than are available elsewhere in the United

States. The CDIF has already been mentioned. The second facility
the Coal Fired Flow Facility (CFFF), is located at the University
of Tennessee Space Institute near Tullahoma, Tennessee. It is
capable of continuous operation at an input of 20 MWt and short
duration testing at large scale.

The ETF is proposed to be a fully-integrated combined-cycle
MHD steam system operating at a power level of 200 MWe. This power
level represents the minimum scale that can demonstrate and verify
the engineering design and operational characteristics of an open-
cycle MHD power plant without incurring unacceptable financial
risk of performance penalties. The ETF will consist of an MHD
system including a coal processing and feed system, MHD combustor
channel, superconducting magnet, power conditioning (including
inverter), bottoming steam plant, slag removal system, seed recovery and regeneration system, cooling system, control systems
and pollution control systems. Oxygen enrichment will be used to
obtain the requisite combustion temperatures.

At present the U.S. programme is experiencing serious difficulties. According to the present administration MHD has developed
so far that industry should take over; consequently the financial
support from the government has been reduced substantially. New
actions are initiated by the utilities (see below).

CLOSED CYCLE MHD

In the closed-cycle MHD concept the working medium of the MHD
generator is a noble gas. After electrical power generation in the
MHD duct the noble gas can be compressed and recirculated through
the heat source (see figure 4). The work in the period up to 1971
led to a number of conclusions which were formulated at the Fifth
International Conference on MHD Electrical Power Generation
(Munich, 1971)[4]: *"Closed-cycle plasma MHD, using seeded noble gases
as the acting medium, can work in a two-temperature regime. The
electron temperature can be elevated over the gas temperature. In
this way the minimum required gas temperature can be approximately
700 K lower in a closed-cycle MHD power plant than for an open-
cycle MHD plant."*

*"Overall conversion efficiencies were calculated
for a high-temperature gas-cooled reactor combined with closed-
cycle MHD power conversion. From the calculated cycle efficiencies
44-54% it was concluded that closed-cycle MHD in combination with
a high-temperature gas-cooled reactor should be further investigated. Large scale, 5-25 MW thermal closed-cycle MHD loops should be
built."*

Further work on closed-cycle MHD, particularly in Western Europe, was at that time oriented on the development of an ultra-high-
temperature reactor (UHTR). A large closed-cycle MHD installation
was built in Jülich; new large installations were planned in Gar-

253

ching and Frascati. Although the development of coated particles as fuel elements for high-temperature gas-cooled nuclear reactors progressed substantially, in 1972 it turned out to be unlikely that a commercial UHTR with outlet temperatures up to 2000 K would become available within 15 years. Thus, the strongest argument for building and operating large-scale closed-cycle MHD installations had fallen.

The large difference of 700 K in the minimum required plasma temperature in open- and closed-cycle MHD warrants further research and development in the field of closed-cycle MHD power plants fired by fossil fuels. In these plants a series of regenerative heat exchangers, in cyclic operation, are heated by fossil fired fuels. The heat is transferred to a cesium seeded argon gas which drives the MHD generator. The remaining heat in the argon gas produces steam. The argon is compressed and reheated in a closed loop: this is the closed-cycle MHD - steam concept (compare figure 4).

In 1975, at the Sixth International Conference on MHD Electrical Power Generation (Washington, 1975), large enthalpy extraction was reported in shock tunnel MHD generator experiments. The experiments were performed on a small scale, 10-40 liter channels, at rather high stagnation temperatures up to above 3000 K. In the presence of strong fluctuations, power densities over 100 MWe/m^3 and enthalpy extraction over 20% were achieved[5,6,7]. Conclusions drawn at the Washington conference in the field of fossil-fired closed-cycle MHD read among others[8]:

"A blow down facility can be considered as a physical demonstration experiment for a fossil-fired closed-cycle MHD plant. In order to approach an isentropic efficiency of 50%, the power level of this facility should be at least 5 MWt."

A MHD blow down facility in the 5 MWt range, designed and built at the Eindhoven University of Technology(THE), has been brought in full operation[9,10]. The facility has been designed and constructed for fossil fuel operation with a thermal input of about 5 MWt, initially to test cold-wall heat-sink type channels (see figure 5). Further under the agreement between the United States of America and the Netherlands a preheat loop was designed and constructed in the U.S. and still has to be brought in full operation in Eindhoven. This preheat loop will enable the testing of preheated hot-wall channels.

A key element of the 5 MWt closed-cycle MHD blow down facility, built at the THE, is the regenerative heat exchanger. A ceramic heat exchanger has been designed by Fluidyne Engineering Corporation, Minneapolis, Minnesota. The heater has been designed for an argon mass flow rate of 5 kg/s, an outlet pressure of 7 bar and an outlet temperature at the beginning of the 60 s run of 2000 K. The maximum temperature drop over the 60 second period is 60 K. The bed is made out of alumina bricks and has a diameter of 0.68 m and a height of 4 m. There are only three layers of insulation to minimize the pump down time. The diameter of the vessel is 1.14 m. See figure 6.

The values for mass flow, stagnation temperature and pressure predicted by the design study have been reached in a number of runs. It has been demonstrated that an alumina brick heater designed to heat gases for closed-cycle MHD generators can deliver an argon plasma with a stagnation temperature of 1970 K. With a proper temperature gradient in the bed the difference between the maximum bed temperature and the argon stagnation temperature can be as low as 90 K.

The obtained values for the impurity levels of CO_2 and N_2 in the argon flow are well below the limits which are acceptable for MHD power generation. The H_2O concentration has been large in initial runs. A strong decrease during consecutive runs has been observed and the H_2O concentration has also become below the limit of acceptable levels for MHD power generation. During the power runs the observed impurity levels of H_2O, N_2 and CO_2 were all well below 50 ppm.

With the present THE installation, including the 5 T cryogenic magnet built by Brown Boveri Company in Switzerland, electrical power up to 360 kWe has been generated. The power extraction is possible through non-equilibrium ionisation. Electron temperatures far above 3500 K have been reached with a stagnation temperature of 1900 K.

Extremely important in this field of ceramic regenerative heat exchangers for closed cycle MHD is the work done at GE[11]. The test facility at GE includes a direct coal fired wet bottom ceramic regenerative heat exchanger. The heat exchanger is fired by a regeneratively air cooled cyclone coal combustor and has a high temperature bottom support structure allowing slag drainage. The test facility has been developed under the GE closed-cycle MHD program to simulate operation of argon blow down and combustion gas reheat modes at up to 4 bar. However, the high temperature bottom support construction also has application to the directly fired air preheaters ultimately desired for the highest efficiency open-cycle MHD configuration and this work is integrated with the development of that component, as well. Tests have been designed to simulate full scale, programmed, and cyclic operations of the combustor/heat exchanger facility and includes measurements of NO_x emission levels as a function of the catalytic effects of alumina core bricks, as well as the effects of flue gas recirculation.

The heat exchanger system is two-staged with a high temperature heat exchanger and a preheater. The high temperature heat exchanger, which is a highly insulated hot bottom heat exchanger, has a 14 inch diameter brick matrix with 1/2 inch diamter flues. This heat exchanger has been designed to operate with a top core brick temperature of approximately 1900 K, with a bottom brick temperature in the range of 1000-1400 K and to permit off-line clean-out of accumulated slag by continuous heating to 1650 K. The preheater functions to preheat a working fluid, argon or air, up to 1100 K as well as to cool the combustion products before entering the stack.

CONCLUSIONS, PROBLEM AREAS AND RECOMMENDATIONS

1. From the work performed with the MHD blow down facility in Eindhoven it can be concluded:
   - Substantial electrical power up to 360 kWe can be extracted in an MHD generator from an argon plasma heated by a fossil fuel.
   - Spectroscopic measurements show non-equilibrium ionization in the MHD generator plasma. Electron temperatures and population temperatures well above 4000 K have been measured at a stagnation temperature of 1900 K.
   - With a regenerative ceramic heat exchanger, using a bed made out of alumina bricks, impurity levels of H_2O, N_2 and CO_2 well below 50 ppm are achieved.
   - Further increase of electrical power output can be achieved with the present facility by increasing the stagnation pressure up to 9 bar and by adapting the hot flow train.

2. Given:
   - the 700 K difference in the required temperature between open and closed cycle MHD;
   - the lower minimum plant size (50-400 MWe for a closed-cycle MHD steam plant with regard to 400-1000 MWe for and open cycle MHD steam plant[12]); and
   - the electrical power extraction achieved in the blow down facility in Eindhoven,
   a reappraisal of MHD work is done in a number of organisations.

3. The Southern California Edison Company is considering[13]) to retrofit the Etiwanda Power Station in California, U.S. with a closed-cycle MHD topping cycle. The key problem in such a program is the heat exchanger. A reduction of the minimum required stagnation temperature by another 200 K will bring recuperative heat exchangers into the picture. Thus work, a.o., should be carried out along the following lines:
   - Electrical power generation experiments at stagnation temperatures down to 1700 K using argon cesium seeded plasma. The minimum magnetic field to be applied in such experiments is considered to be 5 T.
   - Development of ceramic heat exchangers to be fired by coal. Two lines have to be followed: regenerative heat exchangers in cyclic operation working at 1900 K, and recuperative ones working at 1700 K or higher. There is a strong preference to use recuperative heat exchangers in closed-cycle MHD steam power plants.

REFERENCES

1. L.H.Th. Rietjens, G. Rudins and M.M. Sluyter, Position Paper WG 7: "The present status of MHD for large-scale electrical power generation", 11th World Energy Conference, Munich, September 1980.

2. G.R. Seikel, et al., A summary of the ECAS, 15th Symp. on Eng. Aspects of MHD, University of Pennsylvania, Philadelphia, U.S., III. 4, 1976.

3. MHD Power Generation, A report to the Congress, MHD Industrial Forum, Washington, D.C., 1982.

4. 5th Int. Conf. on MHD Electrical Power Generation, April 1971.

5. J.H. Blom,et al., High power density experiments in the Eindhoven shocktunnel MHD generator, 6th Int. Conf. on MHD Electrical Power Generation, Vol. III, p. 73, Washington D.C., 1975.

6. C.H. Marston, E. Tate and B. Zauderer, Large enthalpy extraction results in a non-equilibrium MHD generator, 6th Int. Conf. on MHD Electrical Power Generation, Vol. III, p. 89, Washington D.C., 1975.

7. C.S. Cook, Current experimental results from operation of the G.E. closed cycle ceramic regenerative heat exchanger, 15th Symp. on Eng. Aspects of MHD, University of Pennsylvania, U.S., VIII, 4, 1976.

8. Status Report, MHD Electrical Power Generation, Paris, OECD, 1976.

9. H.J. Flinsenberg, et al., Instability analysis of the first power runs with the Eindhoven MHD blow-down facility, 20th Symp. on Eng. Aspects of MHD, Irvine, Cal., p. 12.1, 1982.

10. P. Massee et al., Gasdynamic performance in relation to the power extraction of the Eindhoven MHD blow-down facility, 20th Symp. on Eng. Aspects of MHD, Irvine, Cal., p. 7.4, 1982.

11. S. Omori, E.S. Fleming and C.S. Cook, A direct coal fired wet bottom ceramic regenerative heat exchanger for closed cycle MHD, 7th Int. Conf. on MHD Electrical Power Generation, Vol. I, p. 379, 1980.

12. F.N. Alyea, et al., Comperative Analysis of CCMHD Power Plants, 19th Symp. on Eng. Aspects of MHD, Tullahoma, Tenn., 1981.

13. Coal fired closed cycle MHD retrofit study, General Electric, SCE Report no 81-RD-32, May 1981.

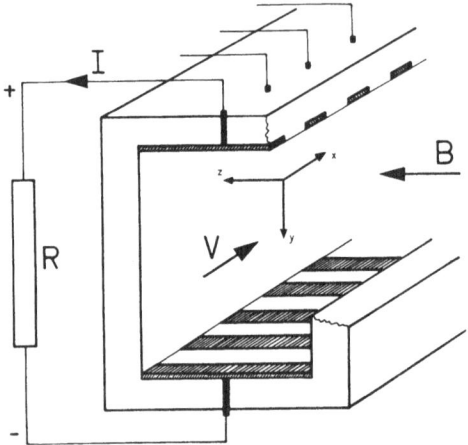

*Fig. 1. The principle of a segmented Faraday generator.
v is the velocity of the plasma, B the magnetic inducion,
I the induced current through a generator segment and
R the load of this segment.*

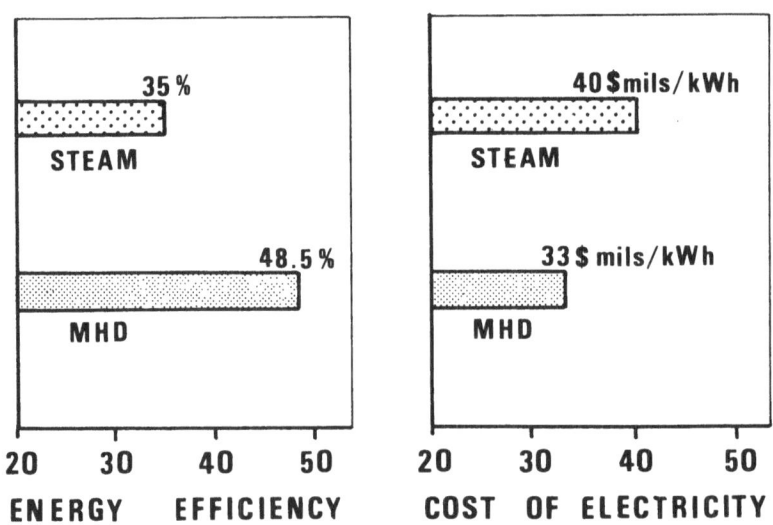

*Fig. 2. A combined MHD steam bottoming cycle plant will
be capable of producing 50% more power from a ton of
coal than a modern steam plant, leading to significant
annual savings in coal resources.*

LOW SO_X NO_X

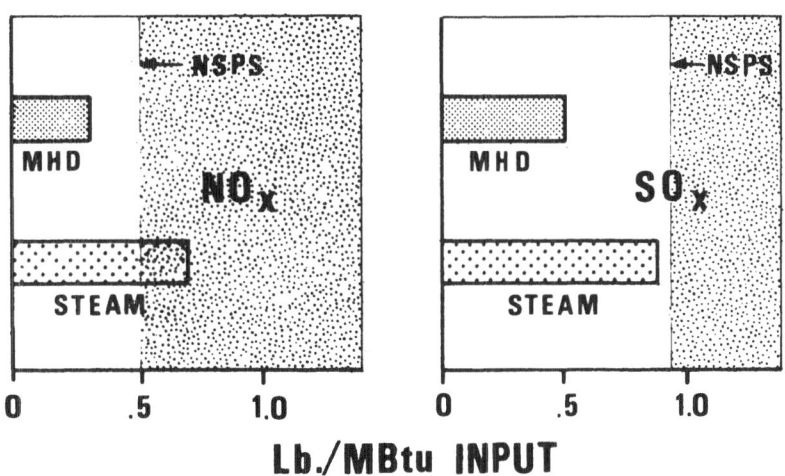

Lb./MBtu INPUT

Fig. 3. An MHD plant will produce less SO_x and NO_x than a modern steam plant with identical power generating capacity. MHD's SO_x and NO_x levels will be well below the EPA's New Source Performance Standards.

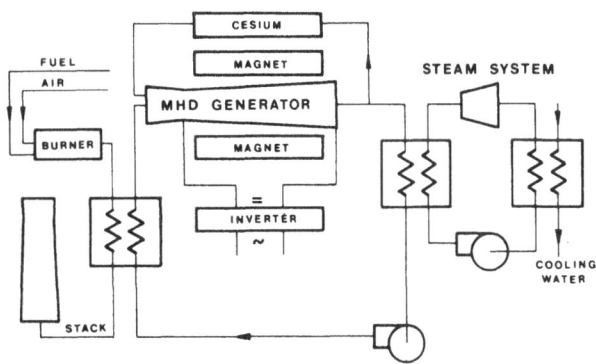

Fig. 4. Scheme of a closed-cycle MHD steam plant. Heat obtained from fossil fuels, solar energy or nuclear power can be transferred to an MHD generator through a heat exchanger.

Fig. 5. MHD channel of the Eindhoven blow down facility.
The walls are made out of boron nitride; the electrodes
out of stainless steel.

Fig. 6. Top view of the ceramic heat exchanger. The bed
is made out of alumina bricks.

DISCUSSION

J.R. Brook: What is the current position with respect to the use of ceramic electrodes in MHD systems?

L. Rietjens: In Eindhoven, we have no experience in ceramic electrodes. Further, it can be stated that no ceramic electrodes are used in large scale MHD facilities. For details I refer to the MHD activities at the Energy Research Centre in Petten (R.A. van der Laken).

H.M. Verhoog: What problems have you encountered in your operation of the heat exchangers?

You have indicated that one of the main objects for further development is the heat exchanger. What are the aspects of this development?

L. Rietjens: Our heat exchanger, designed by Fluidyne Engineering Corporation, has been working without any serious problems.
The further development should concentrate on:
1. Optimization of a full set of regenerative heat exchangers in cyclic operation;
2. Design, build and test of a silicon carbide heat exchanger (5 MWt)
3. Use coal as primary energy.

K.H. Lim: From an operating point of view I have the impression that this system is not easy to operate on a big scale.
Do you have any comments?

L. Rietjens: It will not be easy to build and to operate the first plant. The first full scale closed cycle MHD-steam plant, however, can be as small as 150 MWt.
Recuperative heat exchangers will substantially simplify the operation.

B.C.H. Steele: If a closed cycle MHD system is used then the projected efficiencies (40-42%) only compare with existing technologies using a gas turbine and a conventional steam-raising plant. Can you comment please?

L. Rietjens: Gas turbines require gas as fuel. Efficiencies can only be compared if you start from the same primary fuel and from equal emission requirements for the system considered. Starting from coal and applying high emission standards, calculations show higher efficiencies for fully developed closed cycle MHD steam systems than the gas turbine or the conventional steam plant.

F. Starr: What about the relative magnitude of slag attack between the closed and open cycle systems. Is not slag attack going to be a major limitation in governing the success of MHD?
Could the author indicate what size advantage (i.e. reduction in size) results from operating regenerators and recuperators at higher pressures than ambient?

L. Rietjens: In closed cycle MHD the slag attack is limited to the high temperature heat exchanger (temperature 2000 K). In open cycle MHD slag provides for a protection of wall materials in the channel. Substantial decrease in size of heat exchangers are foreseen at higher pressures. I have no numbers available.

H.M. Verhoog: It is well known that alumina is vulnerable to alkali attack. What is the effect of the seed material on the alumina refractories of the heat exchangers?

L. Rietjens: The seed material, cesium, will be extracted downstream of the channel, purified and injected again upstream. Therefore the seed concentration in the heat exchanger will be limited.

P. Popper: You mentioned the use of boron nitride. Could you say more about this? Presumably it was hot-pressed BN. In what geometrical form was it used and who supplied it? How was the channel constructed?

L. Rietjens: The boron nitride is hot-pressed in cylinders and machined into plates (max. 155x80x10 mm3). It is supplied by Duramic Products, Inc., U.S., it has been further machined. The channel has been built up by elements. The elements are kept in place by the electrodes.

REFRACTORIES IN COAL GASIFICATION PLANTS

R. Dürrfeld

Ruhrkohle, Oel und Gas GmbH
Gleiwitzer Platz 3
D - 4250 Bottrop, FRG

1. Introduction

Coal is a natural raw material, beautiful to look at, and containing a tremendous variety of physical laws and phenomena as well as a variety of organic and anorganic chemical matters. For this reason, only a descriptive definition can be given for coal which up to present remains the object of comprehensive research work all over the world.

If coal is just burnt, the technology of its utilization is comparatively simple. There is, however, a number of quite good reasons to consider coal as valuable enough to use it as feed stock for higher-developed processes, viz.:
- coking
- gasification, and
- liquifaction.

With steadily increasing knowledge about the properties of coal, the coal utilization technology underwent steady changes. Hardly any other raw material gave more impetus to the creativity of engineers. The paper therefore concentrates just on one aspect of coal conversion, i.e. gasification. This contribution is meant to be an introduction for the experts of the refractory industry into the technology of coal gasification and its requirements with respect to refractory materials. The conclusions to be drawn from this lecture with respect to the development of suited refractories will then be left to the Colloquium.

2. Coal gasification technologies.

The universally applicable gasification process suited for any feed coal does not exist yet. Accordingly, distinction is made between 4 different processes:

- fixed bed gasification
- fluidized bed gasification
- entrained flow gasification
- molten iron bath gasification.

The last-mentioned process looks farther ahead into the future than the other ones. All the processes mentioned may be sub-divided into various process configurations allowing for particular requirements of the feed coal used.
Basically, however, process-specific requirements need to be met by the feed coal with respect to avoiding agglomeration as well as with respect to particle size and ash melting temperature. (Fig. 1). A preliminary selection of a gasification process therefore is based on coal specification. The parameters to be considered are listed on Fig. 2. In the present context the term "proximate analysis" stands for the analysis shown on Fig. 3, and the term "ultimate analysis" stands for the one shown on Fig. 4. The complexity of the parameters is obvious from these figures.

2.1. Conventional technology

Gasification processes commercialised in the past are the ones of: (Fig. 5):
- Lurgi (fixed bed gasification)
- Koppers-Totzek (entrained flow gasification) and the one of
- Winkler (fluidized bed gasification).
Commercial plants were constructed all over the world until eventually they lost importance under the tough competition of oil and natural gas, the use of which was much easier.

As you may see on Fig. 5, the technological requirements are quite tough. In all cases, gasification temperature is > 800°C. One has therefore to speak about high temperature processes, the control of which requires particular measures.

The Lurgi approach to mastering these problems was watercooling of the thermally highly stressed plant components. The steam produced is no disadvantage for the process since it can be used directly in the process as gasification agent.

A Koppers-Totzek process is run at a gasification temperature above the ash melting point, i.e. > 1.100°C in a reaction chamber lined with a refractory mass which, however, requires additional cooling in order to avoid excessive wear of the refractory liner.

The Winkler gasifier, operating safely at a temperature below the ash melting point, is also refractory-lined. Compared with Koppers-Totzes gasification, however, the requirements of the Winkler process with respect to refractory materials are definitely lower.

2.2. Change of process requirements

The technologies discussed above are those of the first generation. As a consequence of the feared energy scarcity a new development phase was started in the early seventies in view of a second generation of processes incorporating improvements of known technologies, viz.:

- Pressurised gasification
  The energy needed in atmospheric gasification systems to compress the raw gas to the pressure required for the down-stream processing steps, can be saved in pressurised gasification.

- Efficiency increase
  High efficiency is compulsory due to the imminent scarcity and resulting price increase. For maximizing carbon utilization the gasification temperature is raised to the limits of controllability.

- Higher flexibility
  Load changes must not affect efficiency.

- Wider coal base
  Low-grade coal, high-ash and high-sulphur coal as well as feed coal with high fines content - a consequence of modern coal winning technologies - will have to be used as well.

The following two illustrations (Figs. 6 and 7) show details of the process development for the second generation. We should still mention that coal gasification in molten iron bath is a very recent development which may be categorized under the heading "third generation" of coal gasification processes. This process is run at approx. 1.400°C in molten iron and accordingly implies particular requirements for the materials used in plant construction.

2.3. Required specifications of refractory materials

Plant construction materials for gasification processes
undergo operational stresses which vary widely as a function
of the various process configurations and operation modes. By
providing for the combination of all parameters we obtain a
confusingly wide range of stresses. The following text
therefore is necessarily limited to basic essentials. In
addition, we should mention that an analysis of performance
requirements should be set up for each individual
gasification reactor.

First of all, (Fig. 8) the function principles of the basic
variants should be explained. In a Lurgi fixed-bed gasifier
the coal bed rests on a rotary grate. The gasification agent,
a blend of O_2 and steam, is led into the bed from the bottom
upwards. Due to the counter-current configuration a
temperature profile, the peak of which comes close to the ash
melting point, is obtained in the longitudinal axis of the
reactor. The gas leaves the reactor at a temperature of
approx. 400°C and entrains – besides the gaseous components –
condensing components as e.g. higher hydrocarbons and steam.
This reactor is not refractory-lined.

A further development of this configuration by British Gas
and Lurgi, the slagging gasifier, on the other hand, is
refractory-lined (Fig.9). The slag withdrawal system of the
reactor contains an oxygen injector by means of which
solidification of slags in case of operation trouble is
prevented.

The fluidized bed (Fig. 10) is characterized by the movement
of the individual particles. These particles move in
suspension within the reactor since they are borne by the
gasification agent streaming through the fluidized bed from
the bottom upwards. A fluidized-bed temperature which
naturally is homogeneous over the whole reactor is lower than
the ash melting point in order to avoid agglomeration. The
impact effect of the particles in suspension is negligible so
that erosion in the reactor needs not to be catered for. The
operation pressure is of max. 10 bar according to the present
state of the art.

A special application of this technology is nuclear steam
gasification (Fig. 11). In this case, the fluidized bed is
heated from outside via an immersed heat exchanger. The
stresses the materials have to resist are comparable with
those of other applications.

For entrained-flow gasification coal may be fed to the reactor either
- from the top,
- from the side, or
- from the bottom.

The function principle of the Texaco gasifier (Fig. 12),where a coal-water suspension is injected by means of oxygen from the top may be shown as example. The temperature is higher than the ash melting point so that liquid slag runs down the reactor walls and drips off from an opening in the bottom. The reactor is refractory-lined.

In Koppers-Totzek or Shell-Koppers gasifiers (Fig. 13) two burners are arranged opposite to each other in the lateral walls. The flames are deflected upwards without the reactor walls being touched. While the slags are withdrawn by gravity from the reactor bottom the gas leaves the reactor via the top. The reactor is fitted with a cooled refractory lining.

In the Saarberg-Otto process (Fig. 14) coal is gasified in a slag bath in the reactor bottom. Coal and oxygen are injected slightly excentrically by a burner. In this way, the slag bath is set in rotary motion. While the gas flows upwards, the slags flow downwards. The reactor is lined with cooled refractory materials although the raw gas is cooled in the cylindrical section of the reactor. The gasifier dome is protected against high temperatures by an uncooled brick lining. In this dome the hot raw gases are blended with a cold recycling-gas stream. It possibly sounds trivial to experts that refractory materials are to protect reactor materials against temperature-induced, localized or general, reduction of mechanical strength. I mention this, nevertheless because reliability of these components is of particular importance when looking at pressurized gasification. An analysis of the stressing mechanisms, is therefore, the prerequisite for a choice or development of reliable materials and for safe operations of gasification plants.

(Fig. 15). In principle, one has to differentiate between
- chemical
- mechanical and
- operational attacks and stresses, respectively.

Each kind of stress and attack on its own as well as the complex interaction of several of these factors is to be assigned specific importance. Certain phenomena as e.g. purpose-serving reactions must be accepted without any limitation. Other mechanisms, however, can be influenced and may be used for approaches in view of reducing wear.

The reactions shown in Fig. 16 take place synchroneously during gasification and produce a gas comprising CO, CO_2, H_2, H_2O and other components. The concentration of the individual components is a function of the chemical equilibrium which, in turn, is controlled by temperature, pressure, and starting conditions. Each gasification process,therefore produces a gas of typical composition. This fact should be underlined because also reactions of gas components with the refractory materials used can occur. I should like to mention in this context that for instance free silicic acid can react according to the operation conditions prevailing with steam or hydrogen, and that iron contained in the refractory material may act as catalyst for methane formation.

In addition, the mineral substance of coal may possibly react with the refractory material. This attack probably is the most serious one, since refractory materials as well as coal are natural materials. This implies that almost an unlimited number of reaction possibilities exists. Fig. 17 shows the frequency of minerals found in German coal qualities according to the total ash content. The question concerning the quality of the feed coal therefore influences the choice of sufficiently resistent refractory materials to a considerable extent.

In coal gasification also mechanical stresses of the construction materials occur. These stresses necessarily need to be controlled in view of safe operation. The intensity of attack by erosion is least with fluidized-bed gasification, stronger with fixed-bed gasification and strongest with entrained-flow gasification. This is conditioned partially by different feed systems accelerating the individual coal particles more or less. Attack by erosion can be reduced to some extent by design modifications.

Temperature and pressure shocks occur mostly in case of operation trouble. Temperature shocks are of particular high intensity in entrained-flow gasification since, due to the small mass of reaction products in the gasifier, no damping effect is given. Rapid temperature variations by several hundred degree centrigrade must be tolerated.

Pressure shock occurs always if intentionally or inadvertently pressure relief of the reactor takes place. In such a case, forces are activated which may cause destruction by implosion if these phenomena were not considered adequately in the design phase.

Also condensation induced locally by temperatures below the dew point can have destructive effects on the materials. Careful consideration is therefore to be given to the concentration profiles in the reactor. Furthermore, refractory materials are jeopardized by penetration by molten slag. This is a matter of porosity of the refractory material and of the viscosity of the slags. Accordingly, high material density and viscous slags are desirable.

Restrictions imposed by the manufacturers of refractories for protecting their materials are frequently reason for concern to the plant operators because operation conditions of start-up and shut-down as well as in continuous operation or in case of load changes can be adapted to the requirements of the refractory manufacturers only to a limited extent. This applies also to drying after initial lining of the reactors, and in particular when catering for operation troubles. After all, prophetic gift is required for predicting possible operation troubles prior to the introduction of new technologies.

Fig. 18 shows that by combination of simultaneous measures, examplified by a development of refractory material for the pressurized coal dust gasification according to the Texaco process, substantial success was achieved within less than 5 years. While, in the initial phase of the project, the wear of the refractory material was stated to be of 0.2 mm/h, the respective value today is rated to 0.01 mm/h.

To sum up, the Fig. 19 summarizes the conditions of normal operation for the three processes discussed. The range of figures quoted results from the multitude of the individual process variants.

3. The way to solution of problems

Development and introduction of new technologies today touches the majority of disciplines of technology and natural sciences.Obviously, the multitude of the tasks to be tackled can only be dealt with in close cooperation with competent institutions. This also applies to refractory materials for coal gasification plants. The history of the Texaco gasification process has shown that a logically scheduled development can be carried out successfully by joint efforts of industry and research institutions. Within the framework of subsequent discussions we may expect contributions from companies involved. Just 4 1/2 years were required for increasing the durability of a refractory material to a value which now possibly guarantees at least one year of continuous operation of a commercial plant. The costs incurred were high. When considering the funds contributed by developing companies the overall costs to be quoted amount to several million DM. The work done for this development covers preliminary testing of approx. 50 commercially available materials, 20 of which – approximately – were tested under simulative gasification conditions. According to the present state of development, only two material qualities are suited for commercial operation.

4. Summary

Pressurised high temperature coal gasification uses varying technologies depending on the available feed coal and the desired product range. Refractory materials are in most cases absolutely necessary for these processes. The requirements which these refractory liners have to meet vary for the individual processes with respect to chemical stability, mechanical strength, and with the capability to withstand operational stresses. The development of a universally applicable refractory material does not seem feasible, from our present-day viewpoint, both for technical and financial reasons. Preferably, the particular conditions of a process should be defined, in order to enable the development of specifically tailored materials. The possibility that technologically promising processes would founder due to a lack of durability in commercially available materials, makes the merits of joint actions in the development of refractory materials for coal gasification obvious.

Requirements for Physical Coal Properties

Fixed Bed   : weakly agglomerating
                   lumpy
                   highly melting ash

Fluidized Bed   : not agglomerating
                   coarse ground
                   highly melting ash

Entrained   : finely ground

Molten Iron Bath: crushed

Fig. 1 Coal Gasification
        Applicability of Principles

- Proximate Analysis
- Ultimate Analysis
- Fusion Temperature of Ash
  . Oxidizing Atmosphere
  . Reducing Atmosphere
- Hardgrove Grindability
- Swelling Index
- Size Distribution

Fig. 2 Coal Gasification
        Specification

Content of

        Moisture

        Fixed Carbon

        Ash

        Volatiles
              ————————

                100 %

        Heat Value

Fig. 3 Coal Gasification
       Proximate Analysis

Content of

        Carbon

        Hydrogen

        Nitrogen

        Chlorine

        Sulfur

        Oxygen

        Ash

        Water
              ————————

                100 %

Fig. 4 Coal Gasification
       Ultimate Analysis

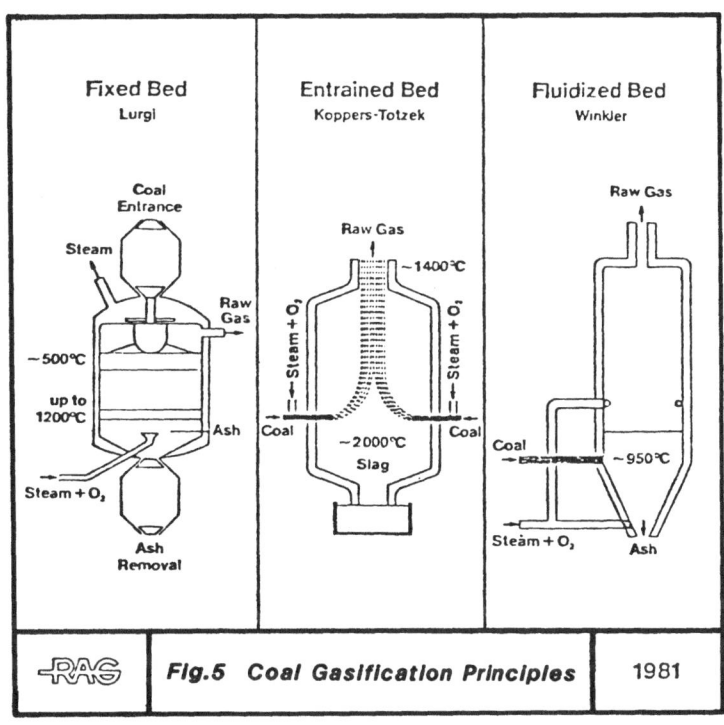

Fixed Bed
Lurgi

Coal Entrance

Steam

~500°C

up to 1200°C

Raw Gas

Ash

Steam + O₂

Ash Removal

Entrained Bed
Koppers-Totzek

Raw Gas

~1400°C

Steam + O₂

Steam + O₂

Coal

Coal

~2000°C

Slag

Fluidized Bed
Winkler

Raw Gas

Coal

~950°C

Steam + O₂

Ash

RAG **Fig.5 Coal Gasification Principles** 1981

Projects	Partners	Total Costs Period	Process Data	Products
Texaco Oberhausen-Holten	Ruhrkohle AG Ruhrchemie AG	48 Mio DM 1975–1981	Entrained bed pressure gasification according to Texaco 150 t/d 40 bar 1450 C	290 000 m³/d Synthesis gas
Ruhr 100 Dorsten	Ruhrgas AG Ruhrkohle AG/ Steag AG	150 Mio DM 1975–1983	Fixed bed pressure gasification according to Lurgi 70–170 t/d 100 bar 700–1000 C	40–95 000 m³/d SNG 100–236 000 m³/d Town gas
PNP Prototype Plant Nuclear Process Heat	Bergbau-Forschung GmbH Gesellschaft fur Hochtemperatur-reaktor-Technik GmbH Hochtemperatur-Reaktorbau GmbH Kernforschungsanlage Julich GmbH Rheinische Braunkohlenwerke AG	1200 Mio DM 1975–1984	Fluidized bed gasification 1500 t hard coal/d 4000 t lignite/d Hydrogasification: 80 bar, 820–930 C Steam gasification: 40 bar, 630–800 C	1 000 000 m³/d SNG from hard coal 640 000 m³/d SNG from lignite Synthesis gas Reduction gas
-RAG	***Fig.6 Coal Gasification Projects***			1981

Project	Partners	Total Costs	Process Data	Products
Shell-Koppers Hamburg-Harburg	Shell International Deutsche Shell AG Krupp-Koppers GmbH	100 Mio DM	Entrained bed gasification according to Shell: 150 t/d 30 bar 1150–1600 C	Synthesis gas Reduction gas
Saarberg/ Otto Fürstenhausen/Saar	Saarbergwerke AG Dr. C. Otto & Co., GmbH	71 Mio DM	Slag bath gasification 250 t/d 30 bar 1450–1650 C	Synthesis gas Reduction gas
High temperature **Winkler-Process** Frechen	Rhein. Braunkohlen-werke AG	32 Mio DM	Fluidized bed gasification 25 t/d 10 bar 870–1070 C	Synthesis gas Reduction gas
KGN-Plant Hückelhoven	PCV Gewerkschaft Sophia Jacoba	25 Mio DM	Fixed bed gasification 35 t/d 6 bar 920–1120 C	Low BTU gas Synthesis gas
VEW-Process Dortmund	Vereinigte Elektrizitätswerke Westfalen AG	25 Mio DM	Pilot Plant 24 t/d 1 bar	Low BTU gas for Electric Power Generation
⊣RAG	*Fig.7 Coal Gasification Projects*			1981

275

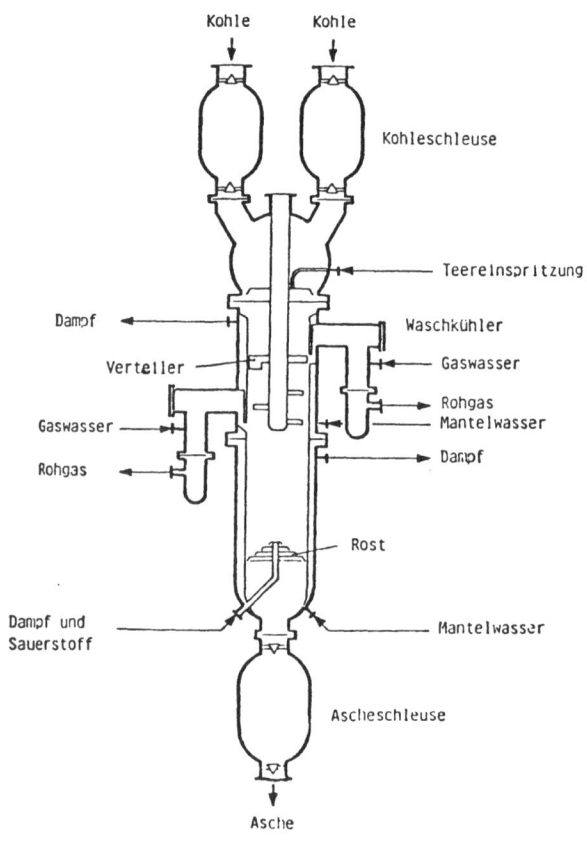

Kohle   Kohle

Kohleschleuse

Teereinspritzung

Dampf

Waschkühler

Verteiler

Gaswasser

Gaswasser

Rohgas
Mantelwasser

Rohgas

Dampf

Rost

Dampf und
Sauerstoff

Mantelwasser

Ascheschleuse

Asche

RUHRKOHLE OEL UND GAS GMBH	*Fig.8 Festbett-Druckvergaser RUHR 100*	1 9 8 1

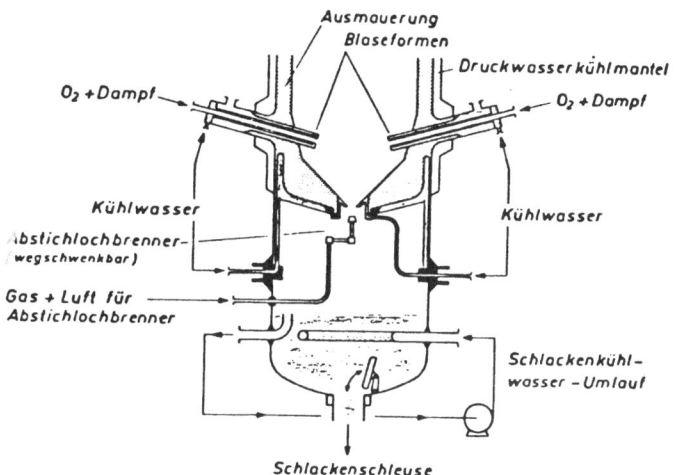

Fig. 9 Coal Gasification
British Gas

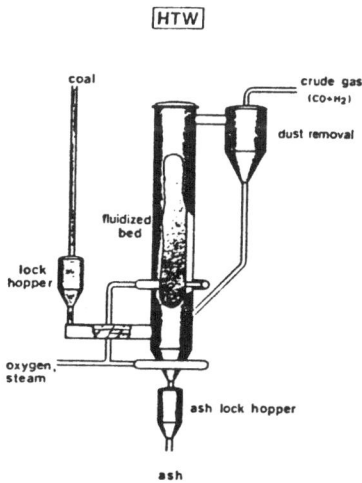

Fig. 10 Coal Gasification
High Temperature Winkler

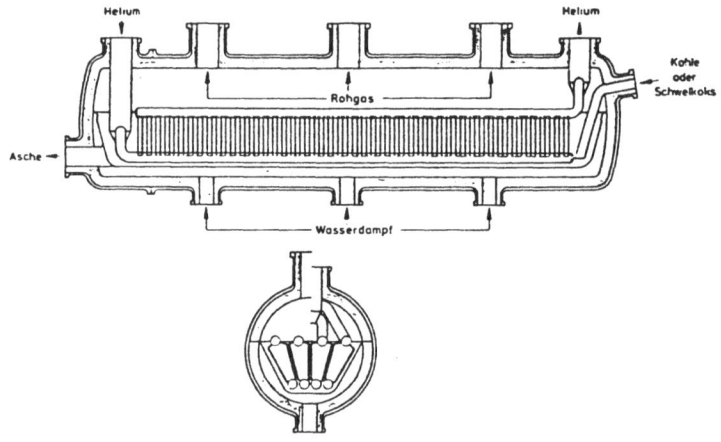

Fig. 11 Coal Gasification
Steam Gasification (PNP)

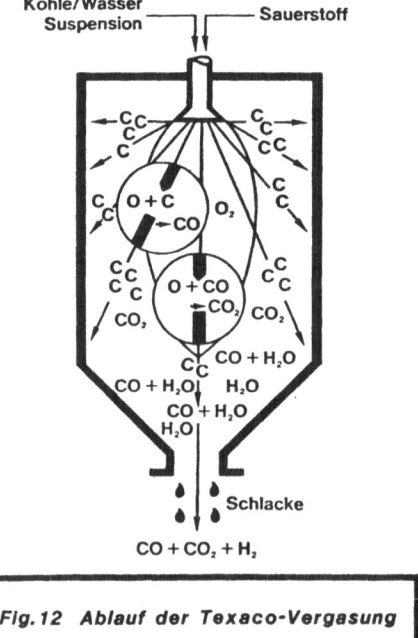

Kohle/Wasser
Suspension ——————— Sauerstoff

$O + C$

O_2

$\rightarrow CO$

$O + CO$

$\rightarrow CO_2$

CO_2

CO_2

$CO + H_2O$

$CO + H_2O$

H_2O

$CO + H_2O$

H_2O

Schlacke

$CO + CO_2 + H_2$

| *Fig. 12 Ablauf der Texaco-Vergasung* | 1981

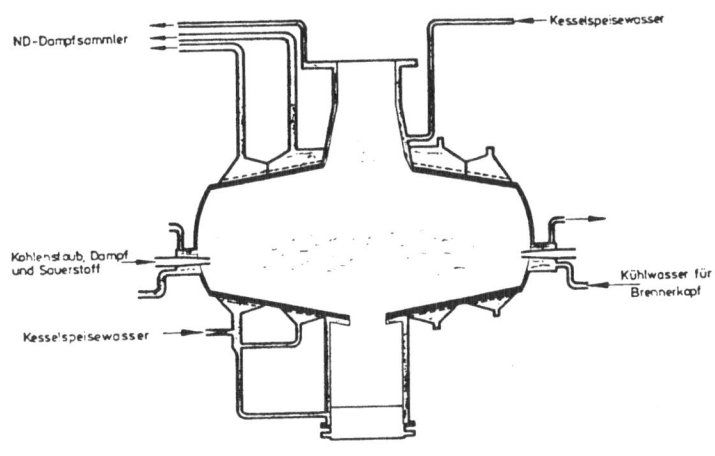

Fig.13 Coal Gasification
Koppers-Totzek

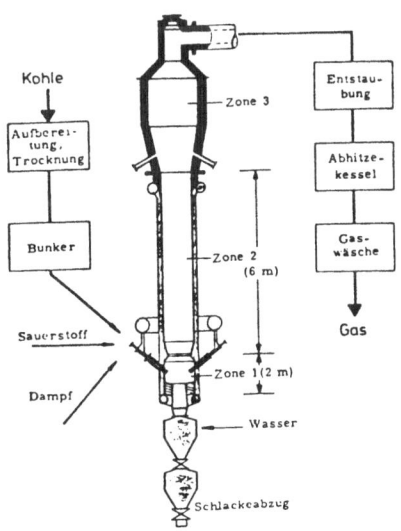

Fig.14 Coal Gasification
Saarberg-Otto

```
Chemical      - Temperature
              - Pressure
              - Interaction Gas/Refractories
              - Interaction Slag/Refractories

Mechanical    - Erosion
              - Thermal Shock
              - Pressure Shock
              - Slag Penetration

Operational   - Start up/Shut down
              - Steady Operation
              - Load Change
              - Failure
```

Fig. 15 Coal Gasification Stress of Refractories

$$C + 1/2\ O_2 = CO \qquad \Delta H = -123,1\ kJ/mol$$
$$C + O_2 = CO_2 \qquad \Delta H = -393,6\ kJ/mol$$
$$C + CO_2 = CO \qquad \Delta H = -159,9\ kJ/mol$$
$$C + H_2O = CO + H_2 \qquad \Delta H = -118,5\ kJ/mol$$
$$C + 2H_2 = CH_4 \qquad \Delta H = -\ 87,5\ kJ/mol$$
$$CO + H_2O = H_2 + CO_2 \qquad \Delta H = -\ 40,9\ kJ/mol$$
$$CO + 3H_2 = CH_4 + H_2O \qquad \Delta H = -205,9\ kJ/mol$$

Fig. 16 Coal Gasification Basic Reactions.

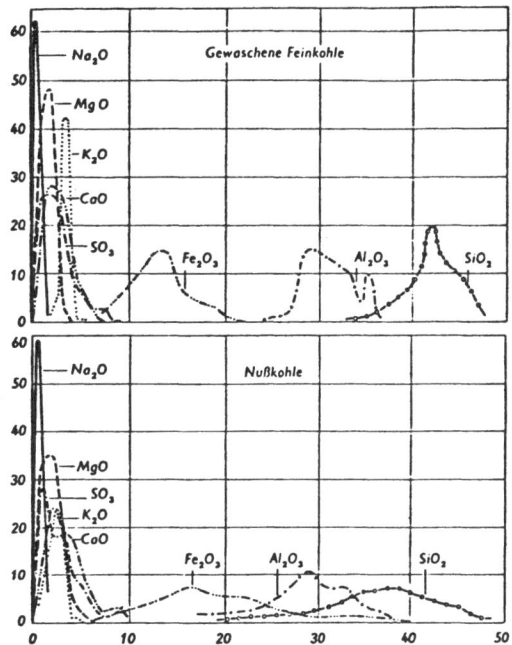

Fig. 17   Frequency of Ash Components
vs. Ash Content in wt%

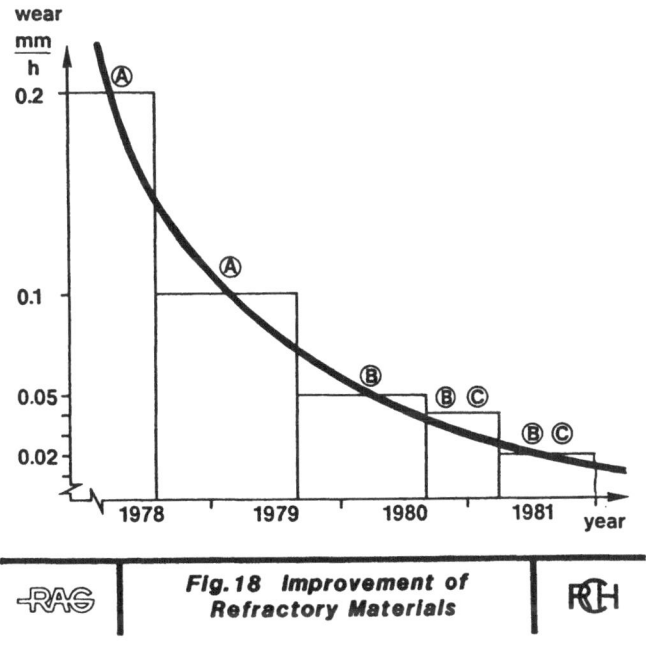

Fig. 18   Improvement of
Refractory Materials

Technology	Fixed Bed	Fluidized Bed	Entrained
Pressure (bar)	30- 100	1- 10	1- 40
Temperature (^{0}C)	760-1400	800-1000	1300
Gas:			
CO_2 (%)	3-29	13-25	0,5-12
CO (%)	18-56	30-50	55-71
H_2 (%)	29-38	35-46	21-32
CH_4 (%)	7- 6	1- 2	0,1
CnHm (%)	0,6	----	------
N_2 (%)	4	0,5-1,5	1
Byproducts	Tars Phenols	----	----

Fig. 19 Coal Gasification Operational Data

REFRACTORIES FOR COAL GASIFICATION.

THE STATE-OF-THE-ART IN THE U.S.

C. R. Kennedy and P. E. Schlett

Materials Technology Division
Exxon Research and Engineering Co.
Florham Park, NJ, USA   07932

ABSTRACT

        The state-of-the-art of refractories for coal
gasification is reviewed. Both dry ash and slagging coal
gasification process conditions and materials are discussed.
Among the potential failure mechanisms to be dealt with
include: (1) corrosion and erosion due to molten coal slag,
iron, or both; (2) alkali attack; (3) carbon monoxide induced
disintegration; (4) reduction of silica by H_2; (5) iron oxide
bursting; (6) steam related reactions; (7) thermoelastic
stresses; and (8) particulate erosion. Results from pilot
plant exposures and a variety of DOE-sponsored laboratory
experiments indicate that dense fireclay castables containing
approximately 50% alumina and having a low iron content are the
preferred materials for use in dry ash gasifiers (temperature
<1200°C). In slagging gasifiers operating at intermediate
temperatures (1200-1600°C), coal slag corrosion/erosion may be
minimized through the judicious selection of materials (high
density, high-chromia refractories), coals, fluxes, and process
temperatures, and through careful design of water-cooled
linings. For slagging gasifiers operating in excess of 1600°C,
the slagging boiler design consisting of a thin sacrificial
refractory lining on water-cooling tubes is preferred. Current
DOE-sponsored research and development programs are reviewed
and areas for future research are identified.

I. INTRODUCTION

As a result of sharp increases in oil and natural gas
prices in the 1970's, a large number of different coal-
gasification systems are currently being considered for
commercialization in the U.S. Several reviews of these
processes are available in the literature.[1-4] For the purposes
of discussion, it is convenient to divide these systems into
two broad categories: dry-ash and slagging gasifiers. Dry-ash
gasifiers generally operate at temperatures of <1200°C, whereas
slagging gasifiers may have temperatures as high as 1750°C.
The composition of the gaseous atmosphere can vary widely
(Table 1), depending on the process, the feedstock, and the
desired output, e.g. high-Btu pipeline gas, low-Btu gas, or
process gas (H_2, CO) for chemical industries. Coal slag
compositions are also highly variable (Table 2). Although the
designs of the gasification plants can differ significantly,
refractory linings are generally utilized in the main reaction
vessel, as well as in some transfer lines and cyclones, to
reduce heat losses and protect the metallic components from
erosion and corrosion.

The purposes of this paper are to delineate potential
refractory degredation mechanisms in coal gasification, review
research relevant to these problems, and to identify preferred
materials based on the "state-of-the-art". Recommendations for
areas of future research will also be presented.

II. POTENTIAL REFRACTORY PROBLEMS IN COAL GASIFICATION

Based on the limited operation of pilot plant
gasifiers and from extensive experience with refractories in
related service conditions, such as in blast furnaces, it is
possible to anticipate the following potential failure
mechanisms for refractories in dry ash gasifiers: (1) alkali
attack; (2) carbon monoxide disintegration; (3) silica
volatilization; (4) steam-related reactions; (5) thermoelastic
stresses; and (6) erosion due to solid particulates.

Since many slagging gasifiers have lower temperature
slag-free sections (Figures 1a, 1c), refractories in these
gasifiers may be subject to all of these degradation
mechanisms, with the addition of (7) corrosion and erosion due
to molten coal slag, iron, or both, and (8) iron oxide
bursting.

A. Corrosion and Erosion by Molten Coal Slag, Iron, or Both

Of all the potential failure mechanisms, corrosion by coal slags appears to be the most serious and most difficult to solve effectively. Operating experience at the Combustion Engineering pilot plant has demonstrated[5] that at temperatures of $1750^{\circ}C$, the lifetimes of a variety of thick (185 mm) alumina and alumina-chromia refractory bricks were only of the order of hundreds of hours. Also included was a test panel consisting of a thin layer (19 mm) of high-alumina refractory on studded water-cooled tubes, typical of the design of many slagging boilers. Although the refractory was quickly corroded and replaced by a frozen slag layer, little corrosion of the metallic tubes was observed after 4000 h. High-purity alumina, chrome-magnesia, alumina-zirconia-silica, zirconia, and silicon carbide hearth plates in the Grand Forks Energy Technology Center (GFETC) gasifier have undergone severe corrosion in as little as 10 h of operation at $1550^{\circ}C$.[6] Refractory lining lifetimes on the order of only hundreds of hours were observed[7] initially in the Ruhrchemie Texaco gasifier, operating at temperatures up to $1600^{\circ}C$.

The relative corrosion resistances of about 60 commercial and experimental refractories have been evaluated in a series of laboratory tests at Argonne National Laboratory (ANL) using a slag-bath furnace. Both acidic ($CaO/SiO_2 < 1$) and basic ($CaO/SiO_2 > 1$) slags have been utilized. Test conditions were chosen to approximate those found in the Bigas slagging coal gasification plant. In most tests, bricks of different lengths (228 mm and 114 mm) were included and were water-cooled on their cold faces to produce different thermal profiles. The shorter the brick, the lower the hot face temperatures. Detailed accounts of the apparatus, experimental procedures, test conditions, and results of these tests have been presented elsewhere.[8-10]

The corrosion data from these ANL tests should only be used to provide a semiquantitative, relative ranking of the performance of the refractories. The data cannot be used directly to predict lining lifetimes in an actual gasifier. Of the three major mechanisms of the corrosion process, i.e., dissolution, penetration and disruption, and erosion, these tests tend to emphasize the first two due to the low velocity of the slag at the interface with the refractory. In an actual gasifier with higher velocity slag, the rate of corrosion can be anticipated to increase significantly since the rates of dissolution and/or erosion might increase. This could be particularly pronounced for those refractories that exhibited substantial intergranular attack due to the increased erosion of entire grains.

In general, dense high-chromia (Cr_2O_3) content
refractories have demonstrated superior corrosion resistances
to a wide variety of coal slags ($CaO/SiO_2 = 0.2$-1.7). As shown
in Figures 2a and 2b, a chrome-spinel (primarily $MgCr_2O_4$)
refractory (No. 22) exhibited the least corrosion and
penetration of ten refractories when exposed to a high-iron
oxide acidic coal slag at $1575^\circ C$. As the chromia content of
the refractory increased, the corrosion generally decreased.
This is due to the extremely low solubility of Cr_2O_3 and
$MgCr_2O_4$ in SiO_2-Al_2O_3-CaO liquids, as shown by Muan[11] at Penn
State University. Low porosity was also an important factor in
minimizing the amount of corrosion and penetration.
Refractories containing > 30% Cr_2O_3 reacted with all types of
coal slags to form complex spinels which dissolved slowly.
High alumina refractories were intermediate in performance in
acidic slags (Figures 2 and 3) and poor in basic slags (Figure
4). Thin (<150 mm) silicon carbide bricks have shown good
resistance to acidic coal slags at $1500^\circ C$ when heavily water
cooled (data not shown). However, the corrosion rate increased
rapidly when the temperature was raised to $1575^\circ C$. Attack was
also rapid at $1500^\circ C$ in basic slags when uncooled (Figure 4).
The mechanism of corrosion of these refractories is the
reaction of the silicon carbide with iron oxide to produce a
low-melting ferrosilicon alloy. Magnesia-chromite refractories
containing free MgO, i.e. numbers 400, 20 and 18, performed
somewhat better in basic slags than in acidic slags (Figures 3
and 4). Preferential dissolution of the MgO was observed in
all cases.

The effect of water cooling on the performance of the
various refractories is evident from the differences in the
corrosion and penetration of the 228 mm and 114 mm length
bricks (Figure 2); the shorter bricks suffered considerably
less corrosion and penetration because of their lower hot face
temperature. In general, an increase in the hot face
temperature of $\sim 100^\circ C$ (going from 1450 to $1550^\circ C$) resulted in
an increase in the corrosion of from 400 to 1300%.

The importance of limiting penetration by the slag
has been described by Bakker and Stringer[12], who observed
failure of a 90% Al_2O_3-10% Cr_2O_3 refractory lining by
pyroplastic deformation (Figure 5). This particular lining was
heavily insulated, which resulted in a shallow thermal profile
(Figure 6). Therefore, the fluid slag was able to penetrate
deeply into the porous ($\sim$ 16% open porosity) structure of the
refractory, attack the bonding phase, and significantly reduce
the creep resistance of this material. To alleviate this
problem, they suggest that the thermal gradient be increased
near the hot face so that the slag will "freeze" after
penetrating only a short distance into the brick. This can be

accomplished by utilizing a material with high thermal conductivity at the cold face (Figure 7) or by water cooling. In addition, refractory lifetime can be maximized by operating the gasifier at as low a temperature as possible, without plugging the slag outlet (resulting in the formation of potentially destructive molten iron pools) or sacrificing too much efficiency.

Although dense chrome-based refractories (composed primarily of $MgCr_2O_4$) have demonstrated the highest resistance to corrosion by coal slags, potential problems still exist; as will be discussed subsequently, these refractories possess poor thermal-shock resistance and may be susceptible to iron oxide bursting.

For gasifiers operating at temperatures in excess of $1600^{\circ}C$, the slagging boiler design, consisting of a thin sacrificial refractory lining on studded water-cooling tubes (Figure 8), is the only practical solution. For gasifiers operating at lower temperatures ($1200-1500^{\circ}C$), thick refractory linings, utilizing chrome-based refractories at the hot face, offer the promise of reasonable corrosion resistance with minimum heat losses. Indeed, lining lifetimes in the Rurchemie Texaco gasifier are now estimated to be in excess of one year.[7] The use of fluxes, coals with low-melting slags, or both, will allow many gasifiers to operate at lower temperatures, thus maximizing lining lifetime. In some cases, water cooling of these thick linings may also be necessary.

B. Thermal Shock Resistance of Brick Linings

Few data are available on the thermal shock resistance of the high-chromia refractory brick that have demonstrated potential for use in slagging gasifiers. Results of limited water quench tests[13a] have indicated that six commercially available dense high-chromia (~80 wt%) brick have significantly lower thermal shock damage resistance than a more conventional sintered 90 wt% Al_2O_3-10 wt% Cr_2O_3 brick (Table 3). Thus, unless the thermal shock resistance of the high-chromia refractories can be improved by microstructural alteration, heating and cooling rates in slagging gasifiers will have to be carefully controlled to avoid spalling.

C. Iron Oxide Bursting

As discussed in Section IIA, chrome-spinel ($MgCr_2O_4$) refractories have demonstrated the highest resistance to

corrosion by all types of coal slags. However, failure of
refractories containing spinel phases can occur when iron
oxides are absorbed.[13b] Shown in Figure 9 are some of the many
different pure spinel compounds and the variation in their
unit-cell sizes. In general, the ferrite spinels have
significantly larger unit-cell sizes than do the chromites or
aluminates. Thus, the reaction of aluminate or chromite-based
refractories with magnetite ($Fe^{+2}Fe_2^{+3}O_4$)-containing slags may
result in internal stress which can cause spalling. Change in
the partial pressure of oxygen can cause further embrittlement
by altering the Fe^{+2}/Fe^{+3} ratio, and thus the size of the unit
cell. In the extreme case, $FeCr_2O_4$ has been found to
dissociate into a solid solution of Fe_2O_3 and Cr_2O_3 after only
2 h at $850^\circ C$ in air. When this product was reheated in H_2 at
$1050^\circ C$ for 2 h, the spinel phase reformed, but the reaction was
accompanied by a devastating increase in the volume.

The use of refractories with low porosity will limit
the penetration of iron oxides from the slag and, thus, confine
the spalling to a thin surface layer. However, since high-
chromia refractories with low porosity have poor thermal shock
resistance, tradeoffs may have to be considered.

D. Alkali Attack

Alkali attack is a chronic problem in blast furnaces,
but usually manifests itself only after many thousands of hours
of operation. Destruction of the refractory can occur by the
formation of low-melting low-viscosity liquids, or, more
usually, by the formation of dry expansive alkali-alumino-
silicate compounds that result in chemical spalling. Since
most coal slags contain significant amounts of alkali (1-10%),
it is reasonable to anticipate such problems in the nonslagging
regions of gasifiers.

The only documented alkali-related failure in a
gasifier occurred in a non-slagging section of the GFETC
slagging gasifier.[14] In this case, a mullite refractory (76%
Al_2O_3, 24% SiO_2) failed after an intermittent exposure of only
~125 h at a hot-face temperature of ~$1000 \pm 200^\circ C$. The coal
that was being gasified during these particular runs was a
lignite coal with high sodium content, and chemical analysis of
the coal ash and slag revealed substantial volatilization of
the sodium. The volatile sodium species probably reacted with
the moisture in the atmosphere to form NaOH, which then
attacked the refractory. An examination of the failed
refractory revealed that it had reacted with NaOH to form
carnegieite ($NaAlSiO_4$) and beta-alumina ($NaAl_{11}O_{17}$). This
reaction was accompanied by an ~30% volume expansion which

caused the refractory to fail by chemical spalling. Experiments of short duration (500 h) in coal-gasification atmospheres saturated with alkali (NaOH and KOH) vapor at 980°C revealed little or no degradation of a variety of aluminous refractories ranging from 45 pct Al_2O_3 superduty firebrick to 95 pct Al_2O_3 concrete.[15] Nevertheless, after 500 h in contact with molten NaOH at 980°C, these refractories exhibited severe deterioration. Thus, the presence of molten NaOH at some time in the operating history of the GFETC gasifier is indicated.

Other laboratory experiments[16] have shown that the amount of $NaAlSiO_4$ formed when ground-up aluminous refractories were mixed with alkali compounds and heated to 900-1400°C decreased as the alumina content increased; 70 pct (high-fired) and 90 pct alumina produced no $NaAlSiO_4$. The reaction of aluminous refractories with alkali compounds to form beta-alumina was observed only at temperatures in excess of 1100°C. Despite the decreased tendency of high-alumina compositions to form $NaAlSiO_4$ in the above tests, high-silica (~60 pct) refractories performed better in alkali cup tests at 950°C. This was attributed to the ability of the high-silica refractories to react more rapidly with the alkali and contain its attack at the surface; microstructure was cited as an important factor in these results. However, Rigby and Hutton[17] have shown that at temperatures in excess of ~ 1000°C, a specimen with a composition of 40 wt% Al_2O_3 and 60 wt% SiO_2, typical of a superduty fireclay refractory, can react with alkali to form a liquid.

Several options exist for minimizing the alkali attack problem in the nonslagging region. Obviously, the use of low-alkali coals and lower process temperatures (perhaps with an associated penalty of decreased efficiency) will result in lower alkali volatility. High-density high-alumina refractories, which limit the penetration of the sodium species, or high silica (60%) refractories, which react with the alkali to produce a glass that seals off the surface from further attack, might be utilized. However, the use of these materials incurs trade-offs involving either a loss in thermal-shock resistance or a maximum operating temperature of <1000°C, respectively, which may be unacceptable. For use at temperatures of 1000-1300°C, dense magnesia-chrome or magnesium-aluminate spinel ($MgAl_2O_4$) refractories may give suitable service.

E. Carbon Monoxide Induced Disintegration

In certain refractories, the deposition of carbon, as the result of the decomposition of carbon monoxide according to

the equation $2CO = C+CO_2$, can cause internal stresses leading
to failure. This reaction typically occurs in reducing
atmospheres in the temperature range of 400-700°C.[18] Although
the exact nature of catalyst is open to debate, carbon monoxide
disintegration has been shown[18,19] to be greatly accelerated by
the presence of metallic iron, free iron oxides that can be
reduced to metallic iron, or iron carbides. In blast furnace
applications, CO attack has been controlled by the use of high-
fired fireclay refractories, in which the iron oxides have
reacted to form the stable compounds fayalite (Fe_2SiO_4) or
hercynite ($FeAl_2O_4$). Although no carbon monoxide-related
refractory failures in U. S. pilot plant gasifiers have been
reported to date, laboratory experiments[20] have demonstrated
that the rate of attack increases rapidly as the pressure
increases (Figure 10). The results clearly show that even very
small amounts (0.25%) of iron impurities can substantially
affect the rate at which alumina castables lose strength in
pure CO. Tests have also been conducted to evaluate the
effects of alkali, NH_3, H_2S, CO_2, and H_2O at elevated
pressures. Alkali compounds appear to increase the rate of
attack, whereas H_2S generally tends to retard attack. Current
experiments are focusing on stainless steel fiber-reinforced
castables, whose performance depends on the iron content of the
castable, the surface condition of the fibers, and the
prefiring treatment of the samples.

Dense alumina castables with low iron contents or
high-fired fireclay or alumina refractories will probably
provide reasonable lifetimes in the low temperature regions of
gasifiers. At present, a metal fiber-reinforced lining is
undergoing trial at the Hygas Gasifier (see Section IIIB)

F. Reduction of Silica by H_2

The loss of silica due to the formation of volatile
compounds has been observed in both reducing[21] and steam-
containing atmospheres.[22] Upon exposure to hydrogen-rich
atmospheres volatile SiO can be produced by the reaction

$$SiO_2 + H_2 \rightarrow SiO + H_2O.$$

As an example, after several years of service, an alumina-
silicate refractory in a secondary ammonia reformer was found
to have lost 50% of its silica at the hot face due to reduction
by hydrogen.[23] Nevertheless, the silica content was
essentially unchanged at a depth of ~10 mm from the hot face,
indicating that the diffusion rate of SiO is extremely slow

below 1200°C. Thus, it is unlikely that the reduction of silica by H_2 will be a serious mode of degradation in gasifiers.

G. Steam-Related Reactions

Sadler et al.[24] have performed an extensive series of tests to evaluate refractories for nonslagging service. These tests exposed a variety of refractories to both pure steam and steam-containing low- and high-Btu atmospheres. Their results are applicable to the low-temperature (1000-1100°C) sections of most gasifiers. The data show that calcium aluminate-bonded refractory concretes (40-50% Al_2O_3) made from fireclay calcine aggregate, and refractory brick containing 50-90% alumina, perform better than refractory concrete (90% Al_2O_3) made from tabular alumina and high-purity cement (Figure 11). Many of these materials actually increased in strength after exposure.

Although silica has been observed[22] to react with steam to form volatile phases such as $Si_2O(OH)_6$, exposures of up to 1000 hours to a simulated coal gasification atmosphere (containing both steam and hydrogen) produced no evidence of serious silica loss in a wide variety of refractories. Thus, it appears that silica volatilization will not be a major degradation mechanism for alumina-silicate refractories operating at temperatures below 1200°C. At temperatures above 1200°C, it is likely that the refractory will be sealed with a layer of molten coal slag. However, fouling of downstream components, such as heat exchanger, may occur when the volatile silicates from the slag condense.

Silicon carbide refractories (both oxynitride and nitride bonded) disintegrated completely when exposed to coal gasification atmospheres containing high partial pressures of steam.[24] Other experiments[25,26] in saturated gasification atmospheres at temperatures <300°C, indicate significant strength losses in phosphate-bonded refractories, but no degradation of cement-bonded castables.

H. Thermomechanical Degradation of Monolithic Linings

At Babcock and Wilcox Co. (B&W), Anderson et al.[27] conducted a series of eighteen heat-up tests on various lining types and anchor configurations with the objective of reducing cracking during initial dryout and heat-up of monolithic refractory linings. A 1.52m diameter by 4.27m high pressure vessel/test furnace lined with a 0.31 m thick 1.52m high refractory test cylinder was utilized. The unit was operated with air or steam pressures up to 1.725 MPa. The test linings were monitored by use of thermocouples and high temperature strain gauges imbedded in the castables, attached to the refractory anchors, and on the metal shell. Acoustic emission monitoring was used to detect the times at which cracking occurred in the refractory.

From these tests, the following guidelines for improved lining integrity during heat-up of refractory castables were established: (1) a 50% Al_2O_3 dense refractory castable with a low cement content, very low shrinkage, good fracture toughness, and superior creep resistance (as compared to a conventional 50% Al_2O_3 dense refractory castable) was incorporated as the hot face material. The lower coefficients of thermal expansion and thermal conductivity (as compared to 90% Al_2O_3 dense refractories) reduced the thermal stresses generated in the lining and provided better insulation for the shell; (2) four weight % of 310 stainless steel melt extracted fibers were incorporated in the 50% alumina dense component; (3) a low strength, low thermal conductivity insulating castable was used as a backup; (4) anchor spacings were increased to 0.6 - 0.9 meters from the originally tested 0.15 - 0.30 meters; (5) anchors were coated with asphalt based burn-out tape which left an expansion gap around the anchor; (6) bonding barriers were used between the lining components and between the lining and the shell; (7) a corrosion resistant material was attached to the shell and acted as a compliant layer between the shell and the lining; and (8) a continuous, slow, heat-up rate of 13.9 to 27.8°C/hr was used.

A numerical analysis model, based on the mechanical property data gathered by Anderson et al.[27] and Bray et al.[28], has been developed at the Massachussetts Institute of Technology (MIT) to predict the stress/strain behavior of dual component refractory linings.[28] This model has been implemented in a three dimensional finite element computer program. Although the details of this model are beyond the scope of this paper, it contains a temperature and stress-dependent constituative law, failure criteria, creep behavior, thermal expansion-contraction behavior, and shrinkage characteristics of refractory concrete. Also included in the model is a heat transfer analysis, which was based on the CONBEC (Connected Block and Effective Conductivity) program developed by Battelle Columbus Laboratories.[30] Verification analysis of the model is limited to the tests performed at B&W. Results obtained from parametric studies indicate that the mechanical reliability of the lining can be improved by the following: (1) minimizing the amount of linear shrinkage of the refractory; (2) elimination of long hold periods during the heating and cooldown; (3) maintaining the vessel shell temperature as close to ambient as possible; (4) using incompressible bond barriers; and (5) using anchor spacings greater than 1.5 times the lining thickness. In general, the results of the parametric studies confirm the guidelines developed in the experimental program at B&W. However, additional data (both short and long term) on the mechanical behavior of monolithic refractories under multiaxial stress are needed to improve the accuracy of the model.

I. Erosion of Refractory Materials

The erosion of refractories for gasifier applications has been studied [31] by direct-impingement, fluidized-bed, and impingement-tube tests. Both castables and bricks have been evaluated.

The direct-impingement device consisted of a regulated air supply, particle feed, venturi nozzle, rotatable specimen holder, and enclosure. These tests utilized various dolomite and sand particles sizes propelled at various velocities and impingement angles. All tests were at ambient temperature and only a few materials were tested. A chrome castable was found to be more erosion resistant than a high-alumina and a lightweight castable.

The fluidized-bed experiments were carried out at ambient temperature and at 810°C with dead-burned dolomite particles as the erodant. The facility consisted of a combustion chamber for natural-gas firing, a fluidized-bed test section, a disengaging section, a cyclone, and a high-temperature exhaust system. The specimens were located in racks within the bed. Prior to testing, all specimens were heat treated in air for 24 h at 1066°C to thoroughly decompose the hydraulic bonds. The 810°C data are given in Table 4. Contrary to the results of the direct impingement tests, the high- and intermediate-alumina castables exhibited better erosion resistance than the chrome castable.

The pneumatic impingement tube test facility was designed to simulate hot-gas transfer lines found in gasification and combustion systems carrying high-velocity particulate-laden gas streams. The facility consists of a horizontal leg where dolomite particles are introduced and the natural gas is combusted, a vertical section with a 3.05 m long acceleration zone and a 1.53 m long test zone, a cyclone, and a high-temperature exhaust system. The test specimens were hollow cylinders with 51 mm ID and 102 mm OD. The results of the elevated temperature tests are shown in Table 5. The least amount of erosion occurred on the fired shapes. In these tests, the 50% alumina castable containing fireclay aggregates performed as well as, or better than, the high-alumina castables. The erosion losses were found to vary directly with kinetic energy of the particles and inversely with the compressive strength and surface area of the refractory. In most materials the erosion occurred primarily in the softer matrix, as evidenced during visual examination of protruding aggregates and recessed matrices.

III. SELECTED EXAMPLES OF DEPARTMENT OF ENERGY-SPONSORED DRY-ASH GASIFIER PILOT PLANT EXPERIENCE

A. Refractory Lining Performance in the CO_2-Acceptor Dry-Ash Gasifier

Initially, a 3-component lining was utilized.[32] A lightweight (384 Kg/m^3) silicate material was installed next to the shell, 432 mm thick in the boot section and 178 mm thick in the rest of the vessel. An intermediate density (800 Kg/m^3) insulating castable was then installed over the block, followed by 75 mm of a dense calcined fireclay castable composed of about 40% SiO_2, 50% Al_2O_3 and 10% CaO as the hot face material. The inside diameter of the gasifier was ~1m in the upper portion and 508 mm in the boot section.

The initial lining was inspected after 6 months and found to have cracked in several places exposing anchor tips. Three vertical cracks had formed about one-third the length of the boot and were 25 mm wide. Two cracks, 25 and 30 mm wide, were observed in the bottom head refractory. The inside diameter of the boot had increased from 508 to 533 mm. Several anchor tips were exposed by the cracking, an abnormal event. During the refractory removal, some anchor bolts were found which had 38 mm bows in the studs.

The refractory damage was attributed to the thickness and low compressive strength of the block insulation which could not support the initial vessel pressurization or pressure surges. The force exerted on the hard-faced refractory with the soft block insulation behind it caused the refractory cracking compression of the insulation, and buckling of the anchor studs. Gas channeling was also noted in the insulation as a result of cracks in the hot face refractory.

Repairs after 6 months operation involved complete removal of the refractory from the bottom head boot and transition section and the replacment of it with 457 mm of insulating castable and 152 mm of dense fireclay castable. Four vapor stops were also installed in the new lining to minimize the possibility of gas channeling.

Numerous repairs were made to the new refractory lining; however, these repairs may well have been due to modifications. The refractory appeared to be in very good condition when examined for chemical attack by the gasifier atmosphere. No carbon had penetrated the refractory and physical properties testing of the refractories after service showed marked increases in compressive strength after exposure to the gasification atmosphere. The gains in strength were due to the reaction of calcium aluminate cement with silica in the presence of steam to form anorthite, $CaAl_2Si_2O_8$. The silica involved in this reaction had been leached from the mullite-cristobalite fireclay aggregate.

Nepheline $[(Na_{1.5}K_{0.5})Al_2Si_2O_8]$, was detected at the hot face of all specimens taken from all levels in the gasifier. This compound is formed by the reaction of alkalis (present from the lignite) with the refractory material. A black crust (1 to 3 mm thick) had formed at the hot face of the lining due to formation of liquid on the surface during gasifier operation; however, the effect of alkalies on the refractory was not considered detrimental to the lining.

Considering the operating environment to which the refractories were subjected, overall performance was rated as very good.

B. Refractory Lining Performance
    in the Hygas Dry-Ash Gasifier

The original refractory lining was composed[33] of 102 mm of a dense 90% alumina calcium aluminate bonded castable backed up by 279 mm of a lightweight low iron insulating castable. The original lining was considered adequate; however, spalling and cracking were observed in the dense castable and appeared to be due to the gunnited installation technique rather than to atmospheric attack. In subsequent relines of stages 2 (maximum temperature of 930°C) and 3 (maximum temperature of 1040°C), the 90% castable was reinstalled by casting (instead of gunning) and the cracking and spalling observed earlier was significantly reduced.

Based on the findings[27] at B&W, stage 1 (maximum temperature of 815°C) of the gasifier was recently relined with a stainless steel fiber-reinforced high density, high strength, low cement 50% alumina castable. This lining has reportedly performed very well.

Refractory performance in this pilot plant unit has been described as very good from an atmospheric interaction standpoint. The only significant refractory problems in the gasifier resulted from the wrong installation technique, i.e., gunning rather than casting. This installation technique was considered improper because of the space limitations in the unit and because of the critical service requirements of the refractory.

C. Performance of Refractory Coupons in Various Dry-Ash Gasifiers

A wide variety of refractory coupons have been exposed[34] in the CO_2-Acceptor, Synthane, Hygas, Bigas, U-Gas and Battelle coal gasification pilot plants. Three groups of refractories were evaluated. Group A consisted of monolithic materials including dense and lightweight castables, mortars, patching mixes and ramming mixes, with alumina contents ranging from 50% to over 90%. Group B consisted of fireclay and high alumina bricks which were either phosphate, vitreous or self-bonded. Group C was composed of refractories which were specially prepared for the tests and which are special order products from refractory vendors. These included fused-cast alumina, carbide and nitride bonded silicon carbides, tar-impregnated alumina-chrome, zircon, and hot-pressed boron carbide.

Samples of Group A materials were prepared according to manufacturer's recommendations and fired to $1500^{\circ}F$ before putting them in the pilot units. All groups, after preparation, were tested similarly for apparent porosity, apparent specific gravity and bulk density, before and after exposures. Abrasion resistance was evaluated before and after exposure using the rotating rod hot abrasion test.

In general, most specimens were mounted on racks along with metal specimens, placed into the gasifiers and exposed to the atmospheres on all sides. Problems involved in these tests included cracking, possibly due to thermal cycling and/or interaction between the supporting metal racks and the specimens. Although no mechanisms of deterioration were identified, all of the lightweight and medium density materials were reported to have exhibited highly variable performance, with internal penetration and bond deterioration after only short exposures. The carbide and nitride bonded refractories

were oxidized after only 700-1000 hours of exposure. Most
dense alumina-silica products containing 60% alumina and above
exhibited no evidence of significant deterioration. No carbon
monoxide induced disintegration was observed since the exposure
temperatures for these specimens ranged from 760 to 1040°C,
well above the temperature range of 425-700°C for carbon
deposition.

IV. CURRENT RESEARCH PROGRAMS

Summaries of the progress in 1981 of most of the
research programs on materials for coal gasification funded by
the Department of Energy (DOE), the Electric Power Research
Institute (EPRI) and Gas Research Institute (GRI) can be found
in reference 35.

Many questions about the performance of refractories
in slagging gasifiers remain to be answered. Although
laboratory corrosion tests at ANL have demonstrated that six
dense high-chromia refractories exhibit better corrosion
resistance than any other types of refractory when in contact
with coal slags, the tests to date have not been able to
separate the performances of these six refractories despite
significant differences in their microstructures and bonding.
Thus, more aggressive dynamic accelerated tests are being
developed. These tests will emphasize the synergism between
the corrosion and erosion aspects of degradation. Leading
candidate refractories will be exposed to a flowing slag in a
small-scale, rotating-drum, accelerated-corrosion test
furnace. The furnace will have overnight unattended operating
capability with an automatic slag-feeding system and a reducing
atmosphere. The data from these tests will yield a relative
ranking of the corrosion resistance of the leading candidate
refractories and lend guidance with respect to improving their
performance.

Research into the effects of coal slag (and flux)
compositions on the flow and corrosion properties of slags is
also underway at ANL. To date, the compositional effects of
the slags on the corrosion rate of refractories have been
evaluated at an arbitrarily chosen temperature. Future tests
will be conducted in a dynamic environment such that the
viscosity (not the temperature) of the slag is constant
throughout the corrosion test matrix; thus, the temperature in
each individual corrosion test will be set according to the
viscosity-temperature relationship for each slag (and flux)
being evaluated. The rationale for such an approach is related
to several of the critical operating parameters of many

slagging gasifiers, i.e., the gasifier must convert a high percentage of the coal into the desired product gas, the slag must be tapped continuously, and the refractory lining must have a reasonable (> 1 year) lifetime. Since some types of slagging coal-gasification systems can operate at rather low (1400-1500°C) temperatures without a significant loss in efficiency, and since the corrosion rates of the refractories decrease as temperature decreases, the limiting factor in operating at relatively low temperatures in these processes is often the plugging of the slag tap hole. Because plugging of this hole is directly related to the flow properties of the slag, the criteria of maintaining the viscosity of the slag < 25 Pa·s (generally accepted as the point at which slag starts to flow under its own weight) may guide the selection of the process temperature. Thus, the effect of slag (and flux) composition on the corrosion rates of the various candidate refractories will be evaluated in a series of tests at a constant slag viscosity, e.g. at 5 Pa·s. It is anticipated that significant increases in refractory lining lifetimes can be achieved through the proper choice of coals with low melting slags (or through the use of fluxes) provided the process temperatures are lowered concomitantly.

As part of the overall effort to understand the corrosion processes, the determination of phase diagrams relevant to coal slag-refractory interactions is continuing at Penn State University, with the focus on systems containing various forms of Cr_2O_3.

In addition to the evaluation of corrosion resistance, experiments are in progress at ANL to evaluate the thermal shock damage resistance of candidate refractories using the ribbon test method.[36] A program to fabricate experimental high-chromia refractories with improved thermal-shock damage resistance is also underway. Both commercial and experimental refractory bricks will be subjected to a thermal shock similar to what is expected in an actual plant during normal startups, shutdowns, and expected operating transients. Specific guidelines for heating and cooling rates will be generated. To compliment this experimental effort, models to predict thermal stress and strain in brick linings are being developed in programs at B&W and at MT. Research at Iowa State University and National Bureau of Standards is providing some basic high temperature materials property data on both brick and monolithic linings.

The environments found in dry-ash gasifiers (temperature <1200°C) are significantly less demanding of refractory materials than are those of slagging gasifiers. To date, no critical problems have been identified from pilot

298

plant exposures. However, some questions remain to be resolved. Results at B&W have indicated that the use of metal fiber-reinforced monolithic linings results in less cracking upon heatup. However, laboratory tests at the Virginia Polytechnic Institute have indicated that carbon monoxide disintegration can occur in metal fiber-reinforced monolithics as well as in materials with as little as 0.1% metallic iron. Thus, further research is underway to characterize CO disintegration, expecially in metal fiber-reinforced refractories.

V. FUTURE RESEARCH NEEDS

As previously described, DOE-funded research programs are currently addressing many of the remaining questions related to refractory linings in coal gasification. However, budgetary restrictions (both current and future) hamper the pace and scope of these investigations. In certain areas (particularly with respect to slagging gasifiers), parallel efforts, funded by sources other than DOE, are needed. These areas include: (1) the optimization of the microstructure and phase assemblage of high-chromia refractories (considering the tradeoffs between corrosion, thermal shock, mechanical properties, and cost); (2) the acquisition of a suitable data base on the best candidate refractories (both brick and metal fiber-reinforced monolithics) including such properties as strength, creep, fracture toughness, elastic modulus, and thermal conductivity; and (3) the evaluation of refractory/slag/flux systems in engineering scale tests conducted in pilot plants or at a large-scale special test facility such as that which was built at IIT Research Institute (but which has been mothballed due to lack of funding). These engineering scale tests are necessary to provide an accurate prediction of the corrosion rates of various refractories as a function of such variables as temperature, temperature profile (affected by lining design), composition of slag and/or flux, slag throughput, slag flow rate, and atmosphere. They would also offer the opportunity to assess tradeoffs between operating variables (e.g. temperature, pressure, type of coal), lining design (e.g. uncooled vs. cooled, two component vs. single component), and process efficiency (e.g. heat loss vs. time between repair). Finally, they could be utilized as bench marks for the calibration of stress/strain and thermal conductivity prediction codes.

VI. SUMMARY

A great deal of public information on the behavior of refractories in coal gasification atmospheres is now available. Results of pilot plant exposures and DOE-funded research programs indicate dense fireclay castables containing approximately 50% alumina and having a low iron content are the preferred materials for use in non-slagging ($< 1200^\circ$C) conditions. In slagging gasifiers operating at intermediate temperatures (1200-1600°C), the results suggest that coal slag corrosion/erosion may be minimized through the judicious selection of materials (high density, high-chromia refractories), coals, fluxes, and process temperatures, and through careful design of water-cooled linings. For slagging gasifiers operating in excess of 1600°C, the slagging boiler design consisting of a thin sacrificial refractory on water-cooling tubes is the only practical design.

300

References

1. R. E. Balzhiser: EPRI J., 1979, Vol 4, No. 1, p37-40.

2. B. Barker, editor: EPRI J., 1979, Vol. 4, No. 3, p6-14.

3. K. Chandra, B. McElmurry, and S. Smelser: Electric Power Research Institute Final Report AF-782, 1978.

4. R. A. Bradley and R. R. Judkins: Oak Ridge National Laboratory Report ORNL/TM-7612, July, 1981, p41-66.

5. W. T. Bakker, S.L. Darling, and W. C. Coons: Amer. Ceram. Soc. Bull., 1982, Vol. 61, No. 3, p395.

6. R. C. Ellman, B. C. Johnson, and M. M. Fegley: "Studies in Slagging Fixed-Bed Gasification at the Grand Forks Energy Technology Center," presented at the Tenth Synthetic Pipeline Gas Symposium, Chicago, IL, October 30-31,1978.

7. W. Konkol, P. Ruprecht, B. Cornils, R. Durrfeld, and J. Langhoff: Hydrocarbon Processing, 1982, Vol. 61, No. 3, p97-102.

8. C. R. Kennedy: J. Materials for Energy Systems, 1981, Vol. 3, No. 3, p39-47.

9. C. R. Kennedy: J. Materials for Energy Systems, 1980, Vol. 2, No. 2, p11-20.

10. C. R. Kennedy and R. B. Poeppel: InterCeram, 1978, Vol. 27, No. 3, p221-226.

11. A. Muan: Fifth Annual Conference on Materials for Coal Conversion and Utilization, Gaithersburg, MD, October 7-9, 1980, pIV:38-IV:41.

12. W. T. Bakker and J. Stringer: Sixth Annual Conference on Materials for Coal Conversion and Utilization, Gaithersburg, MD, October 13-15, 1981, p6-26.

13a. G. Bandyopadhyay, J. Chen, C. R. Kennedy, and D. R. Diercks: to be submitted to J. Materials for Energy Systems for possible publication, 1982.

13b. J. H Chesters: Refractories: Production and Properties, p213-261; 1978, London, The Iron and Steel Institute.

14. C. R. Kennedy: J. Materials for Energy Systems, 1981, Vol. 3, No. 1, p27-30.

15. H. Heystek and N. S. Raymon: Fourth Annual Conference on Materials for Coal Conversion and Utilization, Gaithersburg , MD, October 9-11, 1979, CONF-791014, pIV:11-IV:13.

16. R. E. Farris and J.E. Allen: Iron and Steel Engineer, 1973, Vol. 50, No. 2, p67-74.

17. G. R. Rigby and R. Hutton: J. Amer. Ceram. Soc., 1962, Vol. 45, No. 2, p68-73.

18. T. F. Berry, R. N. Ames, and R. B. Snow: J. Amer. Ceram. Soc., 1956, Vol. 39, No. 9, p308-318.

19. W. R. Davis and G. R.Rigby: Trans. Brit. Ceram. Soc., 1954, Vol. 53, p511-523.

20. J. J. Brown, Jr.: Fifth Annual Conference on Materials for Coal Conversion and Utilization, Gaithersburg, MD, October 7-9, 1980, pIV:18-IV:24.

21. M. S. Crowley: Amer. Ceram. Soc. Bull., 1970, Vol. 49, No. 5, p527-530.

22. R. E. Dial: Amer. Ceram. Soc. Bull., 1975, Vol. 54, No. 7, p640-643.

23. M. S. Crowley and J. F. Wygant: Amer. Ceram. Soc. Bull., 1973, Vol. 52, No. 11, p828-831.

24. L. Y. Sadler III, N. S. Raymon, K. H. Ivey, and H. Heystek: J. Amer. Ceram. Soc., 1979, Vol. 58, No. 7, p705-714.

25. S. F. Rahman and D. E. Day: J. Materials for Energy Systems, 1979, Vol. 1, No. 3, p34-44.

26. S. F. Rahman and D. E. Day: J. Materials for Energy Systems, 1980, Vol. 1, No. 1, p3-23.

27. E. M. Anderson, R. P. Glasser, M. A. Schroedl, R. W. Sheriff, R. S. Williams, and C. E. Zimmer: Fifth Annual Conference on Materials for Coal Conversion and Utilization, Gaithersburg, MD, October 7-9, 1980, pIV:25-IV:33.

28. D. J. Bray, J. R. Smyth, T. D. McGee: J. Amer. Ceram. Soc., 1980, Vol. 59, No. 7, p706-710.

29. O. Buyukozturk and T. Tseng: J. Amer. Ceram. Soc., 1982, Vol. 65, No. 6, p301-307.

30. J. R. Schorr: Fifth Annual Conference on Materials for Coal Conversion and Utilization, Gaithersburg, MD, October 7-9, 1980, pIV:32-IV:33.

31. T. Vojnovich, D. L. Keairns, W. C. Yang, W. G. Vaux, E. P. Weaver, D. P.Stiles and G. Derge: Electric Power Research Institute Final Report AF-1151. July 1979, p3:1-3:59.

32. R. C. Dobbyn, H. M. Ondik, W. A. Willard, W. S. Brower, I. J. Feinberg, T. A. Hahn, G. E. Hicko, M. E. Read, C. R. Dobbins, J. H. Smith, and S. M. Wiederhorn: unpublished work at the National Bureau of Standards, Gaithersburg, MD, 1979.

33. W. Bair, Institute of Gas Technology, personal communication, 1981.

34. G. H. Criss, R. F. Firestone: Procedings of Properties and Performance of Materials in the Coal Gasification Environment, Pittsburgh, PA, September 8-10, 1982, p. 97-123.

35. Sixth Annual Conference on Materials for Coal Conversion and Utilization, Gaithersburg, MD, October 13-15, 1981.

36. C. E. Semeler, Jr., T. H. Hawisher, and R. C. Bradt: Amer. Ceram. Soc. Bull, 1981, Vol. 60, No. 7, p724-729.

Table 1

Typical Range of Product

Gas Compositions

Component	Vol.%
H_2	10-55
H_2O	2-35
CO	7-65
CO_2	2-25
CH_4	1-30
Miscellaneous (H_2S, NH_3, N_2)[a]	1-5

[a] Up to 60% N_2 for low-Btu gas

Table 2

Typical Compositions of Coal Slags

Component (wt%)	Acidic Slag	Basic Slag
SiO_2	30-60	10-25
CaO	2-20	20-45
Al_2O_3	10-30	10-20
Fe_2O_3 (Eq.)	3-30	3-20
MgO	1-10	1-10
$Na_2O + K_2O$	1-10	1-10
TiO_2	1-2	1-2
SO_3	1-20	1-20

[a]$CaO/SiO_2 = 2.0 - 0.5$

Ferritic content $= \dfrac{\text{wt\% } Fe_2O_3}{\text{wt\% } Fe_2O_3 + 1.11 \text{ wt\% } FeO + 1.43 \text{ wt\% } Fe} = 0\text{-}60\%$

Table 3

Retained Strength of Test Refractories Before and After
Five Quenches from 1000 to 100°C (From Reference 13a)

Number/ Type[a]	Composition	As-Received Tensile Strength (MPa)	% Retained Strength After Five Quenches
16-S	Al_2O_3 (90), Cr_2O_3 (10)	14.8 ± 1.3	40.2
23-S	Al_2O_3 (90), Cr_2O_3 (10)	30.6 ± 2.0	20.4
600-S	Al_2O_3 (60), Cr_2O_3 (20), ZrO_2 (12)	30.0 ± 3.4	7.5
22-FC	Cr_2O_3 (80), MgO (8), Fe_2O_3 (6)	31.1 ± 6.8	5.6
800-S/FC	Cr_2O_3 (76), Al_2O_3 (9), SiO_2 (11)	27.6 ± 3.1	7.4
255-S/FC	Cr_2O_3 (79), Al_2O_3 (6), MgO (5)	115.9 ± 15.1	1.5
812-S/FC	Cr_2O_3 (78), MgO (18)	7.3 ± 3.0	24.6
251-S/FC	Cr_2O_3 (72), MgO (26)	11.3 ± 1.3	25.5

[a] S = sintered
FC = fused cast
S/FC = rebonded fused-cast grain

Table 4

Average Erosion Loss Per Unit Area for Alumina
Castable Refractories Tested in a Fluidized Bed[a]
(From Reference 31)

Refractory	Average Erosion Loss (10^{-4} m^3/m^2)
96% Alumina (Mfg. A)	1.02
96% Alumina (Mfg. B)	1.49
60% Alumina	1.26
Chrome	1.95

[a]Test temperature = 810°C; duration of test = 50.5 h; bed
material was −20 + 70 mesh dead-burned dolomite particles;
fluidizing velocity of combustion gases = 1.30 m/s.

Table 5

Average Erosion Loss Per Unit Area Per Particle[a] for
Alumina Castable Refractories Tested in an Impingement Tube
(from reference 31)

Refractory	Type[b]	Erosion Loss (10^{-14} $m^3/m^2/P$)	
		Test HTP-3[c]	Test HTP-4[d]
96% Alumina (Mfr. A)	C	2.9	4.1
95% Alumina (Mfr. B; Fine aggregate)	C	1.9	6.7
96% Alumina (Mfr. B; Coarse aggregate)	C	6.2	10.7
90% Alumina (Phosphate bond)	P	1.5	1.1
90% Alumina (Mullite bond)	FS	0.3	0.6
85% Alumina (Phosphate bond)	FS	0.3	0.1
60% Alumina	C	3.8	11.2
50% Alumina	C	0.7	5.6
Lighweight 40% Alumina	C	48.3	97.7
Chrome	C	1.8	4.4
Silicon Carbide (Calcium Aluminate Bond)	C	1.9	2.1

[a] Mean particle size = 250 μm; particle velocity = 12.0;
particle loading = 1.13-1.18 wt. solids/wt. gas.
[b] C = castable, FS = fired shape, P = plastic
[c] Average temp. = 715°C; peak temp. = 740°C; total mass of
impinging particles = 2.26 x 10^3 kg.
[d] Average temp. = 807°C; peak temp. = 924°C; total mass of
impinging particles = 1.21 x 10^3 kg.

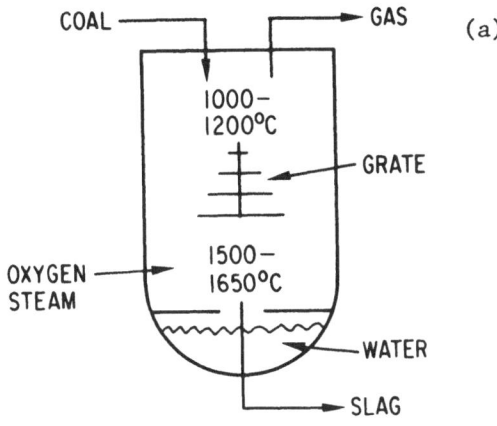

(a)

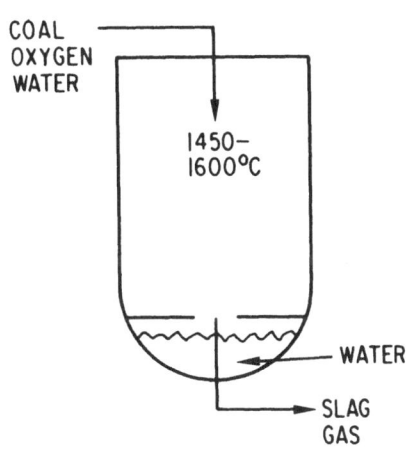

(b)

Fig. 1. Schematic design of
(a) fixed-bed gasi-
fier, (b) down-fired
single-stage en-
trained gasifier,
and (c) up-fired
two-stage en-
trained gasifier

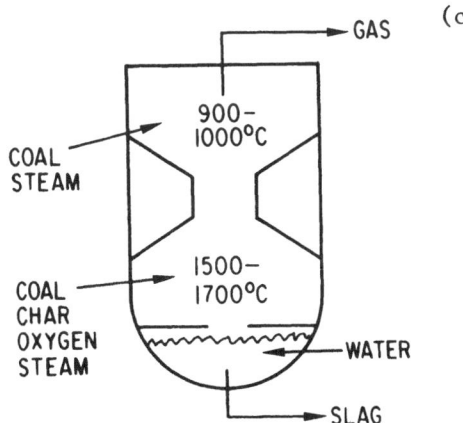

(c)

309

HIGH IRON OXIDE (20% FeO) ACIDIC COAL SLAG (B/A = 0.55)
WITH WATER COOLING (500 h) at 1575°C

DEPTH OF CORROSION (mm)

NUMBER/TYPE	POROSITY (%)	COMPOSITION (wt %)
2-FC	1	Al_2O_3 (99)
190-S	17	Al_2O_3 (92), Cr_2O_3 (7.5), P_2O_5 (0.5)
16-S	16	Al_2O_3 (90), Cr_2O_3 (10)
86-C	21	Al_2O_3 (85), Cr_2O_3 (10), P_2O_5 (4.5)
109-C	23	Al_2O_3 (67), Cr_2O_3 (32), P_2O_5 (1)
260-S	25	Al_2O_3 (67), Cr_2O_3 (32)
400-S	13	MgO (42), Cr_2O_3 (27), Fe_2O_3 (16)
38-FC	5	Al_2O_3 (60), Cr_2O_3 (27), MgO (6)
280-FC	7	Al_2O_3 (65), Cr_2O_3 (32)
22-FC	6	Cr_2O_3 (80), MgO (8), Fe_2O_3 (6)

0 — 5 — 10 — 15 — 20 — 25 — 40

LEGEND
228 mm LENGTH
114 mm LENGTH

Fig. 2a. Relative corrosion resistances of water-cooled refractories exposed to a high iron oxide acidic coal slag for ∿500 h at 1575°C. The depth of corrosion was measured from the original position of the hot face. FC = fused-cast, S = sintered, and C = chemically bonded.

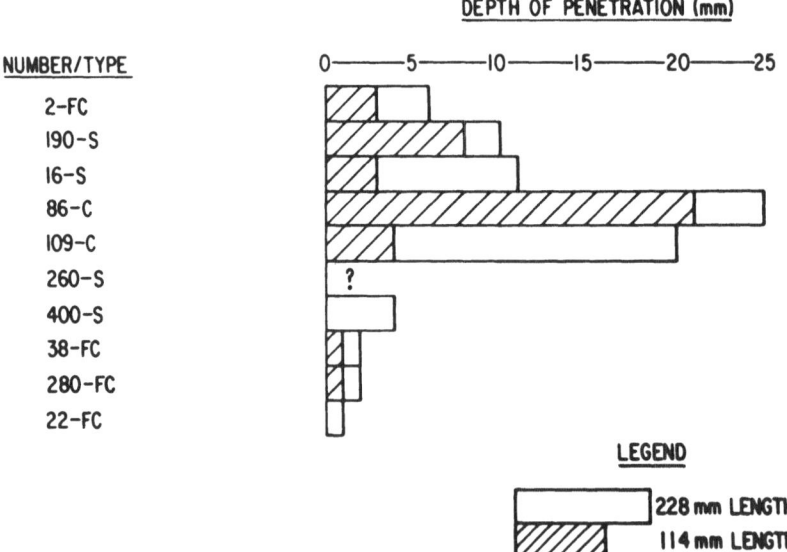

DEPTH OF PENETRATION (mm)

NUMBER/TYPE

0 — 5 — 10 — 15 — 20 — 25

2-FC
190-S
16-S
86-C
109-C
260-S
400-S
38-FC
280-FC
22-FC

LEGEND
228 mm LENGTH
114 mm LENGTH

Fig. 2b. Approximate depths of penetration by the slag into the refractories (as measured from the final position of the slag-refractory interface).

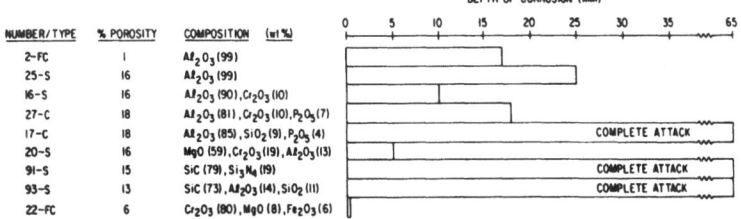

ACIDIC SLAG, 1500°C WITHOUT WATER COOLING (200 h)

DEPTH OF CORROSION (mm)

NUMBER/TYPE	% POROSITY	COMPOSITION (wt %)
4-FC	2	Al_2O_3 (99)
1-S	16	Al_2O_3 (99)
15-S	14	Al_2O_3 (92), SiO_2 (8)
16-S	16	Al_2O_3 (89), Cr_2O_3 (10)
20-S	16	MgO (59), Cr_2O_3 (19), Al_2O_3 (13)
8-C	25	Al_2O_3 (94), CaO (5)
10-C	22	Al_2O_3 (93), P_2O_5 (6)
17-C	18	Al_2O_3 (84), SiO_2 (9), P_2O_5 (4)
18-C	18	Cr_2O_3 (40), Fe_2O_3 (23), MgO (10)
22-FC	6	Cr_2O_3 (80), MgO (8), Fe_2O_3 (6)

Fig. 3. Relative corrosion resistances of refractories exposed to an acidic coal slag at 1500°C for 200 hours.

BASIC SLAG, 1500°C WITHOUT WATER COOLING (500 h)

DEPTH OF CORROSION (mm)

NUMBER/TYPE	% POROSITY	COMPOSITION (wt %)
2-FC	1	Al_2O_3 (99)
25-S	16	Al_2O_3 (99)
16-S	16	Al_2O_3 (90), Cr_2O_3 (10)
27-C	18	Al_2O_3 (81), Cr_2O_3 (10), P_2O_5 (7)
17-C	18	Al_2O_3 (85), SiO_2 (9), P_2O_5 (4) — COMPLETE ATTACK
20-S	16	MgO (59), Cr_2O_3 (19), Al_2O_3 (13)
91-S	15	SiC (79), Si_3N_4 (19) — COMPLETE ATTACK
93-S	13	SiC (73), Al_2O_3 (14), SiO_2 (11) — COMPLETE ATTACK
22-FC	6	Cr_2O_3 (80), MgO (8), Fe_2O_3 (6)

Fig. 4. Relative corrosion resistances of refractories exposed to a basic coal slag at 1500°C for 500 hours.

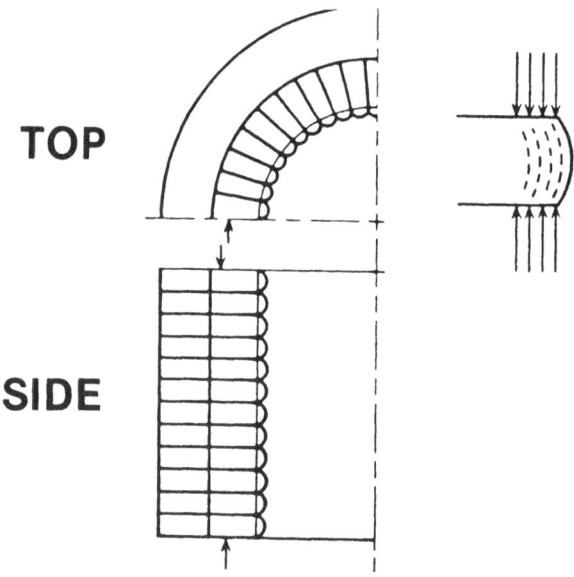

Fig. 5. Pyroplastic deformation of refractory lining after slag penetration (from Reference 12).

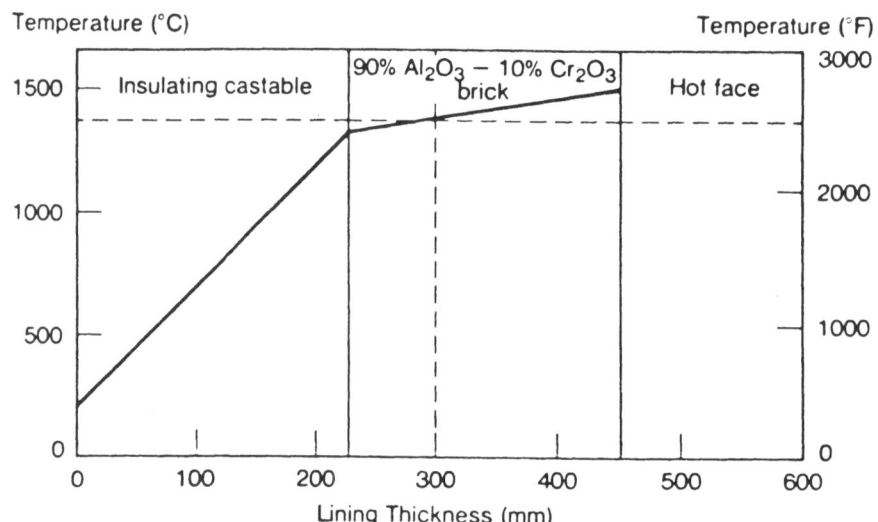

Fig. 6. Thermal profile of insulated refractory lining (from Reference 12).

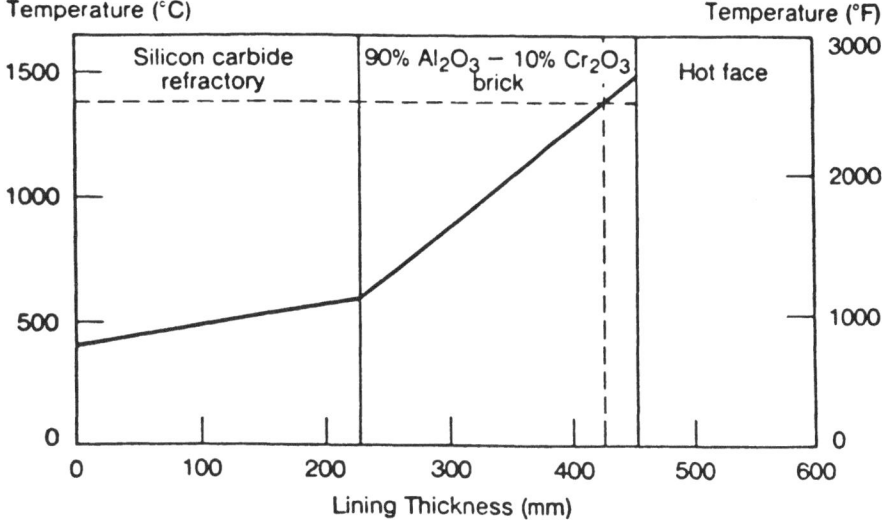

Fig. 7. Thermal profile of dense refractory lining (from Reference 12).

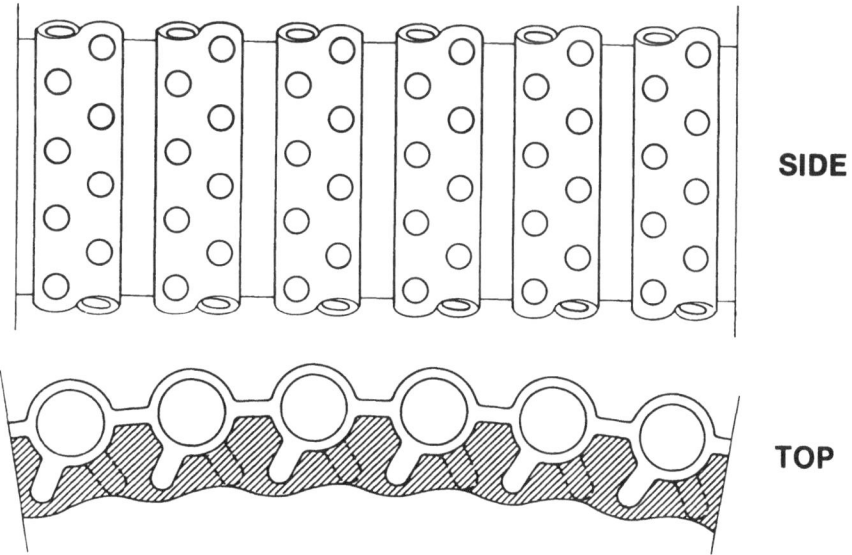

Fig. 8. Conceptual design of a studded membrane wall of a slagging coal gasifier.

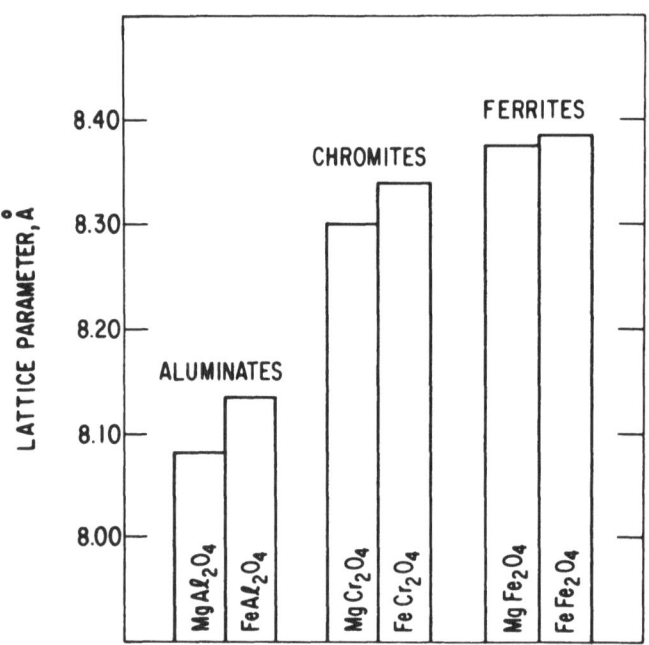

Fig. 9.   Lattice dimensions of synthetic spinels.

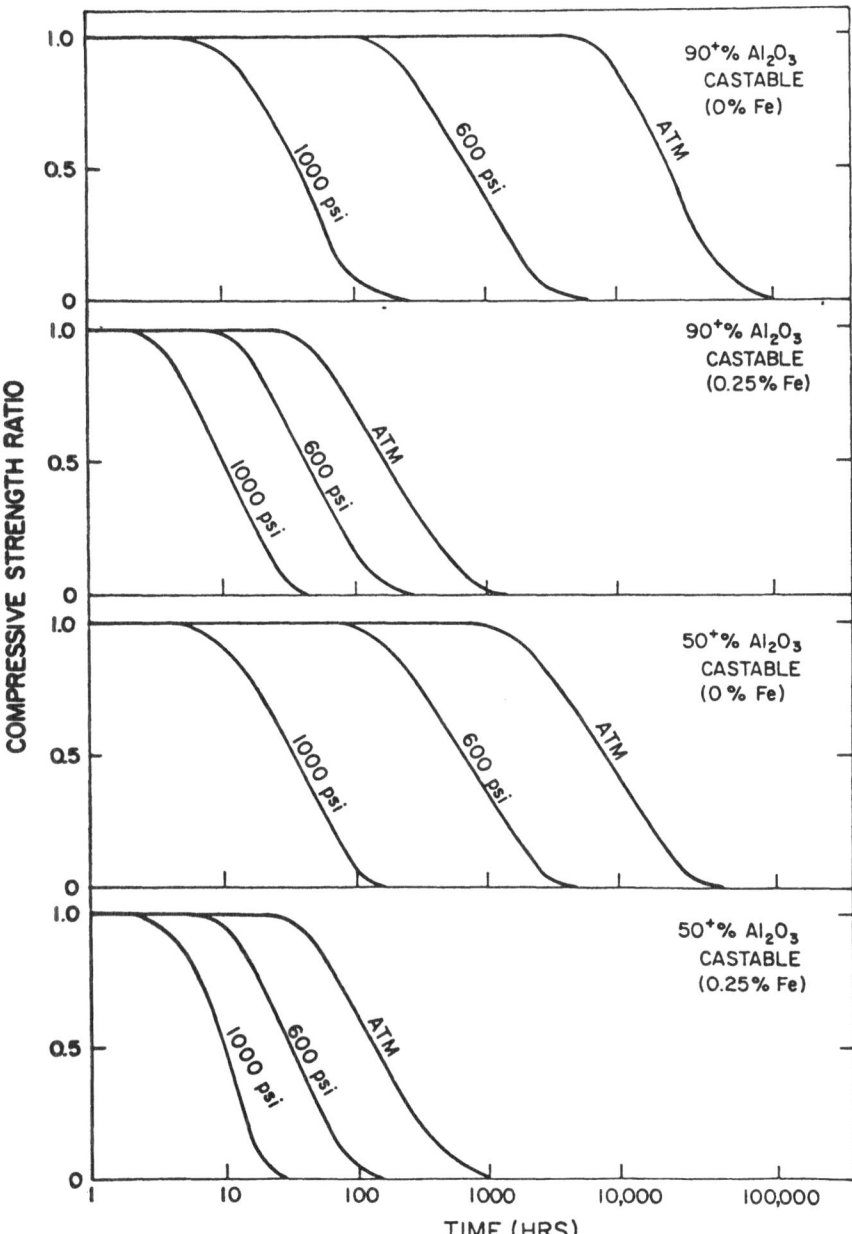

Fig. 10. Summary of CO disintegration behavior of 50+
Al$_2$O$_3$ castable up to 1000 psig and 0.5 wt.% iron
addition. (Pure CO, 500°C) (from Reference 20)

315

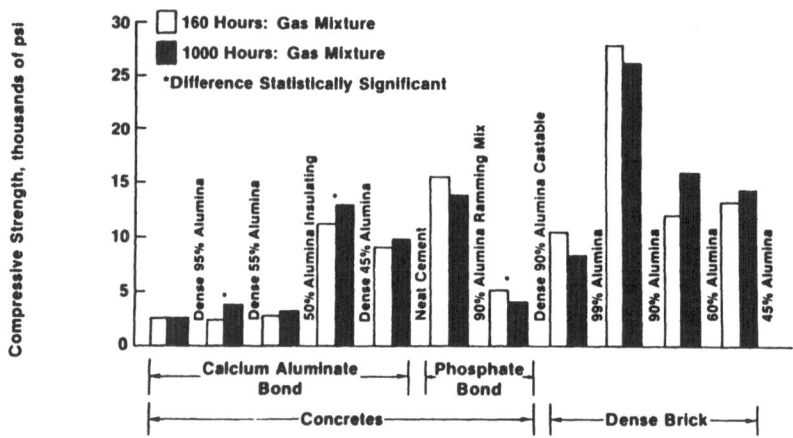

Fig. 11.   Effect of gas mixture on refractory strength - 1,000
hours (from Reference 24).

DISCUSSION

D. van der Giessen: Most tests on slag corrosion have been
done with molten slag. Would test results be different if
tests were done with slag vapour?

C.R. Kennedy: Yes. At temperatures < 1750°C, the volatility of
alkali and sulfur are much higher than that of the other
components of the slag. Therefore, the primary concern with
respect to slag vapour is alkali resistance (see Section
II-D). In our corrosion tests at ANL, the bottom half of the
bricks were submerged in a pool of slag, whereas the upper
half were exposed to the vapour. The $MgCr_2O_4$ type refractories
exhibited excellent resistance to both vapour and slag attack.

D. van der Giessen: What would be the effect of operating
pressure on slag penetration?

C.R. Kennedy: No data exist to answer this question. Assuming that no pressure differential exists across the slag/refractory interface, I would expect the effect would be small. See, however, the comment of G. de With concerning the corrosion of alumina by sodium.

G. de With commenting on the previous question: we did some corrosion tests on alumina with sodium and found a severe corrosion increase upon increasing the pressure. Therefore, I expect you would have the same in the gasifier.

F.K. van Dijen: Concerning the behaviour of bricks normally employed in glass ovens, I would like to inquire about the possibility of Cr carbides or Cr nitrides formation.

C.R. Kennedy: Thermodynamic calculations based on slagging gasification conditions indicate that the formation of Cr oxide is preferred over that of Cr carbides or nitrides.

F.K. van Dijen: Could you comment on the formation of SiALON or SiO_2 in the case that the system Si_3N_4-Al_2O_3 is subjected to a reducing environment.

C.R. Kennedy: I have not worked with this system. Perhaps J.P. Kiehl could answer this question.

REFRACTORIES PROBLEMS IN COAL GASIFIERS

Jean-Pierre KIEHL
Lafarge Réfractaires
Vénissieux - France

Due to the numerous coal gasification processes it is obvious that
refractory applications are very complex in themselves and each is
presenting a particular case.

They can, however, be roughly divided into 3 groups:

1. Refractories for gasifiers with fluidised bed and temperatures
   below 1200°C, said: "dry" ash reactors.

2. Refractories for high temperature gasifiers (1500/1600°C) with
   liquid ash running down the reactor walls.

3. Refractories used for very high temperature reactors reaching
   temperatures equal or above 1700°C in the burning zone.

1. FLUIDISED BED GASIFICATION WITH DRAY ASH
   (temperature below 1200/1300°C).

This type of gasifiers is more or less derived from the old
Winkler gasifier.

The refractories used require the following principal properties:
good mechanical erosion resistance together with sufficient
resistance to alkali vapours derived from the coal and,
eventually, resistance to sudden temperature changes which might
occur:

- from burner failure

- or from mechanical incidents in the installation.

These refractories must, of course, be CO/Boudouart's reaction, hydration and carbonation resistant.

If working temperature remains inferior to 1100/1200°C refractories with silica content combined as mullite can be employed. But above 1200°C, it is recommended to use a refractory without silica content, this being later reduced to volatile SiO which causes a weakening of the mechanical resistance and, as a result, a weakening of the erosion resistance.

A first selection could be done according to usual refractory controls and classifications: bulk density, open porosity, cold crushing strength, hot crushing strength at 1200°C and modulus of rupture at 1200°C, plus standard CO resistance test.

In order to select the more suitable refractories, it is recommended to carry out the Morgan Marshall's test of hot erosion as described by H. Aldred, Elliott & Cowling (1) with an improved high temperature version (Fig. 1).

This is an easy test up to 1200°C, but it is not possible to simulate erosion for a refractory with a surface undergoing changes by alkali attack.

A certain percentage of alkaline salts could be added to the abrasive powders (silicon carbide or fused alumina), but because of the very short duration of contact it is unlikely that the refractory will react with these alkalies.

Thus a separation of the tests in two parts is preferred:

a) preparation of a brick (110 x 110 mm) by coating the surface to be exposed to erosion by a potassium carbonate paste

b) firing of this sample at 1200°C during 24 hours.

Visual examination:

- rejection of all refractories with a bloated and friable surface after first treatment

- refractory retaining an apparent hard and compact surface: the sample is tested by the Morgan Marshall method.

Startling varying results are obtained from the abrasion test despite apparent soundness after corrosion by alkalies. A difference of erosion from 1 to 2 can be observed, but in some cases only a loss of 10% or less is observed.

2. GASIFICATION AT 1500/1600°C WITH LIQUID ASH RUNNING ALONG THE WALLS

The TEXACO process is the prototype for this kind of reactor.

Problems encountered with the refractory linings are mainly of chemical nature. Liquid ash reacts with the refractory forming a glass. The problem would be solved if this glass is viscous.

The working conditions of these reactors are all the more severe due to high insulation which results in a low temperature gradient through the working lining: often inferior to 150/200°C.

Depending on the refractory type observation should include the corrosion phenomena whether or not the refractory is easily impregnated by molten ash. This impregnation may be purely physical, filling of the pores, or physico-chemical, i.e. combined with a diffusion of the liquid ash through the refractory compounds and, as a consequence, modification of the structure of the refractory matrix.

3 types of corrosion tests are currently applied:

a) slag resistance by the cup method

b) "finger" slag test

c) slag resistance by horizontal or vertical rotary slag test.

In these 3 cases the tests have a significant value only when they can be made in a reducing atmosphere.

Advantages and disadvantages of these methods:

a) Cup method:

Usually a half brick (110 x 110 mm) drilled to form a cavity of 50/60 mm diameter and 40/50 mm depth. The cup thus created is filled with the ash and covered by a lid. Once prepared, this half brick is placed in a SiC saggar filled with graphite and topped with a SiC cover.

The test is carried out at 1650°C during 24 hours which gives a temperature of about 1550°C at the inside of the cup.

Once the test is completed, the samples are cut in two parts and corrosion phenomena examined with an optical microscope, a scanning microscope and X rays.

Advantage: simple method, no interaction of one refractory with another

Disadvantage: static test, no renewal of the corrosive agents.

b) "Finger" test: See apparatus fig. 2.

25x25x150 mm test bars are usually used in a graphite crucible forming the inductive element of an induction furnace. The coal ash to be examined is introduced into the crucible. Once the material is melted, the test bars can be rotated in the liquid bath to increase slag attack.

In order to avoid a too rapid combustion of the graphite crucible, the whole installation is flushed by Argon + Hydrogen.

Advantages: - Easy temperature control and possibility of attaining temperatures of 1800/1750°C in the liquid slag,

- A reducing atmosphere is easy to maintain,

- It is also possible to test either one product alone or to compare several products with each other.

Disadvantages: - Testing conditions are severe as full thickness of the refractory is fired to the testing temperature,

- Limited duration of the test.

Fig. 3: example of results obtained with this method.

c) Rotary furnace slag test:

The rotary furnace used can either have a horizontal or a vertical axis.

The number of tested products can vary from 4 products in the simplest case up to 10 in some trials i.e. as practiced by Kennedy.

Samples are often bricks of standard size or parts of bricks specially shaped or cut for this test.

The quantity of the corrosive agent is usually important : 40-50 kg, it can be:

- either continuously renewed

- or emptied at the end of a day of testing and completely

renewed the next morning.

Advantages: - The refractories can be employed in conditions
rather similar to real service conditions
because of the large quantity of molten ash.

- Different gradients and sudden temperature
variations of temperature can be simulated.

Disadvantages: - Expensive test, requiring a lot of bricks. If
bricks are not of the same class, they might
react with each other through the molten ash.

- Difficulties in maintaining the reducing
atmosphere necessary for the desired
reactions.

In addition all these corrosion tests have to be completed by a
bursting test, i.e. refractories must be fired up to 1650°C in
contact with Fe_2O_3. This very simple test shows in a striking way
the difference in attack which might occur in the chrome oxide
based refractories due to the FeO/Cr_2O_3 and FeO/Al_2O_3 SPINEL
forming (fig. 4). This, however, occurs only partially with the
previous testing methods, because the lime-silica glasses dissolve
the iron oxides. This reaction can be observed in pilot plant
installations after a working period of 600 to 1000 hours.

3. VERY HIGH TEMPERATURE GASIFICATIONS (< 1650°C) WITH HIGH
MECHANICAL EROSION AND CHEMICAL CORROSION

These gasifiers can be classifed as follows:

- gasifiers with a coal combustion taking place first in conical
burners and then running into a larger reaction chamber

- gasifiers with fluidised bed but with liquid ash at very high
temperature.

In addition to the phenomena of corrosion seen above, mechanical
erosion is caused by high speed coal particles, partially burned.

This test is particularly difficult to simulate in the laboratory
and would in practice require a special burner for correct
testing.

The refractories which are likely to be employed in practice can
be selected firstly by a corrosion test of the "finger" type and
secondly, after a slight rectification, hot modulus rupture tests
at 1550/1600°C must be performed on briks.

With both, the study of the corrosion phenomena and the hot
modulus of rupture, a classification can be established which
provides criteria for the possible usage of each product.

Laboratory simulation of gasifier's working conditions are hardest
to achieve.

Further study of the different types of refractories employed for
coal gasification applications is, after this quick account of the
specific selection tests, presented in the following chapter:

A. GASIFIERS WITH FLUIDISED BED AND TEMPERATURE < TO 1200°C
==
Up to now, the best results obtained are:

- either with fused alumina refractories in which the mullite
  bond is partially vitrified,

- or with fused alumina thixotropic castables with low hydraulic
  cement content. These castables require a low water addition
  (< 4%). They have cold crushing strengths already superior to
  1000 kg/cm^2 at 110°C, remaining practically unchanged up to
  1200°C.

The main characteristics of these 2 series production type
products are given in the table fig 5.

Castables with 10% of chrome oxide can be employed for this type
of reactors when the temperature exceeds 1200°C (see
specifications given in fig. 6). These castables are in addition
very successfully used for blast furnace applications (3).

B. HIGH TEMPERATURE GASIFIERS WITH LIQUID ASH RUNNING ALONG THE
==
WALLS
=====

Coal ash is mainly acid and can lie in the following typical
composition ranges:

Al_2O_3	25 /	35%
SiO_2	31 /	53%
Fe_2O_3	12 /	18%
CaO	4 /	5%
MgO	4 /	7%
Na_2O	1 /	2%
K_2O	1 /	2%

In response to experience acquired in basalt furnaces, research
has been orientated towards refractories with a high Cr_2O_3 content
(4) (5).

Two families of refractories seem to be outstanding here:

- a basic one, based on electrofused picochromite, reagglomerated by sintering

- an acid one, based on Cr_2O_3/Al_2O_3 and protected from metallic iron or from FeO by a high viscosity zirconia silica based glass phase.

Various studies are being carried out in Europe as well as in U.S. and it is difficult for the time being to give a typical composition.

In both cases the reacting agents in the system ash/refractory form complex spinels of the $(Fe-Mg)O-Cr_2O_3$ and $(Fe-Mg)O-Al_2O_3$ or Fe_2O_3 types.

C. VERY HIGH TEMPERATURE (> 1650°C) GASIFIERS WITH FLUIDISED BED

The gasifiers of KORF belong to these types of reactors. In the reactors of the second generation the interacting refractory/liquid ash layer is continuously renewed and passes through the main combustion chamber from where it drips into the bottom of the reactor.

At these operating temperatures, and the previously mentioned working conditions, the basic or acid refractories with a high Cr_2O_3 content are no longer practicable. They dissolve in less than 1000 hours.

The only refractories on which hopes can be focussed are nitrides or sialons as binders for a corundum matrix.

The principal properties of these refractories are given in table fig. 7.

These refractories have excellent hot mechanical properties and are hardly altered by coal ash, even at 1700°C.

In practice, they show noticeable less wear than the above mentioned chrome oxide refractories.

However, it is too early to form a precise opinion as long as several thousand working hours are not achieved.

Even if their use as brick or precast pieces, fired under N_2 atmosphere, is well known, it is not so for plastic or ramming specialities due to difficulties with the bond system. Prospects in this field cover a vast range.

It is not to be concluded that nitride based and chrome oxide based refractories are incompatible. Chrome oxide and silicon nitride form compounds melting at temperatures inferior to 1500°C. This makes field trials more complicated.

Another tricky point is the resistance of the silicon nitride in reducing atmosphere and aqueous vapour, laboratory results do not always agree with those obtained in pilot plant installations (6). When used at high temperatures, Si_3N_4 and SIALON seem to be stable from 1400°C up to 1800°C.

CONCLUSION
==========

. Fluidised bed gasifiers with dry ash do not give any particular problem for the refractory industry.

Subject to a good knowledge of the volatile components existing at this temperature, the problem of selecting a refractory is above all a problem of mechanical erosion.

Although several product types can be employed due to the moderate temperature, high quality refractories (e.g. fused alumina) are not excluded when high physical or chemical resistance is required due to mechanical erosion and attack by flux even at temperatures as low as 1200°C.

. Gasifiers with molten ash. Subject to a good control of their working conditions and to temperatures not exceeding 1600°C – at the very most 1650°C – they could be industrially operated with refractories of a high Cr_2O_3 content which have been especially studied for this type of reactor.

. As for the various gasifiers working at temperatures superior to 1650°C where erosion and chemical corrosion problems are encountered: refractories with a pure oxide based are unlikely to be a solution. On the other hand, nitride based refractories seem to have possibilities. Long running field tests are still to be done to confirm this position.

REFERENCES :
==========

(1) F.H. ALDRED, A. ELLIOTT & K.W. COWLING
    Trans. Brit. Ceram. Soc. 1955 - 54 - 239.

(2) C.R. KENNEDY & R.B. POEPPEL
    Interceram 27 - 221 - 226 (1978)

    J.A. BONAR, C.R. KENNEDY & R. SWAROPS
    Am. Ceram. Soc. Bull. 59 (4) 473-478 (1980)

(3) J.P. KIEHL & Y. LE MAT
    84th Meeting American Ceramic Society 1982.

(4) C.R. KENNEDY
    90th Nat. Meeting of the American Institute
    of Chemical Engineering
    April 1981 - Houston.

(5) A. MUAN
    Fossil Energy Material Program - March 31 - 1981.

(6) E.R. FULLER & Alias.
    Fossil Energy Material Program - March 31 - 1981.

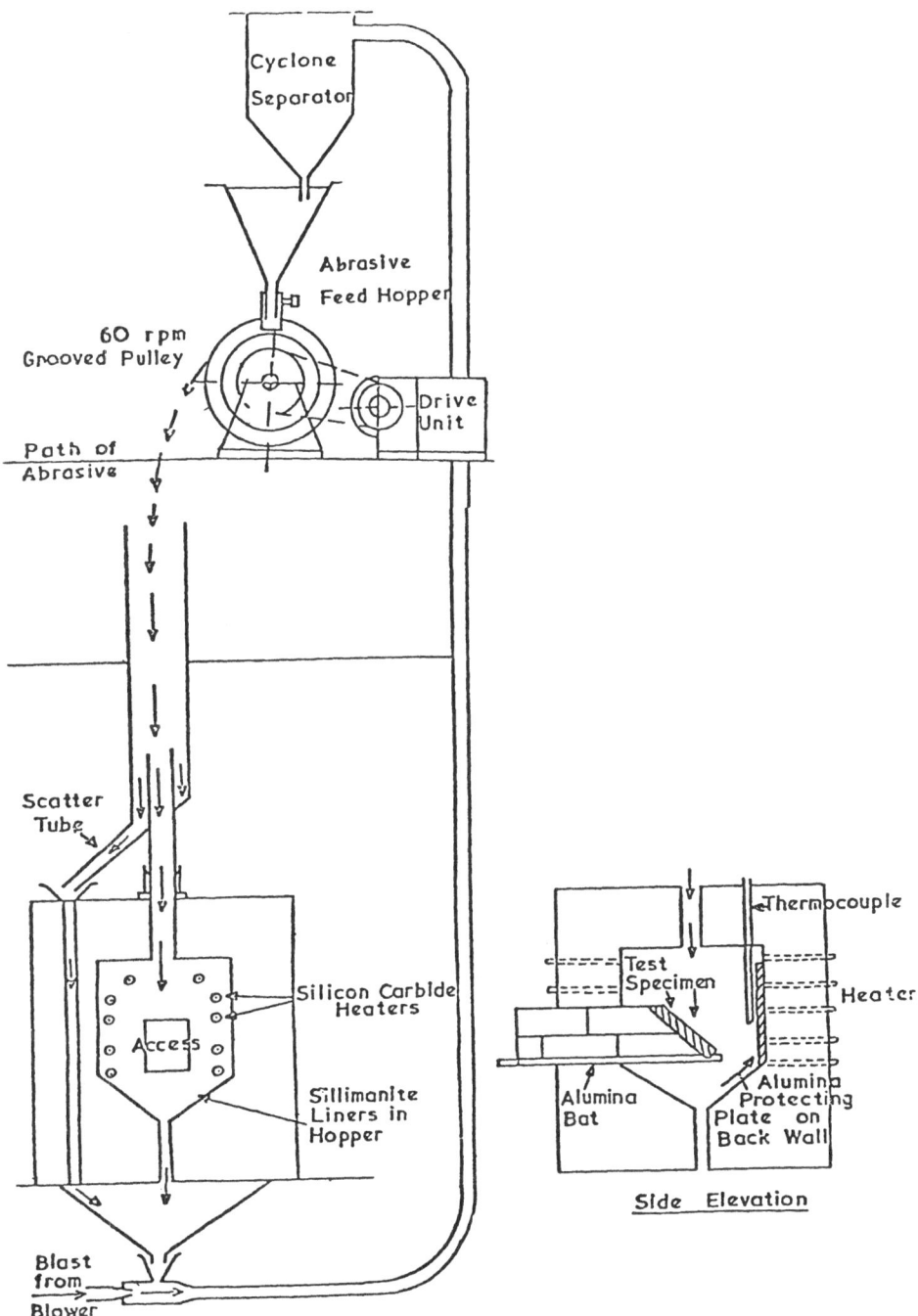

Fig. 1   Improved high temperature version of Morgan Marshall test
apparatus.

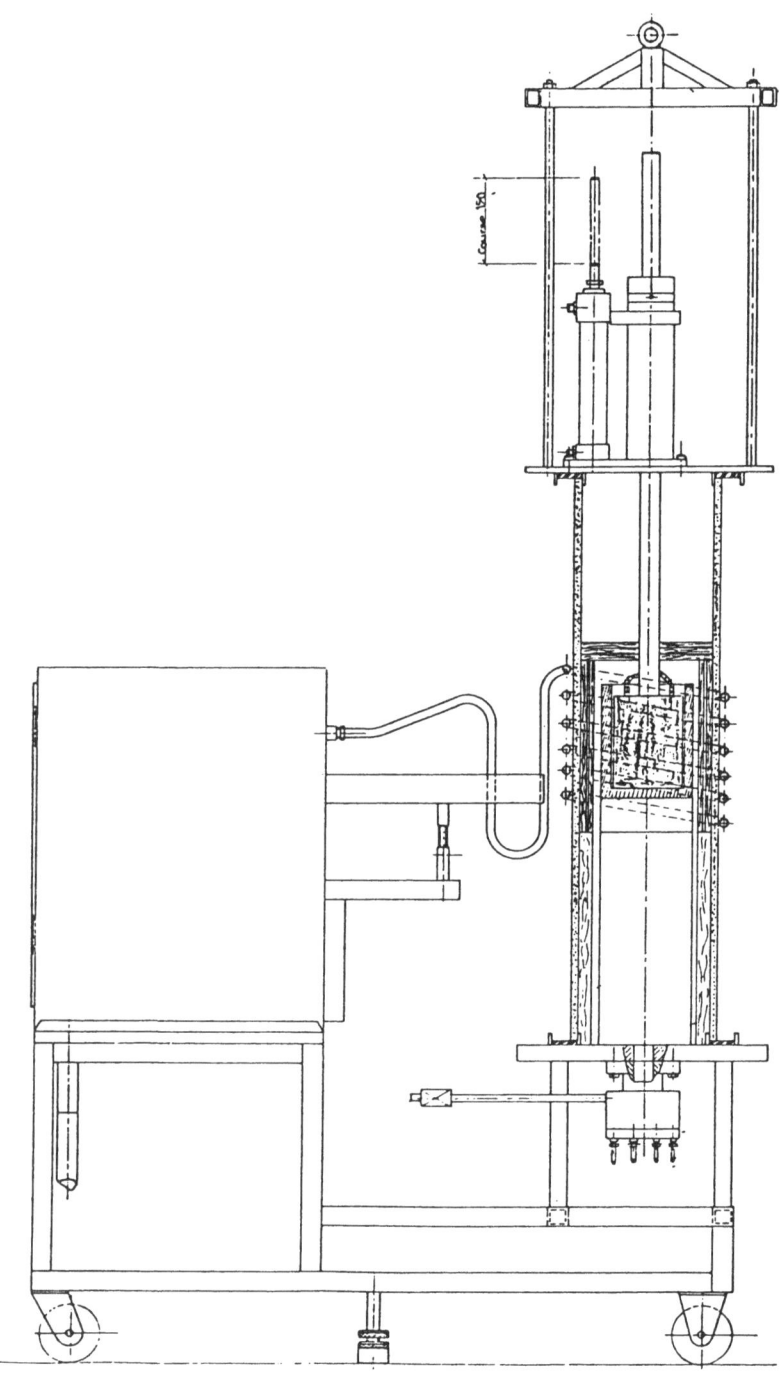

Fig. 2 Finger Test Apparatus.

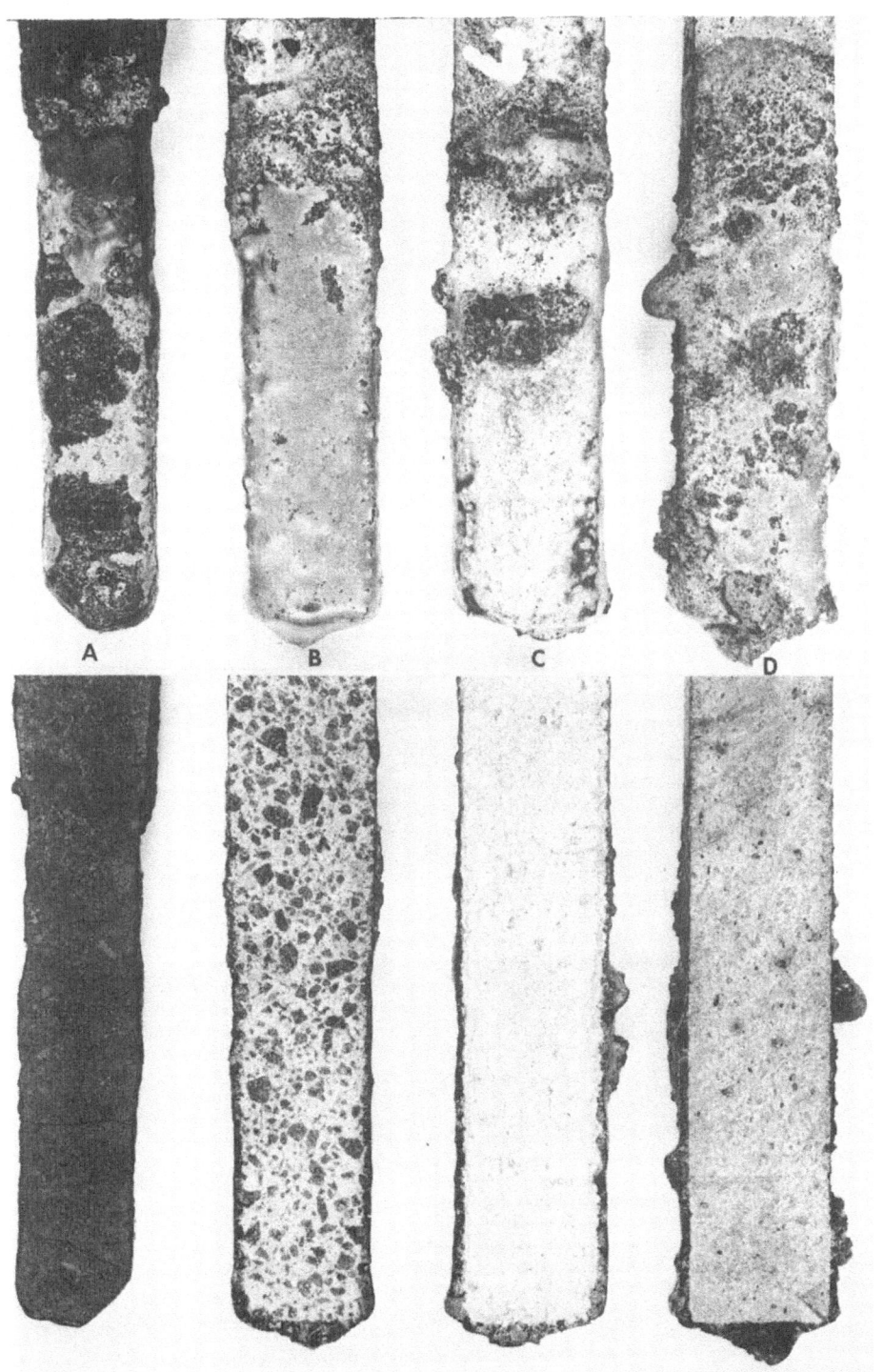

A   B   C   D

Fig. 3   "Finger" test samples: Coal ash corrosion at 1650°C, 4 hrs.

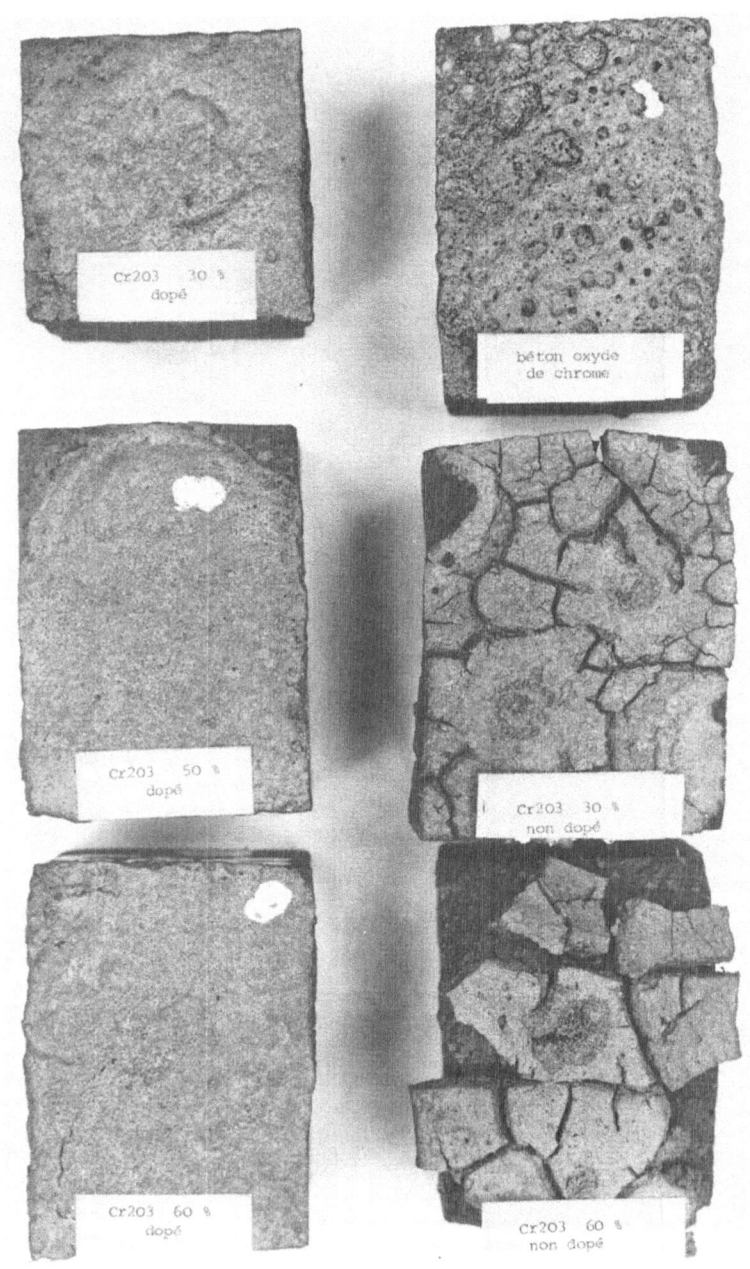

Fig. 4   Results from rotary furnace slag tests.

	CORUNDUM/ MULLITE	LC-CASTABLES
Chemical analysis : SiO2	10 / 12	5 / 6
A1203	85 / 87	90 / 92
Fe2O3	0.1	0.1
TiO2	0.1	-
CaO	0.1	1.2
MgO	-	-
Na2O	0.3	0.3
Bulk density (g/m3)	2.8 / 3.0	3.1 / 3.2
Apparent porosity (%)	15 / 18	10 / 15
Crushing strength (N/mm2) :		
20°C	110	110
1500°C	15 / 20	7 / 10
Thermal expansion	$7 . 10^{-6}$	$6 . 10^{-6}$
Thermal conductivity (W/mm) :		
800°C	1.9	1.8
1000°C	1.8	1.9
1200°C	1.4	2.0

Fig. 5   Main characteristics of mullite bonded and thixotropic castables fused alumina.

```
CHEMICAL ANALYSIS : Al2O3          84   / 86

                     Cr2O3          9    / 10

                     SiO2           0.5 /  0.6

                     Fe2O3          0.2 /  0.3

                     TiO2           2.4 /  2.6

                     CaO            1.1 /  1.2

                     MgO            0.2 /  0.3

                     Na2O           0.1 /  0.2
```

PHYSICAL PROPERTIES :

After firing at :	110°C	1000°C	1600°C
. Apparent porosity (%)	3.55	3.50	3.45
. Cold crushing strength (kg/cm2)	6/8	8/10	10/12
. Permeability (np)	0.1	–	0.1

HIGH TEMPERATURE PROPERTIES :

. Crushing strength at 1700°C (kg/cm2)   400

. Mean thermal expansion : 1500°C

$8.6 \cdot 10^{-6}$

. Permanent linear change (%)
   after firing at 1650°C   + 0.6

. Creep test 24 h – 1550°C – 5 b.   < 0.2

. Thermal conductivity in
   Kcal.m/m2.h.°C at 1000°C   2

Fig. 6   Chrome Oxide Castable.

	Si3N4 / Al2O3	Si3N4
	75 % Al2O3 25 % Si3N4	70 % Si3N4 β 30 % Si3N4 α
Bulk density (g/cm3)	3.0 / 3.1	2.4
Apparent porosity (%)	16 / 18	18 / 20
Cold crushing strength (N/mm2)	80 / 100	200
Thermal expansion	$4.7 \cdot 10^{-6}$	$2.7 \cdot 10^{-6}$
Thermal conductivity (W/mm) :		
500°C	2.0	12
1000°C	2.4	10
1500°C	2.8	8
Modulus of rupture at 1500°C under N2 (kg/cm2)	25	40

Fig. 7   Principal properties of refractories with nitride and sialon binders in a corundum matrix.

Investigations and design considerations for
refractories for slagging coal gasification

Lim Kong Hoa
Uhde GmbH
Dortmund, FRG.

As an introduction to the comments on the three
papers presented, let me briefly review our
activities in this field, which date back to the
late seventies. As part of its contribution to
making the industrial-scale engineering of the
advanced coal gasification technology a successful
enterprise, Uhde launched a refractory research
programme of its own. We have been involved in the
development of two types of coal gasification
processes starting from the planning stage of the
pilot plant, namely the Texaco process in
cooperation with Ruhrchemie/Ruhrkohle and the
fluidized-bed HT Winkler process in cooperation
with Rheinische Braunkohlenwerke AG. These comments
give a brief summary of our paper entitled
''Investigations and design considerations for the
refractory lining of coal gasifiers'', which is to
be published.

The coal slag

Table 1 and Table 2 show the composition of the
coal contaminants prior to and after gasification.
The ternary system shown in Fig. 1 may be used for
slag with a low iron content. The behaviour of
those slags with increasing calcia-silica ratio is
changing from the anorthite towards the gehlenite
region. With an increasing iron oxide content of
the slag, the anorthite-gehlenite-wustite system
becomes significant. The melting and fluid flow
behaviour of coal slag depends to a great extent on
the composition of the slag, the gaseous
environment as well as the temperature, as can be
seen in Fig. 2. Curve no. 3 shows the viscosity of
coal slag with a very high iron oxide content,
which was measured in a gaseous nitrogen
environment.

The refractories

Fig. 3 shows the stability of relevant refractory
oxides. The solubility of these oxides in the
molten slag is presented in Fig. 4. These
thermodynamic considerations lead to the following
three conclusions: firstly only a refractory with a
high Cr_2O_3 content could be expected to have a high
corrosion resistance to molten coal slag, secondly,
corrosion cannot be precluded completely, and
thirdly, the brick microstructure is likely to be
complex.

The refractories investigated are shown in Table 3.

The investigations

Our investigations were carried out in cooperation with TH-Aachen (5, 6). The crucible test and pellet reactions methods were used. The types of slag and the reaction conditions are shown in Table 4 and Table 5, respectively. Interpretations were done by means of x-ray diffractometer, microscopy and x-ray dispersive energy analysis. The results indicate that the molecular or ionic migrations from and to the refractory are the main source of corrosion on the boundary surface between the slag and the refractory. The migration of sesquioxide components, i. e. Cr_2O_3, Al_2O_3, Fe_2O_3, as well as MgO, FeO and ZrO_2 is critical to the brick. The main corrosion mechanism is summarized in Table 6.

We also tested the relative corrosion of a brick in a rotary furnace at a temperature of 1500 °C in the presence of H_2, CO, CO_2 and H_2O vapour. The results confirm the thermodynamic prediction that a high Cr_2O_3 content is responsible for the corrosion resistance to low-iron coal slag. Other investigations (7, 8) reveal a similar trend. Cornils and others (9) have reported on experience with picrochromite bricks in the pilot plants of Ruhrchemie-Ruhrkohle, West Germany. A corrosion rate of 0.01 mm/h is claimed. We also tested and measured quantitatively the thermoshock resistance of the bricks.

Design Considerations

As corrosion cannot be precluded completely, the design considerations must be strongly guided by the kinetical aspects in order to keep the corrosion rate as low as possible. The rate of corrosion is $r = f(\frac{1}{\eta}, S, T.V)$
where η = viscosity, S = solubility of refractory in slag,
T = temperature, V = slag quantity. For a given coal and a given gasifier capacity, the rate depends to a great extent on the viscosity and the temperature. The basic idea of any design consideration is therefore to keep the viscosity at the refractory/slag surface boundary at a sufficiently high level.

In order to achieve a long service life of the refractory lining it is necessary to have a precise refractory specification, proper aerodynamic and temperature distribution within the gasifier, optimum definition of process parameters and correct engineering of the bricklining. We expect an overall service life of about two years in our design of the Texaco gasifier.

In conclusion, I would like to point out how important it is for engineers dealing with slagging coal gasification to have a thorough understanding of the science of refractory materials.

References:

1. Schulze, R. :
    Energie 5, (11), 1953

2. Albrecht, W. and Pollmann, S.:
    VGB Kraftwerkstechnik, 60, 90 - 97 (2) 1980

3. Levin, E. M., et. al.:
    Phase diagram for ceramists, 1964

4. Richardson, F.D. and Jeffes, J. H. E.:
    J. Iron Steel Inst., 261 - 270, Nov. 1948

5. Dietrich, P. and Krönert, W.:
    Uhde-Auftrags-Nr. 0198 10041, Bericht Inst.
    f. Gesteinshüttenkunde, TH-Aachen 1981

6. Dietrich, P.:
    TH-Aachen, private communications, 1981

7. Kennedy, C. R. and Poeppel, R. B.:
    Interceram, 221 - 226, Nr. 3, 1978

8. Bonar, J. A., Kennedy, C. R. and Swaroop, R. B.:
    Ceramic Bulletin, 473 - 478, Vol. 59 No. 4, 1980

9. Cornils, B., et. al.:
    Radex-Rundschau, 707 - 723, Heft 1/2, 1982

10. Chaurey, K. H. and Sharma, K. C.:
    Proceeding of TVA's symposium ammonia from coal,
    1978

338

1. Clay minerals:
   Kaolinite: $Al_2O_3 \cdot 2SiO_2 \cdot 2H_2O$
   Illite: $K_2O \cdot 3 (Al, Fe)_2O_3 \cdot 16\ SiO_2 \cdot 4H_2O$

2. Carbonates:
   Calcite: $CaCO_3$
   Dolomite: $CaMg (CO_3)_2$
   Siderite: $FeCO_3$

3. SiO_2:
   Quarz: SiO_2
   Chalcedon: SiO_2

4. Sulfides:
   Pyrites: $Fe\ S_2$
   Marcasite: $Fe\ S_2$

5. Sulfates:
   Gypsum: $CaSO_4 \cdot 2H_2O$

6. Other minerals:
   Feldspar: $(K, Na)\ AlSi_3O_8$
   Hematite: Fe_2O_3
   Salt: $NaCl$
   Rutile: TiO_2

UHDE	TYPICAL MINERALS IN COAL (2)	TABLE 1

Compositions (Wt. %)	Texaco (9) Various coals	K-Totzek (10) Talcher / India	HT-Winkler Brown coal
SiO_2	37 − 60	61.2	6.6
Al_2O_3	16 − 33	18.2	0.9
TiO_2	0.9 − 1.9	1.5	0.3
Fe_2O_3	4 − 25	11.9	10.8
CaO	3 − 11	5.4	54.7
MgO	1.2 − 2.9	1.4	6.8
K_2O_3	0.3 − 3.6	−	0.2
Na_2O	0.2 − 1.9	−	1.2
P_2O_3	0.1 − 2.4	−	−
SO_3			18.7

UHDE	SLAGS OF GASIFIED COALS	TABLE 2

Composition Wt. %	1 fused	2 sintered.	3 sintered.	4 fused	5 fused	6 sintered.	7 sintered.
Cr_2O_3	80	76 – 80.0	60	28	27	8 – 10	99.8
Al_2O_3	4.5	1.0	40	28	60	90 – 92	
MgO	8	18.0	0.1	–	7		
ZrO_2	–	–	11 – 12	28	–		
SiO_2	1.3	0.50	6 – 7	6 – 7	1.8		
Fe_2O_3	6.1	1.50	0.1	0.1	4.2		
TiO_2	–	–	0.7 – 0.8	0.7 – 0.8	–		
Na_2O	0.1	–	0.5	0.5	0.3		
CaO	–	0.5	0.1	0.1	–		
Mineral, %							
$(Cr, Al, Fe)_2O_3$	52	95	75	53	60 – 65	Korund s.s.	Korund
$(Mg, Fe) O (Cr, Al, Fe)_2O_3$	40		–	–	40 – 45		
ZrO_2	–		15	27			
Glassy phase	5		10	20	2		
Metallic phase	3	–	–	–	3		
Density, (g/cm^3)	4.09	3.70	3.65	4.0	3.1	3.2	3.2
Open porosity, %	5.4	15.0	10 – 15	3	4.2	15 – 20	18

INVESTIGATED REFRACTORIES

TABLE 3

UHDE

Components	Wt. %	Wt. %	Wt. %
SiO_2	50,9	26,9	11,8
Al_2O_3	28,3	15,0	6,5
CaO	12,6	53,8	2,8
Fe_2O_3	3,9	2,1	77,8
$MgO + TiO_2$	3,5	1,8	0,9
$K_2O + Na_2O$	0,8	0,4	0,2
	100,0	100,0	100,0
$\dfrac{CaO}{SiO_2}$ — Ratio	0,25	2,0	0,24
$\dfrac{FeO}{100}$ — Ratio	0,02	0,01	0,40

UHDE	SLAG COMPOSITIONS	TABLE 4

Gas component	Vol. %
H_2	22
CO	28
CO_2	17
H_2O	33
Pressure, Atm	1
Temperature, °C	1500

UHDE	REACTION CONDITIONS	TABLE 5

Refractory grain Cr₂O₃ - rich in refractory	New boundary layer formation	Migration: from Refractory into slag / from slag to refractory
Escolaite (Cr_2O_3)	FeO. $(Cr, Al, Fe)_2O_3$	Cr^{3+} → $Al^{3+}, Fe^{2+}, Fe^{3+}$ →
Sesquioxide solid solution ($Cr, Al, Fe)_2O_3$	FeO. $(Cr, Al, Fe)_2O_3$	Cr^{3+} → Fe^{2+} → Al^{3+}, Fe^{3+} → ←
Complex spinel $MgO (Cr, Al, Fe)_2O_3$	$(Mg, Fe)O. (Cr, Al, Fe)_2O_3$	Mg^{2+}, Cr^{3+} → Fe^{2+} → Al^{3+}, Fe^{3+} → ←
Picrochromite $MgO. Cr_2O_3$	$(Mg, Fe)O. (Cr, Al, Fe)_2O_3$	Mg^{2+}, Cr^{3+} → $Al^{3+}, Fe^{2+}, Fe^{3+}$ → ←

UHDE	SLAG—REFRACTORY INTERACTIONS	TABLE 6

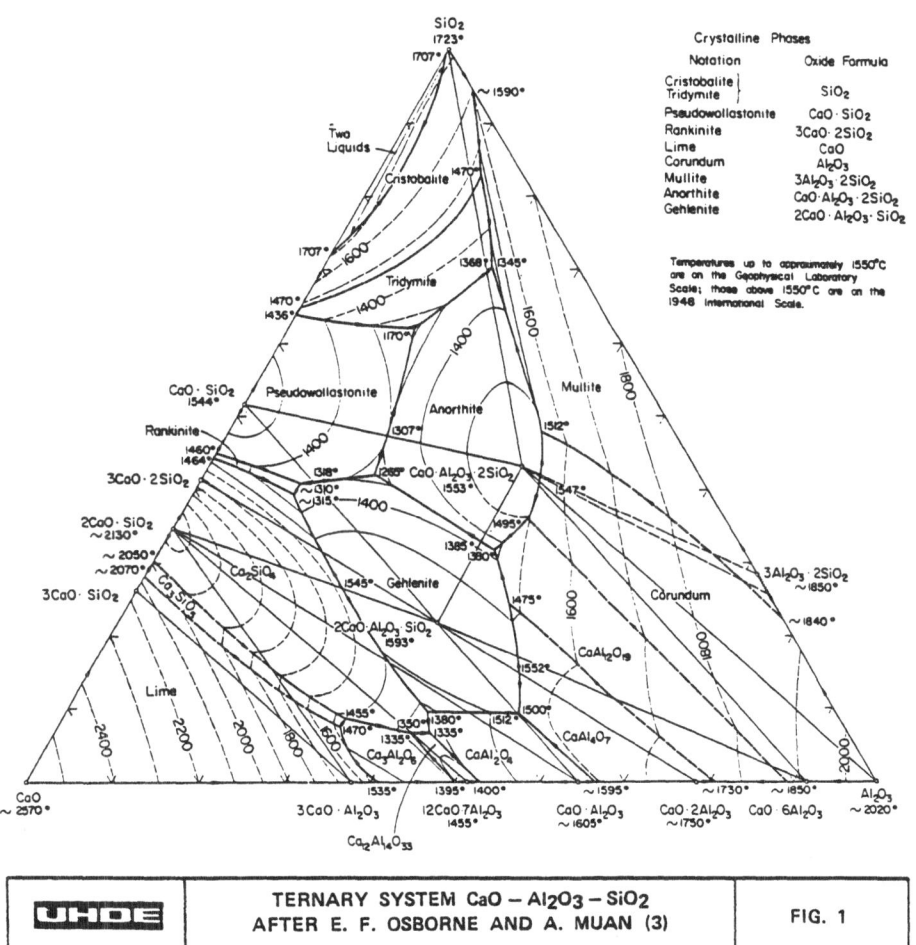

SiO₂
1723°
1707°

~1590°

Crystalline Phases

Notation	Oxide Formula
Cristobalite / Tridymite	SiO₂
Pseudowollastonite	CaO · SiO₂
Rankinite	3CaO · 2SiO₂
Lime	CaO
Corundum	Al₂O₃
Mullite	3Al₂O₃ · 2SiO₂
Anorthite	CaO · Al₂O₃ · 2SiO₂
Gehlenite	2CaO · Al₂O₃ · SiO₂

Temperatures up to approximately 1550°C are on the Geophysical Laboratory Scale; those above 1550°C are on the 1948 International Scale.

Two Liquids

Cristobalite

1707° · 1600
1470° 1436°

Tridymite

1368° 1345°

1600

1400

1170°

CaO · SiO₂
1544°

Pseudowollastonite

Anorthite

Mullite

1307°

1512°

Rankinite
1460°
1464°

3CaO · 2SiO₂

1318°
~1310°
~1315° 1400

1265° CaO · Al₂O₃ · 2SiO₂
1553°

1547°

2CaO · SiO₂
~2130°

~2050°
~2070°

Ca₂SiO₄

1495°

1385°
1380°

3Al₂O₃ · 2SiO₂
~1850°

3CaO · SiO₂

Ca₃SiO₅

1545°

Gehlenite

Corundum

~1840°

Lime

2CaO · Al₂O₃ · SiO₂
1593°

1475°

CaAl₂O₁₉

1552°

2400 2200 2000 1600

1455°
1470° 1350° 1380°
1335° 1335°
Ca₃Al₂O₆ CaAl₂O₄

1512°

1500°

CaAl₄O₇

1800

2000

CaO
~2570°

1535°

3CaO · Al₂O₃

1395° 1400°
12CaO 7Al₂O₃
1455°

~1595°

CaO · Al₂O₃
~1605°

~1730°

CaO · 2Al₂O₃
~1750°

~1850°

CaO · 6Al₂O₃

Al₂O₃
~2020°

Ca₁₂Al₁₄O₃₃

	TERNARY SYSTEM CaO – Al₂O₃ – SiO₂ AFTER E. F. OSBORNE AND A. MUAN (3)	FIG. 1
UHDE		

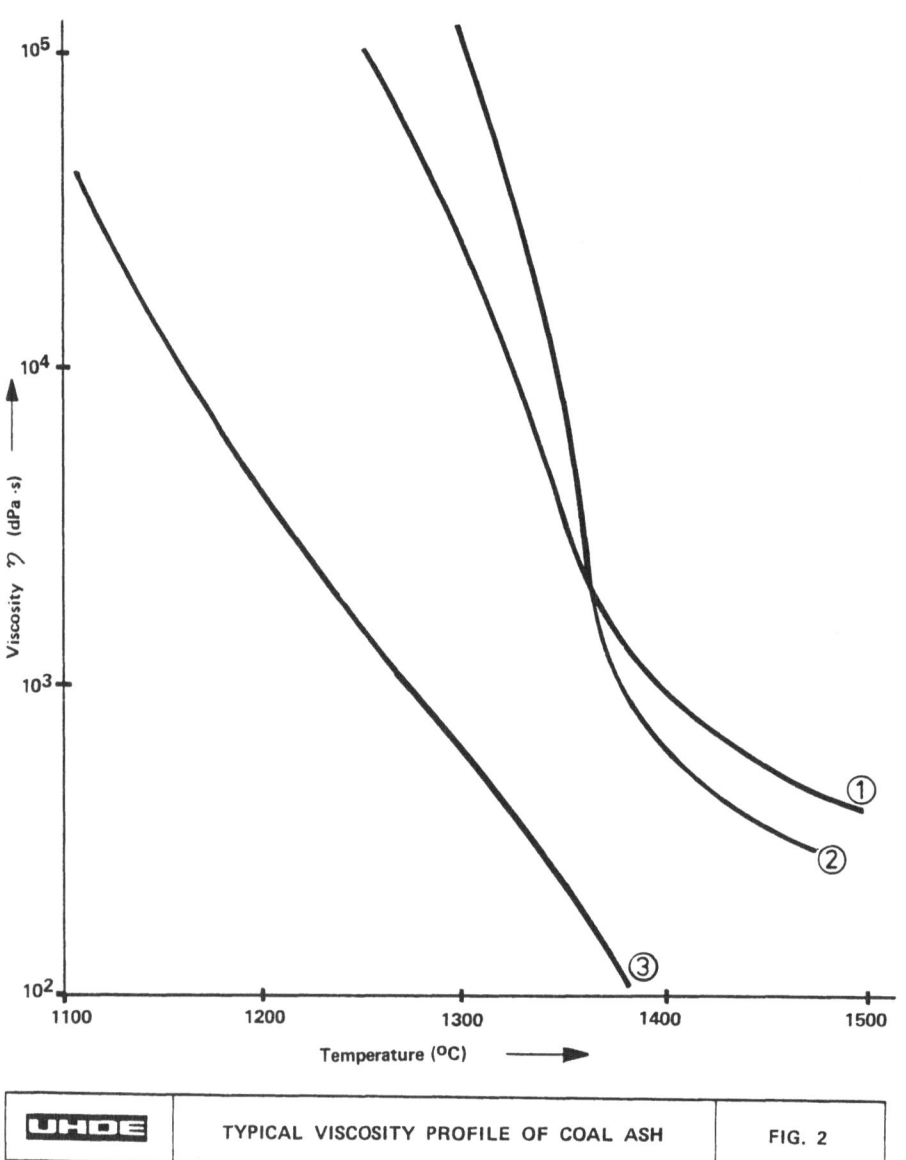

TYPICAL VISCOSITY PROFILE OF COAL ASH — FIG. 2

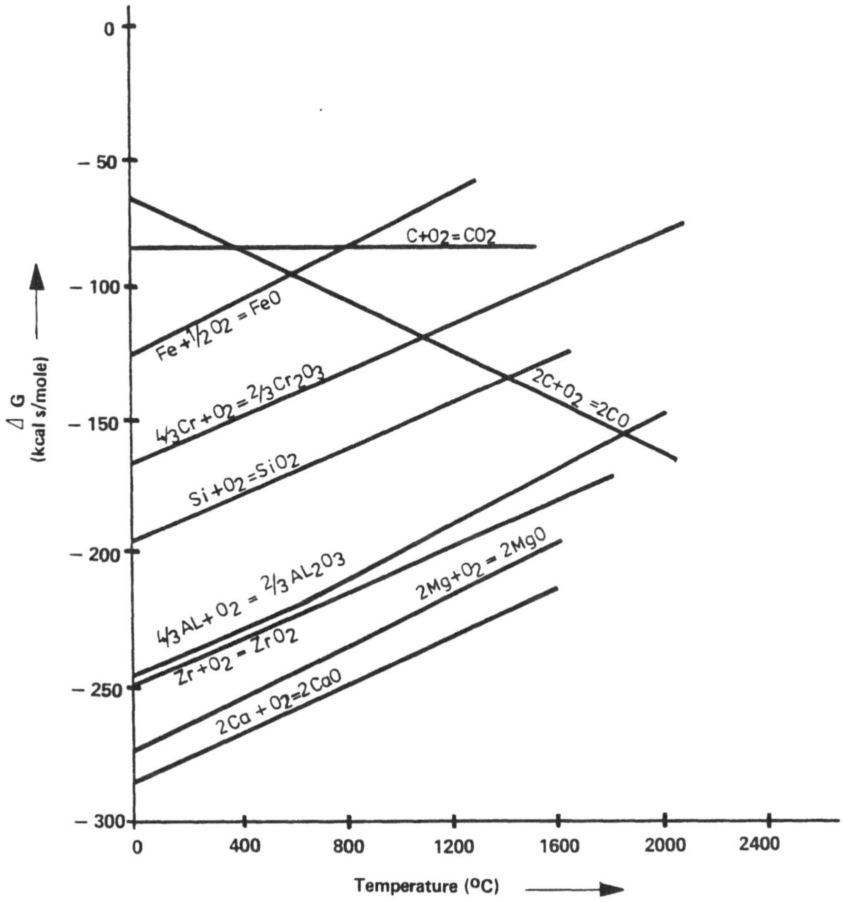
0
- 50
- 100
- 150
- 200
- 250
- 300
Δ G
(kcal s/mole)
C+O2=CO2
Fe+1/2O2 =FeO
4/3Cr+O2=2/3Cr2O3
2C+O2 =2CO
Si+O2=SiO2
4/3AL+O2 = 2/3AL2O3
Zr+O2 = ZrO2
2Mg+O2 = 2MgO
2Ca + O2=2CaO
0
400
800
1200
1600
2000
2400
Temperature (°C)

UHDE	STANDARD FREE ENERGY OF REFRACTORY OXIDES	FIG. 3

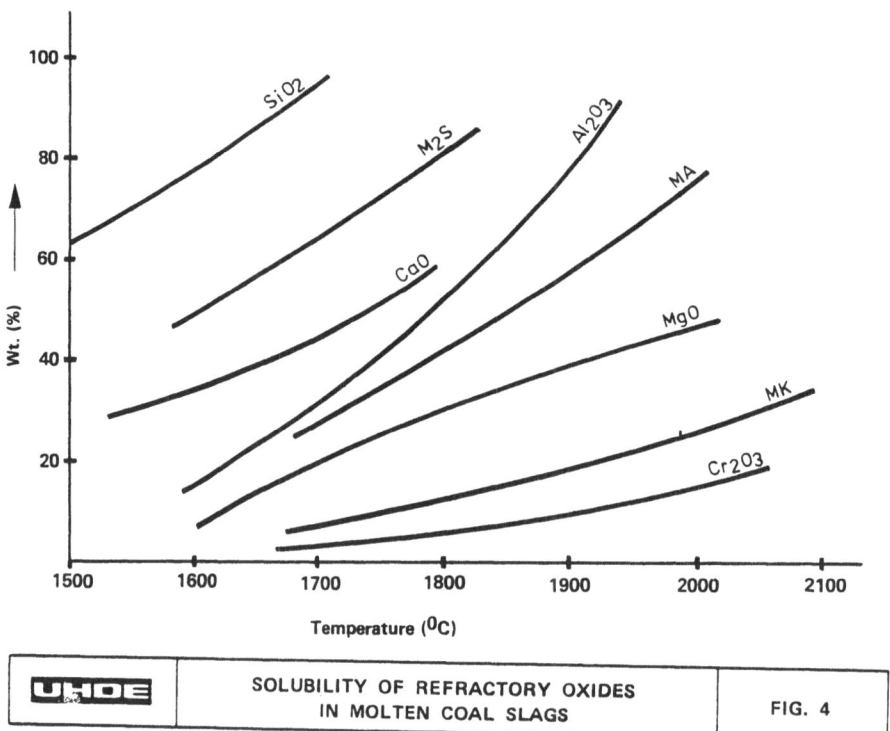

Wt. (%)

Temperature (⁰C)

SiO2 M2S Al2O3 MA CaO MgO MK Cr2O3

| UHDE | SOLUBILITY OF REFRACTORY OXIDES IN MOLTEN COAL SLAGS | FIG. 4 |

DISCUSSION

F. Starr: I have a question for Dr. Lim who presented work on slag attack of refractories. Could he give the constants K, a, b and c and units in the equation

$$R = K . \frac{1}{\mu a} . S \ T^b . V^c$$

where
- R = rate of slag attack
- K = constant
- μ = viscosity of slag
- S = solubility of refractory in the slag
- V = slag volume
- a, b, c = constants.

K.H. Lim: It is a correlation on rate of slag attack on refractory where $a \sim 0,89$, $b \sim 0,67$, $c \sim 0,11$, $k \sim 1,0$
- R = mm/h
- S = solubility, weight fraction
- μ = viscosity, poise
- V = slag volume per unit area per unit time
- T = temperature, °K.

The correlation has empirical character, investigated and derived mainly from experiences in steel processing furnaces.

THERMODYNAMIC ASPECTS OF THE APPLICATION OF CERAMICS/REFRACTORIES

IN ADVANCED ENERGY TECHNOLOGIES

Arnulf Muan

The Pennsylvania State University
College of Earth and Mineral Sciences
University Park, Pennsylvania, USA, 16802

SYNOPSIS

The present paper is concerned with equilibrium relations in refractory materials and their reaction products with coal-ash slags in slagging coal gasifiers. Phases containing chromium oxide have shown exceptional resistance to corrosion in such gasifiers. High-temperature equilibrium data on systems involving chromium oxide as a component at various oxygen partial pressures are discussed. Particular attention is given to the presence of Cr^{2+} in silicate melts under strongly reducing conditions. It is shown that the latter may cause large decreases in liquidus and solidus temperatures compared to those prevailing under more oxidizing conditions.

Great progress has been made in recent years in determining and interpreting thermodynamic properties of oxide and silicate solid solutions at high temperatures. Activity-composition relations in several refractory oxide phases involving chromium oxide as a component have been determined experimentally, and are discussed in the present paper. General conclusions are drawn regarding the importance of volume differences in determining the degree of deviation from ideality of oxide and silicate solid solutions.

INTRODUCTION

Advanced energy technologies are placing increasingly severe demands on the properties of the materials used in these technologies. Commonly, the materials are exposed to high temperatures and corrosive environments. Slagging coal gasifiers are examples of such environments: The refractory linings of the vessels are subjected to the corrosive action of coal-ash slags, in a temperature range which may lead to partial or complete melting of the refractory, and in atmospheres which may lead to precipitation of a metal phase (notably iron) from the slag, reduce some of the ions of slag and refractory phases to unusual oxidation states, or cause high volatilization losses of some of the oxide constituents.

The oxygen potentials of the atmospheres associated with common combustion practices in slagging coal gasifiers are commonly in the range of $\sim 10^{-10}$ atm. at 1400-1500°C, although locally, such as in the vicinity of unburned carbon particles, the conditions may be much more reducing. What are the implications of these levels of oxygen pressures as far as refractory behavior is concerned? The oxygen pressures are too high to make carbides, nitrides and oxynitrides thermodynamically stable under the prevailing conditions, yet are sufficiently reducing to cause reduction of some components into unusual oxidation states or precipitation of a metal (e.g. iron) from the oxide phases. By and large, however, it appears that oxides are the most promising refractory materials for use in slagging gasifiers unless ways can be found to protect carbides and nitrides from continued oxidation in the gasifier environment.

In the present paper, we will first consider briefly the thermodynamic relations attending the possible use of SiC as a refractory material in slagging coal gasifiers and report on some results obtained in laboratory experiments involving the presence of SiC with various oxide phases under atmospheric conditions approximating those commonly encountered in slagging coal gasifiers. This will be followed by a consideration of phase relations in some oxide and silicate systems of importance for an understanding of reactions between coal-ash slags and refractory oxide materials. Among the latter, those containing chromium oxide as a component will be given particular attention. Finally, activity-composition relations in oxide and silicate phases will be discussed, dealing with general aspects of such relations as well as specific examples pertaining to chromium oxide-containing phases of particular importance in coal gasifiers.

EQUILIBRIUM ASPECTS OF THE USE OF CARBIDES AND NITRIDES AS REFRACTORIES IN SLAGGING COAL GASIFIERS

Silicon carbide and similar materials have some attractive properties relative to their potential use as refractory materials in coal gasification processes, e.g. they have high thermal stability, and are relatively "inert" to most silicate melts, unless the latter contain significant amount of transition-metal oxides such as iron oxide. There is, however, at least one serious chemical problem relative to the possible application of these materials as refractory linings, viz. their poor oxidation resistance. Some simple thermodynamic data illustrate this point. First, note that the standard free energy of formation of SiC from its constituent elements is a relatively small negative value, ~-10 Kcal. Hence, its susceptibility to oxidation is nearly as large as those of the constituent elements, Si and C. Silicon carbide, as well as silicon nitride (Si_3N_4) and silicon oxynitride (Si_2O_2N), are therefore far from being equilibrium phases at the oxygen potentials prevailing in typical coal gasification atmospheres. This is illustrated in Figs. 1 and 2, showing interrelations between oxygen potentials of typical coal gasification atmospheres and stability ranges of the various carbide-, nitride- and oxynitride phases as a function of temperature. The first diagram (Fig. 1) illustrates relations for SiC. The solid curve in the lower part of the diagram represents the equilibrium between SiO_2 (stable in the area above the curve) and SiC (stable in the area below the curve) at a CO pressure of 1 atm. The dash curves above and below the solid curve depict the same equilibrium (i.e. coexistence of SiO_2 and SiC) at CO-pressures of 50 and 0.02 atm., respectively. It is seen that over the entire temperature range covered (~800-1600°C), and at the three CO pressures used as examples, the SiO_2/SiC equilibrium curves are located far below the shaded area representing approximate oxygen potentials (in terms of the function R'T log_{10} P_{O_2} as plotted along the vertical axis) of typical coal gasification atmospheres.

Similar relations apply in the case of silicon nitride and oxynitride, as demonstrated in Fig. 2. The solid curves in the lower part of the diagram show, from top to bottom, the equilibrium curves for SiO_2/Si_2N_2O, Si_2N_2O/α-Si_3N_4 and α-Si_3N_4/β-Si_3N_4, in that order, at an N_2 pressure of 10^{-2} atm. The dash and dash-dot curves represent the SiO_2/Si_2N_2O equilibrium at N_2 pressures of 1 atm. and 10^{-4} atm., respectively. It is seen that the SiO_2/Si_2N_2O curves, at all three N_2 pressures considered, are located far below the shaded area representing approximate oxygen potentials of typical coal gasification atmospheres. Hence, the typical coal gasification atmospheres are strongly oxidizing relative to these nitride- and oxynitride phases.

The suitability of SiC, Si_3N_4 and Si_2N_2O as refractory materials in coal gasification processes, therefore, depends on whether the rates of oxidation can be maintained at an acceptably low level, either by using cooling, or by otherwise inhibiting the reaction.

Reactions between SiC or Si_3N_4 and simplified slags involving the components CaO, MgO, Na_2O, Al_2O_3, SiO_2 have shown that only small to moderate depressions of the known liquidus temperatures of the oxide phases takes place.[1] This is an indication that relatively small amounts of C and N are dissolved in the largely oxidic liquids present. At elevated pressures (e.g. 50 atm.), these effects are undoubtedly somewhat larger than at 1 atm., but they are still small. There is one component of coal-ash slags, however, which behaves quite differently, viz. iron oxide (FeO and/or Fe_2O_3). This oxide is much less stable relative to its constituent elements than are the oxides previously mentioned (CaO, MgO, Al_2O_3, SiO_2), as illustrated by the standard free-energy curves ($\Delta G°$) shown in Fig. 3.[2] Hence, in the case of FeO, this component of the slag will have a tendency to react with SiC according to the simplified equation

$$SiC + 3(FeO) = SiO_2 + 3[Fe] + CO \qquad (1)$$

where parenthesis and brackets indicate species dissolved in the silicate slag and in a metal phase, respectively. The change in standard free energy for this reaction is relatively small ($\sim$−20 Kcal). Hence equilibrium between SiC, the oxidation product SiO_2, metallic iron and FeO of the slag is reached at a moderately low FeO-concentration of the slag. Metallic iron will tend to be precipitated from the slag at or near the SiC refractory-slag interface until the FeO activity has been lowered to the equilibrium value required by equation (1).

That this reaction actually takes place at the SiC-slag interface has been shown in testing of SiC brick corrosion by actual coal-ash slags[3] and by laboratory reactions between SiC and FeO-containing silicate melts,[4] as sketched in Fig. 4.

The reaction between SiC and FeO of the slag, as described above, serves to remove part of the FeO component from the slag. The equilibrium relations in the remaining oxide system may then, as a first approximation, be discussed in terms of the system CaO-Al_2O_3-SiO_2. This is shown in Fig. 5,[5,6] illustrating the liquidus surface of this system and showing as a black circle an example of the approximate composition of a FeO-depleted coal-ash slag in terms of the components used in this diagram. The solidus temperatures for a mixture of this composition, and consisting mainly of anorthite, with lesser amounts of silica and wollastonite, is $\sim$1100°C. It is inferred that it is desirable to cool the SiC-slag interface to temperatures below this level in order to provide better protection for a SiC refractory against oxidation-corrosion by coal-ash slags.

Inasmuch as the reaction between SiC and the FeO of the slag removes only a part of the FeO component of the slag, a more accurate picture of equilibrium relations attending the reaction is afforded by consideration of phase relations in parts of the system CaO-"FeO"-Al_2O_3-SiO_2, as illustrated in Figs. 6-7. The first of these (Fig. 6)[2,7] is a perspective drawing of a tetrahedron representing this system and showing the various composition volumes. The parts of the tetrahedron of greatest interest in the present research are the volumes and the bounding triangles in the vicinity of the SiO_2 apex, i.e. $CaAl_2Si_2O_8$-$CaSiO_3$-FeO-SiO_2. Phase relations in one of the ternary systems bounding this quaternary system are portrayed in Fig. 7[7]. The solidus temperatures are of the order of $\sim$1080-1120°C, thus establishing an upper limit for the temperature that may be tolerated without severe fluxing-oxidation of a SiC refractory by coal-ash slags.

EQUILIBRIUM RELATIONS IN CHROMIUM OXIDE-CONTAINING SYSTEMS OF IMPORTANCE IN SLAGGING COAL GASIFIER TECHNOLOGY

It is clearly desirable to use as refractory linings of coal gasifiers only those materials that are thermodynamically stable in the prevailing atmospheres. Hence, very stable oxides, i.e. those with a large negative value of the standard free energy of formation of the oxide from its elements (see Fig. 3), are primary candidates for such linings. The choice must be further limited to those oxides which are compatible with or show good resistance to corrosion by coal-ash slags.

Testing of various commercial refractory materials in slagging coal gasifier environments has shown[8] that chromium oxide-containing materials display particularly strong resistance to corrosion by coal-ash slags. This is hardly surprising in view of the well-established fact that the solubility of chromium oxide in silicate melts is very small. We will show, however, that this low solubility applies only when conditions (oxygen pressure, temperature and basicity of the melts) are such as to keep most of the chromium as Cr^{3+}. The solubilities may increase manifold at very low oxygen pressures, where substantial proportions of the chromium may be present as Cr^{2+}, or in basic melts at relatively high oxygen pressures, for instance in air, where substantial proportions of the chromium may be present as Cr^{6+}.

In the following, we will review briefly* phase relations in some chromium oxide-containing systems in order to understand the behavior of this oxide under various atmospheric conditions, from very reducing ($\sim$10^{-14} atm P_{O_2}) to relatively oxidizing (air, 0.2 atm. P_{O_2}).

Consider first phase relations in simple binary and ternary chromium oxide-containing systems under two extreme conditions of oxygen potentials, viz. air and contact with metallic chromium, as shown in Figs. 8[2,9-11] and 9,[2,12,13] respectively.

It is seen that liquidus and solidus temperatures in these systems are drastically lower at the very low oxygen potentials,
*Detailed descriptions of the various systems will be published elsewhere.

where substantial proportions of the chromium is present as Cr^{2+} in the liquid phase.

Experimental studies of these systems have also been carried out at intermediate oxygen potentials, based on the observation that the presence of Cr^{2+} in amounts large enough to change significantly the phase relations observed in Cr_2O_3-containing systems is restricted to a range of very low oxygen pressures. It was considered essential that the extent of this range be established for at least a couple of systems. Toward this end, the solidus temperatures as a function of oxygen pressure were determined experimentally for the systems chromium oxide-SiO_2 and chromium oxide-Al_2O_3-SiO_2. The results are summarized in Fig. 10.[14] Superimposed on this diagram is a shaded area indicating the range of oxygen potentials usually encountered in coal gasifiers. It is clear that the oxygen partial pressures of the gas phase in slagging coal gasifiers usually is higher than those at which the presence of Cr^{2+} in the slag phase suppresses solidus (and liquidus) temperatures of the refractory significantly. However, Cr^{2+} may play a major role in the chemistry of coal-ash slags at the C-slag interface, especially at the highest temperatures ($\sim1600°C$) involved.

Among the most promising refractory materials for slagging coal gasifiers are Cr_2O_3-Al_2O_3 solid solutions and various chromium oxide-containing spinels, for instance $MgCr_2O_4$-$MgAl_2O_4$ solid solutions. In view of the fact that coal-ash slags as a first approximation may be represented within the ternary system CaO-Al_2O_3-SiO_2 (see previous section of this paper), a sound approach to developing an understanding of the chemistry of refractory-slag interactions in coal gasifiers is to combine the components of this system with Cr_2O_3 or $MgCr_2O_4$ as an additional component. This approach is illustrated in Fig. 11 showing a tetrahedral model for the system CaO-Al_2O_3-SiO_2-Cr_2O_3. Considering that compositions of coal-ash slags commonly fall approximately in the middle of this diagram, i.e. the range between the composition points of gehlenite ($Ca_2AlSi_2O_7$) and anorthite ($CaAl_2Si_2O_8$) (compare Fig. 5), the composition volume within the tetrahedron of Fig. 11 of particular interest in the present work is $Ca_2Al_2SiO_7$-$CaAl_2Si_2O_8$-Cr_2O_3-Al_2O_3. This volume, as well as the volume $Ca_2Al_2SiO_7$-$CaAl_2Si_2O_8$-$MgCr_2O_4$-$MgAl_2O_4$ in the system CaO-MgO-Al_2O_3-Cr_2O_3-SiO_2, are portrayed as regular tetrahedra in Fig. 12.

Phase relations in a number of planes through these tetrahedra have been determined in our laboratories, or are available in previous literature. A few examples are shown in succeeding diagrams. The first of these (Fig. 13) shows the systems $Ca_2Al_2SiO_7$-$CaAl_2Si_2O_8$-Cr_2O_3[15] and $Ca_2Al_2SiO_7$-$CaAl_2Si_2O_8$-Al_2O_3.[5,6,16-19] Attention is directed toward the very low solubility of chromium oxide (~0.5 wt.% Cr_2O_3) in the silicate liquid and the relatively low depression of liquidus temperatures in Fig. 13a. It is to be noted that the solubility of Al_2O_3 in the corresponding silicate liquids in the system $Ca_2Al_2SiO_7-CaAl_2Si_2O_8-Al_2O_3$ (Fig. 13b) is considerably higher.

Phase relations in the systems $Ca_2Al_2SiO_7-CaAl_2Si_2O_8-MgCr_2O_4$ and $Ca_2Al_2SiO_7-CaAl_2Si_2O_8-MgAl_2O_4$ are shown in Fig. 14.[15,20] Again, the lower solubility of chromium oxide compared with that of Al_2O_3 is the most noteworthy feature of the diagrams.

ACTIVITY-COMPOSITION RELATIONS IN OXIDE AND SILICATE PHASES OF IMPORTANCE IN SLAGGING COAL GASIFIERS

In order to treat quantitatively the equilibria among refractory and slag phases in coal gasification processes, it is necessary to know the thermodynamic properties of the phases involved. Such data also establish an important basis for future quantitative treatment of kinetic aspects of such reactions.

Great strides have been made in recent years toward better experimental data and better theoretical treatment of the thermodynamics of oxides and silicates. In the present paper we will limit the discussion to refractory phases of the type showing particular promise as materials for lining of coal gasification vessels, viz. $Al_2O_3-Cr_2O_3$ and chromium oxide-containing spinel phases.

A key to the determination of activities of chromium oxide in silicate melts has been the availability of thermodynamic data for Mo-Cr alloys, as determined in our laboratories recently.[21] With such data available, activity-composition relations for chromium oxide in various oxide and silicate solutions (solid or liquid) may be determined from equilibria of the type

$$(Cr_2O_3) = 2[Cr] + 3/2O_2(g) \qquad (2)$$

where parenthesis designates chromium oxide dissolved in an oxide or silicate phase, brackets designate chromium in a Mo-Cr alloy and g indicates species in a gas phase. The equilibrium expression may be written as

$$\frac{a_{Cr}^2 \cdot P_{O_2}^{3/2}}{a_{Cr_2O_3}} = K = \exp. \frac{\Delta G^0}{RT} \qquad (3)$$

where ΔG^0 is the standard free energy of formation of Cr_2O_3 from the elements, $\underline{a}$ with appropriate subscripts are activities of the various species in the alloy and oxide (silicate) phase(s), K is the equilibrium constant and P_{O_2} is partial pressure of oxygen. If a_{Cr} is known (from known activity-composition relations in Mo-

Cr alloys, see above), ΔG^0 is known from previous work in the binary system Cr-O, and P_{O_2} is measured experimentally, the activities of chromium oxide may be calculated as

$$a_{Cr_2O_3} = a_{Cr}^2 \cdot P_{O_2}^{3/2} \cdot \exp. - \frac{\Delta G^0}{RT} \qquad (4)$$

Caution must be exercised in formulating the most appropriate expressions for the ideal activity of the various species of ionic solid solutions. It has been shown[22,23] that for a solution $(A,B)_u Z_w$ involving mixing on only one atomic site, the equations become

$$a_{A_u Z_w} = (X_{A_u Z_w})^u \qquad (5)$$

and

$$a_{B_u Z_w} = (X_{B_u Z_w})^u \quad . \qquad (6)$$

When applied to an ideal sesquioxide solid solution, for instance Al_2O_3-Cr_2O_3, the expressions would be

$$a_{Cr_2O_3} = X_{Cr_2O_3}^2 \qquad (7)$$

or, if the equations are re-cast into a formulation involving only one atom of the exchangeable cations, a linear relationship of activity with composition for the ideal solution is obtained

$$a_{CrO_{1.5}} = X_{Cr_2O_3} \quad . \qquad (8)$$

Results obtained for Al_2O_3-Cr_2O_3 solid solutions at 1500 and 1600°C[24] are shown in Fig. 15. It is seen (Fig. 15) that the system displays considerable positive deviation from ideality. This can be correlated with the size difference between Al^{3+} and Cr^{3+} ions, analogous to relations which have been observed previously for oxide solid solutions of periclase-type structure.[23,25]

Knowledge of the thermodynamic properties of sesquioxide solid solutions can be used to determine activity-composition relations in more complex oxide phases, for instance spinels. Consider as an example the system MeO-Al_2O_3-Cr_2O_3, where MeO represents an oxide of periclase-type structure (e.g. MgO, FeO). Experimentally determined conjugation lines between coexisting spinel and sesquioxide solid solutions, in combination with known activity-composition relations for the Al_2O_3-Cr_2O_3 solid solutions, may be used to derive approximate activity-composition relations for the spinel solid solution.

In general, simple relations with modest positive deviation from ideality is displayed, at high temperatures, in spinels in which the solid solution involves either mainly the substitution of approximately equal-sized divalent cations on tetrahedral sites or mainly the substitution of approximately equal-sized trivalent cations on octahedral sites. In the former case, the ideal activity-composition relations may be expressed (using $MgCr_2O_4$-$FeCr_2O_4$ as example) as

$$a_{MgCr_2O_4}^{(IV)} = X_{Mg^{2+}} \tag{9}$$

and

$$a_{FeCr_2O_4}^{(IV)} = X_{Fe^{2+}} \quad . \tag{10}$$

In the latter case the ideal activity-composition relations may be expressed (using $FeAl_2O_4$-$FeCr_2O_4$ as example) as

$$a_{FeAl_2O_4}^{(VI)} = (X_{Al^{3+}})^2 \tag{11}$$

and

$$a_{FeCr_2O_4}^{(VI)} = (X_{Cr^{3+}})^2 \tag{12}$$

because two trivalent cations are involved in the substitution, for each formula unit of each of the spinel end member(s). Examples of experimentally determined activity-composition relations in spinel solid solutions, viz. $FeAl_2O_4$-$FeCr_2O_4$[26] and Fe_3O_4-Cr_3O_4,[27] are shown in Figs. 16-17. In the former solution, a moderate positive deviation from ideality is shown to exist, which may be correlated with the size difference between Al^{3+} and Cr^{3+}. In the second solution, Fe_3O_4-"Cr_3O_4" (Fig. 17), it may be assumed that the solution involves mainly the substitution of Cr^{2+} for Fe^{2+} on tetrahedral sites in the composition range between $FeCr_2O_4$ and Cr_3O_4. This situation results from the strong octahedral site preference energy of Cr^{3+}. The solution is shown to display close to ideal behavior in this composition range.

ACKNOWLEDGMENTS

Most of the ideas and results presented in this paper were derived during research on refractory phases sponsored by the United States Department of Energy.

REFERENCES

1. K. Kitayama and A. Muan: unpublished.

2. A. Muan and E. F. Osborn: Phase Equilibria Among Oxides in Steelmaking, 236 pp., 1965, Reading, Massachusetts, Addison-Wesley Publishing Co.

3. C. R. Kennedy: personal communication.

4. N. Toker, J. DeVilliers and A. Muan: unpublished.

5. G. A. Rankin and F. E. Wright: The Ternary System $CaO-Al_2O_3-SiO_2$, Am. J. Sci. (4th series), 1915, v. 39, pp. 1-79.

6. J. W. Greig: Immiscibility in Silicate Melts, Am. J. Sci. (5th series), 1927, v. 13, pp. 1-44, 133-154.

7. J. F. Schairer: The System $CaO-FeO-Al_2O_3-SiO_2$. I. Results of Quenching Experiments on Five Joins, J. Am. Ceram. Soc., 1942, v. 25, pp. 241-274.

8. C. R. Kennedy: The Corrosion of Refractories in Contact with Siliceous Mineral Residues at High Temperature, Proc. Symposium on The Chemistry of Materials at High Temperatures, September 1981, Harwell, England.

9. E. N. Bunting: Phase Equilibria in the System $Cr_2O_3-SiO_2$, J. Res. Nat. Bur. Standards, 1930, v. 5, RP203, pp. 325-327.

10. M. L. Keith: Phase Equilibria in the System $MgO-Cr_2O_3-SiO_2$, J. Am. Ceram. Soc., 1954, v. 37, pp. 490-496.

11. H. Collins and A. Muan: The System Chromium Oxide-SiO_2 in Contact with Metallic Chromium, unpublished.

12. P. L. Roeder, F. P. Glasser and E. F. Osborn: The System $Al_2O_3-Cr_2O_3-SiO_2$, J. Am. Ceram. Soc., 1968, v. 51, n. 10, pp. 585-594.

13. K. Kitayama and A. Muan: Phase Relations in the System Chromium-Oxide-$Al_2O_3-SiO_2$ in Contact with Metallic Chromium, unpublished.

14. K. Kitayama and A. Muan: Melting Relations in the Systems Chromium Oxide-SiO_2 and Chromium Oxide-$Al_2O_3-SiO_2$ as a Function of Oxygen Potential, unpublished.

15. A. Muan and E. Lopuski: unpublished.

16. N. E. Filonenko and I. V. Lavrov: Investigations of Equilibrium Conditions in the Al_2O_3 Corner of the System $CaO-Al_2O_3-SiO_2$, Zhur. Priklad. Khim. (U.S.S.R.), 1950, v. 23, pp. 1040-1046; J. Appl. Chem. (U.S.S.R.), 1950, v. 23, pp. 1105-1112 (English Translation); Chem. Abstr., 1952, v. 46, p. 786a.

17. S. Aramaki and R. Roy: The Mullite-Corundum Boundary in the Systems $MgO-Al_2O_3-SiO_2$ and $CaO-Al_2O_3-SiO_2$, J. Am. Ceram. Soc., 1959, v. 42, pp. 644-645.

18. F. C. Langenberg and J. Chipman: Determination of 1600°C and 1700°C Liquidus Lines in $CaO \cdot 2Al_2O_3$ and Al_2O_3 Stability Fields of the System $CaO-Al_2O_3-SiO_2$, J. Am. Ceram. Soc., 1965, v. 39, pp. 423-433.

19. A. L. Gentile and W. R. Foster: Calcium Hexaluminate and Its Stability. Relations in the System $CaO-Al_2O_3-SiO_2$, J. Am. Ceram. Soc., 1963, v. 46, pp. 74-76.

20. G. W. Morey: Phase-Equilibrium Relations of the Common Rock-Forming Oxides except Water. Chapter L of Data on Geochemistry, M. Fleischer, Technical Editor, 1964, Washington, D.C., U.S. Government Printing Office.

21. R. A. Snellgrove, T. Tsai, L. S. Darken and A. Muan: Activity-Composition Relations in Mo-Cr Alloys at 1500 and 1600°C, in preparation.

22. K. Schwerdtfeger and A. Muan: Activities in Olivine and Pyroxenoid Solid Solutions of the System Fe-Mn-Si-O at 1150°C, Trans. AIME, 1966, v. 236, pp. 201-211.

23. D. M. Kerrick and L. S. Darken: Statistical Thermodynamic Models for Ideal Oxide and Silicate Solid Solutions, with Application to Plagioclase, Geochim. Cosmochim. Acta 39, 1975, pp. 1431-1442.

24. T. Tsai and A. Muan: Activity-Composition Relations in $Al_2O_3-Cr_2O_3$ Solid Solutions at 1500 and 1600°C, in preparation.

25. A. Muan: Equilibrium Distribution of Elements Between Coexisting Solid-Solution Phases, in Origin and Distribution of the Elements, 1968, Oxford and New York, Pergamon Press, L. H. Ahrens, Editor.

26. T. Tsai and A. Muan: Activity-Composition Relations in $FeAl_2O_4-FeCr_2O_4$ Solid Solutions at 1500 and 1600°C, in preparation.

27. N. Toker, L. S. Darken and A. Muan: Equlibria in the System Fe-Cr-O in the Temperature Range 1500-1800°C, in preparation.

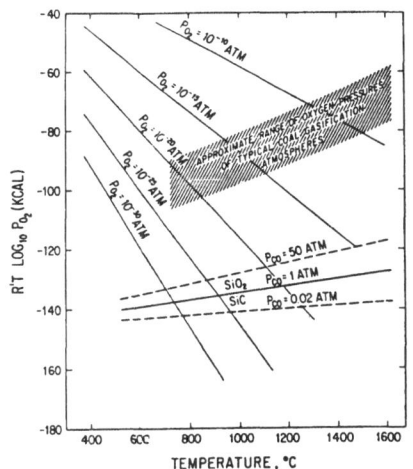

Fig. 1

Curves showing the SiO_2/SiC equilibrium at CO-pressures of 1 atm. (solid line), 50 and 0.02 atm. (dash line) as a function of temperature.

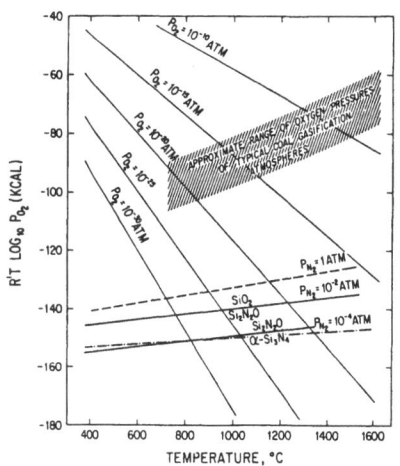

Fig. 2

Curves showing the SiO_2/Si_2N_2O and $Si_2N_2O/\alpha-Si_3N_4$ equilbria at N_2 pressures of 10^{-2} atm. (solid line), 1 atm. (dash line) and 10^{-4} atm. (dash-dot line).

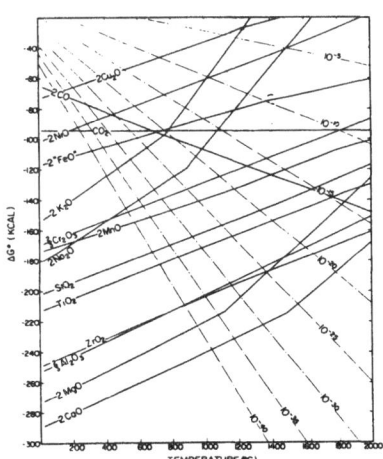

Fig. 3

Standard free energies ($\Delta G°$) of formation of various oxides from their elements.[1]

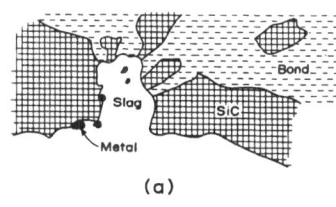

(a)

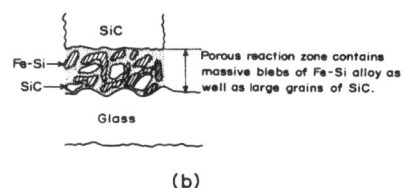

(b)

Fig. 4

Sketches showing typical features of reactions between SiC and FeO-containing coal-ash slags, as observed in testing of commercial refractories[3] (a) and in laboratory-scale experiments[4] (b).

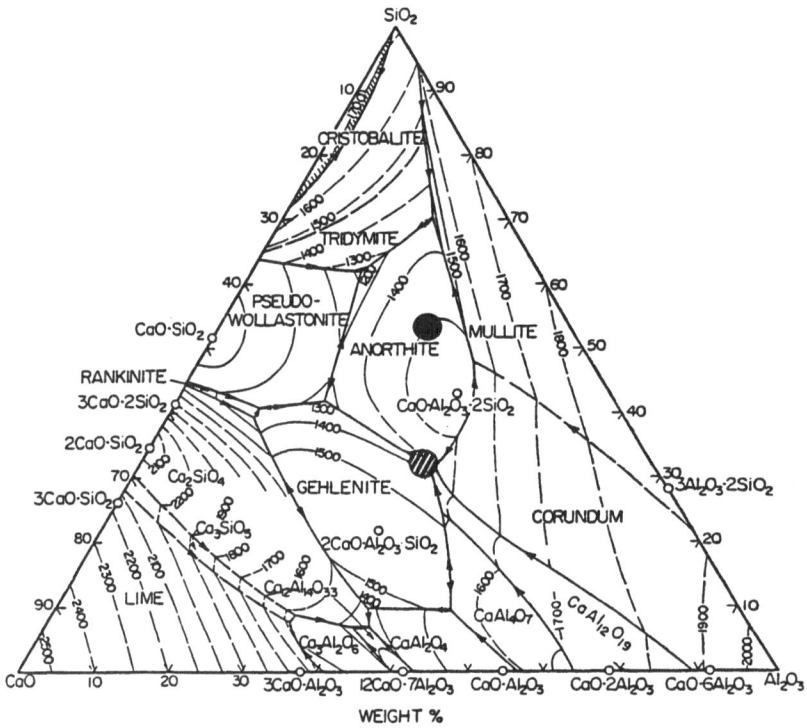

Fig. 5 Liquidus surface of the system CaO–Al$_2$O$_3$–SiO$_2$, reproduced from Muan and Osborn,[1] based on data available in previous literature.[5,6] Solid black and crosshatched circles show examples of compositions of coal-ash slags, as projected onto this plane.

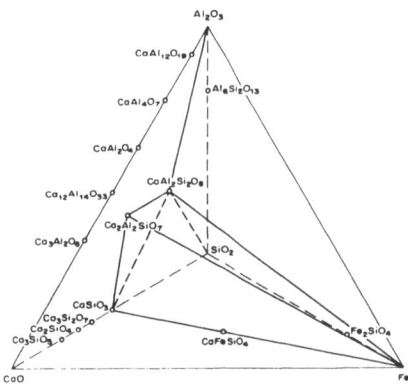

Fig. 6
Sketch of tetrahedron representing the system CaO–FeO–Al$_2$O$_3$–SiO$_2$, showing the composition volumes of main interest in conjunction with coal-ash slags.

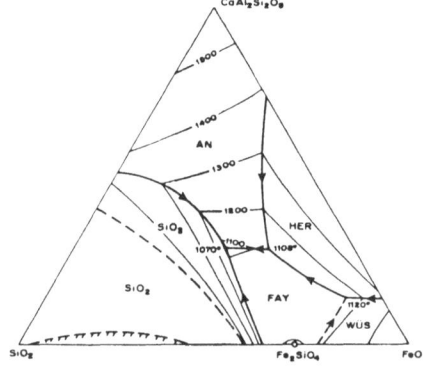

Fig. 7
Liquidus phase relations in part of the system CaO–FeO–Al$_2$O$_3$–SiO$_2$, based on data of Schairer.[7]

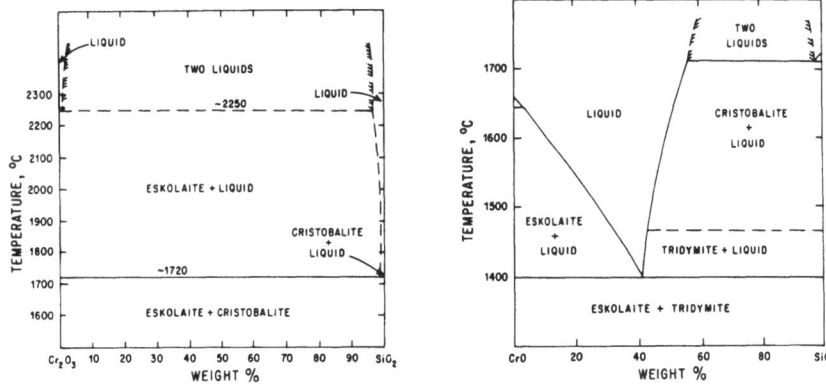

Fig. 8 Phase relations in the system chromium oxide-SiO$_2$ in air (a)[9,10] and in contact with metallic chromium (b).[11]

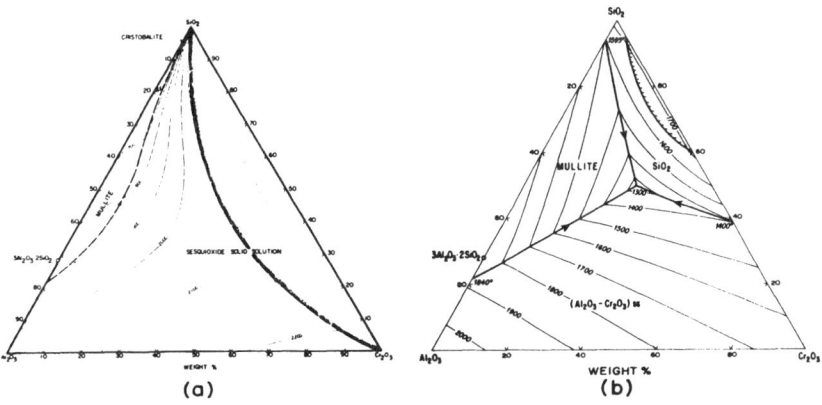

Fig. 9 Liquidus phase relations in the system Al$_2$O$_3$-chromium oxide-SiO$_2$ in air (a)[12] and in contact with metallic chromium (b).[13]

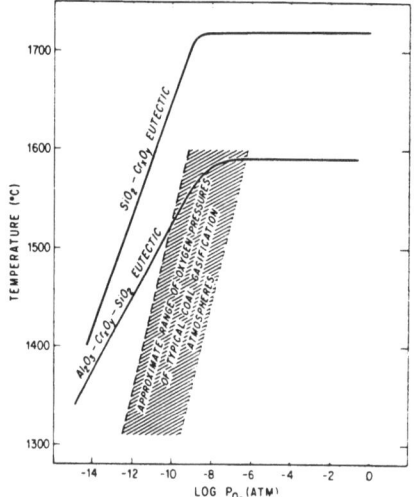

Fig. 10

Solidus temperatures in SiO_2-chromium oxide and Al_2O_3-chromium oxide-SiO_2 mixtures as a function of oxygen potential.[14]

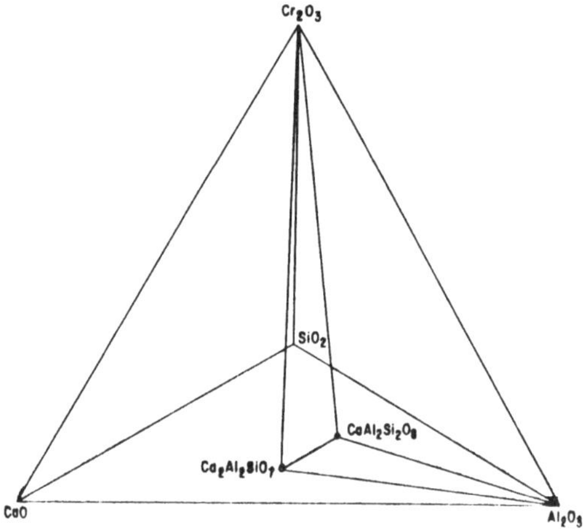

Fig. 11

Perspective drawing of tetrahedron representing the system $CaO-Al_2O_3-Cr_2O_3-SiO_2$.

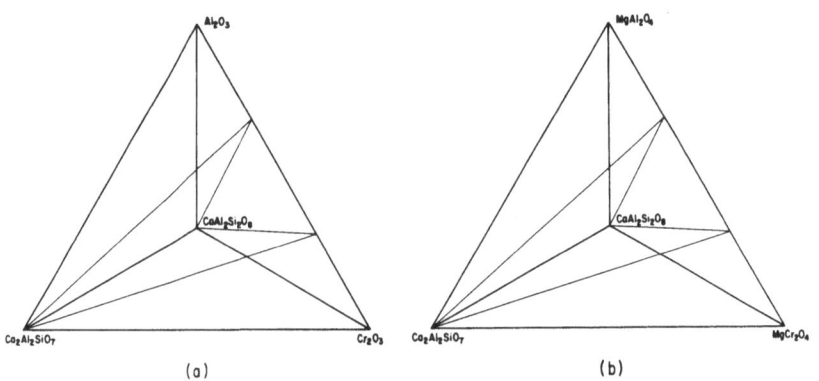

Fig. 12   Tetrahedra representing the systems $Ca_2Al_2SiO_7-CaAl_2Si_2O_8-Al_2O_3-Cr_2O_3$ and $Ca_2Al_2SiO_7-CaAl_2Si_2O_8-MgAl_2O_4-MgCr_2O_4$.

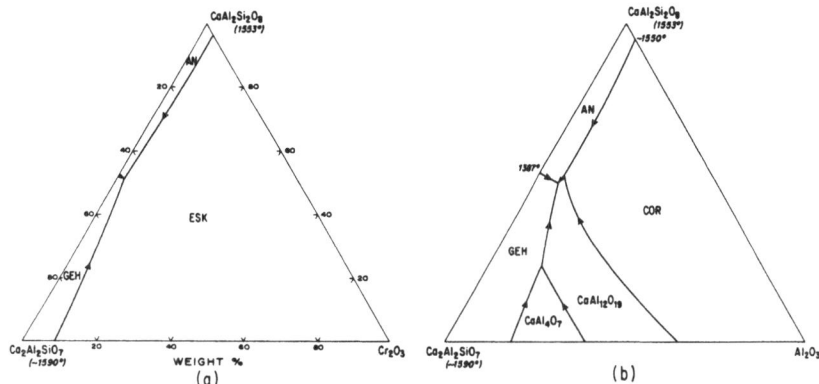

Fig. 13   Sketches showing main features of liquidus phase relations in the systems $Ca_2Al_2SiO_7$-$CaAl_2Si_2O_8$-Cr_2O_3[14] (a) and $Ca_2Al_2SiO_7$-$CaAl_2Si_2O_8$-Al_2O_3[5,6,16-19](b).

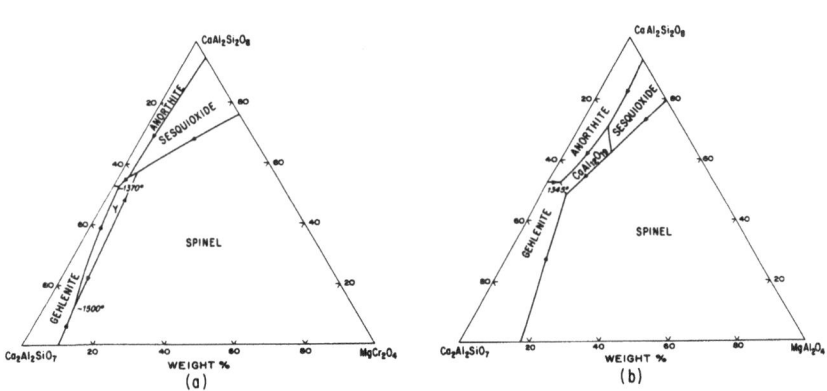

Fig. 14   Approximate liquidus phase relations in the systems $Ca_2Al_2SiO_7$-$CaAl_2Si_2O_8$-$MgCr_2O_4$[14] (a) and $Ca_2Al_2SiO_7$-$CaAl_2Si_2O_8$-$MgAl_2O_4$[15,20] (b).

363

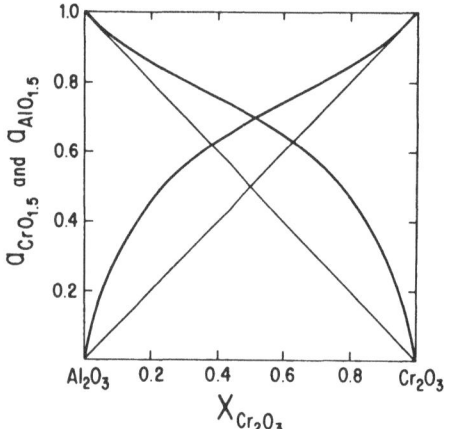

Fig. 15

Activity-composition relations in Al_2O_3-Cr_2O_3
solid solutions at 1500 and 1600°C.[24]

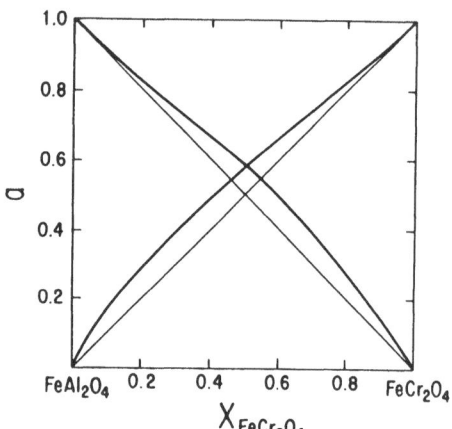

Fig. 16

Activity-composition relations in $FeAl_2O_4$-$FeCr_2O_4$
solid solutions at 1500 and 1600°C.[26]

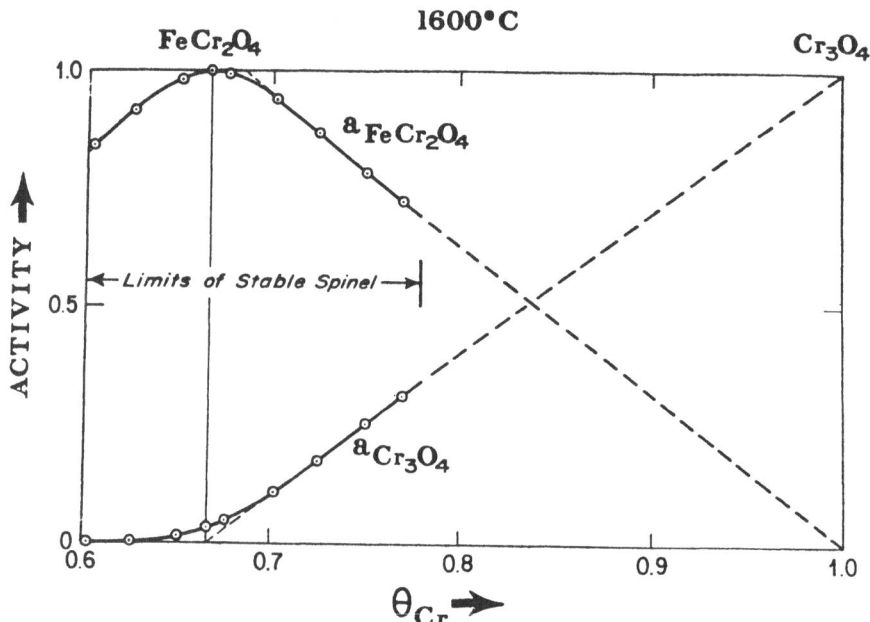

Fig. 17 Activities of $FeCr_2O_4$-Cr_3O_4 for the spinel region
of the Fe-Cr-O system at 1600°C (in equilibrium
with Fe-Cr liquid alloy).[27]

DISCUSSION

P. Popper: Thermodynamics predict that Si_3N_4 and SiC are unstable in an oxidizing environment. We know that in practice it can be used - so what use is thermodynamics? I know the answer is 'kinetics' and the protection of silica layers etc. In other words, you have to investigate the behaviour experimentally and my question remains.

A. Muan: In most high temperature reactions, of which the slagging gasifier is an example, the reaction rates are usually sufficiently large to make thermodynamic data extremely valuable guides for predicting materials' behaviour.

During the course of the experimental work aimed at delineating the equilibrium reactions described in the present paper, qualitative observations were also made in general regarding the rates of the reactions involved. In addition, quantitative observations were made on the rates of oxidation of SiC at various oxygen pressures of the atmospheres, and in the presence and absence of various liquid or solid silicate phases. The evidence obtained in all these studies suggests that it would be difficult to maintain SiC, for extended periods of time, without serious oxidation, if the SiC body is in contact with a silicate phase assemblage that is not saturated with SiO_2. It is our contention that the prospects for adequate protection against destructive oxidation/erosion of SiC refractories are greatly enhanced if the SiC/slag interface is maintained at a temperature below the solidus temperature of the coal-ash slag.

F. Starr: Does Professor Muan know of the existence of a CrO/Cr_2O_3 phase diagram?

Professor Muan requested ideas for future avenues of work. In view of what has been said about steam and gaseous transport of silica, K_2O, Na_2O etc., it would be most helpful to have a greater understanding of the thermodynamics of these processes in the temperature range 500-1500°C at pressures of up to 100 bar. Eventually this understanding should be extended to mixed gas environments.

A. Muan: Recent data obtained in our laboratories for equilibria in the system Cr-O at high temperatures (~1500-1750°C) (Toker, Darken and Muan, unpublished) strongly suggest that Cr_3O_4 is a stable phase in the temperature range of 1665 to 1705°C. We have not been able to preserve this phase by quenching to room temperature, nor has the existence of the phase been confirmed yet by high temperature X-ray diffraction. The lowest liquidus (and solidus) temperature in the system (oxide in equilibrium with metallic chromium) was found to be 1650°C, in good agreement with data reported in previous literature (Y.I. Ol'Shanskii and V.K. Shlepov, Dokl, Akad, Nauk. SSSR, 91, 561 (1953), and R.E. Johnson and A. Muan, J. Amer. Ceram. Soc., 51, 430 (1968)).

In my oral presentation, I presented some of my personal thoughts regarding future avenues of research. I very much welcome the ideas put forth by Dr. Starr. While we do not plan, in my laboratories, to study the thermodynamics of K_2O- and Na_2O-containing phases at temperatures as low as 500°C, where attainment of equilibrium within a reasonable period would pose some serious problems, we are in the process of adding alkali oxides (K_2O and/or Na_2O) to our simplified, synthesized $CaO-Al_2O_3-SiO_2$ coal-ash analogs and studying systems such as $Ca_2Al_2SiO_7-CaAl_2Si_2O_8-NaAlSi_3O_8-Cr_2O_3$ in a temperature range ($\sim1100-1300°C$) and under atmospheric conditions ($CO_2/H_2 = 10/1$) of interest in slagging coal gasifiers.

INORGANIC, POROUS MEMBRANES.

PREPARATION, STRUCTURE AND POTENTIAL APPLICATIONS

K. Keizer, A.F.M. Leenaars and A.J. Burggraaf

Twente University of Technology, Department of
Chemical Engineering, Laboratory of Inorganic
Chemistry and Materials Science,
P.O.Box 217, 7500 AE Enschede, The Netherlands

SYNOPSIS

In this paper a survey is given about possibilities for gas and
liquid separation using ceramic membranes.

Some application fields connected with energy saving or production
are indicated below.
Hydrogen, obtained from water or hydrogen sulfide decomposition,
or from coal-gasification plants, has to be separated from other
gases at high temperatures.
Oxygen-enrichment of air is very useful in medical applications
and combustion processes. The exhaust gases of coal gasification
and power plants should be purified from sulfurdioxide in order
to prevent air polution. Also phenolic compounds in waste water
have to be removed.

Except for hydrogen separation, interaction between pore surface
of the membrane and molecules leading to surface diffusion, is
necessary to obtain reasonable separation factors. This interaction phenomenon has to be investigated because data are required for adsorption and diffusion of gases and liquids along
the pore surface on different porous materials with very small
pore sizes. Research of this type can lead to (partial) separation
of e.g. CH_4 from CO and/or lower hydrocarbons, of NO from air,
Ar from NH_3 etc.

Separation problems, connected with the above mentioned applications, can be solved in principle with inorganic, porous
membranes having uniform pore diameters smaller than 5 to 10 nm.

These types of microstructures are realized in Vycor-type glass membranes and in ceramic membranes produced by a sol/gel process as described in this paper.

Ceramic membranes are especially useful in cases where it is advantageous to perform the separation process at relatively high temperatures and/or aggressive circumstances (corrosion, erosion). Further research is needed in order to decrease the pore diameter to below 4 nm, to decrease the membrane thickness to 1μm or lower and to modify the pore surface characteristics.

1. LIST OF SYMBOLS

b a parameter depending on the geometry of the membrane $(m^2 mol^{-1})$

c_A molar concentration of component A $(mol\ m^{-3})$

c_f molar concentration in the feed of a component A in a mixture A/B $(mol.m^{-3})$

c_p molar concentration in the permeate of a component A in a mixture A/B $(mol.m^{-3})$

c_{sA} molar surface concentration of component A $(mol.m^{-2})$

Δc_{sA} difference in molar surface concentration of component A at the inlet and outlet of the membrane $(mol.m^{-2})$

D_{KA} Knudsen diffusion coefficient of component A $(m^2 s^{-1})$

D_p pore diameter (m)

D_{sA} surface diffusion coefficient of component A $(m^2 s^{-1})$

f fraction of molecules with a diffuse reflection on the pore wall (-)

M_A molecular mass of component A $(kg\ mol^{-1})$

N_{AD} molar diffusive flux of component A $(mol\ s^{-1}m^{-2})$

ΔP pressure drop over a membrane (Nm^{-2})

P critical vapour pressure in a pore system (Nm^{-2})

P_o saturated vapour pressure at a temperature T (Nm^{-2})

R rejection = $1 - c_p/c_f$ (-)

R_g gasconstant = 8.317 $(J\ mol^{-1}K^{-1})$

T temperature (K)

t thickness of an adsorbed layer (m)

V molar volume of condensed vapour $(m^3 mol^{-1})$

v_s diffusion velocity along the surface (ms^{-1})

Δx thickness of the membrane (m)

α   separation factor $(= C_p(1 - C_f)/(1 - C_p)C_f)$ (-)

γ   surface energy of the condensed vapour (Nm^{-1})

φ   angle of contact between the liquid and the wall of the pore

2. INTRODUCTION

In the literature the problem of separation of gases, liquids and solids is broadly discussed[1]. Distillation, extraction, filtration and washing processes are commonly used. Membrane processes are relatively new and their potentials have not yet been fully explored. For special cases and for economic reasons such as energy-saving, membrane processes can certainly be useful.

A membrane is defined as a barrier between two homogeneous phases through which transport can take place under the influence of a pressure gradient, a chemical or an electrical potential difference. Separation of a mixture occurs if the transport coefficients are different for the components of the mixture. Membranes are dense or porous, can be solid as well as immobilized liquid phases, are constituted of organic or inorganic materials and can have catalytic properties.

Recently Lonsdale[2] published an extended review paper on organic membranes concerning micro- and ultrafiltration, gas separations and applications. Especially the prospects of polymeric membrane processes and 'hollow fiber' membranes are reviewed. For inorganic membranes no such review has been published because the study of inorganic membranes is relatively new and they are not widely used except for the enrichment of U^{235} by gas diffusion. Recently literature review reports have been made concerning separation mechanisms, liquid and gas separations[41].
In this paper a survey of possible applications of inorganic membranes is given together with a short discussion of underlying mechanisms of the separation processes. From these applications and mechanisms requirements are derived for the desired micro-structure of inorganic membranes. Preparation possibilities of these types of structures are described and examples of micro-structures are given.

New developments are suggested to improve the properties of the membranes to make them useful for new separation processes.

3. STATE OF THE ART AND PROSPECTS OF INORGANIC MEMBRANE PROCESSES

3.1. Gas separation

Some twenty-five years ago Barrer et al.[3,4,5], Carman and Raal[6] and Gilliland et al.[7] investigated transport of various gases through several inorganic, porous plug systems such as carbon, Vycor glass, silica-alumina and silica. They found a difference in transport behaviour for the various gases. Light gases such as He, H_2 and CH_4 have a larger diffusion velocity through the gas phase (Knudsen diffusion) compared with heavier gases such as CO_2, C_3H_8, iC_4H_{10} and SO_2, but the surface diffusion contribution to the total transport is larger for the latter gases. This is due to a higher surface coverage of the solid with gas molecules because the surface-diffusion coefficients seem to be about the same for all molecules mentioned.

Kammermeyer[8] investigated and reviewed the permeation of the gases He, Ne, H_2, D_2, O_2, N_2, CH_4, CO_2, C_2H_2, C_3H_8 and C_4H_{10} through a Vycor-glass system. He found a considerable amount of surface diffusion especially for the last four gases mentioned and he concluded that selectivity improvement for separation with microporous barriers would seem to be dependent on surface-diffusion manipulation. This is a system-dependent variable, which is just beginning to be recognized as a promising operation variable.

Barrer[9] investigated the separation of mixtures of H_2/SO_2, N_2/CO_2 and N_2/Ar at temperatures near the boiling point of the heavier gas with porous Vycor glass, carbon and alumina/silica. In the H_2/SO_2 system he found enrichment factors for SO_2 from 10–1000 at -20°C depending on the partial pressure of SO_2. In this case a capillary condensation process takes place of the SO_2 molecule which permeate as a liquid at the high pressure side and as a vapour at the low pressure side of the membrane. The hydrogen does not dissolve much in the SO_2-liquid and therefore does not permeate.

Ash et al.[10] investigated the transient state separation factors with the same membranes as Barrer for the systems H_2/Ar, H_2/N_2 and $^{235}UF_6/^{238}UF_6$ and found separation factors of 2 for the latter gas pair and up to 10 or more for the other gas pairs. For the separation process they suggest a fractionation technique in which pulses of strongly enriched component are collected and combined by switching the mixture successively to each of a series of membranes.

In the literature there is much more wide-spread evidence for the possibility of separation of gases through inorganic systems[11,12].

In the last three years Toei and Okazaki et al.[13-16] published on surface diffusion and capillary condensation of C_2H_4, C_3H_6, iC_4H_{10} and SO_2 in Vycor glass and high-surface alumina. The results again indicate that separation of these types of gases is possible. Even

a solar-powered dehumidifier based on the surface diffusion phenomenon through an activated alumina[17] membrane has been developed.
The dehumidification mechanism is based on a difference in temperature and therefore on a difference in surface coverage at the
two sides of the membrane. This concentration gradient of water on
the surface initiates a diffusion along the surface.

Technical application of inorganic membranes in gas separation is
realized for the purification of hydrogen through palladium-alloy
membranes. Recently Japanese scientists[19,20,21] started to use
Vycor glass and alumina ceramics for the production of hydrogen
with two types of processes.

Kameyama, Fukuda and co-workers[19,20] developed a catalyst for the
direct conversion of hydrogen sulfide into sulphur and hydrogen at
temperatures of 500°C-1000°C. To obtain hydrogen advantageously,
it is desirable that the hydrogen formed is continuously or at
leat intermittently separated from the reaction gas. Scrubbing
with a selective solvent requires cooling of large quantities of
reactants. Therefore a membrane process operating at high temperature is advantageous. Kameyama et al. used Vycor glass, alumina-
coated Ni-metal and alumina-coated alumina. The last membrane is
a new development of a ceramic composite membrane which presents
a lot of possibilities. However, the technology of making these
types of membranes needs improvement. Separation factors α of 1.5
are achieved, but this is still too low for an economic feasible
separation process.

Shindo et al.[21] studied the Magnesium-Iodine cycle for the development of a process for the thermochemical production of hydrogen
from water. The Magnesium-Iodine cycle is as follows:

$$6/5 \text{ MgO} + 6/5 \text{ I}_2 \xrightarrow{150°C} \text{MgI}_2 + 1/5 \text{ Mg(IO}_3)_2 \quad \ldots\ldots\ldots\ldots (1)$$

$$1/5 \text{ Mg(IO}_3)_2 \xrightarrow{650°C} 1/5 \text{ MgO} + 1/5 \text{ I}_2 + 1/2\text{O}_2 \quad \ldots\ldots (2)$$

$$\text{MgI}_2 + \text{H}_2\text{O} \xrightarrow{400°C} \text{MgO} + 2\text{HI} \ldots\ldots\ldots\ldots\ldots (3)$$

$$2\text{HI} \xrightarrow{<400°C} \text{H}_2 + \text{I}_2 \ldots\ldots\ldots\ldots\ldots\ldots (4)$$

The hydrogen produced in step (4) has to be separated from the
mixture at temperatures between 150°C - 400°C and this is performed
by a Vycor-glass membrane. As a model experiment mixtures of hydrogen or helium with nitrogen are permeated through a membrane and
separation factors α of 2 to 3 are obtained which are in the order
of those according to Knudsen diffusion.

3.2. Liquid separation

The most frequently used inorganic membranes in liquid separation
processes are the dynamic membranes formed on the basis of hydrous
Zr(IV)-oxide and of polyacrylic acid[22,23]. They are suitable for

the removal of sodium chloride or phenolic compounds from water.
These phenolic compounds occur in waste water of Lurgi coal
gasification plants which have been built in the USA and the
German Federal Republic. These dynamic membranes cannot be made
very reproducible and chemically stable and other types of membranes such as Vycor glass have been investigated by Ballou[24-26]
and Littman[28] and more recently by Schnabel[27] and McMillan[42].
Schnabel[27] produced porous Vycor-like glass capillaries with an
outer diameter of 0.6 mm and an inner diameter of 0.4 mm and tested
them for the desalination of water. With a modified membrane he
found a value for the rejection coefficient R of salt of 90 to
99% at salt concentrations of 3.0 to 0.1 wt%.
Ballou et al.[24-26] used porous glass-capillaries of 1.6 mm outer
diameter and 1.0 mm inner diameter with pores of 4 nm diameter.
R-values for sodium chloride were 40 to 90% depending on the
sodium chloride concentration and 40% for urea nearly independent
of the urea concentration. Littman et al.[28] found a rejection of
70% for phenol-poluted water with the same membrane as Ballou et
al. used.
A problem with all the glass membranes is a relative lack of chemical stability in sodium chloride, phenol and urea-solutions. The
silica of the membrane dissolves slowly and there is a pore growth
with increasing test time. The addition of aluminium chloride can
retard this solution process of silica.

The Union Carbide Corporation developed an inorganic ultrafiltration
composite membrane[29]. This membrane consists of a carbon tube
support with pores of 10 nm to 10 µm and an inorganic top layer
with pores of 2 to 100 nm. The latter layer is probably a zirconia
compound.
This membrane can be used for waste water control especially for
the removal of emulsified oil or other agents, detergents etc.,
in the pulp and paper industry etc.
A French firm developed the same type of inorganic membranes[30]
with suggested applications such as oil/water separation, cheese
whey protein recovery, latex and milk concentration etc. These types
of membranes can be used at relatively high temperatures (>100°C)
and at pH-values ranging from 1 to 14.

4. MECHANISMS OF SEPARATION

In the previous paragraph a number of examples of separation processes and possible applications are given. A systematic description
of separation mechanisms will now be given with emphasis on surface
diffusion and Knudsen diffusion. In the separation process a mixture of liquids or gaseous components is in all cases pressed through
a porous membrane with a driving force ΔP (Fig.1). The component A
with a relative low transport coefficient through the membrane
accumulates on the feed side of the membrane. A continuous flow

along the feed and the effluent side of the membrane yields a
mixture, which is enriched with A at the feed side and with B at
the effluent side (at the outlet of the system).
We can distinguish five separation mechanisms for microporous
membranes:

-1 geometric exclusion
-2 Donnan exclusion
-3 adsorption/diffusion mechanism
-4 Knudsen diffusion
-5 capillary condensation mechanism

With geometric exclusion separation takes place by size effects i.e.
the pore size has to be smaller than the particle size of one of
the components of the mixture to be separated (Union Carbide membrane).
The separation of charged particles or molecules, for instance
sodium chloride from water, is achieved by the Donnan exclusion
mechanism. The membrane develops a charged pore surface at the
start of the process. At small pore sizes the interaction between
charged molecules and pore surfaces gives a rejection of the charged
molecules.

In Fig.2 an example of diffusion af adsorbed particles through a
membrane is given. If the thickness of the adsorbed layer of one
of the components is equal to t, the pore size D_p has to be smaller
than 2t in order to inhibit transport of the non-adsorbed component through the membrane. If the pore size D_p is larger than 2t,
the non-adsorbed component is also transported through the membrane
and the separation factor α decreases.
For gases, pure Knudsen diffusion appears if the mean free path of
molecules is much larger than the average pore size (50 to 100
times).
In this case the total diffusion is determined by Knudsen diffusion
according to equations (5) and (6).

$$N_{AD} = - \text{constant}.D_{KA} \cdot \frac{dC_A}{dx} \dots\dots\dots\dots\dots\dots\dots\dots\dots\dots\dots\dots\dots\dots\dots (5)$$

where N_{AD} is the molar diffusive flux of component A

$\quad\quad D_{KA}$ is the Knudsen diffusion coefficient

$\quad\quad C_A$ is the molar concentration of component A

$$D_{KA} = 1/3 \; D_P \; (\frac{8R_g T}{\pi M_A})^{\frac{1}{2}} \cdot \frac{2-f}{f} \dots\dots\dots\dots\dots\dots\dots\dots\dots\dots\dots\dots\dots (6)$$

where M_A is the molecular mass of component A

$\quad\quad R_g$ is the gasconstant

$\quad\quad T$ is the temperature

$\quad\quad f$ is the fraction of molecules with diffuse reflection on
the pore wall

In this case separation of gases is possible. The separation efficiency is proportional to the ratio of the square root of the mass of the molecules. High separation factors α are obtained only with light gases such as hydrogen and helium in mixtures with heavier gases. The mean free path for these molecules at 0.1 MPa is about 200 nm and at this pressure pore diameters of about 5 nm are necessary to obtain pure Knudsen diffusion. Larger pores introduce the possibility of slip flow and viscous flow and the separation factor α decreases.

A combined transport by Knudsen diffusion and adsorption/diffusion is possible for gases. As an example assume that a gas is adsorbed in a monolayer on a surface of pores with a diameter of 5 nm. At 0.1 MPa the number of molecules in this layer is over 200 times larger than in the gas phase of the 5 nm pore itself. Even if the surface diffusion coefficient is somewhat smaller than that of the gas phase there can be a considerable amount of transport along the surface compared with that in the gas phase. Specific adsorption of one of the components then results in an enrichment of the effluent gas with that component.

Capillary condensation can occur according to the Kelvin equation:

$$R_g T \ln(p/p_o) = - \frac{4\gamma V}{D_p} \cos \Phi \dots\dots\dots\dots\dots\dots\dots\dots\dots\dots (7)$$

where p   is the critical vapour pressure at pore diameter D_p

     p_o is the saturated vapour pressure at a temperature T

     γ   is the surface energy of the condensed vapour

     V   is the molar volume of the condensed vapour

     Φ   is the angle of contact between the liquid and the pore wall.

Take for example the mixture N_2/SO_2 where the SO_2-vapour condenses at the high pressure side of the membrane as is shown in Fig.3. In the membrane, evaporation of the condensed liquid occurs at the low pressure side.
Because nitrogen is not very soluble in liquid SO_2, only SO_2 passes the membrane and high separation factors are achieved. Other mixtures can also be separated by this kind of mechanism such as for instance alcohol/water. However, at relatively low partial pressures of the component with the highest boiling point, high pressures or low temperatures are necessary for condensation of that component and there is a change that both components may condense. In this case the capillary condensation mechanism ceases to operate. In conclusion it appears that not only geometric exclusion makes separation possible but that diffusion and interaction with the pore wall can provide considerable effects too. In this case the ratio between the pore diameter and the molecule (particle) dia-

meter should not be larger than 10. This means that pore diameters should not exceed 10 nm or even better 5 nm. The pore size distribution must be kept very small because pores larger than 10 nm should be absent. If they are larger than 10 nm transport by viscous flow contributes to the total flux in liquid and gas separation processes and the separation efficiency decreases or vanishes. With pore diameters smaller than 4 nm interaction effects with the solid surface become very important.

5. REALIZATION AND CHARACTERISATION OF VERY FINE PORE STRUCTURES

Porous materials with all pores smaller than 5 or 10 nm are reported in the literature in only a few cases. The only commercially available system which fulfil the above mentioned requirements, is Vycor or Vycor-type glass.
This glass is a sodium borosilicate glass with a composition in a miscibility gap of the phase diagram. Under certain time and temperature conditions, the homogeneous glass separates into two phases, a silica-rich and a alkali borate-rich phase. The latter phase is soluble in a mineral acid and a porous skeleton of substantially insoluble silicon dioxide is left. According to Schnabel[27] an average pore diameter of this matrix can be selected between 2 and 40 nm with a standard deviation of $\pm$ 10% and a porosity of about 30%. These materials can be used up to temperatures of 700°C and pressures of more than 10 MPa without changing or damaging the structure. However, these membranes are not very stable chemically.
The pore size can increase in extreme circumstances by solution of a part of the silica. However, Schnabel claims that his modified glass membranes have an improved stability.
Another structure with very fine pores was developed by Yoldas[31,32]. He started with a sol preparation by hydrolysis of aluminium s-butoxide in water followed by peptization of the aluminium hydroxide with inorganic acids.
From this sol a gel is formed by slow evaporation of the water. The sol-gel transition proceeds without any discontinuity in properties. The dried gel body is calcined very carefully at about 500°C and mean pore diameters of 4 to 10 nm are obtained. The pore diameter distribution is reported to be very narrow ($\pm$ 1 nm) and the pore size can be varied by the kind of acid used and by the acid concentration. The pore size increases with increasing temperatures. Dense, crack-free membranes were not reported.

A modified sol/gel method is used by us[33] to obtain thin, homogeneous and crack-free layers.
These thin layers of γ-AlOOH (boehmite) are dried and calcined at different temperatures. For crack-free calcination the layers should have a thickness less than 20 μm. The microstructural characteristics of heat-treated samples are given in Table 1 and in Figs. 4 and 5.

It can be seen that the pore diameter of the hydroxide is 3.5 nm and the porosity is 34%. At 390°C a phase transition occurs according to the reaction

$$2 \text{ AlOOH (boehmiet)} \longrightarrow Al_2O_3 + H_2O \dots\dots\dots\dots\dots\dots\dots\dots (8)$$

During this reaction new pores are developed. The diameter of the pores depends on the calcination temperature and increases from 4.4 nm at 500°C to 8.2 nm at 780°C. The total porosity in this temperature region is almost constant. The pore size distribution is small as is shown in Fig.4 especially for the samples which are calcined at 500°C and 600°C. The pore size can also be varied with the amount of acid used for the peptization of the hydroxide as is shown is Fig.5. With increasing acid concentration the pore size decreases together with the porosity. This is caused by an increase in the stacking density of the crystallites in the gel with increasing acid concentration. With mercury porosimetry no pores larger than 10 nm could be detected. It should be noted that this sol/gel method can be easily applied to other systems[43]. It is also known that in the case of sol/gel silicas in *powder* form the pore size can be brought down to much smaller values than those obtained by us so far[44].

Porous systems such as carbon, silica-alumina and silica catalysts are reported in literature to contain pores of 2 to 10 nm size but in all cases these systems also have pores of 10 to 100 nm and they are not prepared in the form of thin, crack-free layers. This means that besides surface and Knudsen diffusion slip flow and viscous flow also occurs. The latter mechanisms do not show any separation and therefore porous systems with this morphology are not suited for the separation of mixtures of molecules or small particles in a continuous process.

6. SOME RESEARCH OPPORTUNITIES FOR SEPARATION PROCESSES WITH VERY FINE PORE STRUCTURES

Because very fine pore structures can be realized as is stated in the previous section, a number of mixtures are of interest in separation processes. Three separation mechanisms can be used. Firstly the Knudsen diffusion mechanism can be applied to solve the problems mentioned by Kameyama[20] and Shindo[21] for the production of hydrogen from H_2O cycles or from H_2S decomposition. In these cases no surface diffusion must be present because this phenomenon fastens the diffusion of heavier components and decreases the separation factor for hydrogen and helium. Therefore the separation should take place between 300°C and 700°C with membranes having pores of about 5 nm and at pressures less than 0.5 MPa. At higher temperatures there is a pore growth within the membranes for the Vycor-type glasses and to a lesser extent for

the sol/gel type ceramic membranes. At lower temperatures surface
diffusion can play an important role. These membranes can also be
used for the production of helium from natural gases or for the
production of hydrogen in the coal gasification process. For a
membrane thickness of 10µm hdyrogen fluxes of 3.10^{-5} m^3/m^2.s.Pa
at normal temperatures and pressures are achieved[21,41]. Inorganic
or ceramic membranes have potential applications in nitrogen and
oxygen separation if an adsorption/diffusion mechanism is applied
at moderate temperatures (20-200°C). For this mechanism the amount
of Knudsen diffusion has to be small compared with the diffusion
along the pore surface of the material due to equal diffusivities
of nitrogen and oxygen (about the same molecular mass), therefore
pores smaller than 5 nm are necessary. Furthermore there has to
be a specific interaction of oxygen or nitrogen with the pore wall.
The pore wall of sol/gel produced porous materials with pores of
5 nm can therefore be covered with a material with specific properties for one of the components. The technique of covering
porous materials with monolayers of foreign oxides has been
developed for catalyst applications in our laboratory[38-40] and
can also be used for membrane applications. Specific interactions
as well as a decrease of the pore diameter is now being investigated in our laboratory.
In Table 2 the results of model calculations are presented for an
oxygen enrichment process with air at the inlet of a membrane.
The calculation method and results will be discussed below.
Assumptions are made for a specific interaction of oxygen with the
pore surface of the membrane. These assumptions imply a higher
surface coverage and higher diffusion coefficients for oxygen than
for nitrogen. The diffusion coefficients of oxygen and nitrogen
are based on literature values. Furthermore it is assumed that
Knudsen diffusion does not contribute to the separation because
the molar masses of N_2 and O_2 are almost equal.
The transport velocity along the surface is in first approximation
equal to:

$$V_s = b \ D_{sA} \ \frac{\Delta C_{sA}}{\Delta x} \tag{9}$$

where V_s is the diffusion velocity along the surface

   b   is a parameter depending on the geometry of the porous
        medium

   ΔC_{sA} is the difference in molar-surface concentration between
        the inlet and outlet of the membrane.

   Δx is the thickness of the membrane.

The outlet oxygen concentration is calculated at an inlet concentration of 20% as a function of the pore diameter, the surface
diffusion coefficient of N_2 and O_2 and the surface coverage difference (in fractions of a monolayer). There are two mass fluxes,

one through the gas phase, which does not result in component
separation and one along the pore surface. These two fluxes are
treated in the calculation as acting parallel and independently.
If the pore size decreases, the relative amount of diffusion
through the gas phase decreases because of a decreasing diffusion
coefficient (eq.6) and a smaller volume to surface ratio. Because
of this effect the separation factor increases and the amount of
oxygen at the outlet also increases. This is clearly indicated in
Table 2 for every value of the surface coverage and diffusion
coefficient. With equal surface coverage difference and surface
diffusion coefficients for nitrogen and oxygen the maximum oxygen
concentration at the outlet is 50%. This value is achieved with
2 nm pores, a surface-coverage difference of 0.5 monolayer and a
surface diffusion of $10^{-7}m^2s^{-1}$. At a lower surface-coverage
difference for nitrogen and a lower surface-diffusion coefficient
the oxygen outlet-concentration can become 83% and 96 % resp.
which is found at the same small pore diameter and large diffusion
coefficient.
In these cases the gas-phase diffusion can be neglected compared to
the surface diffusion. At pores of 10 nm and a diffusion coefficient
of $10^{-9}m$ s^{-1} no appreciable separation occurs. For pores of 2 to
5 nm and diffusion coefficients of 10^{-7}-$10^{-8}m^2s^{-1}$ oxygen fluxes
are of the same order of magnitude as mentioned before for hydrogen
fluxes e.g. about $10^{-5}m^3/m^2s.Pa$ at standard temperature and pressure
and for a 10μm thick membrane.
This type of separation due to molecule/surface interaction can
also be applied for mixtures of SO_2 with air and for the separation
of phenol and urea from water as is schematically given in Fig.2.
In these cases very small pores of 5 nm or less are also necessary
to obtain separation factors which are large enough to be of
interest.
Another interesting mechanism for SO_2-separation with high separation
factor α is the capillary-condensation mechanism. At 25°C the surface energy γ of SO_2 is 0.021 Nm^{-1} and the saturated vapour pressure
is 0.34 MPa. To obtain condensation in pores with 4 nm diameter a
partial pressure for SO_2 of 0.23 MPa can be calculated with eq.(7).
This means that at SO_2 concentrations less than 5% the total pressure
has to be at least 5 MPa. A combination of surface adsorption/
diffusion at relative low pressures (<0.1 MPa), which enriches the
SO_2/air mixture to about 10% SO_2 and a capillary-condensation
mechanism at relative high pressures (about 5 MPa), to obtain 90%
or more SO_2, seems to be a possibility for the desulfurization of
exhaust gases of coal gasification and power plants and for the
production of sulfuric acid.

- A typical example of a *dense*, ceramic membrane is presented by
  Browall[36]. Pure oxygen is achieved in relatively small amounts
  (<10 ton/day) with an oxygen pump which makes use of solid
  electrolytes (stabilized Bi_2O_3 or ZrO_2). Solid electrolytes can

transport oxygen as oxygen ions due to an electrical potential difference. These types of pumps can be used to increase the efficiency of combustion processes by oxygen-enrichment of air or even as a sensor for the control of the efficiency of combustion processes[45].

- Other oxygen-enrichment methods are the pressure-swing adsorption on carbon molecular sieves[34,35], and the use of immobilized liquid membranes[37].

In the pressure-swing method oxygen adsorbs faster on carbon molecular sieves than nitrogen does. So air is compressed in an adsorber as shown in Fig.6. The nitrogen desorption pressure is somewhat lower than the adsorption pressure and the oxygen-rich air is desorbed below atmospheric pressures.

Together with cryogenic techniques pressure-swing adsorption is not a membrane technique, but for oxygen-enrichment processes these methods enter into competition with membrane processes.

7. <u>CONCLUSIONS AND RECOMMENDATIONS</u>

Porous ceramic materials and glass can be used for the separation of gas mixtures to produce hydrogen or oxygen or to remove sulphur dioxide from exhaust gases. These materials are also suitable for the removal of sodium chloride, phenolic compounds and urea from waste water. However, to obtain reasonable separation factors for gases and rejections for cleaning waste water, all pore diameters of the membranes should be smaller than 10 nm. This has been realized in Vycor-type glasses and in ceramics prepared by the sol/gel method.

For the hydrogen production an inert membrane material is needed but for other applications interaction between molecules or particles and the membrane material is necessary.

Several systems should be investigated to obtain quantitative data of adsorption of molecules on pore surfaces of different chemical composition and of the mobility of molecules on the pore surfaces. This would give information about which combination of molecules and porous materials can be used for separation and rejection.

A limited number of membranes with different chemical composition should be made with pore sizes smaller than 5 nm. Studies which are directed to modify the chemical composition of the internal pore surface of the membrane, are important to achieve optimal gas or liquid-pore surface interactions.

In principle ceramic sol/gel membranes can be produced in such a way that large crack-free surfaces can be made for the production of for instance membrane modules. The technology for producing membranes with large surfaces has to be further developed.

An economic evaluation is necessary to obtain data about which kind of separation processes or waste rejections ceramic membranes can be used.

8. REFERENCES

1. R.H. Perry and C.H. Chilton: Chemical Engineers' Handbook, fifth edition, 1973, London, McGrawHill Inc.

2. H.K. Lonsdale: J. Membr.Sc.1982, 10 (2+3), 81-181.

3. R.M. Barrer and J.A. Barrie: Proc.Roy. Soc.A, 1952, 213, 250-64.

4. R.M. Barrer and T. Gabor: Proc.Roy.Soc.A, 1959, 251, 353-68.

5. R.M. Barrer and T. Gabor: Proc.Roy.Soc.A, 1960, 256, 267-90.

6. P.C. Carman and F.A. Raal: Proc.Roy.Soc.A, 1957, 209, 38-50.

7. E.R. Gilliland, R.F. Baddour and J.L. Russell: A.I.Ch.E.J., 1958, 4(1), 90-96.

8. K. Kammermeyer in Progress in separation and purification, E.S. Perry, editor, vol.1, pp.335-72, 1968, New York, Interscience N.Y.

9. R.M. Barrer: A.I.Ch.E. - I.Chem.Symposium Series, 1965, 1,112-21.

10. R. Ash, R.M. Barrer and T. Foley: J.Membr.Sci, 1976, 1(4), 335-70.

11. E.R. Gilliland, R.F. Baddour, G.P. Perkinson and K.J. Sladek: Ind.Eng.Chem.Fundam. 1974, 13(2), 95-9.

12. K.J. Sladek, E.R. Gilliland and R.F. Baddour: Ind.Eng.Chem.Fundam. 1974 13(2), 100-5.

13. M. Okazaki, H. Tamon and R. Toei: A.I.Ch.E.J., 1981, 27(2), 262-70.

14. H. Tamon, M. Okazaki and R. Toei: A.I.Ch.E.J., 1981, 27(2), 271-7.

15. H. Tamon, S. Kyotani, H. Wada, M. Okazaki and R. Toei: J.Chem. Eng.Jap., 1981, 14(2), 136-41.

16. M. Okazaki, H. Tamon, T. Hyodo and R. Toei: A.I.Ch.E.J., 1981, 27(6), 1035-8.

17. R. Toei and H. Tamon, private communication.

18. K. Kammermeyer: Chem.-Ing.Tech., 1976, 48(8), 672-5.

19. K. Fukuda, M. Dokiya, T. Kameyama and Y. Kotera: Ind.Eng.Chem. Fundam. 1978, 17(4), 243-8.

20. T. Kameyama, K. Fukuda, M. Fujishige, H. Yokokawa and M. Dokiya: Hydrogen Energy Progress, Proc.3rd World Hydrogen Energy Conference, Tokyo, Japan, T.H. Veziroglu, K. Fueki and T. Ohta, editors, 569-79, 1980, vol.2, Oxford, Pergamon Press.

21. Y. Shindo, K. Obata, T. Hakuta, H. Yoshitone, N. Todo and J. Kato: Adv.Hydr.Energy, 1981, 2, 325-33.

22. D. Freilich and G.B. Tanny: J.Coll.Interf.Sci., 1978, 64(4), 362-70.

23. S.L. Klemetson and M.D. Scharbow: Prog.Wat.Tech., 1978, 10(1/2), 479-91.

24. E.V. Ballou, T. Wydeven and M.I. Leban: Environmental Sc. and Techn., 1971, 5(10), 1032-8.

25. E.V. Ballou and T. Wydeven: J.Coll.Interf.Sc., 1972, 41(2), 198-207.

26. E.V. Ballou, M.I. Leban and T. Wydeven, J.Appl.Chem.Biotechnol., 1973, 23(2), 119-30.

27. R. Schnabel and W. Valont, Desalination 1978, 24(4), 249-72.

28. F.E. Littman, F.D. Kleist and G.A. Croopnick: Saline Water Res. Develop., Progress Rep. 1971, no.720.

29. I.K. Bansal: A.I.Ch.E.-Symposium Series, 1975, 71(151), 93-9.

30. Carbosep* Third generation, inorganic, ultrafiltration membrane, SFEC, 84500 Bollène, France.

31. B.E. Yoldas: J.Amer.Ceram.Soc. 1975, 54(3), 286-90.

32. B.E. Yoldas: J.Mat.Science 1975, 10, 1856-60.

33. A.F.M. Leenars, K. Keizer and A.J. Burggraaf: 2nd Europ.Conf. on Solid State Chem., 1982, Veldhoven, Neth.

34. K. Knoblauch: Chem.Eng., 1978, Nov. 6, 87-9.

35. H. Jüntgen, K. Knoblauch and K. Harder, Fuel, 1981, 60 (Sept.), 817-22.

36. K.W. Browall, Paper 467 presented at the Electrochem.Soc.Meeting, 1978, May 21-6, Seattle, Washington.

37. J.J.M. Snepvangers, Energiespectrum, 1982, 6(4), 102-9.

38. T. Fransen, thesis, 1977, Twente Univ. Techn., Enschede, Neth.

39. F. Roozeboom, thesis, 1981, Twente Univ. Techn., Enschede, Neth.

40. J.G. van Ommen, K. Hoving, H. Bosch, A.J. van Hengstum and P.J. Gellings: to be published.

41. A.F.M. Leenaars, K. Keizer and A.J. Burggraaf: Membrane separation processes with inorganic membranes, Progress Report, 1982, Lab. Inorg.Chem.Mat.Science, Dept. Chem.Eng., Twente Univ. Techn., Enschede, Neth.

42. P.W. McMillan: Process Eng., 1980 (April), 44.

43. 'Glasses and glass ceramics from gels': Proc.Int.Workshop Glasses and Glass ceramics from Gels, V.Gottardi, ed., J. Non-Cryst. Solids, 1982)1), 48, 1-230.

44. R.K. Iler: Colloidal Silica in Colloid and Surface Science,vol.6, E. Matyevic, 1973, 1-100, New York, Wiley Interscience.

45. M.J. Verkerk,thesis,1982,Twente Univ.Techn., Enschede, Neth.

Table 1. Microstructural characteristics of heat-treated samples.

	dry (150°C)	T_s=500°C	T_s=600°C	T_s=780°C
porosity (%)	34	48	48	47
B.E.T. surface area (m^2g^{-1})	256	246	209	129
pore diameter*	3.5	4.4	5.1	8.2
crystallite size** (nm)	–	3.6	3.9	5.4

*   pore diameter belonging to the peak in the size distribution (see fig.4)

** determined by X-ray line broadening

Table 2. The oxygen concentration (%) at the outlet of the membrane as function of the pore size, the surface-coverage difference (as a fraction of a monolayer) and the surface diffusion coefficients of oxygen and nitrogen at a inlet concentration of 20% (air).

	$\Delta C_{sO_2} = \Delta C_{sN_2}$			$\Delta C_{sO_2} > \Delta C_{sN_2}$			$\Delta C_{sO_2} > \Delta C_{sN_2}$		
ΔC_{sO_2}	0.02	0.10	0.50	0.02	0.1	0.5	0.02	0.1	0.5
ΔC_{sN_2}	0.02	0.10	0.50	0.005	0.05	0.1	0.005	0.02	0.1
pore diameter (nm)	$D_{sN_2} = D_{sO_2} = 10^{-7} m^2 s^{-1}$			$D_{sN_2} = D_{sO_2} = 10^{-7} m^2 s^{-1}$			$D_{sO_2} = 5*D_{sN_2} = 10^{-7} m^2 s^{-1}$		
2	46	47	50	69	81	83	64	89	96
5	35	45	49	44	68	80	47	76	91
10	26	37	46	28	48	69	28	50	79
	$D_{sN_2} = D_{sO_2} = 10^{-8} m^2 s^{-1}$			$D_{sN_2} = D_{sO_2} = 10^{-8} m^2 s^{-1}$			$D_{sO_2} = 5*D_{sN_2} = 10^{-8} m^2 s^{-1}$		
2	32	44	48	38	64	79	40	70	91
5	23	30	42	24	35	50	24	36	60
10	21	23	32	21	24	28	21	25	39
	$D_{sN_2} = D_{sO_2} = 10^{-9} m^2 s^{-1}$			$D_{sN_2} = D_{sO_2} = 10^{-9} m^2 s^{-1}$			$D_{sO_2} = 5*D_{sN_2} = 10^{-9} m^2 s^{-1}$		
2	22	28	41	22	32	58	23	32	63
5	20	21	27	20	22	29	21	22	30
10	20	20	22	20	20	23	20	20	23

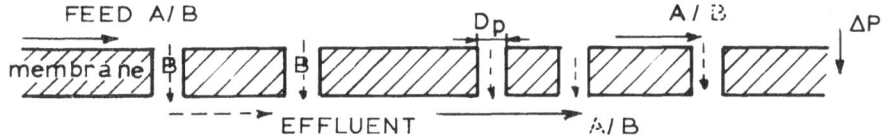

Fig.1. A schematic representation of a membrane separation
process.

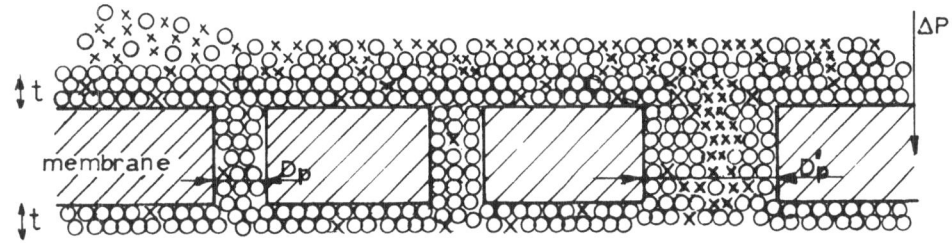

Fig.2. An example of diffusion of
adsorbed particles (0) through
a membrane. At larger pore
sizes (D_p') a transport of
non-adsorbed particles (x)
occurs.

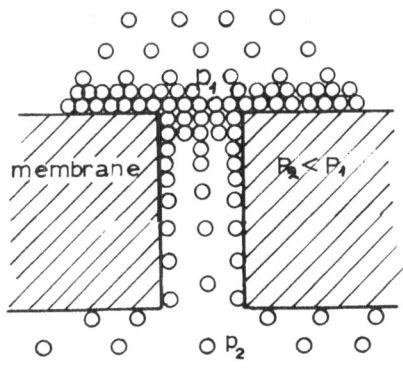

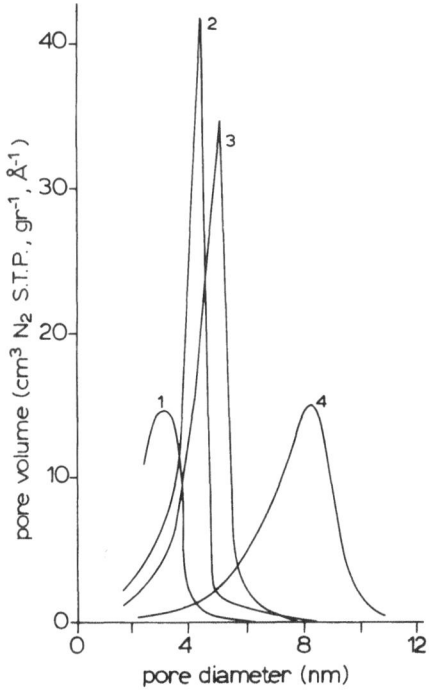

Fig.3. A schematic representation
of vapour condensation and
transport through a microporous membrane.

Fig.4. Pore size distribution
for different tempera-
ture treatments.
($1:150^{\circ}C$; $2:500^{\circ}C$;
$3:600^{\circ}C$; $4:780^{\circ}C$).

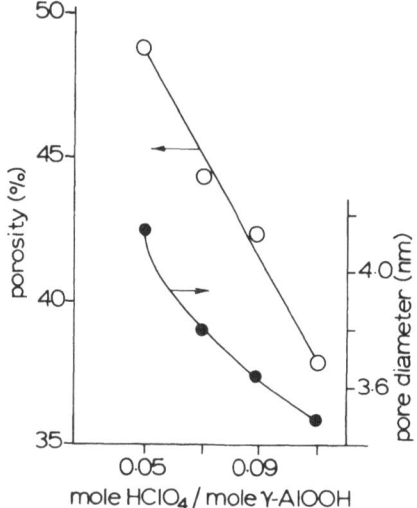

Fig.5. Porosity and pore diameter of calcined
samples (500°C) as a
function of the acid
concentration in the
sol.

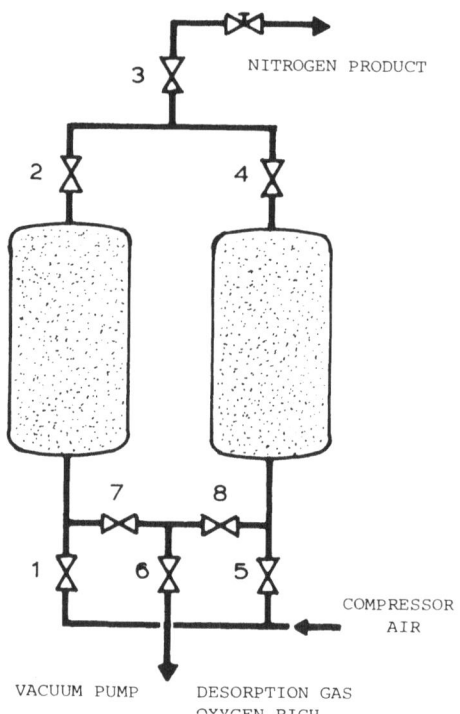

Fig.6. Pressure-swing process
for N_2 recovery using
carbon-molecular sieves.
An adsorption pressure
of 0.4, 0.6, 0.7 of
0.8 MPa is combined
with a nitrogen product
pressure of 0.2, 0.4, 0.6
or 0.7 MPa. The desorption
pressure for O_2 is 0.01-0.1
MPa (according to ref.35).

DISCUSSION

A. Auriol: Organic membranes are generally made 'non-symmetrical' by a surface treatment which results in a very thin layer with much smaller pore size. Are you able to make such a treatment on your alumina membranes?
Would it be possible with your method to make reverse osmotic inorganic membranes?

A.J. Burggraaf: The methods, used for the fabrication of asymmetric organic membranes, cannot be used for inorganic Al_2O_3-membranes because the latter membranes are not soluble in any (inorganic) liquid and have a much higher melting point than organic materials. It is possible to make reverse osmosis membranes with this kind of technique because with these membranes small (salt) molecules or ions can be separated from a liquid solution.

P. Popper: When a ceramic is sintered there must be a stage, just before the open porosity disappears, when there are only small diameter channels. Would such a product be useful for the application you mentioned?

A.J. Burggraaf: Such products are not suitable for the application of gas and liquid separation because
(a) the pore volume is very small and therefore the permeability for gases and liquid is too small;
(b) the pore size or better pore entrance distribution is too large.
Some pore entrances are still larger than 100 nm if other pores are already closed.

G. Grathwohl: How do you measure the pore size distribution in these membranes and which other pore characteristics besides pore size do you need to characterise the structure?

A.J. Burggraaf: The pore size distribution is measured with an adsorption/desorption technique at liquid nitrogen temperature and the pore shape is also important (slit- or cylindrical shape).

CERAMIC MATERIALS FOR ELECTROCHEMICAL ENERGY CONVERSION DEVICES

B.C.H. Steele

Wolfson Unit for Solid State Ionics,
Department of Metallurgy and Materials Science,
Imperial College, London, SW7 2BP.

Synopsis

High energy secondary batteries are briefly surveyed. For systems
incorporating liquid electrolytes the principal ceramic components
are generally limited to inert porous membranes designed to immobilise the electrolyte and separate the electrodes. The provision
of ceramic electrolytes for sodium/sulphur systems operating at
300°C, for moderate temperature (100-180°C) systems incorporating
liquid sodium, and for completely solid state systems is also discussed. A requirement for improved sodium and lithium ion electrolytes is identified with emphasis upon the synthesis of novel
glass electrolytes because of their probable fabrication advantages. Fuel cell systems are next considered and attention is
drawn to their superior conversion efficiency when compared to
heat engines. Porous inert ceramic matrices for immobilising the
liquid electrolyte are the principal contributions made by ceramic
technology except for the high temperature (800-1000°C) fuel cell
incorporating an oxygen ion conducting ceramic electrolyte. This
type of fuel cell is very versatile insomuch as it can operate
with many types of fuels and would have many advantages over other
fuel cell systems if the operating temperature could be reduced to
around 500-600°C. Accordingly the current status of oxygen ion
conducting electrolytes is summarised and it is concluded that research in this area should be expanded. Finally attention is
drawn to the large disparity between the R & D funding for fuel
cells in the EEC compared to the situation in North America.

1. Introduction

Basic research needs on ceramic materials for energy applications
have recently been summarised [1] . With regard to electrochemical
energy conversion systems this publication concludes that substantial work is needed to develop principles of crystal structures,
and solid state chemistry to stimulate the synthesis of novel materials exhibiting superionic transport for such ions as Na^+, Li^+,
O^{2-} and H^+. Developments in this area are contained in relevant
books[2,3,4] and proceedings of recent conferences on Fast Ion Transport published in Solid State Ionics. With the increasing application of ceramic materials in batteries, fuel cells, electrolysers,
and MHD electrodes a relatively new failure mode of electrolytic
decomposition is becoming more frequent. This is an area which has
received very little fundamental attention although electrolytic
effects in ceramics are probably more widespread than is generally
realised. It should be emphaised that the application of only a
few volts is needed to exceed the decomposition potential of even
the most stable ceramics.

The brief preceding comments are included to emphasise the need for
basic studies to underpin the technological developments that are
summarised in this brief review. Although electrochemical energy
conversion systems can be considered as encompassing electrolytic
systems used, for example, in the production of H_2, the comments
in the subsequent sections are restricted to the application of
ceramic materials in those electrochemical energy conversion systems, batteries and fuel cells that are designed to produce electrical energy.

2. High Energy Secondary Batteries

2.1. General Comments

High energy secondary batteries are already used in electric
vehicles and in association with intermittent energy sources such
as wind and solar power generation. In the U.S.A. there is also
a large demonstration programme to examine the application of secondary batteries for load-levelling in electrical power generation
systems. However, it is widely recognised [5,6] that improvements
in battery performance are essential if electrochemical energy
storage is going to have a significant role in future energy

strategies. Recent surveys and conferences[6,7,8,9] have emphasised the crucial role of materials science in the development of improved and alternative secondary batteries. It is in fact a statement of the obvious that the development of batteries is essentially a material science programme.

Taking electric vehicles as an example, the anticipated performance requirements obviously depend upon the type of vehicle and duty cycle but minimum specific energies of 100-200Wh/Kg and power capabilities of 100-200W/ℓ are usually specified (Table 1). Selecting a figure of 100W/ℓ then Fig. 1 indicates that in principle this requirement can be obtained by a variety of cell configurations. The diagonal line (100W/ℓ) locates the relationship between current density (mA/cm^2) and the maximum thickness (cm) that it is possible for a cell having a power density of 100W/ℓ operating with an open circuit voltage of 2.0V and IR losses of 10% (0.2V). The upper axis shows the electrode areas required for different electrode thicknesses if the cell is to have an energy density of 100Wh/ℓ. Also displayed on the diagram are the minimum specific conductivity values required for IR losses less than 0.2V as a function of current density and thickness. Clearly it is possible to have a relatively small electrode area, high current density configuration (region A) which is typified by the high temperature Na/S battery incorporating the Na-β"-Aℓ_2O$_3$ ceramic electrolyte (section 2.2). An alternative configuration is to have a very large electrode area requiring only small operating current densities (region C) which is the approach adopted by the Dow Chemical Company for their Na/S battery incorporating thin glass capillary electrolytes. The large area configuration can pose major fabrication problems although the development of thin film polymeric electrolytes (section 2.4) should give a stimulus to this type of design.

Although a large variety of electrochemical couples have been proposed for advanced systems only a limited number of secondary systems are being developed and their principal characteristics are summarised in Table 2. Most of these systems are ambient temperature batteries incorporating aqueous electrolyte, Zinc negative electrodes are also common as this is the most electro-positive metal that can be used in contact with aqueous electrolytes. In general these systems have little direct relevance to ceramic technology although it should be noted that the successful commercial development of fully-sealed maintenance-free lead-acid batteries was only made possible by incorporating a very porous glass fibre separator which allows oxygen evolved at the positive plate to rapidly permeate to the negative electrode, where it is recombined. The negative plate does not become further charged and no water is lost during overcharge. Moreover many electrodes consist of porous assemblies of non-stoichiometric oxide and hydroxide powders, and the preparation, characterisation and be-

388

haviour of these particulates have many features in common with the technology of ceramic powders. However this brief review will concentrate on those non-aqueous high temperature and solid state battery systems that incorporate ion conducting ceramics as the principal structural component.

2.2. Sodium Based Batteries

Since the announcement of the sodium/sulphur battery incorporating a beta-alumina ceramic electrolyte by the Ford Motor Company in 1967 this system has been the subject of major development pro- grammes[10,11] in the U.S.A., France, Germany, Japan, U.K and more recently China. The system is attractive because of the re- lative abundance and cheapness of the principal battery compo- nents, Na, S, Al_2O_3, but has the disadvantage of requiring a high operational temperature in the range 300-350°C. At this tempera- ture the electrode reactants (Na,S) and discharge products (sodium polysulphides, e.g. Na_2S_5) are liquids so that high cur- rent densities (>100mA/cm^2) can be sustained without excessive polarisation losses. A schematic diagram of the Na/S cell de- sign adopted by Chloride Silent Power in the U.K. is depicted in Fig. 2. The sulphur is contained inside the ceramic electrolyte tube and the sodium electrode is in a narrow outer annulus con- nected by a wick to the sodium reservoir. This configuration pre- vents corrosion of the outer metallic cell case by sulphur and sulphide species and also allows the bulk of the sodium to be con- tained in a reservoir at the end of the cell when fully charged. This arrangement reduces the quantity of liquid sodium that can directly react with molten sulphur in the event of an electrolyte tube failure and is an important safety feature.

Nearly all the battery evaluation programmes have utilised a cera- mic sodium aluminate composition and development of the microstruc- ture to provide the optimal mechanical and electrical properties have been well documented[12,13]. There are two crystallographic forms of beta-alumina, β and β". The latter, stabilised by small addition of magnesium or lithium oxides is the more conductive but is more difficult to process and most laboratories utilise a cera- mic composition that incorporates about 85% β" phase and 15% β phase and which exhibit specific resistivities around 5 ohms cm at 300°C. Production of the beta-alumina tubes is usually accomplish- ed by a rapid zone sintering technique or programmed batch firing method to minimise volatilisation of sodium oxide at the high tem- peratures (1600-1700°C) required for complete densification. Pilot plant production of ceramic electrolyte tubes having diame- ters upto 75mm has shown that these components can be successfully mass-produced and has allowed the construction of multi-cell bat- tery modules for operational evaluation.

Many individual cells with capacities greater than 100 Ah have successfully sustained more than 1000 charge-discharge cycles, and 10 KWh batteries incorporating at least 100 cells in series-parallel connections are currently been evaluated by many development groups around the world. Although the majority of cells constructed with a particular batch of ceramic electrolyte tubes successfully sustain more than 1000 cycles there is, at present, always a small proportion ($\sim$ 5%) of cells failing before 100 cycles have been completed. The most frequent reason for failure is the formation of cracks in the electrolyte tube during cell cycling. It appears that these cracks are initiated on the sodium side of the electrolyte and possibly arise due to stresses caused by local high current densities associated with surface inhomogeneities. High resolution electron microscopy studies[14] have also revealed that cracks can be initiated when sodium ions are displaced from the conducting plane in crystals of the relatively unstable β'' phase. It is obviously still necessary to develop ceramic electrolyte tubes with improved lifetime statistics and this requirement may prove to be particularly difficult to achieve in view of the exacting fabrication procedures required to ensure dense beta-alumina ceramic electrolyte tubes.

Fortunately a variety of alternative sodium ion conducting solid electrolytes have been synthesised using 'molecular engineering' concepts and a selection of these are incorporated in Fig. 3. These crystalline electrolytes are three dimensional framework sodium ion conductors exhibiting almost isotropic mechanical and electrical properties in contrast to the two dimensional anisotropic behaviour associated with the beta-alumina structure. A particular feature of the material $Na_{3.1} Zr_{1.55} Si_{2.3} P_{0.7} O_{11}$ is that dense ceramic electrolytes can be fabricated at relatively low temperatures ($\sim$ 1200°C) compared to those (> 1600°C) required for beta-alumina based electrolytes. At present there is some uncertainty regarding the long term stability of these alternative electrolyte materials in contact with molten sodium at 300°C and relevant investigations are now in progress.

For many applications the relatively high temperature of operation of the Na/S system is a disadvantage and thermal management of the battery assembly can be complicated. Reducing the temperature can minimise these problems and a variety of cells incorporating alternative cathode materials have been investigated[15]. These include the following systems,

Na / Na-β-Al_2O_3 / Na $AlCl_4$, Sb Cl_3 Sb (200°C)[16]

Na / Na-β-Al_2O_3 / Na $AlCl_4$, NiS_2 (165°C)[17]

Na / Na-β-Al_2O_3 / Na^+ organic electrolyte, S (150°C)[18]

Na / Na-β-Al_2O_3 / Na^+ organic electrolyte, TiS_2 (130°C)[19]

The cells incorporating NiS_2, for example, exhibit excellent re-chargeability when cycled between voltage limits of 1.7 - 3.5 V at current densities of 10 mA/cm². The development of moderate temperature systems (100-150°C) has many attractions particularly as materials compatibility problems are also reduced. However IR losses at these lower temperatures will restrict the power density performance unless better sodium ion conductors are synthesised or unless thinner ceramic electrolyte tubes can be used.

Mention should also be made of the large area Na/S configuration developed by Dow Chemical Company[20]. This system also operates at 300°C but the electrolyte consists of thin sodium ion conduc-ting glass fibres (80 μm diam. with walls approximately 10 μm thick) and typical current densities are of the order of 10mA/cm² To attain the required power density requirements it is necessary therefore to have at least 1000 glass capillaries in a typical cell. Ingenious fabrication procedures have been devised to in-corporate these glass capillaries into cell assemblies and cycle lives greater than 1000 have been achieved. The principal prob-lem still to be overcome appears to be associated with cracking of the glass capillaries during thermal cycling.

It should be emphasised that preparation of glass tubes and cap-illaries is very amenable to mass production techniques and the de-velopment of alternative sodium ion conducting glasses should lead to glass electrolytes being incorporated into a much wider range of battery systems than at present. An obvious application is the moderate temperature systems (100-180°C), and recent investi-gations[21] have already shown that sulphide glasses (Fig.3) have the requisite sodium ion conductivity. The redox and long term stability of these glasses obviously requires careful evaluation but it is at least encouraging that it appears to be possible to formulate alternative glass compositions with conductivities com-parable to polycrystalline beta-alumina ceramic electrolytes.

It should be emphasised that the application of glasses in high energy batteries is not confined to the electrolyte phase. Special glasses, for example, have been developed[10] to seal the beta-alumina tubes to alpha-alumina collars and the electrical hermetic seals through the outer case often incorporate glassy components.

Finally it should be noted that ceramic coatings are often used to protect metallic current collectors from corrosion[10]. An ex-ample is provided by the aluminium rod coated with niobium doped rutile which is used for current collection from the sulphur el-ectrode in Na/S cells.

2.3. Lithium Based Batteries

The large energy density theoretically available from lithium battery couples has ensured a continuing high level of interest in these systems. Several primary lithium batteries are now commercially available. These include: Li/MnO_2, Li/V_2O_5, Li/CuS, Li/SO_2, $Li/SOCl_2$, $Li/CF_x)_x$ and Li/I_2 (poly-2-vinylpyridine). Although progress has been made in the development of secondary lithium systems[22,23] many problems still remain. These are generally associated with the lack of suitable organic electrolytes which are stable in contact with lithium. Increasing attention is therefore being given to alternative lithium ion conductors including molten salt, solid, and polymeric electrolytes.

Conceptually, batteries incorporating molten salt electrolytes appear very attractive because of the very high current densities that can be sustained by these phases. However the corrosive nature of the fused salt electrolyte introduces many materials compativility problems, and the Li-FeS secondary battery provides an example of a system in which there appears to be no economic solution to the formidable material problems[6,24] associated with operating at 400-450°C in a molten salt environment. Originally conceived as a Li/Cl_2 couple than a Li/S battery the preferred solution at present is to use a solid Li-Al or Li-Si alloy as the negative electrode, and FeS (or possibility FeS_2) as the positive electrode. Whilst these developments have made it possible to fabricate complete cells and prototype batteries, the specific energy density values have obviously been down graded. Typical prismatic cells have specific energy densities in the region of 75 - 100 Wh/Kg and specific power values[6] of 75 - 100 W/Kg although the average cycle life (500 - 1000) at present appears to be inferior to that attained by current Na/S cells. 40 Kwh battery modules have also been constructed for evaluation tests. The preceding performance characteristics have only been achieved by incorporating relatively expensive ceramic materials in the cell construction as can be seen from the schematic diagram (Fig. 4) of a prismatic Li-Al/FeS cell. The molten LiCl-KCl electrolyte is immobilised within a woven BN cloth separator which provides electrical isolation of the electrodes but permits migration of the lithium ions. The cell also incorporates particle retainers of Y_2O_3 felt for the Li-Al and FeS electrodes. It should also be noted that BN blocks and strips are used for electrical insulation purposes. The selection of BN and Y_2O_3 components is dictated by the requirement for materials that are compatible with the molten halide electrolyte, and also with the active electrode materials which exhibit either high lithium or sulphur activities. Until recently only BN and Y_2O_3 appeared able to satisfy these exacting requirements[6] but more recently the Varta Battery Company in Germany claim to have fabricated an MgO separator which can replace the BN cloth component. There is no doubt that cheaper components are essential if the Li-Aℓ/FeS battery is ever to become a practical system.

Finally it should be noted that because of the higher melting point of lithium (180°C) and its greater reactivity compared to sodium (m.p. 98°C) there have been no investigations into moderate temperature lithium batteries incorporating liquid positive plate components.

2.4. Solid State Battery Systems

All the systems summarised in Table 2, except for Li/TiS$_2$ can be regarded as essentially classical secondary battery systems in that the charge-discharge reaction involves the formation of new phases. Nucleation and growth of these new phases often restricts the magnitude of the current and power density. The associated morphological changes can also lead to a loss of active electrode material and consequent deterioration in overall performances. In contrast, the alternative concept of insertion electrode materials avoids the formation of new phases and also overcomes many of the problems associated with the construction of completely solid state batteries as highly resistive reaction products are not produced at the electrolyte/electrode interface.   The concept of using insertion electrodes was first discussed by Steele[25] and Armand[26], and the practical implementation of these ideas was pioneered by the Exxon group[27] who demonstrated the high rate reversibility of lithium insertion electrodes such as TiS$_2$. The mode of operation of these electrode materials can be explained with reference to Fig. 5, using TiS$_2$ as a specific example. During the discharge reaction, lithium ions are produced at the negative plate electrode. These ions migrate across the electrolyte phase and are then inserted into the TiS$_2$ host lattice thus producing the solid solution Li$_x$TiS$_2$. At the same time an equal number of electrons are transported around the external circuit and are incorporated into the Li$_x$TiS$_2$ insertion electrode. The discharge process thus consists of a double injection of ions and electrons into the TiS$_2$ host lattice. For one equivalent of lithium inserted (i.e. LiTiS$_2$) the associated theoretical specific energy density is 480Wh/Kg which should produce practical cell configurations with specific energy densities greater than the required 100Wh/Kg. Many other insertion electrodes have now been synthesised and some of these have theoretical energy densities greater than 1000Wh/Kg. Relevant details have been summarised in recent reviews[28,29].

Many groups[22] have attempted to incorporate these solid reversible positive electrode into practical cell configurations using liquid organic electrolytes. The best performance so far was exhibited by the Exxon system[27]

Li / dioxolane (LiCℓO$_4$) / TiS$_2$

393

which had excellent cycling characteristics. Although very promising, development work on this system had to be abandoned because of a series of explosions initiated by the decomposition of $LiClO_4$ that occurred during high rate cycling experiments.

The alternative approach of incorporating solid electrolytes to make completely solid state batteries has to overcome the problems of fabricating good solid electrolyte-electrode interfacial contacts capable of withstanding the stresses generated during repeated charge-discharge cycles. Experience within the Wolfson Unit for Solid State Ionics at Imperial College has indicated that it is possible to fabricate appropriate solid-solid interface providing that at least one of the components can exhibit some degree of either elastic or plastic deformation. Measurements[30,31,32] on well designed solid electrolyte-electrode interfaces indicate that interfacial impedances can be very small indeed which implies that the energy barriers to ion transfer are also very small.

Selection of appropriate lithium ion solid electrolytes is severely restricted as few lithium ion conductors exhibit conductivities in the temperature range 25 - 150°C greater than 10^{-3} $(ohm.cm)^{-1}$ which are the minimum requirements for many practical applications. Reference to Fig. 6 confirms that development of superionic crystalline lithium ion electrolytes has been much less successful than the synthesis of alternative sodium ion conductors The values shown in Fig. 6 for $Li_{3.6}P_{0.4}Si_{0.6}O_4$ are representative of a number of crystalline oxide lithium conductors including $Li_4B_7O_{12}Cl$, and it should be noted that minimum specific conductivity of $10^{-3}(ohm.cm)^{-1}$ is not attained until approximately 200°C Although Li_3N appears to satisfy the conductivity requirements it has a relatively low thermodynamic decomposition potential (0.44V at room temperature) which must raise serious doubts[33] about the long term stability of this material when exposed to the exacting environmental conditions associated with secondary high energy batteries. The better oxide lithium ion conducting glasses have conductivities comparable to the oxide crystalline electrolytes (Fig. 6), but replacement of the oxygen by the more polarizable sulphide ion[34] increases the lithium ion conductivity by a significant amount. It has been reported[35] that the stability of sulphide glasses based on $Li_2S-P_2S_5-LiI$ compositions also exceeds 2V, which will ensure that this and other similar glasses will be extensively investigated as promising electrolyte components in batteries, although it must be emphasised that their extreme sensitivity to moisture will inevitably introduce fabrication problems for practical cell configurations.

The material $LiI-Al_2O_3$ is an example of a poly-phase solid electrolyte[35,36]. The intimate mixing of small particles of an ion conducting phase with a non-conducting phase has been found to enhance the total conductivity. Although explanations for the mechanisms of lithium ion transport in these polyphase electrolytes remain controversial, there is general agreement that the mechanism is associated eith the high specific surface area of the interparticle contact regions. It should be noted that the $LiI-Al_2O_3$ polyphase electrolyte attains a specific conductivity of $10^{-3}(ohm.cm)^{-1}$ around 100°C, and at this temperature is sufficiently plastic to ensure good interfacial contact with solid insertion electrode materials. This material has been incorporated into the solid state cell[35]

$$Li - Si / LiI(Al_2O_3) / TiS_2, Sb_2S_3, Bi,$$

which operated at 300°C. Preliminary design studies indicated that practical energy densities of 200Wh/Kg and 500Wh/ℓ could be realised with this cell but only limited cycling data were reported and the high temperature of operation appears to be a disadvantage.

The fabrication of solid state cells could be facilitated by incorporating polymeric lithium ion[37,38] electrolytes of the type depicted in Fig 6. Although these materials, such as polymeric polyethylene oxide (PEO) incorporating appropriate lithium salts, are relatively poor conductors ($10^{-5}-10^{-6}ohm^{-1}cm^{-1}$) at room temperature the specific conductivity can attain values of $10^{-3}-10^{-2}$ $(ohm\ cm)^{-1}$ at 120°C. These polymeric electrolytes can easily be fabricated into thin films and certain of the formulations also exhibit elastomeric properties which are very desirable for the fabrication of completely solid state batteries. Although it is possible to envisage batteries incorporating only polymeric electrolytes and composite electrode components, it is also possible that polymeric electrolytes will be used to provide an accommodating interface between hard ceramic electrolyte and electrode components in solid state batteries.

It will be difficult to design completely solid state batteries incorporating sodium ion conductors because of the problems associated with synthesing sodium insertion electrodes[29] principally due to the larger size of the sodium ion. The structural requirements for interstial channels large enough to accommodate sodium ions invariably means that the transition ions within the framework are usually too far apart for significant d orbital over lap to occur resulting in relatively poor electronic conductivity. An exception are the 5d electrons of tungsten. The metastable hexagonal form[39] of WO_3 can reversibly and rapidly accommodate upto 0.33 equivalents of sodium and the inserted electrons exhibit a good conductivity. However the relatively low capacity and heavy tungsten ions combine to produce an energy density that is unattractive for practical cells.

Finally it should be noted that solid state cells incorporating fluoride ion solid electrolytes have been fabricated[40] but the performance characteristics are not very encouraging principally due to the absence of appropriate insertion electrodes and at present the high resistive interfacial discharge products rapidly reduce the current density.

2.5 Ceramics in Advanced Batteries : Conclusions

The large investment in the high temperature (300°C) Na/S battery will ensure that this battery will continue to receive financial support in the near future so that evaluation of prototype battery modules can be completed. The replacement of beta-alumina by alternative electrolytes such as $Na_{3.1}Zr_{1.55}Si_{2.3}P_{0.7}O_{11}$ needs to be carefully examined, but probably the greatest need is to develop moderate temperature (100-180°C) sodium batteries. Not only will this reduce the exacting requirements for the ceramic electrolyte but lower temperature operation will make commercial exploitation much easier. Although 300°C operation is not necessarily a disadvantage for load-levelling applications there will be little demand in Europe for off-peak storage in the foreseeable future, and yet it is unlikely that a battery operating at 300°C will find widespread application as a power source for electric vehicles. Associated with lower temperature operation is the need to synthesise alternative sodium and lithium ion solid electrolytes with improved performance in the temperature range 25 - 150°C, and particular attention should be given to the development of novel glass electrolytes because of their ease of fabrication.

3. Fuel Cells

3.1. General Comments

The fuel cell is an electrochemical device for converting a fuel directly into electricity without the thermodynamic limitations inherent in heat engines. A comparison of the theoretical efficiencies of a fuel cell utilising the H_2/O_2 reaction, and a heat engine, as a function of temperature are summarised in Fig. 7. The superior performance of the fuel cell throughout most of the temperature range is clearly demonstrated. However it is more realistic to compare actual operating efficiencies as a function of load factor and this is done in Fig. 8 for a variety of car engines (41). A typical load factor for urban driving conditions will be around 10% and the superior performance of fuel cells in this situation is clearly demonstrated in Fig. 8. Only if the load factor approaches 50%, which is very rarely achieved in commercial operation, does the efficiency of automotive heat engines approach that exhibited by fuel cells. As well as superior efficiencies fuel cells in principle offer additional advantages such as, modular and simpler construction, fewer moving parts requiring less maintenance, quieter and cleaner in operation. At present the superior characteristics of fuel cells, however, are offset by higher production costs and uncertainties about operating lifetimes of certain components such as the electro-catalytic electrode materials.

3.2 Principal Types of Fuel Cells

In principle any fuel can be oxidised at an electrode to generate a current but in practice the rates of reaction are often slow resulting in low values for power production. For most fuel cells hydrogen is the preferred fuel. Certain hydrocarbons (e.g. methanol and methane) can be used directly but it is often desirable to reform them to hydrogen first. Similarly, coal and residual oil may be gasified to provide the hydrogen. However the cost of fuel processing can be a critical factor in the economics of fuel cell systems. The principal types of fuel cells under development are summarised in Table 3 with some information about their relevant fuel requirements. More details are given in the review by Kordesch[47]. It is necessary to operate the fuel cells incorporating aqueous electrolytes at elevated temperatures to improve the kinetics of oxygen reduction which is relatively slow at ambient temperatures even in the presence of electro-catalysts such as platinum. Operation at elevated temperatures also improves the conductivity of the electrolyte particularly in the case of phosphoric acid. It should be emphasised that all the fuel cells except the one incorporating a solid oxide electrolyte have severe restrictions on the type and purity of fuel that can be accepted. In this respect the solid oxide electrolyte based fuel cell is the most versatile. The fuel restrictions arise from the need to prevent poisoning of the catalysts by CO and H_2S, or to prevent the formation of insoluble K_2CO_3 due to reaction of CO_2 with the alkaline aqueous electrolyte KOH.

The following brief comments outline the principal features of the different fuel cells incorporating liquid electrolytes. Details of the high temperature solid oxide electrolyte system are considered in section 3.3.

The alkaline electrolyte fuel cell, originally designed by Bacon in 1959, has since been used extensively for the electrical power source in space vehicles. The cell performs best on pure hydrogen and oxygen, and the availability of these fuels on spacecraft has ensured the successful application of this system in this specialised area. In principle the fuel cell can be operated on other fuels, for example, ammonia and reformed hydrocarbons. However the removal of carbondioxide from a hydrocarbon based fuel is expensive, and so the alkaline fuel cell is not a practical choice for use with conventional fuels. However alkaline fuel cells are being developed in Europe by Elenco and Siemens for transport and specialist applications. The hydrogen fuel will be stored in pressure vessels or as a hydride.

The electrolyte used in the phosphoric acid fuel cell is not sensitive to carbon dioxide and so it is suitable for use with conventional hydrocarbon fuels. This system, manufactured by United Technologies Corporation, is now at an advanced stage of development, and 4.5 MW demonstration units have been installed in Manhattan and Tokyo. In addition about fifty 40Kw power units are to be evaluated in on-site industrial combined heat and power applications. Initially the two major problem areas were re-crystallisation of the platinum catalyst producing loss of active surface area and corrosion of the catalyst support. Both of these problems have apparently been solved by the choice of appropriate materials and fabrication processes. A schematic diagram of the phosphoric acid fuel cell is depicted in Fig. 9.

The sulphuric acid/methanol fuel cell has been extensively invest-igated by Shell[41] as the power source for a variety of electric vehicles. Although methanol is not at present regarded as economic fuel it may become so in the future, and it represents an attractive possibility for applications in fuel cells as it exhibits some electrochemical activity even at ambient temperature. However the kinetics of the electrode reactions are still too slow, and successful exploitation will depend upon the development of cheap, effective, electro-catalysts for both the methanol and air electrode.

The molten carbonate fuel cell (Fig. 10) consists of a thin layer of molten carbonate ($Li_2CO_3 + K_2CO_3$) immobilised in an inert mat-rix which is sandwiched between two electrodes made of porous nickel. At the operating temperature of 650°C, the reaction rate at the electrodes and the diffusion rates of gases and ions are so high that polarisation is virtually eliminated even without special catalysts. The cell therefore has a higher efficiency than the phosphoric acid system and can accept a lower quality fuel gas containing high carbon monoxide concentrations. However the molten carbonate cell suffers from a gradual loss of electro-lyte which limits endurance and reliability. Corrosion and de-gradation of the electrodes and other components is also a problem and there is some dispute about the level of H_2S that can be tol-erated by the cell. These problems may be capable of solution but the molten carbonate fuel cell is still some way from commer-cialisation.

Examination of the preceding summaries about fuel cells incorp-orating liquid electrolytes indicates that the role of ceramic technology in these systems is essentially twofold. The charac-terisation and behaviour of the electro-catalytic particles (e.g. Pt, Ni, NiB_2, WC) have many features in common with ceramic powder technology except that in the fuel cell situation it is important to reduce agglomeration and sintering of the particles to maintain a large active surface area for as long as possible. The other relevant area is the provision of inert matrices to immobilise the liquid electrolyte and separate the two electrodes. For example, phenolic and SiC/teflon separators are the preferred material for the phosphoric acid electrolyte, and the molten carbonate electrolyte (Li_2CO_3 and K_2CO_3) is retained in a porous tile of $LiAlO_2$.

3.3. Fuel Cells Incorporating Oxygen Ion Conducting Electrolytes

In principal this high temperature system should have the highest overall efficiency and it avoids the corrosion and electrolyte stability problems encountered with the molten carbonate fuel cell. The preferred electrolyte is usually a solid solution of zirconia with small quantities of yttria which has an oxygen ion conductivity of 10^{-1} (ohm cm)$^{-1}$ inbetween 900 and 1000°C. When thin ($\sim$ 1 mm) ceramic electrolytes of this composition are incorporated into fuel cells current densities of around 150mA/cm² have been sustained for over three years operation at 1000°C. The principle problem still appears to be the life-time of the air electrode (cathode) with plasma-sprayed Sr doped $LaNiO_3$ and SnO_2 doped In_2O_3 exhibiting the most promising results. The thin ZrO_2 based electrolyte is usually supported on a porous tubular substrate and series connected multi-cell assemblies can be fabricated by successively depositing thin layers of anode, electrolyte, cathode and interconnection materials to produce the structure depicted in Fig. 11. The evaluation and performance of 100 multi-cell modules (100 V, 115 W) has been described[42,43] using H_2 as the fuel. However in Europe development work has stopped principally because the fuel cell components and the ancillary structural materials required for operation in the temperature range 900 - 1000°C were considered too expensive for commercial systems.

It should be emphasised that data for the electrical properties of the zirconia based electrolytes incorporated in these high temperature fuel cells have been available for almost thirty years. Theoretical calculations and careful measurements on these oxide fluorite solid solutions have certainly contributed to a better understanding[44,45] of oxygen migration in highly defective materials, but in general very little effort has been made to synthesise novel oxygen ion conductors with improved anionic conductivity values. The conductivity-temperature relationships for a selection of fluorite and fluorite-related oxygen ion electrolytes are summarised in Fig. 12 which also depicts the region of conductivity values required for moderate temperature (500 - 600°C) fuel cell operation. It is evident that at present the electrical conductivity values of oxygen ion electrolytes cannot attain the required performance characteristics. However, only moderate improvements are required, and it should be noted that the ionic conductivity behaviour of δ-Bi_2O_3 indicates, that in principle at least, fast oxygen ion migration is possible in oxide structures with relatively low (0.4 eV) energies for migration. Even lower values (0.35 eV) for oxygen migration energies have recently been reported[46] for $Bi_{12}PbO_{19}$. There is every incentive, therefore, to try to develop alternative oxygen ion conductors using the concepts of molecular engineering.

3.4. Ceramics in Fuel Cells : Conclusions

The fuel cell is more fuel efficient than other electricity generating devices and therefore has great potential application in the area of energy conservation. This is recognised in the U.S.A with fuel cell R. and D. attracting significant funds which were increased in 1982. A partial breakdown of the sources of this U.S. funding ($77.5 million in 1982) is given in Table 4. In addition NASA is still funding fuel cell development for aerospace, with the U.S. Departments of Energy and Defence also supporting work for fuel cells for transport. It is estimated that at least an additional $15 million are also involved in these programmes. In contrast the sums spent in Europe per annum are very small ($2-3 million?). It is suggested that development work on selected fuel cell systems should be expanded. In view of the large funding in North America for the phosphoric acid and molten carbonate systems it is recommended that development work in Europe should concentrate on those alternative systems which could provide the second generation of more efficient and cheaper systems that would be produced commercially towards the end of this century. The sulphuric acid/methanol system designed for transport applications is one such system, and it is possible that the development of novel carbide or boride electro-catalysts could provide the break through required to make this fuel cell a practical reality. The other system is the moderate temperature (300 - 600°C) fuel cell incorporating either oxygen ion or proton conducting ceramic electrolytes. The development of improved oxygen ion conductors is particularly attractive as these materials could also be incorporated into amperometric oxygen monitors designed to operate at much lower temperatures than the conventional potentiometric devices based on zirconia electrolytes. The successful development of a fuel cell incorporating a ceramic electrolyte would obviously provide a major stimulus for the technical ceramic industry in Europe.

4. References

1. H.K. Bowen, Mat. Res. & Engin, 1980, 44, 1-56.
2. S. Geller (ed), Solid Electrolytes, Vol. 21, Topics in Appl. Physics, 1977, Springer-Verlag.
3. P. Hagenmuller and W. Van Gool (eds), Solid Electrolytes, 1978, Academic Press.

4. E.C. Subbarao (ed), Solid Electrolytes and Their Applications 1980, Planum Press.
5. J. Jensen, P. McGeehin and R. Dell, Electric Batteries for Energy Storage and Conversion, 1979, Odense Univ. Press.
6. D.W. Murphy, J. Broadhead and B.C.H. Steele (eds), Mats. for Advanced Batteries, 1980, Plenum Press
7. C. Stein (ed), Critical Materials Problems in Energy Production, 1976, Academic Press.
8. G.G. Libowitz and M.S. Whittingham (eds), Materials Science in Energy Technology, 1979, Academic Press.
9. B.C.H. Steele, Phil. Trans. R. Soc. Lond. A 1981, 302, 361-374.
10. R.M. Dell and R.J. Bones, in P. Vashishta, J.N. Mundy, G.K. Shenoy (eds), Fast Ion Transport in Solids, 1979, 29-37, North Holland.
11. W. Fischer, Solid State Ionics, 1981, 3/4, 413-424.
12. G.J. May, Power Sources, 1978, 3, 1-22.
13. L.C. De Jonghe, L. Feldman and A. Buechele, Solid State Ionics, 1981, 5, 267-270.
14. J.O. Bovin in P. Vashishta, J.N. Mundy and G.K. Shenoy (eds), Fast Ion Transport in Solids 1979, 315-318, North Holland.
15. K.M. Abraham, 'Moderate Temperature Sodium Batteries', To be published in Solid State Ionics.
16. W. P. Shalette, I.S. Klein and J. Werth in N.P. Yao and J.R. Selman (eds), Proceedings of Symposium on Load Levelling, 1977, 77-4, 306-317, Electrochemical Soc. Inc.
17. K.M. Abraham, M.W. Rupich and J. Elliott, to be presented at the Fall Meeting of the Electrochemical Soc. (Detroit, Oct. 1982).
18. G. Weddigen in S. Gross (ed), Proceedings of Symposium on Battery Design and Optimization, 1979, 79-1, 436-445, Electrochemical Soc. Inc.
19. K.M. Abraham, and L. Pitts, J. Electrochem Soc. 1981, 128, 2574-2577.
20. C. Levine, J. Power Sources 1980, 5, 363-364.
21. S. Susman, L. Boehm, K.J. Volin and C.J. Delbecq, Solid State Ionics, 1981, 5, 667-670.
22. K. M. Abraham and S.B. Brummer, Lithium Secondary Batteries, to be published in J.P. Gabano (ed), Lithium Batteries 1982, Academic Press.
23. B.C.H. Steele, Organoelectrolyte and Solid State Batteries, to be published in D.A.J. Rand and B.D. McNicol, Power Sources for Electric Vehicles, 1983, Elsevier Press.
24. J.E. Battles, J.A. Smaga and K.M. Myles, Metall. Trans. A, 1978, 9, 183-191.
25. B.C.H. Steele in W. Van Gool (ed), Fast Ion Transport in Solids, 1973, 103-122, North Holland.
26. M.B. Armand, ibid., 665-673.

27. M.S. Whittingham, U.S. Patent 4, 004, 052, (1973).
28. K.M. Abraham, J. Power Sources, 1981/82, 7, 1-43.
29. See ref. 23.
30. J.M. Shemilt, B.C.H. Steele and J.E. Weston, Solid State Ionics, 1981, 2, 1-11.
31. M. Kleitz, J.R. Ackridge and J.H. Kennedy, Solid State Ionics 1981, 2, 67-72.
32. J.R. Owen, J. Drennan, G.E. Lagos, P.C. Spurdens, and B.C.H. Steele, Solid State Ionics, 1981, 5, 343 -346.
33. J.R. Rea and D.L. Foster, Mater. Res. Bull, 1979, 14, 841-846.
34. J.L. Souquet, Solid State Ionics, 1981, 5, 77-82.
35. C.C. Liang, A.V. Joshi and N.E. Hamilton, J. Appl. Electrochem. 1978, 8, 445-454.
36. J.B. Wagner, Mater. Res. Bull. 1980, 15, 1691-1701.
37. M.B. Armand in E.B. Yeager, B. Schumm, G. Blongren, D.R. Blakenship, V. Leger and J. Ackridge (eds), Lithium Non Aqueous Battery Electrochemistry, 1980, 80-7, 261-275, Electrochemical Society Inc.
38. J.E. Weston and B.C.H. Steele, Solid State Ionics, 1981, 2 347-354.
39. B. Gerand, J. Desseine, P. Ndata and M. Figlary in Proceedings of 2nd European Conf. on Solid State Chem. (Veldhaven, Netherlands, 7-9th June, 1982), to be published.
40. J. Schoonman, Solid State Ionics, 1981, 5, 71-76.
41. R. Glazebrook, J. Power Sources 1982, 7, 215-256.
42. F.J. Rohr in P. Hagenmuller and W. Van Gool (eds), Solid Electrolytes, 1978, 431-450, Academic Press.
43. T.L. Markin, R.J. Bones and R.M. Dell in G.D. Mahan and W.L. Roth (eds), Superionic Conductors, 1976, 15-35, Plenum Press.
44. J.A. Kilner and B.C.H. Steele in O.T. Sorensen (ed), Mass Transport in Anion Deficient Fluorite Oxides, 1981, 233-269, Academic Press.
45. J.A. Kilner and R.J. Brook, Solid State Ionics, 1982, 6, 237-252.
46. J.A. Kilner, J. Drennan, P. Dennis and B.C.H. Steele, Solid State Ionics, 1981, 5, 527-530.
47. K.V. Kordesch, J. Electrochem. Soc. 1978, 125, 77C-91C.

Table 1. Battery Performance Targets for Electrical Vehicles

Peak Power : 15-20 Kw/tonne
Average Energy : 0.1-0.2 Kwh/tonne - km.
Acceptable Battery wgt : <25% Unladen wgt

	Urban Van	Medium Car
Gross Vehicle wgt (tonnes)	3.5	1.5
Unladen wgt (tonnes)	2.0	1.3
Acdeptable Battery wgt (tonnes)	0.5	0.3
Range Desired (Km)	140	160
Peak Power Required (Kw)	50	27
Energy Required (Kwh)	50	27
Recharge Time Available (h)	14	16
Minimum Cycle Life	1000	500

Derived Parameters

	Urban Van	Medium Car
Battery Energy Density (Wh/Kg)	100	90
Battery Peak Power Density (W/Kg)	100	90
Recharge Rate (Kw)	3.5	1.7

Petrol*

	Urban Van	Medium Car
Volume (Litres)	39.2	19.2
Weight (Kg)	33.3	16.3

*0.08 litres (0.068 kg) / tonne - km.

Table 2 Battery Systems

Components				Current performance	
Negative plate	Electrolyte	Positive plate	Open circuit voltage/V	Energy Wh/kg	Power W/kg
Pb	H_2SO_4(aq)	PbO_2	2.05	30 (171)	50
Zn	KOH(aq)	NiOOH	1.7	50 (321)	150
Fe	KOH(aq)	NiOOH	1.3	60 (267)	70
Zn	$ZnCl_2$(aq)	$Cl_2(Br_2)$	2.1	110*(465)	100
Fe	KOH(aq)	Air	1.3	81 (764)	30
Zn	KOH(aq)	Air	1.6	120*(1080)	80
Na	β-alumina (solid)	S	2.0-1.7	180*(664)	220*
Li-Al	LiCl-KCl (molten salt)	FeS	1.6	50 (870)	-
Li	organic/ solid electrolyte	TiS_2	2.4-1.9	132*(480)	-

*Projected values.

Table 3   Principal Fuel Cell Types

Electrolyte	Electrode Catalyst	Operating Temperature	Fuel Requirements
Alkaline	Nickel Compounds	60 - 200°C	H_2(pure), NH_3, H_2H_4
Phosphoric acid	Platinum on carbon	200°C	H_2(free of CO, H_2S)
Sulphuric acid	Pt, Pt alloys	25-60°C	CH_3OH (direct) or H_2 (indirect)
Molten carbonate	Porous nickel	650°C	H_2(free of H_2S)
Solid oxide	Conducting oxide	1000°C	H_2(low purity) Ch_3OH

Table 4   U.S.A. Annual Spending ($M) in 1982

Fuel Cell Technology	Department of Energy	Department of Defence	EPRI*	GRI+	Private Industry	NASA
Alkaline	?	?	0	0	0	?
Phosphoric Acid	21.5	5	7	19	10	0
Molten Carbonate	10.5	0	1.6	0.8	?	0
High Temperature	1.5	0.3	0.3	0.15	0	0

* Electrical Power Research Institure, California

+Gas Research Institute, Chicago.

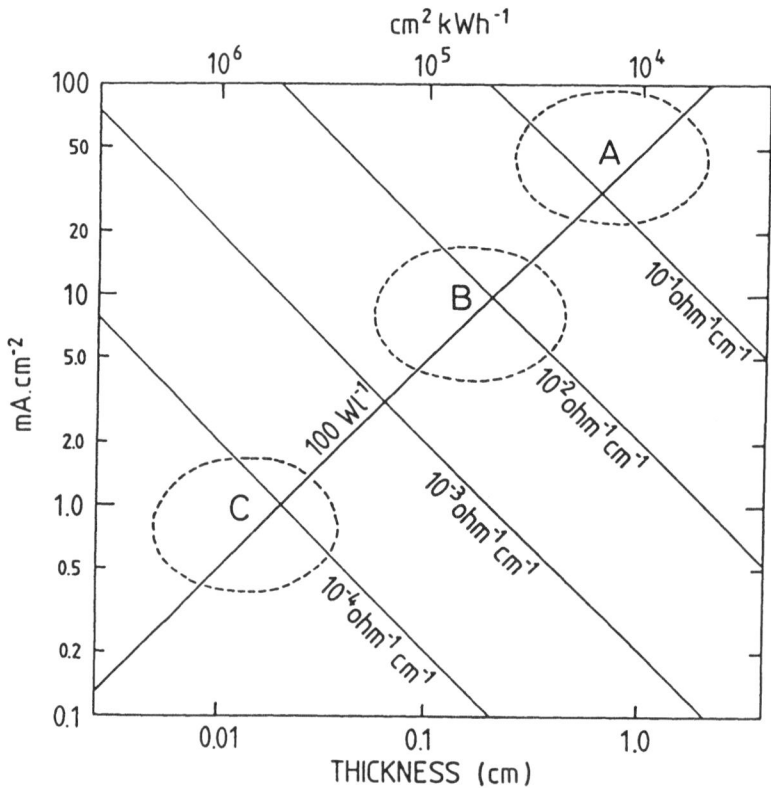

Figure 1 Configuration parameters of 100 W/1 battery cells
(sectn. 2.1.)

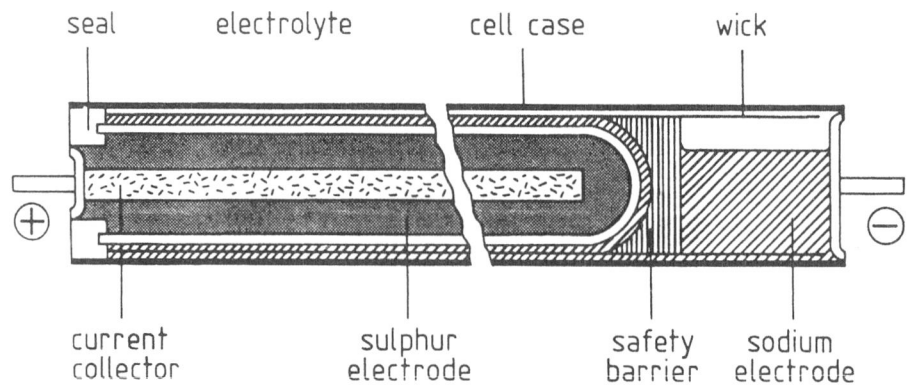

Figure 2   Design scheme of a sodium/sulphur battery cell

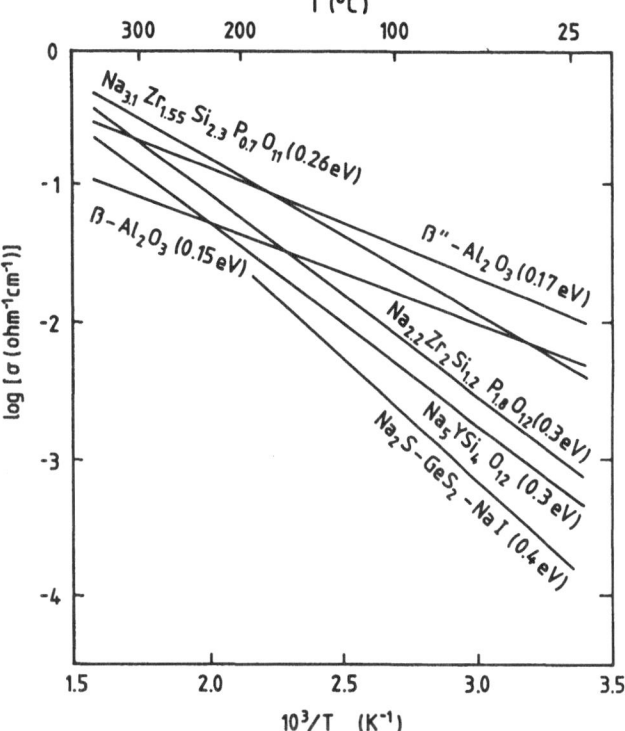

Figure 3   Sodium ion conductivity of some solid electrolytes

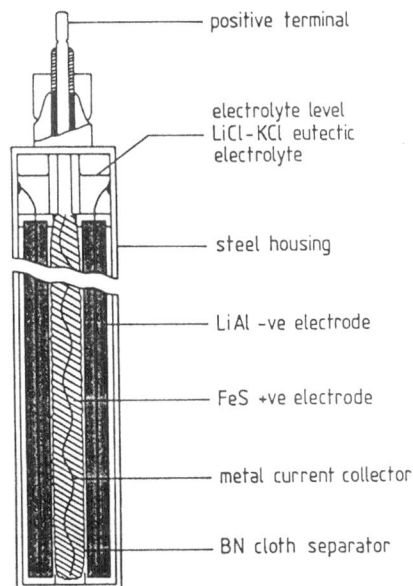

Figure 4   Design scheme of a Li–Al/FeS battery cell

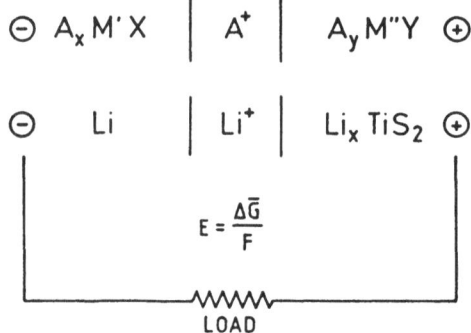

Figure 5   Schematic discharge process using TiS$_2$ as Li
insertion electrode

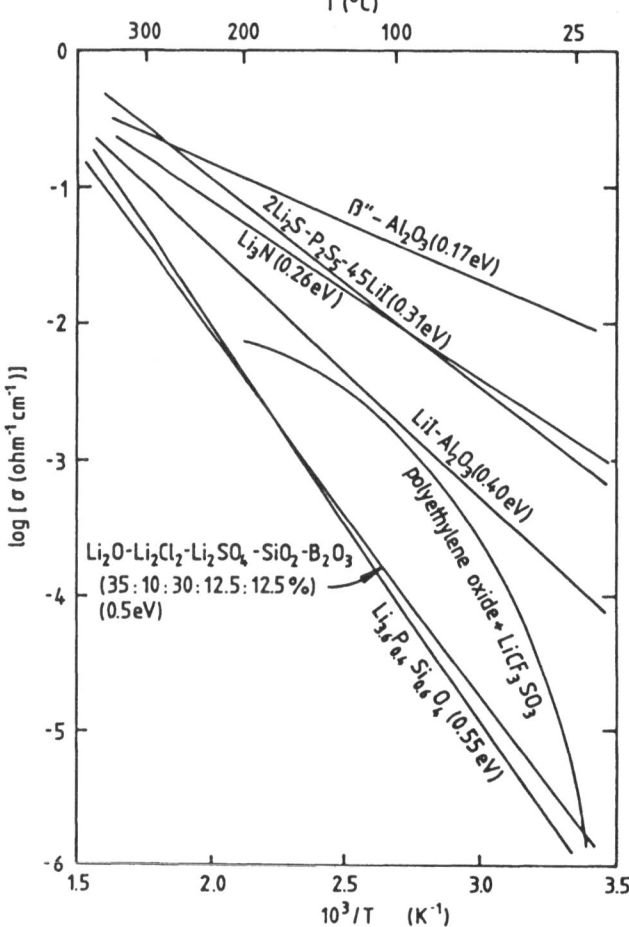

Figure 6   Lithium ion conductivity of some solid electrolytes

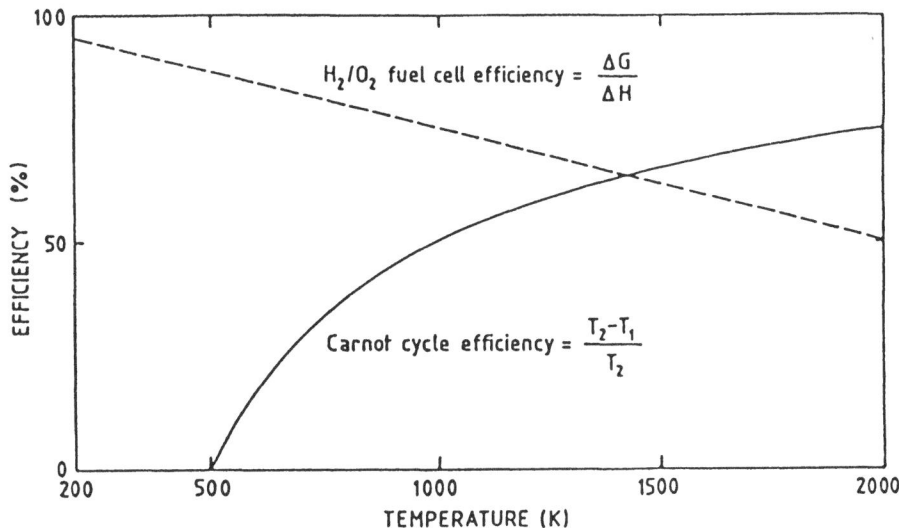

Figure 7

Theoretical conversion efficiency of fuel cells in comparison with heat engine Carnot efficiency

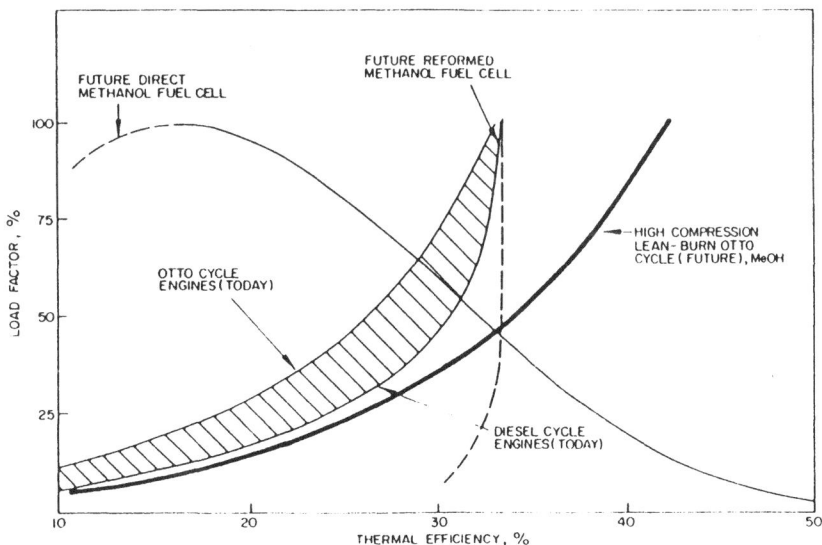

Figure 8

Operating efficiencies of car engines and fuel cells as function of load factor

409

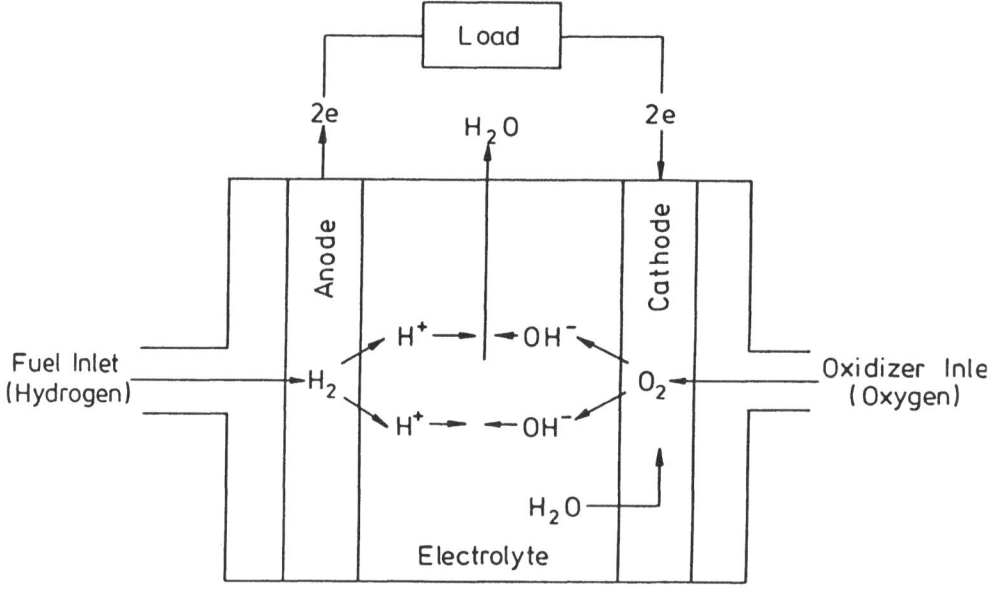

Figure 9   Operation scheme of phosphoric acid fuel cell

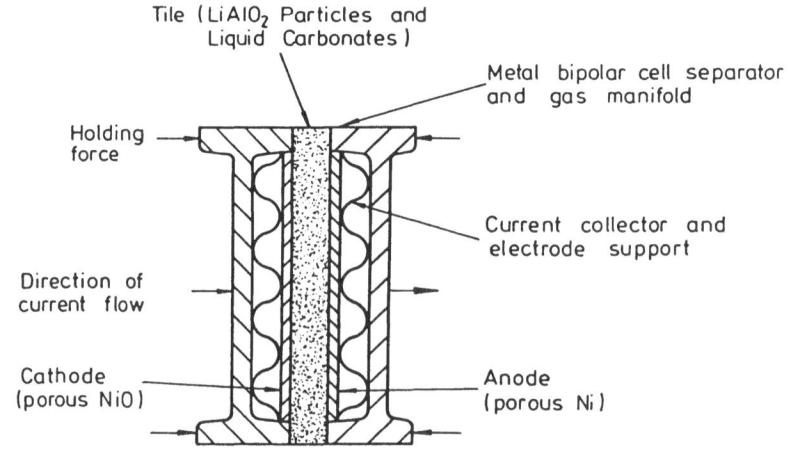

Figure 10   Design scheme of molten carbonate fuel cell

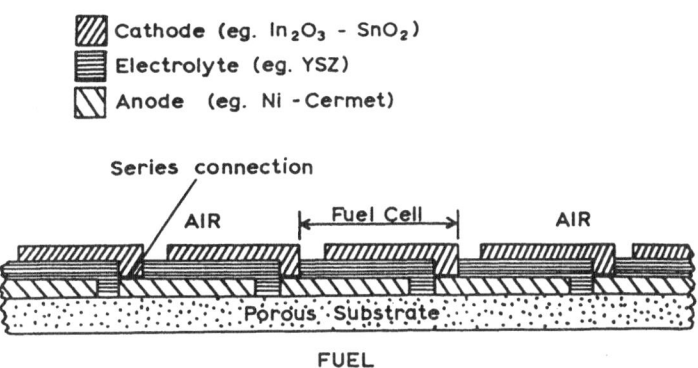

Cathode (eg. In₂O₃ - SnO₂)
Electrolyte (eg. YSZ)
Anode (eg. Ni - Cermet)

Figure 11

Series connected multi-fuel-cell assembly fabricated by thin-layer
deposition

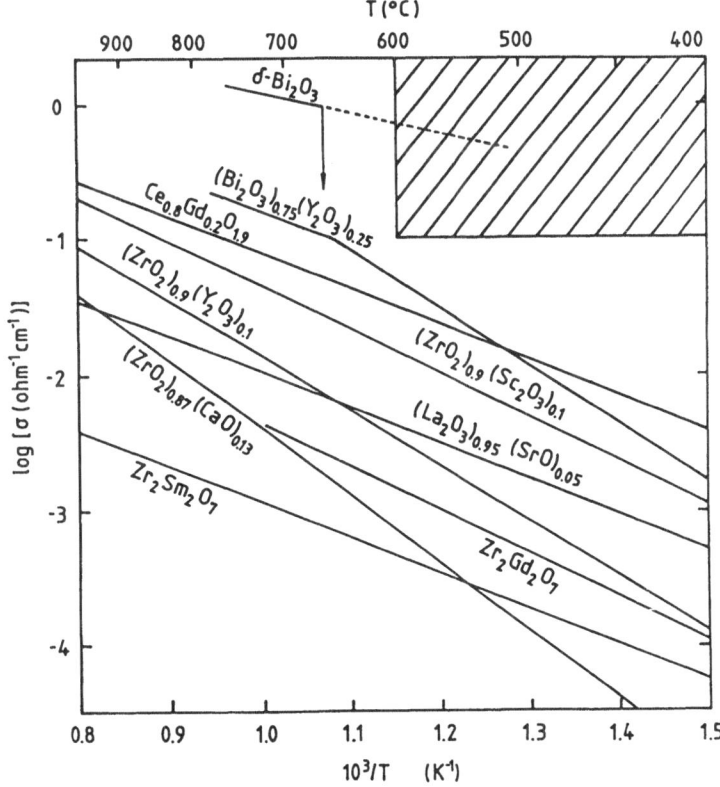

Figure 12   Oxygen ion conductivy of some solid electrolytes

DISCUSSION

R.J. Brook: The very favourable picture of fuel cell
efficiency shown in the Shell study is impressive; however,
with the efficiency of fuel cells falling as the
temperature is raised (the crossover with heat engines
being at about 1400 K), what is the picture for the high
temperature cell? What type of cell was considered in the
Shell study?

B.C.H. Steele: The efficiencies predicted by the Shell
evaluation were based on the direct methanol fuel cell
working at around 80°C. The comparison with Otto and Diesel
engines is so favourable for fuel cells because of the
relatively low load factor involved in transport applications.
The high temperature fuel cell would be used as a combined
heat and power source or in association with a coal
gasification scheme. At low load factors the fuel cell
would still be the more efficient system but the margin
between heat engines and fuel cells is not so great, and
at present is not sufficient to overcome the greater cost
of fuel cell systems which are not mass produced.

BETA-ALUMINA FOR SODIUM-SULPHUR BATTERIES

A. WICKER

Laboratoires de Marcoussis, Centre de Recherches de la
COMPAGNIE GENERALE D'ELECTRICITE
Route de Nozay, 91460 MARCOUSSIS, FRANCE

SYNOPSIS

This paper aims to review the state-of-the-art concerning the use
of beta-alumina ceramics as a solid electrolyte in the sodium-sulphur battery. Design and operating conditions of a sodium-sulphur
cell are first outlined leading to specifications for beta-alumina
tubes. The various fabrication procedures which have been studied
are presented with regard to powder preparation, shaping and sintering. The major problem encountered by the beta-alumina developers
is the in-cell durability of the electrolyte tube. Studies on the
electrical breakdown which is the main cause of failure have been
recently emphasized : failure is usually associated with the penetration or the formation of sodium in the electrolyte or as a general blackening extending from the sodium side. To solve this problem,
further research and development on improving the durability and the
reliability of beta-alumina tubes are needed.

INTRODUCTION

An Anglo-Danish study made within the frame of the Energy Research
and Development programme of the Commission of the European Communities[1] stated that advanced secondary batteries will have a significant role to play as power source for electrical vehicles and as
storage medium for load leveling in electricity utilities. The use
of secondary battery is expected both to conserve energy and to
allow the substitution of other primary energy sources for oil.
Among the advanced batteries under study the sodium-sulphur system,
in which beta-alumina is used as solid electrolyte, is one of the
most promising with respect to the expected high energy density.

THE SODIUM-SULPHUR CELL

The concept of the sodium-sulphur battery was proposed by WEBER and KUMMER in 1967[2]. Since that time sodium-sulphur battery development programmes have been pursued vigorously in U.S.A., Japan, China, Germany, United-Kingdom and France.
The principles of the construction of the sodium-sulphur cell are illustrated in Fig.1.

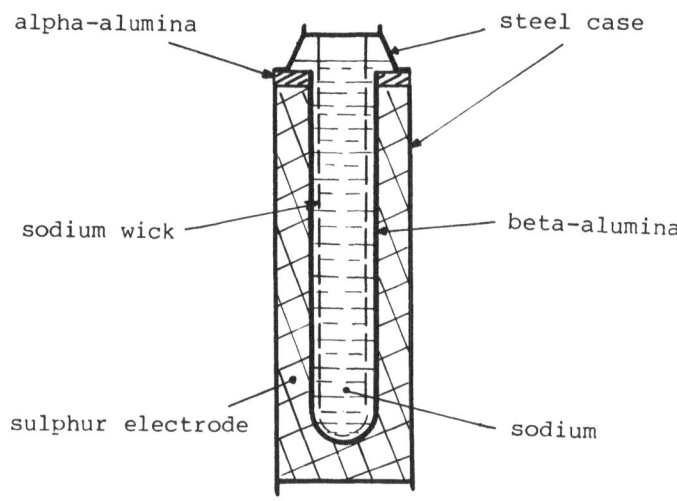

Figure 1 - Schematic of a sodium-sulphur cell (central sodium).

The solid electrolyte in the cell is the sodium conducting ceramic beta-alumina which usually takes the form of a tube. The electrodes, which are liquid sodium and sulphur, are disposed on both sides of the tube. Figure 1 shows the sulphur electrode outside the tube and the sodium electrode inside, but the cell may also be designed with the electrodes interchanged[3]. During operation of the cell, sodium is discharged at the anode and sodium ions migrate through the ceramic to the sulphur electrode. In the first state of discharge, the compound Na_2S_5 is formed giving a constant open-circuit voltage of the cell of 2.08 V. As liquid sulphur is both an electronic and ionic insulation, the sulphur compartment is filled with graphite felt which serves as an electron injector-acceptor. On further discharge lower sodium polysulfides are formed and the voltage falls progressively as the composition changes from Na_2S_5 to Na_2S_3. The cell is operated at between 300 to 380°C so that the discharge products remain liquid.
The sodium-sulphur reaction, based on Na_2S_3 as final product, has a theoretical energy density of 790 Wh/kg. Cells of the central sodium type have energy densities around 200 Wh/kg, depending upon design and size, whereas central sulphur cells have rather lower energy

densities (150 Wh/kg). The utilization of sulphur which is about 80% remains almost constant throughout cell life. Major advances in sodium-sulphur battery technology have been made during the past ten years concerning sealing and corrosion problems. In a sodium-sulphur cell, several seals are needed to prevent for instance, the leaking of molten sodium and sulphur (see Figure 1). These seals can be divided in two groups : ceramic to ceramic and ceramic to metal. A ceramic to ceramic seal must have appropriate thermal expansion, electrical and corrosion resistance characteristics ; borosilicate and aluminoborate glasses can be used to seal alpha-alumina ring to beta-alumina tube. This alpha-alumina ring serves as an electronic insulation between the metallic containers and the electrolyte. For ceramic to metal seal, several technics based on the compression of aluminum gasket between the metal and the ceramic have been developed satisfactorily[4,5]. Liquid polysulphides are very agressive substances, most metal, alloys and other conductive materials are severely corroded but several solutions have been found using chromium plated or alloyed[5,6,7] materials.

BETA-ALUMINA

Sodium beta-alumina, of general formula Na_2O, xAl_2O_3, occurs principally in two crystalline forms : β(hexagonal)[8,9] and β"(rhomboedral)[10,11]. The high conductivity results from the crystal structure which consists of layers in which sodium ions are relatively mobile separated by non-conductive layers of aluminum and oxygen ions (spinel blocks). The β-Al_2O_3 unit cell is made up of two spinel blocks with a minor plane between them, while β" has three blocks which are related by a three-fold screw axis[12]. β" is well-known to possess a lower resistivity than β-Al_2O_3 (3-6 Ω.cm vs 10-15 at 300°C) and is stabilized by small doping additions of lithium and/or magnesium that substitute for aluminum in the spinel blocks.

CERAMIC MANUFACTURING PROCESSES

The manufacturing problem of beta-alumina are more severe than for conventional ceramic tube because the specification which are needed for sodium-sulphur battery operation are very exacting : high ionic conductivity, high density (impervious), straightness, close dimensional tolerances, high mechanical properties (strength, toughness), stable to moisture, long life in cell and reliability. In common with other ceramics beta-alumina is manufactured through three stages : powder preparation, shaping and sintering. Several fabrication procedures have been adopted, they are summarized in Fig. 2.

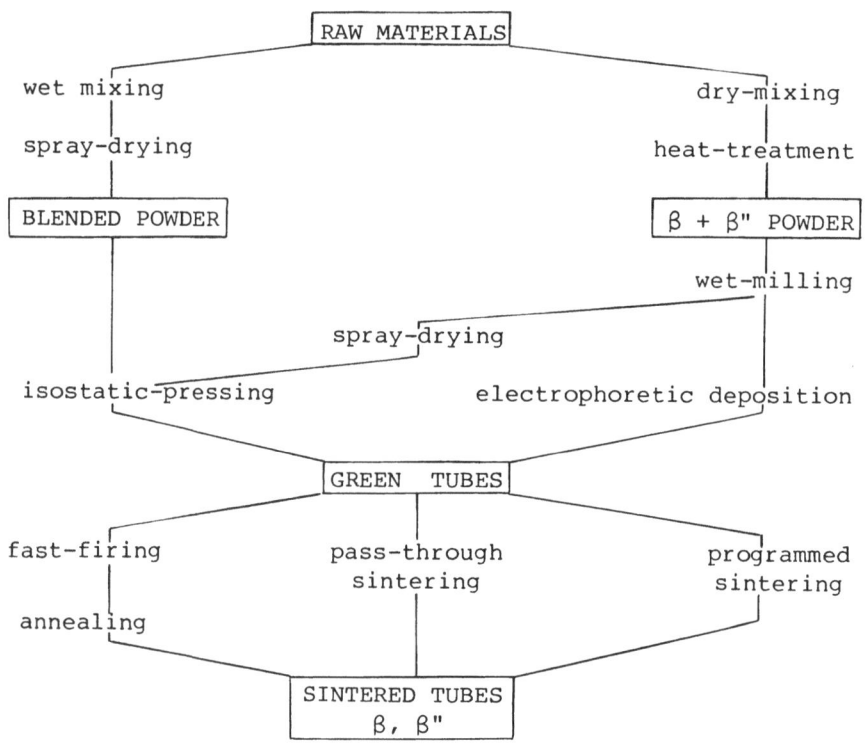

Figure 2 - Manufacturing routes

POWDER PREPARATION

Two main preparation techniques are used according as beta-alumina
is formed before or after the green forming stage. These procedures
aim at producing an homogeneous powder with a high degree of sintering activity. The raw materials are normally α-alumina powder with
soda added as Na_2CO_3, $NaNO_3$, NaOH or $NaAlO_2$, MgO as $MgNO_3$, MgO itself or in the form of magnesium aluminate and Li_2O as Li_2CO_3, LiOH
or Li_2O $5Al_2O_3$. Beta-alumina can be synthetized by heat treating in
the solid state or by direct fusion of the starting materials. Chemical methods have been also investigated : decomposition of nitrate[13],
of oxalates or alkoxides[13,14] ; these processes produce a high homogeneous powder. A gel processing, using citrates[15], leads to an homogeneous powder with a high specific area. Blended powders can be
produced by direct mixing of the starting materials using wet or dry
milling. Spray drying of aqueous solutions or suspensions have been
also investigated[16] in order to obtain homogeneous free-flowing powders.

SHAPING

Several processes have been used for making beta-alumina tubes.

416

. Isostatic pressing, in which powders contained in a rubber bag
with a steel mandrel are compacted by applying a pressure to the
outside of the bag[17] ; by this way high density and good dimensional tolerances are achieved, this technique requires a free flowing
powder and allows the production of close-ended tubes with thin
walls (1 to 3 mm) and lengths over 50 cm.

. Electrophoretic deposition[18,19], the powder suspended in an organic solvent is deposited on a metal mandrel by means of an electric
field. The deposit has good dimensional tolerances and surface
finish, removal from the mandrel requires long drying periods.
Besides these two main processes, other techniques have been investigated : extrusion[20], slip casting[21], but these methods do not appear to have led to large application as yet.

SINTERING

Two main processes are used for sintering beta-alumina tubes : zone
sintering and conventional batch sintering. These processes have
been optimized with regard to specific problems which occur with
beta-alumina : soda evaporation at temperature above 1300°C, difficulty to obtain an homogeneous fine-grained microstructure and differences in the kinetics of densification and β to $\beta"$ conversion.

. Zone sintering[22] is a continuous production process. Tubes are
fed through a high temperature hot zone to provide a short firing
cycling to minimize soda loss. Typical conditions to obtain fine-
grained ceramics are 1700°C and a tube speed of 10-50 mm/mn. For
high speed sintering, a post-sintering annealing is often used at
1400-1500°C which improves the ionic conductivity by increasing the
volume fraction of $\beta"$ phase. This annealing can be made during an
hot isostatic pressing stage[23] which improves the densification and
the mechanical properties.

. Conventional batch sintering allows to fire a stack of tubes, these
tubes are heated to sintering temperature and cooled according to a
programmed cycle[24,25]. In order to prevent soda losses, the tubes
must be enclosed in ceramic containers made of magnesium or beta-
alumina. A firing cycle takes about 10-24 h, with a top temperature
of 1600-1700°C according to the composition. Sintering parameters
must be controlled with accuracy in order to ensure reproducibility
(conductivity, density, strength) and straightness. A two-peaking
firing[26] was developed to control grain growth and electrical resistivity, these two processes have been satisfactorily developed at
different laboratories in the world and they are used routinely to
manufacture a great number of electrolyte tubes with size as large
as 35 mm of diameter and 650 mm length.

CERAMIC PROPERTIES

Electrical, mechanical and thermal properties of beta-alumina ceramics have all been determined, they are tabulated in the review by

MAY[27]. For sodium-sulphur battery application, the ceramic tubes must have suitable thermo-mechanical properties under cell conditions (temperature, voltage, current density, contact with liquid sodium and polysulphides). For example, it has been shown[28] that the strength of beta-alumina measured in sodium at 300°C shows a 25% reduction from that measured in air at 20°C.

At the present time, the principal cause of sodium-sulphur cells failure is the breakdown of the electrolyte. Though many laboratories indicate that 1000 to 2000 charge/discharge cycles have been obtained, a wide dispersion of the performances is noted. Besides mechanical or thermal breakdown which can be eliminated by improving the technology of the cell, the major mode of failure remains the electrical breakdown.

ELECTRICAL BREAKDOWN

This kind of failure is characterized by a variation of the amount of A.h. between charge and discharge which corresponds to a decrease of the faradic efficiency. Failure is associated with the formation or penetration of sodium in the electrolyte, it is evidenced by the growth of sodium-filled cracks and generally speaking by a blackening of the ceramic that spreads from the beta-alumina sodium interface. These cracks have been observed using different methods[29, 30,31]. From these observations several models were suggested to explain the initiation and the growth of the cracks. In one model[31], initial surface cracks, filled with sodium, propagate because the sodium Poiseuille pressure exceeds the mechanical strength of the material at the crack tip. An other model[32] proposes the crack growth by stress-induced dissolution of the ceramic in the liquid sodium. These models have been extended[33,34,35], leading to the concept of critical current density above which sodium penetration can occur and to the influence of the electrochemical aspects of the electrolyte's environment, so that the electrical failure mechanism can be divided into two stages, one with a slow kinetics like an induction phenomenon followed by a rapid one related to crack growth. The first stage might be related to the blackening of the ceramic by molten sodium, which cannot be itself responsible for the electronic conductivity[36] but induces point defects[37,38] and gradients in the electronic/ionic transport number ratio[39,40]. These effects can lead to internal sodium deposition associated with microcracking[41]. Experimental data indicates that the grain boundaries might have a major role in the blackening of the ceramic[36,41] but a clear relationship has not yet been evidenced between the lifetime and the ceramic microstructure[42].

FUTURE RESEARCH

The development of beta-alumina electrolytes for use in sodium-sulphur batteries has reached a critical stage in many laboratories, and several programmes are moving from a phase of scientific research to engineering development. The cycle life of sodium-sulphur cells and batteries is of decisive significance for pratical use, during the past five years the cycle life has been increased up to several thousand cycles, however it is necessary that nearly all the cells of a battery reach the cycle life goal. Therefore, the development of beta-alumina tubes with improved lifetime statistics is still of high priority. The lifetime of the electrolyte depends on its composition and on the manufacturing process. Important progress has been made in this area : ceramics which combine suitable resistivity with satisfactory strength, chemical inertness and durability have been made. However it is not possible, at the present time, to select tubes for a given lifetime. Therefore a great amount of work is needed to characterize the failure mechanism of beta-alumina under condition of electrolysis, this work will help to point out what characteristics will be required to meet the durability goal : purity, level of dopant, microstructure, surface finish... Once these results obtained, work will be done to select the best methods for powder preparation, shaping and sintering, to meet these specifications with regard to reliability. For that,improved techniques for non-destructive testing and quality control are needed. The last step will be the optimization of the manufacturing process, taking account for yield and cost.

The success of the sodium-sulphur battery as an energy storage medium depends mainly on the quality of the electrolyte, but its behavior depends also on the surrounding cell technology ; so that improvements in the other cell components (electrodes, container, safety device) will also contribute to the successful issue of the sodium-sulphur story.

REFERENCES

1 - J. JENSEN, P.Mc GEEHIN and DELL, Electric batteries for storage and conservation, Odense University Press, 1979.

2 - N. WEBER and J.T. KUMMER, Adv. Energy Conv. Eng. ASME Conf., Florida, 1967, p. 913.

3 - Ger. Patent 2, 509, 982 (1975).

4 - Chloride Silent Power Ltd, UK Patent 1, 586, 073 (1977).

5 - Compagnie Générale d'Electricité, French Patent 2,333, 358 (1975).

6 - General Electric Co., US Patent 4, 110, 516 (1978).

7 - UK Patent 1, 582, 845.

8 - W.L. BRAGG, C. GOTTFRIED and J. WEST, Z. Kristallogr. 77 (1931) 255.

9 - C.A. BEEVER and H.A. ROSS, Z. Kristallogr. 97 (1937) 59.

10 - M. BETTMAN and C.R. PETERS, J. Phys. Chem. 73 (1969) 1774.

11 - Y. YAMAGUSHI and K. SUZUK1, Bull Chem. Soc. Japan 41 (1968) 93.

12 - Y. LE CARS, J. THERY and R. COLLONGUES, Rev. Int. Hautes Temp. Refract. 9 (1972) 153.

13 - W. BAUKAL, H.P. BECK, W. KUHN and R. SIEGLER in D.H. COLLINS (Ed.) Power Sources 6, 1976, Academic Press, London (1977) 655.

14 - L.N. GLYZINA, V.I. FADDEVA and Y.D. TRETYAKOV, Inorg. Mat. 11 (1975) 927.

15 - S.E. WEINER, Research on electrodes and electrolyte for the Ford sodium-sulphur battery, report on contract NSF-C 805, July 1975.

16 - D.J. GREEN and S. HUTCHINSON in P. Vincenzi (Ed.) Energy and Ceramics, Elsevier, Amsterdam 1980, p. 964.

17 - P. POPPER, Isostatic Pressing, Heyden, London (1976).

18 - Compagnie Générale d'Electricité, French Patent 2, 401, 119 (1977).

19 - R.W. POWERS, S.P. MITOFF, R.N. KING a,d J.C. BIELANSKI, Solid State Ionics 5, 1981, 287.

20 - Ford Motor Co. UK Patent 1, 541, 850 (1979).

21 - R.T. DIRSTINE in P. VASHISHTA (Ed.), Fast Ion Transport in Solids, 1979, Elsevier, p. 79.

22 - S.R. TAN and G.J. MAY, Sci. Ceram. 9 (1977).

23 - G.J. MAY, S.R. TAN and J.W. JONES, J. Mat. Sci. 15 (1980) 2311.

24 - G. DESPLANCHES, A. WICKER and J.P. DUMAS in New Way to Save Energy (CEC ed.) D. REIDEL publishing Co., 1980, p. 558.

25 - W. HAAR, W. FISHER, H. KLEINSCHMAGER and G.WEDDIGEN, High Temperature Battery Workshop, Argonne National Laboratory, 1976.

26 - J.H. DUNCAN and W.G. BUGDEN, Proc. Br. Ceram. Soc. 31 (1981), 221.

27 - G.J. MAY, J. Power Sources 3 (1978) 1.

28 - R.W. DAIDGE, G. TAPPIN, J.R. McLaren and G.J. MAY, Am. Ceram. Soc. Bull 58 (1979) 771.

29 - C.T.H. STODDART and E.P. HONDROS, Trans. Br. Ceram. Sov. 73 (1974) 61.

30 - J.L. SUDWORTH, M.D. HAMES, M.A. STOREY, M.F. AZIM and A.R. TILLEY in D.H. COLLINS (Ed.) Power Sources 4, Oriel Press, 1972.

31 - R.D. AMSTRONG, T. DICKINSON and J. TURNER, Electrochim. Acta 19 (1974) 187.

32 - R.H. RICHMAN and G.J. TENNENHOUSE, J. Am. Ceram. Soc. 58 (1975) 63.

33 - D.K. SHETTY, A.V. VIRKAR and R.S. GORDON in R.C. BRADT, P.P. HASSELMAN and F.F. LANGE (Eds.) Fracture mechanics of ceramics, Plenum Press, New-York 1977, p. 651.

34 - A.V. VIRKAR and L. VISWANATHAN, J. Am. Ceram. Soc. 62 (1979) 528.

35 - M.P.J. BRENNAN, Electrochim. Acta 25 (1980) 621 and 629.

36 - R.D. AMSTRONG, D.P. SELHICK and S.R. TAN, Solid State Ionics 6 (1982) 203.

37 - D. GOURIER, A. WICKER and D. VIVIEN, Mat. Res. Bull. 17 (1982) 363.

38 - D. GOURIER and A. WICKER, The Electrochem. Soc., 161st Meeting, Montreal, May 1982, Vol. 82-1, p. 1156.

39 - L.C. De JONGHE and A. BUECHELE, J. Mat. Sci., 17 (1982) 885.

40 - L.C. De JONGHE, J. Electrochem. Soc. Vol.129, 4 (1982) 753.

41 - L.C. De JONGHE, L. FELDMAN and A. BUECHELE, J. Mat. Sci. 16 (1981) 780.

42 - J.R. RASMUSSEN, G.R. MILLER and R.S. GORDON, The Electrochem. Soc., 161st Meeting, Montreal, May 1982, Vol. 82-1, p. 1153.

CERAMICS FOR SENSORS AND MONITORS

P. McGeehin and D.E. Williams

Materials Development Division, AERE Harwell,
Didcot, Oxon, OX11 ORA, U.K.

Abstract

The development of electrical ceramics for use in sensors for
industrial process control and environmental monitoring has been
underway for many years. Research to improve devices currently in use
has among its objectives greater specificity of devices to particular
gases, greater sensitivity, less suceptibility to ageing and poisoning,
lower temperature operation and lower cost. Combustion control,
humidity measurement in industrial drying, inflammable gas detection
are particular areas where intensive research to improve on existing
devices continues.

The paper is organized by device type. It also includes a
discussion of combustion processes which is used to illustrate
particular aspects of device performance. Solid electrolyte galvanic
cells, gas sensors utilizing conductivity changes, both surface and
bulk, catalytic gas detectors and semiconductor junction devices are
treated.

1.   INTRODUCTION

The objective of this paper is to give an overview of the use
of electrical ceramics in sensors. The state of the art is briefly
reviewed, problem areas are identified and recommendations are made
on areas for further research and development.

The theme of the conference of which this contribution forms a part is "Ceramics in Advanced Energy Technologies". Many of the applications and potential applications of ceramics to be considered in the programme may require the deployment of sensors to be efficient and possibly also successful. In many cases, currently available devices are impractical, too expensive, suffer from durability problems and/or are insufficiently specific. A particular example is internal combustion engine fuel control, where devices for lean-mixture control are not available. Although sensors are available for control at the stoichiometric air-fuel ratio, they suffer from durability and other problems which will be described.

Control of combustion processes - in boilers, vehicles, coal gasification etc. - can be made more efficient and therefore possibly more cost effective by the deployment of improved instrumentation, and ceramic-based devices are important in providing one of the more robust approaches. This is particularly so for monitoring high temperature processes, for example those related to the oil industry - crackers and catalytic reactors - where monitoring gas composition together with temperature and pressure can give improved yield. Production of metals, e.g. steel, aluminium and copper, and metallic components by high temperature processes can be improved by monitoring melt composition and in particular the levels of alloying additions. In the case of steel, the oxygen potential is important, whilst when casting aluminium, hydrogen content can be critical if porous castings are to be avoided.

The interpretation of the term 'advanced energy technologies' in this paper can therefore be seen to be quite wide, and it includes also lower temperature processes which are important in energy conservation. A particular example is industrial drying, where there is a strong requirement for humidity sensors with improved sensitivity and better durability at temperatures close to or above 100°C.

Emphasis is given to sensors for detecting gases, with some reference to the elemental composition of dissolved metals. It is not possible to do justice to the wide range of other possible measurement techniques which might be employed, some of which, based on, say, the use of fibre optics, may well provide strong competition to electrical ceramics in the future. Ceramic fabrication aspects, covered by others during the Conference, are also excluded except where particularly relevant. These somewhat artificial restrictions are justified on two grounds:

(i)   measurement of gases and in-situ monitoring of molten metals are both important,

(ii) the electrical ceramics sensor can be cheap and apparently simple in operation.

In the applications areas considered, the specifically energy-motivated facets of current research and development are complemented by other important aspects, notably improvement in product quality and reduction in environmental pollution.

The plan of this paper is as follows. We begin by outlining the salient features of combustion processes before considering solid electrolyte sensors, covering such topics as thick film devices having reduced electrolyte resistance and the importance of electrodes to the overall operation of the sensor. Next, gas measuring devices utilising conductivity changes are reviewed. These can conveniently be subdivided into bulk and surface conductivity devices, including in the latter sensors for both humidity and reducing gases. Devices working primarily because of temperature changes resulting from the presence of a particular species or class of substances are considered prior to a brief summary of devices incorporating ceramic-semiconductor junctions. Finally, some general conclusions are drawn, and recommendations for further areas of R & D made.

2. COMBUSTION PROCESSES

The importance of combustion processes in advanced energy technologies is such that they justify fairly detailed attention. This is not least because they represent an important area for demonstrating electrical ceramic sensors. Two different situations will be considered, in outline, firstly internal combustion (as in petrol engines), and secondly boiler (external) combustion.

2.1 Internal Combustion Engines

The fuel taken into the combustion chambers is a hydrocarbon, which we can represent as C_xH_y. Given enough time it should burn with the oxygen from the air which is sucked into the combustion chamber with it, according to the standard equation:

$$C_xH_y + (x + \frac{y}{4})\ O_2 \rightarrow xCO_2 + \frac{y}{2}\ H_2O \qquad \ldots (1)$$

At the flame temperatures experienced, however, some atmospheric nitrogen reacts with the oxygen to produce oxides of nitrogen,

$$N_2 + \frac{x}{2}\ O_2 \longrightarrow NO_x$$

and because of the complex nature of the combustion occurring after the spark ignition, e.g. the expansion of the gases and the quenching of reactions on the cylinder walls, the fuel does not burn completely. Residual hydrocarbons and carbon monoxide in particular are discharged along with the CO_2 and H_2O. In poorly equilibrated exhaust gases, the emissions from an engine look like those shown schematically in Figure 1 when plotted as a

424

function of input air-fuel ratio. The CO_2 and NO levels peak near the stoichiometric air-fuel ratio (the mixture at which just sufficient oxygen is taken into the engine to burn the fuel to completion), whereas the hydrocarbon, CO and H_2 levels fall going from low air-fuel ratios (rich mixture) to high ratios (lean or weak mixture). The exhaust oxygen level remains at a constant low value on the rich side of the stoichiometric point, and rises monotonically on the lean side (when an excess of oxygen is being taken into the engine).

The lack of equilibration of the exhaust gases typically found from internal combustion engines can be judged by a comparison of the oxygen level in Figure 1 with the thermodynamically calculated, equilibrium value shown in Figure 2 as a function of air-fuel ratio. In the rich region (fuel excess) the effective oxygen partial pressure, $p(O_2)$, lies in the range $10^{-13}-10^{-26}$ atmospheres depending on temperature and does not vary greatly. In the lean region there is a similarly gradual change as the air-fuel ratio increases, with $p(O_2)$ being approximately 10^{-2} atmospheres. As the air-fuel ratio goes through the stoichiometric mixture point there is a sharp change in $p(O_2)$ of many orders of magnitude, and the resultant S-shaped curve is typical of that which is observed when one reagent (air) is titrated against another (fuel).

Figures 1 and 2 have important implications for the use of oxygen sensors to control the input air-fuel ratio of engines. Firstly, the lack of equilibration illustrated in Figure 1 must be overcome if a sensor is to work satisfactorily, because if it is not, the measured oxygen level will be an unstable, non-thermodynamic quantity which does not truly represent the state of combustion. Secondly, the variation in thermodynamic $p(O_2)$ at the stoichiometric point is very large, so if equilibration of the products of combustion can be achieved, the detection of this point should be relatively easy. Thirdly, and in contrast, the variation in $p(O_2)$ with air-fuel ratio in the lean region is quite modest, making the detection of some arbitrary air-fuel ratio in this region much more difficult.

Examination of the chemistry of the reactions between the various products of combustion shown in Figure 1 has identified the dominant reactions which are involved in the equilibration of the gases. These are the burning to completion of the carbon monoxide produced in the unstable situation, together with the oxidation of hydrogen to produce water in the water-gas shift reaction:

$$CO + \tfrac{1}{2}O_2 \rightarrow CO_2 \qquad \qquad \dots (2)$$

and
$$CO_2 + H_2 \rightarrow CO + H_2O \qquad \qquad \dots (3)$$

425

2.2 Boiler Combustion[2]

The main features of external combustion are similar to the previous case, though the control philosophy is somewhat different.

In Figure 3 the percentage of the heat of the fuel lost in flue gases for a medium/light fuel oil and the volume percentage of the constituents of the flue gas are plotted against the proportion of air supplied to the burner. The latter parameter ranges from a deficiency of combustion air of around 40% to an excess of combustion air up to about 100%. Whilst this diagram has been derived specifically for the combustion of medium/light fuel oil, the general features of it are widely applicable to almost any external combustion situation.

The major features of the figure are a series of approximately V- shaped curves which show the heat value of the fuel which is lost as a function of flue gas temperature in the range 100-400°C. In the rich region, that is the region where the fuel is being combined with a deficiency of air, we can see that there is a very rapid increase in the amount of heat which is lost resulting from the incomplete combustion of the fuel. In the lean region the rate of loss of heat value from the fuel is rather less as the mixture becomes increasingly lean, but gets worse as the flue gas temperature increases. The reason for the loss in efficiency in the lean region results from the fact that excess air is being heated up and this effectively dilutes the heat available from the fuel.

At the point at which there is just sufficient air present combined with the fuel to burn it completely, the heat loss curves show a broad minimum. Clearly if the combustion is to be as efficient as possible, the amount of air combined with the fuel must be maintained close to this point. If an error in mixture of any kind is made, it is clearly preferable that the mixture be on the lean side rather than on the rich side to avoid carbon monoxide production, although this is accompanied by a fairly rapid increase in the ratio of the production of SO_3 to SO_2. The former can result in rapid corrosion of materials in the flue if the temperature is allowed to fall below the temperature at which sulphuric acid will condense out (approximately 150°C).

Considering now the chemical composition of the flue gas, we can see from figure 3 that the level of CO_2 increases when the mixture moves less rich, reaching a peak at the stoichiometric point, before then falling quite rapidly out into the lean region. This CO_2 curve is in fact quite sensitive to the fuel which is being burnt, and therefore cannot really be used as a variable on which the combustion process can be controlled. In contrast, the level of carbon monoxide, which is very high in the rich region, falls quite rapidly towards the lean region, and at a percentage excess air of about 10% has fallen to very low levels. It does however begin to rise slowly as the mixture becomes excessively

lean. The level of carbon monoxide in a flue is a quite attractive
variable to detect, and indeed a number of infra-red based systems
are available.

It is evident from figure 3 that perhaps the best variable in
the combustion gases with which to control the process is that of
oxygen. Generally speaking a boiler will err on the side of lean
mixture, owing to the undesirability of producing excess carbon
monoxide. The control philosophy involves drawing the combustion
gases closer to the stoichiometric composition, thereby reducing
the wastage of heat through excess gas being present. This is in
fact also accompanied by a reduction in the production of oxides of
nitrogen, as well as SO_3, because of the lower flame temperatures
which result.

Many of the device types described below have been used or
proposed in some form for combustion control. The most successful
and widely used devices have employed solid electrolyte galvanic
cells (Section 3). Devices using the variation in bulk
conductivity of a semiconducting oxide are also important in this
use (Section 4).

3.   SOLID ELECTROLYTE GALVANIC CELLS

In this and subsequent sections, the behaviour of various
device types is illustrated. The wide application of solid
electrolyte galvanic cells dictates a rather detailed examination
of their behaviour.

3.1 General

In these devices, a solid electrolyte separates a reference
compartment and a test compartment. A potential difference is
established between the two sides of the membrane dependent on the
concentration difference across the membrane of species which will
equilibrate with the conduction ions in the membrane[3,4,5].

Thus, oxygen can be measured using oxygen ion conductors such
as yttria-doped zirconia, thoria or ceria, by virtue of the
interface reaction

$$2e^- + \tfrac{1}{2}O_2 \rightarrow O^{2-} \qquad \qquad \ldots (4)$$

and the transport of O^{2-} through the conductor. Similarly,
fluorine can be measured with fluoride ion conductors such as
CaF_2 or LaF_3.

The oxygen content of molten steel can be determined using a
zirconia probe: the probe lasts just long enough for a stable
voltage measurement to be obtained. Probes of yttria-thoria have

been used to measure oxygen in molten sodium. Again, mechanical and chemical durability of the probe and seals is the major problem. Similarly, the hydrogen content of molten aluminium can be determined using a membrane of hydrogen–β–alumina sealed into an α–alumina tube[6]. In this case, the sensor was stable for months but suffered from an interfering response to sodium. Sodium–β–alumina can be used to measure sodium in aluminium, and other metal substituted β–aluminas can be used to measure alloying elements in molten lead[7]. The mechanical durability of β–aluminas used for sensor probes can be improved by incorporation of a dispersion of metal particles[8]. Many patents relating to the sodium–sulphur battery teach how to improve durability by control of microstructure and how to fabricate seals.

Less obviously, oxygen partial pressure can be sensed using a CaF_2 membrane[9]. Oxygen ions substitute on fluorine lattice sites and migrate over fluorine vacancies[10]. The electrode reaction is[10]

$$\tfrac{1}{2}O_2 + V_F' + 2e^- = O_F' \qquad \ldots (5)$$

The calcium fluoride membrane needs to be saturated with CaO in the vicinity of the interface otherwise the response is rather slow; this can be achieved by preheating in oxygen containing a trace of water vapour.

Similarly, the CaF_2 membrane will also generate a response to S_2 vapour[11], by virtue of the notional interface reactions

$$CaF_2 + \tfrac{1}{2}S_2 \longrightarrow CaS + F_2 \qquad \ldots (6)$$

$$e^- + \tfrac{1}{2}F_2 \longrightarrow F^- \qquad \ldots (7)$$

Again, a stable response requires the electrolyte gas interface to be saturated with CaS. The notional equilibrium partial pressures of fluorine are obviously small but the potential changes resulting are easily measured. The detailed mechanism by which the electrode equilibrium is established is not really known, but may perhaps be expressed by

$$\tfrac{1}{2}S_2 + 2e^- + V_F' = S_F' . \qquad \ldots (8)$$

Sulphur oxides can be measured using as membranes sulphates which are metal ion conductors, such as Li_2SO_4[12], $NaSO_4$[13] or K_2SO_4[14], and carbon dioxide can be measured similarly using carbonates such as K_2CO_3. For example, the cell[12]

$$\text{(gas) Pt/23}^m\text{/o Ag}_2\text{SO}_4, 77^m\text{/o Li}_2\text{SO}_4 - \text{Ag}_2\text{SO}_4, \text{Li}_2\text{SO}_4, \text{Ag/Ag} \quad .. (9)$$

operated at 530°C makes a good meter for sulphur oxides. The conducting species is Li^+, the reference electrode is Ag metal equilibrated with the electrolyte. The reference gas partial pressure is determined by the reaction

$$2Ag + SO_2 + \tfrac{1}{2}O_2 = Ag_2SO_4 \qquad \cdots (10)$$

and the equilibration with the electrolyte is established by the notional reaction

$$2Li + SO_2 + \tfrac{1}{2}O_2 = Li_2SO_4 \qquad \cdots (11)$$

The notional electrode reaction is

$$Li^+ + e^- = Li \qquad \cdots (12)$$

The mechanism by which the electrode equilibrium is established is in fact not known, but may be expressed as either

$$2e^- + SO_2 + \tfrac{1}{2}O_2 = 2V'_{Li} + (SO_4)_{SO_4} \qquad \cdots (13)$$

or $$2e^- + 2Li^+ + SO_2 + \tfrac{1}{2}O_2 = 2Li_{Li} + (SO_4)_{SO_4} \qquad \cdots (14)$$

The cell potentials are in agreement with those calculated from the overall cell reaction

$$2Ag + SO_2 + O_2 = Ag_2SO_4 \qquad \cdots (15)$$

A determination of the partial pressure of SO_2 requires a simultaneous measurement of the oxygen partial pressure. A zirconia electrolyte cell can be used to do this. The two cells can be combined to produce a single voltage output[15]. Equilibration amongst the components of the gas atmosphere can be achieved by placing a V_2O_5 catalyst bed in the furnace in front of the cell[12].

This brief description of some solid electrolyte galvanic cells serves to highlight particular features which can become problems in the operation of the devices. Firstly, electrode reactions have to come to equilibrium. Electrodes may be poisoned or the reactions may be slow. Lower limits are thus set on the operating temperature and environment. Secondly, a suitable reference electrode must be provided, which will establish an equilibrium with the electrolyte at the temperature of operation. Thirdly, the solid electrolyte must be sufficiently conducting lest the internal resistance of the device is too high for use with available voltmeters. This too implies high temperature operation. The electrolyte also needs to be stable in the environment. Potassium sulphate electrolytes, for example, can decompose to potassium pyrosulphate[13]. The electronic conductivity of the electrolyte must be low otherwise the cell will be short-circuited by the electrolyte and the output voltage will fall. This consideration sets a limit on the lowest oxygen partial pressure which can be measured by oxygen ion conductors before they begin to be reduced and their electronic conductivity rises. The lower limit[5] is about 10^{-20} atm for ZrO_2[16], 10^{-12} atm for CeO_2[17], less than 10^{-22} atm for ThO_2 but theoretically less than about 10^{-44} atm for CaF_2[9].

The output of a galvanic cell is also logarithmic in the gas partial pressure. Many applications require a linear response over a more restricted range of gas composition.

Many applications involve very rigorous environments for the sensors. Fabrication procedures are of great importance. The conventional measurement technique requires an impervious seal and non-porous electrolyte dividing reference and test compartments. Any gas permeation between the two compartments effectively shorts out the cell and reduces the output voltage. As well as the attention paid to methods of fabricating sufficiently dense materials and sufficiently light seals, there has been considerable effort expended on means of measurement which do not require non-porous electrolytes - for example thermogalvanic measurements[18] and the techniques described in sections 3.4, 3.6 and 3.7.

In the following section the particular application of Nernst sensors to combustion control is described in more detail in order to illustrate the problems outlined above, then subsequent sections will describe recent advances.

3.2 An Example - Solid Electrolyte Devices for Engine Control

In Section 2 combustion processes occurring in engines were described phenomenologically. Currently, the development of sensors to achieve satisfactory combustion control across the whole field of air fuel ratios is widespread and vigorous, and it is appropriate to concentrate on this. Attention will be given to the problems of implementation, from which arise many problems requiring further research and development effort.

A typical configuration of solid electrolyte cell for use in vehicle exhausts is illustrated schematically in Figure 4. In Figure 5 the variation in the output voltage of the sensor as a function of engine air-fuel ratio at various temperatures is illustrated. This indicates that control at the stoichiometric point is relatively straightforward without temperature measurement or stabilisation, providing a set-point of around 500 mV (half the extreme values) is chosen. We should note from Figure 5 that the approximately one volt difference measured between the sensor output from the extreme rich and extreme lean regions is observed independent of the reference oxygen partial pressure; changes in this reference merely move the S-curve up and down the voltage axis, with the step-function change still occurring at the stoichiometric air-fuel ratio point (around 14.7:1).

All the devices developed for automotive applications have been based on zirconia. Stabilisation of the cubic phase with magnesia, calcia or yttria is most common, with the latter two predominating. The level of stabilising addition is generally in the range 5-15%, and a number of other additives are often made to ease manufacture and improve the properties of the material with regard to thermal and mechanical shock in particular. For example, lime stabilised zirconias normally sinter to the required full density only at relatively high temperatures, which may be above 1700°C. By the addition of silica, alumina and other silicates at levels greater than 1% (e.g. say 2% china clay containing ~95% kaolinite) the sintering reaction can be noticeably accelerated. Full density can then be achieved at temperatures below 1600°C. However, great care in optimising the precise composition and firing conditions must be taken, because the microstructure of the ceramics can undergo considerable change as a result of thermally induced stresses occurring in the temperature range between 900-1300°C, and especially when the cubic zirconia grains become partially destabilised. These changes affect both the mechanical and electrical properties of the ceramics, resulting in device ageing.

The physical dimensions of the devices developed for use in vehicle exhausts are compatible with spark plugs; the ceramic tube is a few centimetres in length, with walls a few millimetres thick, and around 1 cm in diameter.

The detailed characteristics of the devices which have been developed, and their ageing behaviour, have been described in a number of papers[19-25]. The latter is particularly important because the working environment is very demanding, and a lifetime of several thousand hours is required. For example, the temperature fluctuates widely and rapidly, the gases are moving with high velocity and often contain fine, abrasive particles. Lead compounds capable of poisoning the catalytic electrodes are also present even in nominally lead-free fuel.

The most important characteristics of the Nernst sensors used in vehicle exhausts can be summarised as follows:

(i) The voltage of the Nernst sensors, whilst still retaining its step-function like character, does not always follow the ideal curve calculated from thermodynamic considerations[20]. The voltage at rich extremes is perhaps 15% lower than expected, and at lean mixtures is ~10% higher; the S-shape becomes compressed. This is also found if the sensors are not operated with a porous plasma-sprayed ceramic protective overcoat on the exhaust electrode to give a structure like that shown in Figure 6. The overcoating acts as a gas diffusion barrier, but is also thought significantly to increase the catalytic activity of the electrode[20].

(ii) Nernst sensor voltages depart significantly from the calculated values as the temperature falls[21]. The voltage measured at the rich mixture extreme (around 1 volt at high temperatures) gets quenched out below 350°C; the voltage measured at the lean extreme is also significantly disturbed by falling temperature. This behaviour is attributed to a decrease in the catalytic activity of the electrodes as the temperature falls.

(iii) The "switch point" of the sensor occurs at different air-fuel ratios depending on whether the mixture of the engine is changing from rich to lean or lean to rich[20]. The rich to lean switch always occurs at leaner air-fuel ratios (by around 0.1 units of air-fuel ratio at around 500°C). Furthermore, for a given direction of air-fuel ratio change, the switch point drifts to richer mixtures as the sensor temperature increases; the drift is ~ 0.3 units of air-fuel ratio on increasing the temperature from 500-700°C. Although these switch point drifts seem quite small they are very important to the overall system performance.

(iv) The response times of the devices are typically around 10-30 m sec, and although a difficult parameter to measure, the response time of the lean to rich transition is always shorter than the rich to lean transition, except perhaps at very low temperatures[21]. When put on an Arrhenius plot, the switching time of the Nernst sensor below 350°C yields a straight line with around 1 eV activation energy. A transition region occurs at around this temperature, for above it the points fall away from a simple linear relationship with reciprocal temperature.

Attempts to understand these details of device behaviour have been restricted to the phenomenological modelling work of Fleming[20,22]. He puts together a physically based model which uses six fitting parameters (which have a physical basis) to reproduce theoretically observed sensor behaviour. His six parameters are:

(i) The equilibrium constants for the oxidation of CO and the water-gas shift reaction which take values different from their thermodynamic values because of the relative lack of catalytic activity at the electrodes. Numerically, the CO oxidation equilibrium constant comes out some ten orders of magnitude lower than the thermodynamic value, and is always lower in the rich region than the lean by about an order of magnitude. This, he claims, correlates with the well known relative slowness of CO oxidation in rich conditions as established by catalysis studies[23]. With an overcoated electrode it is the enhancement of this constant by an order of magnitude which improves the definition of the voltage transition going through the stoichiometric point. The water-gas equilibrium constant takes a numerical value about one order of magnitude higher than expected, and does not vary very much with oxygen-rich or oxygen deficient conditions (lean or rich mixture), or on whether the exhaust electrode is overcoated or not.

(ii)   The adsorption constants for CO and O_2 adsorption on the electrode. Fleming found that the adsorption constant for CO is thermally activated, with an activation energy of around 1 eV, which he ascribes to strong chemisorption of CO on the electrode. In contrast, the oxygen constant did not correlate well with temperature and had an activation energy near zero. He concludes that oxygen is therefore probably only physisorbed on the surface. This, he claims, accounts for both temperature and directional drifts in the voltage curve; going lean the CO is strongly bound and oxygen cannot displace it, so cannot oxidise adsorbed CO readily; as the temperature increases the CO leaves the electrode more readily, allowing oxygen in and so the transition point drifts rich.

(iii)   Standard electrode potentials for the CO fuel cell reaction and for the oxygen concentration cell. These two electrode reactions are:

$$2CO_2(exh) + 4e^- \rightarrow 2CO(ads) + 2O^= (solid) \quad \cdots (16)$$

$$O_2(exh) + 4e^- \longrightarrow 2O^= (solid) \quad\quad\quad \cdots (17)$$

Theoretically the first should be around 1 V and the second zero. Fleming allows these to depart from the ideal values, owing to lack of reversibility and catalytic activity of the electrodes, and reactions with impurities. These quantities do not play a significant role in the model, except in the case of the comparison of electrodes with and without overcoat. In this case, the CO cell voltage comes out at almost half the expected value.

In his later work[24,25] directed towards the development
of an equivalent circuit model for the sensor, Fleming concludes
that:

(iv) The non-ideal characteristics of cells is accounted for
      by a mixed potential obtained from a combination of
      oxygen and carbon monoxide cells in both lean and rich
      regions (see Section 3.4).

(v)   The degree of catalytic conversion of oxygen when rich or
      carbon monoxide when lean is, relatively, very small.

(vi) The observed voltage transition step is predominantly
      controlled by the change in the fraction of carbon
      monoxide cells contributing to the sensor voltage. The
      fraction of very effective carbon monoxide cells is
      controlled by the exhaust electrode CO level, which
      Fleming links directly to the chemical activity of the
      exhaust gas electrode.

3.3 Reducing Electrolyte Resistance - Thick Film Devices

The performance of solid electrolyte galvanic gauges can be
improved in two obvious ways. Firstly, decreasing the thickness of
the electrolyte and secondly providing some integral means of
heating and controlling the temperature[26,27] of the device.
Many recent patents relate to these two aspects of sensor design.
There is also a continuing search for electrolytes of higher
conductivity. For example, oxygen ion conductors such as doped
ceria[17] or δ-Bi_2O_3[18,28] have higher conductivities and
lower activation energy for conduction than doped zirconia but are
more easily reduced and more difficult to fabricate.

Radiofrequency sputtering of oxides[29,30,31] provides a
very thin electrolyte, impervious to gas. Screen printing[32] is
a much simpler technique, suited to mass production. It produces a
thicker layer (~10 μm) and requires great attention to detail in
the preparation of both powders and paste in order to obtain a
pore-free coating. The successful printing of an electrolyte opens
the way to the fabrication of a complete device, comprising heater,
reference electrode, electrolyte, working electrode and protective
coating, all printed onto a flat ceramic substrate. Should this be
achieved it will represent a considerable advance. Some recent
patents describe approaches to this goal[33,34].

As suggested above, many patents on thick film devices have
been published arising from work in the USA, Japan and Europe. To
the authors' knowledge, such devices have not however yet reached
the market. This tends to suggest that there are some very
difficult materials technology problems - especially at interfaces
- yet to be overcome.

434

3.4 Non-Equilibrium Phenomena at Electrodes

3.4.1. Relevant Effects

In Section 3.2 it was shown that the output of a solid electrolyte combustion sensor became insensitive to variations in combustion gas composition as the temperature was reduced. The response will of course be lost if the resistance of the electrolyte rises to values comparable with the input impedance of the measurement circuit. Apart from this, there are two other principal effects which result in a loss of response[22]. The first is a lack of equilibration of the constituents of the gas phase[35,36,37], and the second is the development of a mixed potential (defined below) rather than an equilibrium potential at the electrode-electrolyte interface. The relative importance of these two effects is different for different electrode materials and for different gas mixtures[38]. For example, a solid electrolyte gauge with silver electrodes will indicate the actual (non-equilibrium) oxygen partial pressure in CH_4/O_2 mixtures[39] whereas the use of Pt electrodes results in complete or partial equilibration of the gas phase components, with mixed potentials developing at temperatures below about $350°C$[22,40,41].

When exposed to non-equilibrium gas mixtures, the sensor surface electrode serves both as a catalyst promoting reactant equilibration in the surface boundary layer and as a monitor of the emf developed across the concentration cell. Boundary layer oxygen partial pressures approach values corresponding to equilibrium in the bulk gas under conditions of mass transfer limited kinetics but do not do so if surface reaction rates are slow or comparable to gas diffusion exchange rates[35,36,37]. Under the latter circumstances, the sensor develops a non-equilibrium output voltage which may vary with time if the catalytic activity of the electrodes degrades. Such non-equilibrium output voltages depend on the relative molecular weights of the reactive gases[35], and the response to a mixture of changing composition depends on the direction of the composition change[22,35,36]. The boundary layers may be better defined by coating the sensor with a porous ceramic layer. If this is not done then the sensor output can be dependent on gas flow rate[35].

A mixed potential is obtained when an electrode adopts a potential which is determined by the relative rates of a number of different electrode reactions. This potential is termed a mixed potential because the oxidation and reduction reactions which determine it are not necessarily the reverse of one another. The oxidation reaction might be

$$O^{2-} \rightarrow \tfrac{1}{2}O_2 + 2e^- \qquad \qquad \dots (18)$$

whilst the reduction reaction might be

$$2e^- + CO_2 \rightarrow CO + O^{2-} \qquad \qquad \dots (19)$$

The potential determined by the balance of these two reactions will not be the same as that determined by the redox equilibrium

$$O^{2-} = \tfrac{1}{2}O_2 + 2e^- \qquad \qquad \dots (20)$$

if O_2, CO_2 and CO are not in chemical equilibrium at the electrode surface. The behaviour of silver[38,39] and poisoned Pt[38] electrodes in indicating the actual non-equilibrium oxygen partial pressure can be explained by the assumption that on these materials, electrode reactions such as (19) are very much slower than those of oxygen (18 and 20) so, although the electrode rate may be slow, the electrode potential still depends only on (20).

As noted above, the problem can be alleviated to some extent by coating the working electrode with a porous ceramic layer impregnated with an oxidation catalyst – typically finely divided Pt. If this is not done, the effect is noticeable at temperatures as high as 650°C[20]. Numerous patents relate to the fabrication of this coating: sensor response time is closely related to its microstructure[42]. It has also been noted recently[43] that mixed potential effects on Pt electrodes on stabilised zirconia can be considerably decreased by doping the surface of the electrolyte with fluorine.

3.4.2. Mixed-Potential Devices

The observation of non-equilibrium phenomena at relatively low temperatures on solid electrolyte devices has led to the proposal of a new type of device which uses the voltage difference observable between two electrodes having different electrode kinetics to provide an indication of gas composition[39,40,44,45].

In one type of device (Figure 7a)[44] two electrodes, one of platinum, the other gold, are mounted on a heated piece of solid electrolyte (stabilized zirconia) in a common gas stream. The voltage difference recorded between the two electrodes alters markedly when the gas changes from predominantly O_2 to predominantly CO. The device contrasts with the normal solid electrolyte oxygen gauge in which an impermeable oxygen-ion-conducting electrolyte separates the gas to be measured from a reference gas, one electrode being exposed to the test gas, the other to the reference gas (Fig. 7b).

The variation of mixed potential with varying gas composition is also shown by a version of the device having two identical electrodes placed parallel to one another across the direction of gas flow in a motor vehicle exhaust[46]. When the combustion changes from, say, rich to lean, a gas front passes over the electrodes so that momentarily one electrode sees a CO-rich gas whilst the other still sees an O_2-rich gas. The result is that the sensor produces a transient voltage peak whose sign depends on whether the gas is changing from CO-rich to O_2-rich or vice versa; the voltage peak arises because one electrode sees a different gas mixture to the other, and adopts a different mixed potential.

In another type of device, described recently[40], (Fig. 7c), a piece of oxide ion conductor had on one face exposed to the test gas, a Pt electrode. On the other face, another Pt electrode was separated from the test gas by a porous membrane supporting an oxidation catalyst (finely divided Pt). The voltage difference between the two electrodes varied with the CO content in the air under test. In the CO/air mixture, a mixed potential involving reactions (18) and (19) is developed at the electrode exposed to this mixture; at the electrode covered by the oxidation catalyst, the CO has been removed (hardly affecting the oxygen partial pressure since this is much greater than the CO pressure) and the potential developed by this electrode is the equilibrium potential for reaction (18) alone; the potential difference between the two electrodes therefore reflects the CO concentration. The sensitivity of the device to small concentrations of CO can be explained by noting that CO adsorbs strongly at Pt and inhibits the kinetics of the oxygen electrode reaction (18). At higher temperatures the CO adsorption decreases in strength and the output voltage did indeed fall as the temperature was raised above 350°C, reaching zero at 450°C. The output also saturated for CO concentrations greater than about 500 ppm – the Pt surface had been completely covered by CO so no further inhibition of the oxygen reaction was possible.

A device described much earlier[39], using silver electrodes, was similar in design but slightly different in the working principle. As noted above, Ag electrodes indicate the actual oxygen partial pressure in a non-equilibrium mixture of CH_4 and O_2. If, in a device with two Ag electrodes, one is covered with an oxidation catalyst, then the output voltage will give the difference in oxygen concentration between the equilibrated and un-equilibrated gas, and hence will give a measure of the methane concentration. A device using Pb-poisoned Pt as the electrodes[38] would give a measure of the total concentration of combustible gases in the stream.

Mixed potential phenomena can be observed on a number of different solid electrolytes – the electrode reactions do not need to couple the gas being measured to the major conducting species in the electrolyte. Devices using β-aluminas[45] and β-PbF_2[47] have been described.

3.5 Reference Electrodes

A reference electrode is required to fix a stable reference
gas partial pressure at one electrode. A formal way of expressing
this is to require that the exchange current density at the
reference electrode be sufficiently large that the reference
potential does not vary significantly when currents induced by the
measuring process flow through the cell. These currents are a
steady state current drawn by the input circuit of the measuring
amplifier and transient currents which flow when the test gas
concentration changes.

There are two general ways of forming a reference electrode.
The first is to use a gas stream, air for example in oxygen gauges,
and the second is to define an equilibrium pressure by reaction
between two solids, silver and silver sulphate in the cell
described in Section 3.1 or a metal and its oxide[48] for example.
The current in the measuring circuit will be minimized and its
sensitivity maximized if the cell output is close to zero – reference
partial pressure close to the partial pressure to be measured.

The same problems of electrode kinetics which affect working
electrodes can also affect gaseous reference electrodes – the
reference potential becomes unstable and sensitive to reference gas
flow rate at low temperatures and is also sensitive to
contamination. There have been several studies of the electrode
kinetics of metal–metal oxide couples[49]. In mixtures of metal
and metal oxide powders, one contribution to the overvoltage arises
from steady state oxygen concentration gradients in the metal
particles[3]. It is important to have and to maintain intimate
contact between metal particles and both the electrolyte and the
oxide particles, as well as other metal particles to retain
electrical continuity. It has been found[3] that $Cu-Cu_2O$
electrodes can adjust to small cell currents much more easily than
either the commonly used Fe-FeO or Ni-NiO electrodes. Copper-
copper oxide electrodes will function at temperatures as low as
$100°C$[50]. Metal-metal oxide electrodes also have an upper limit
of temperature defined where the metal is so unstable with respect
to its oxide that there is insufficient metal left at equilibrium
for electrical continuity. There are also practical limits defined
by the need to seal the mixture adequately against the
environment.

A device in which a metal/metal oxide reference is encased
completely within a solid electrolyte body resulting in a compact
and well-sealed construction has been described in the patent
literature[51]. In another device[52], a metal wire is coated
with an oxygen conducting solid electrolyte and fired. Sufficient
oxide must be formed on the metal before or during the processing
to provide a reference system. Although the device is small and
strong its durability and the stability of the reference system
must be questionable.

Recently[38,53,54], oxygen reference systems have been
described in which a reference gas pressure is generated
electrochemically. In one example of the method[38] two zirconia
membranes are used to define a sealed cavity (figure 8). One
membrane is a sensing cell, the other a pump cell. A current
passed through the pump cell removes oxygen from the cavity until
the sensing cell indicates effectively zero oxygen pressure in the
cavity, then the current is reversed and oxygen pumped back in
until the sensing cell output is zero. From the charge passed
through the pump cell and the cavity volume, the cavity partial
pressure, equal to the oxygen partial pressure in the environment,
is calculated. This cavity partial pressure now provides a
reference against which the sensing cell can monitor changes in the
environment. A second embodiment utilizes a dynamic reference
electrode[53,54]. A constant current flow through a porous
electrolyte layer establishes an oxygen partial pressure at the
inner electrode such that the flux of oxygen gas through the pores
of the electrolyte just balances the current flow. The elements of
the reference potential are then the overvoltage at the electrode,
determined by the current flowing, and the potential set by the
oxygen partial pressure at the electrode surface (familiar to
aqueous electrochemists as concentration polarization). This
latter part of the reference potential is determined by the current
flowing, by the oxygen partial pressure in the surrounding gas
atmosphere and by the thickness and porosity of the layer covering
the electrode. A device with a dynamic reference electrode can
have either two[53] or three[54] electrodes (figure 9). In the
device with two electrodes, the measured voltage across the device
is the sum of the overvoltages at the two electrodes, the voltage
drop due to the resistance of the electrolyte and the concentration
polarization. Only the latter varies with the external gas
composition. In the three electrode device, the electrode for
supplying the current is separate from that measuring the exterior
potential. No current flows between the measurement electrode and
the dynamic reference electrode so the potential between the two is
simply the sum of the concentration polarization and the
overvoltage at the reference electrode.

Another mode of operation of dynamic devices of this type is
to measure the current required for a constant potential. Such
devices are described in the next section.

3.6 Amperometric Devices

One difficulty with solid electrolyte concentration cells is
that the output voltage is a logarithmic function of the gas
concentration. Measurement of relatively small changes in
concentration, within one decade say, therefore require voltage
measurements of high precision and very good temperature control.
Devices with a linear output would be more useful for such
applications. Amperometric devices rely on the measurement of a

current flow. A situation with a defined concentration gradient is set up and the current is then proportional to. the diffusion flux. If the current is such that the concentration at the measurement electrode is essentially zero then the diffusion flux is directly proportional to the concentration in the external medium. Several methods of realizing this principle in the measurement of oxygen partial pressure have been described. Depending upon the way in which the principle is realized, the resulting device is more or less sensitive to poisoning of the electrodes and sintering or clogging of the gas diffusion path. Thus the diffusion path can be a leak into a cavity[55]; the partial pressure in the cavity is kept essentially zero by a pump cell, the partial pressure difference between the inside and outside of the cavity is indicated by a measurement cell and the pumping current required to keep this potential difference constant is measured. It is not in fact necessary to form a separate cavity as such; a three-electrode configuration such as that shown in figure 9 can be used. The porous electrolyte layer acts as both cavity and leak.

Two electrode devices such as that shown in figure 9 can work equally well in an amperometric mode: a constant potential is maintained across the device and the current resulting is measured. Various ways of forming the diffusion path have been described[56].

Yet another method of addressing a pump cell device[57] is to allow the cavity to equilibrate with the atmosphere then measure the time required to pump it down to essentially zero pressure, as determined by a measurement cell. Time measurement, using counters, is extremely simple and accurate. An improved method is to set two voltage limits on the measurement cell and reverse the pumping current when the measurement voltage reaches either limit. Oxygen partial pressure in the ambient can then be determined from the period of oscillation of the pump current[58].

3.7 Solid Electrolyte Transistors

Transistor action at elevated temperature has been reported[59,60] for a ceramic structure comprising two zirconia electrochemical cells separated by an enclosed volume (figure 10). One cell emits oxygen from an ambient atmosphere into the volume (base region) where it is collected and returned to the ambient by the opposing cell. Electrical operation can be understood in terms of the current limitation presented by the Pt cathode of the collector cell. Small signal voltage, current and power amplification are observed with a frequency response limited by the double layer capacity of the platinum-zirconia interface. These devices are developments from the devices mentioned in the previous section.

An absolute pressure sensor can be fabricated using a zirconia transistor structure[61]. If the zirconia transistor is used as a feedback impedance in an oscillator circuit, then the resulting oscillation period is inversely proportional to the diffusion coefficient of oxygen in the gaseous ambient. Because the diffusion coefficient of oxygen in a host gas varies inversely with absolute pressure, the oscillation period varies directly with absolute pressure.

4. GAS SENSORS USING CONDUCTIVITY CHANGES

4.1 Bulk Conductivity Changes in Semiconducting Oxides

The change in stoichiometry of semiconducting oxides as a function of the oxygen activity of their environment particularly at elevated temperatures is becoming increasingly well known, as is the effect this change in stoichiometry can have on the electrical resistivity of the materials[62]. The change in resistance can be represented by the relation:

$$\rho \sim \rho_o \; (\frac{E_A}{kT}) \cdot p(O_2)^{\frac{1}{n}} \qquad \qquad \ldots (21)$$

where k and T have their usual meaning, E_A is an activation energy and the sign and value of n depend on the nature of the point defects arising when oxygen is removed from the lattice. The activation energy observed for the change in resistance with temperature can be broken down into a contribution arising from the thermal activation of charge carriers into the conduction band and the energy required to form the defect directly linked to the change in the number of charge carriers in the conduction band. By contrast, the value and sign of n varies only according to the defect chemistry of the oxide of interest, and in particular on whether electrical neutrality of the material is preserved when oxygen is removed from the lattice by the formation of oxygen vacancies, oxygen interstitials or their metal-ion counterparts. If n is negative the material is a p-type semiconductor, and if n is positive, it is n-type.

In the context of the automotive industry a number of materials have been examined in great depth. Cobalt oxide[63] was found to have a strong $p(O_2)$ dependence to its resistivity with an n value of approximately −4 and an activation energy of around 0.3 eV; cobalt vacancy formation accounts for the observed $p(O_2)$ dependence of the conductivity. However, at relatively low levels of reducing $p(O_2)$ around the stoichiometric point, the material tends to reduce to the metal in the temperature range of interest. It could not therefore be considered even as a lean mixture sensor, owing to the fact that the vehicle exhaust is naturally very reducing under normal conditions, for example during a start-up from cold using the choke, which will deliberately enrich the mixture.

Attempts to overcome this by doping CoO with MgO to produce $Co_{1-x}Mg_2O$ with x ~0.85 met with some success[64]. Doping suppresses reduction to the metal to oxygen partial pressures around 10^{-20} atmospheres and decreases the tendency to form spinel phases at lower temperatures under highly oxidising conditions. The activation energy of the material changes slightly to around 0.55 eV even at high dopant levels and it retains an n value of approximately −4. This combination of properties is superficially quite attractive but the relative instability of these materials at the extremes of oxygen content and temperature encountered in vehicle exhausts led to the search for alternatives.

In fact, whilst many Nernst sensors have been developed around the world, only one based upon a semiconducting oxide has reached an advanced stage. The material in this case is TiO_2[65]. When pure and stoichiometric, titania has a very high room temperature resistivity. Loss of oxygen on heating, particularly in reducing conditions, results in the generation of lattice defects such as oxygen vacancies or titanium interstitials which act as electron donors. As oxygen loss increases, more electrons contribute to the conduction process and the resistivity falls further. Over a wide range of temperature from around 300–1000°C the material when 99.5% pure has an n value of 4, and it is an n-type semiconductor except at very high oxygen pressures when the conductivity begins to rise with increasing $p(O_2)$, showing that the material is becoming p-type. The results are thought to be consistent with the presence of doubly ionised oxygen vacancies, with concentrations affected by aluminium and iron electron acceptors which are the major impurities present.

Whilst stable in very reducing conditions, TiO_2-based ceramics do have one major disadvantage. Their activation energy is around 1.6 electron volts, resulting in a 4% per °C temperature coefficient of resistance at 400°C and 2% per °C at 800°C compared with (at the latter temperature) values of around 0.5% per °C for the cobalt oxide-based materials mentioned above. Both these temperature coefficients are rather large for precise control even at the stoichiometric point, and they can be compared with the temperature coefficient of the voltage of the Nernst sensor, which at 400°C is 0.15% per °C, and at 800°C is 0.09% per °C. However, the workers at Ford in the USA came up with a most elegant solution[66]. The probe comprises two ~3 mm diameter, ~1 mm thick pellets of the oxide material with embedded platinum leads identical except that one is porous and the other dense. This is achieved by firing pellets made from material of the same chemical composition under different conditions; as is well known, the tendency to densification of ceramics is associated with higher firing temperatures and finer starting powders. Whilst the resistance of the porous element varies both with exhaust oxygen level and temperature, the dense pellet has only a very slow $p(O_2)$ dependence on its conductivity and responds primarily only to temperature fluctuations. This is because there is insufficient surface area for the change in ambient oxygen level to lead to a change in bulk resistivity of the material in the times of interest. By careful design of the electrical circuit, the dense element can be used to subtract-out the temperature sensitivity of the porous element, allowing its change in resistance with oxygen pressure to be determined accurately.

As well as the problems caused by lack of equilibration in the gaseous boundary layer at the sensor surface, resistive devices also show phenomena caused by slow equilibration of the surface of the semiconductor with the gaseous mixture. In recent work by Logothetis[67] on modelling resistive devices, the situation in which bulk processes in the semiconductor are in equilibrium, but the surface processes are not is considered. This is assumed to be the situation for the TiO_2 sensor as the temperature is lowered and is assumed to give rise, for example, to the oscillations in resistivity observed. Oscillations like these are well known for chemical reactions far from thermodynamic equilibrium, and they require flow of energy or matter, and (non-linear) kinetic processes including autocatalytic reaction steps to provide the required positive feedback. The model concerns only the adsorption and desorption of oxygen and carbon monoxide and the reaction between these species adsorbed on the solid surface.

The model gives a good rationalization of the observed behaviour of titania resistors in vehicle exhaust streams, and gives insight into the causes of problems of shift of the control point and loss of sensitivity. Thus, if the surface reaction rates are insufficiently large then the device can become insensitive to variations in gas composition. The surface reaction rates can be increased by impregnating the pellet with a finely divided Pt catalyst[67]. If the adsorption and desorption rate constants are insufficiently large then the device indicates the rich-lean transition at air-fuel ratios shifted from the stoichiometric point.

4.2 Surface Conductivity Devices

4.2.1. Moisture Detection

Gas phase moisture (humidity) is an important process variable and in principle many devices[68] are available for detection of it. It is a critical variable for a number of industry sectors because of the need to dry materials in a controlled but demanding environment. We choose to describe in this paper a ceramic device which it is believed could be suitable for development into a sensor capable of operating at high temperatures (close to 100°C and above) and high humidities (95-100%), which is typical of the industrial drying situation. Few other devices[68] have this potential.

This sensor is known as the impedance hygrometer, and has been found to be ideal for use in moderate temperatures and humidities.

It generally comprises a porous aluminium oxide layer which has been formed onto an aluminium metal substrate by anodisation. Typical anodising conditions might be 3% oxalic acid at 20°C for one hour at a current density of 1 amp per decimeter[2]. The anodised oxide layer might subsequently be treated with ammonium borate at around 100 Volts potential in order to grow between the columnar pores in the oxide and the metallic substrate a dense and impermeable layer of oxide. On top of the porous oxide material is deposited, usually by evaporation, a gold or otherwise conducting material which will also allow diffusion of moisture through it. The device operates by the diffusion of moisture from the gas phase into the pores of the oxide. Here it probably condenses out on small pores adjoining the large columnar pores, and this results in a change in the dielectric constant and surface conductivity of the material. A device using a porous ceramic element made from a mixture of $MgCr_2O_4$ and TiO_2 has also been described[69].

The devices are formed in a capacitor structure, and it is usual to interrogate it with ac impedance techniques. Either the capacitance or the change in surface conductivity of the material can be used, and this type of device gives a good signal which is roughly proportional to the middle ranges of relative humidity at moderate temperatures. The devices do however have some disadvantages.

Fleming[68] considers that the most significant problem impeding the development of an inexpensive humidity sensor is change of electrical characteristics due to aging, especially that due to thermal aging. Sensors exhibit aging effects which are caused by gradual increases of surface resistances. It is thought that the surface resistances are controlled by the electrostatic field strengths associated with cations on the sensor surface, i.e., by the reactivity of the surface. The more reactive the surface, the more responsive the sensor and, unfortunately, the more susceptible the sensor to aging problems.

Three different physical processes contribute to the aging problem.

(a) Contamination – The sensors are exposed to a multitude of contaminants such as dust, oil vapours, fuel vapours, etc. and any of these contaminants might preferentially chemisorb on cation sites which would otherwise adsorb water. This process chemically "poisons" the sensor surface. Highly polar molecules, such as ammonia or amines, are known to be particularly poisonous. Poisoned surfaces of aged, salt-doped sensors have been cleaned by glow discharge bombardment, and fresh sensor characteristic curves have been recovered.

(b) Loss of Surface Cations – Due to thermal aging, surface cations can either be lost by vaporization or lost by

annealing of ionic crystal surfaces into new structures
of lower reactivity. Furthermore, whenever water
condensate is formed on sensor surfaces, cations can be
lost by dissolution into surface water which might be
subsequently drained off or, upon evaporation, cations
might precipitate out on the surface in segregated
clusters.

(c) Migration of Cations Away From The Surface - Diffusion
processes, accelerated by thermal aging, can cause
cations to migrate away from the surface, thereby forming
a less reactive surface structure. For example, addition
of large dopant ions, which tend to resist the migration
aging process, has been shown to help stabilize the
anodized-aluminium capacitive sensor.

Researchers have worked without success trying to solve sensor
aging problems[70,71]. The main difficulty is that, in order to
detect humidity, the surface of the sensor must possess high
adsorption affinity for water molecules. Since the sensor must
also be exposed to ambient air, surface contamination by molecules
other than water and/or thermally induced changes of surface
structure are unavoidable consequences. Thus, aging of solid
state, impedance-type humidity sensors is a significant problem.

4.2.2. Detection of Reducing Gases

Semiconductor materials whose conductance is modulated
directly by interaction with an active gas have been studied for
over 20 years, since it was discovered in the 1950's that the
reversible chemisorption of reactive gases at the surfaces of
certain metals, oxides, and other materials could be accompanied by
reversible changes in conductance. Unlike metal thin films, whose
conductance modulation by gas adsorption is small and is caused by
changes in the mobility owing to changes in surface scattering, the
conductance changes in the semiconductor materials are large and
are caused primarily by changes in the conduction band electron (or
valence band hole) concentration brought on by charge exchange with
the adsorbed species from the gas phase. The conduction band
electron concentration in the semiconductor sensors can vary more
or less linearly with pressure over a range of up to eight decades
while variations in mobility are generally less than a factor of
two over the same pressure range. It is this large and reversible
variation in conductance with active gas pressure that has made
semiconductor materials attractive for the fabrication of
gas-sensing electronic transducers.

From a practical point of view, the sensors may be in the form of
dense, thin films, screen-printed layers or as a porous ceramic
element. Film devices using both WO_{3-x}[72] and SnO_x as the
sensing element have been described. Pb-doped SnO_2 has been used to
detect CO and ethanol using a device described by Oyabu[73].

Windischmann and Mark[74] present a model for the operation of a thin-film SnO_x carbon monoxide sensor. The model must account for the three known features of such sensors, namely

(a) the dependence of the sensor conduction on the square root of the CO partial pressure;

(b) the existence of a temperature window outside of which the sensor will not function, and

(c) the need for trace oxygen to be present in the background.

The model is very similar to that described above for the titania oxygen sensor. It comprises the following steps:

(i) the surface reaction between associatively adsorbed CO from the ambient gas with chemisorbed oxygen to produce chemisorbed CO_2^-;

(ii) followed by the return of the electron from the CO_2^- to the conduction band, owing to the exothermic reaction energy of the CO oxidation; and

(iii) the subsequent thermal desorption of the neutral sorbed CO_2.

The key feature is again the reaction of adsorbed CO with adsorbed oxygen on the surface of the semiconductor. Thin film detectors such as these respond to any gas which will react with adsorbed oxygen – CO, H_2, hydrocarbons, SO_2, H_2S[75].

The optimum temperature for response to the different gases is different[75] and this is the basis of the somewhat limited selectivity of the devices. This observation can be clearly interpreted using the surface reaction model in terms of the rate constants and activation energies of the various surface oxidations. Oscillations of resistance of films of thoria-doped SnO_2 exposed to CO in air have been observed[77,78], dependent on the temperature and CO concentration.

As noted above for the TiO_2 sensor, the surface reaction model provides a framework for discussing such oscillations. As might be expected, the selectivity, sensitivity, response time, aging and poisoning characteristics of these devices are dependent on the method of preparation[76,79,80], particularly the particle size and dopant concentration – foreign atoms and surface defects such as surface oxygen vacancies can act as inhibitors or activators of the surface reactions. They can provide adsorption sites and surface electronic states which can mediate in electron transfer processes. Certainly the selectivity of SnO_x detectors towards different gases can be changed fairly markedly by doping with other oxides[80],[81]. Devices utilizing WO_{3-x} seem more selective to gases such as H_2S and NH_3[72] which may be dissociated to hydrogen. In this case there is a possibility that the conductance modulation arises because of a surface doping by hydrogen.

Porous, electrical ceramic surface conductivity gas detectors have been developed primarily for monitoring environmental gases. They have been developed mainly in Japan, and are sometimes known as "Taguchi" detectors. In figure 11 earlier types of Taguchi sensor are shown. They comprise generally a metal heating element, and the porous ceramic material, in this case tin oxide, which has appropriate connection leads attached. Generally speaking the porous tin oxide material will be doped with transition metal ions to make it into an n-type semiconductor and adsorption of electron donor gases will decrease the resistance of such a material. The mechanism of operation is assumed to be essentially the same as that of the thin-film devices, since the conductance of a bulk pellet is dominated by the conductivity of the surface layers at the necks between the grains of the porous material[82]. Sintering of the grains therefore constitutes another aging mechanism.

These devices are used for detecting the lower explosive limit of for example propane gas, solvents and petroleum fumes primarily for explosion risk monitoring. Their specificity is determined by the temperature of operation of the sensor, typically in the range 100-200°C. At the high temperatures they may need flame arresters to avoid promoting an explosion. They also have a number of disadvantages in use, in that their signal depends on relative humidity, they are subject to burn-in effects and drift and can be poisoned relatively easily.

A later type of Taguchi sensor is shown schematically in figure 12. It comprises a ceramic tube which has inside it a coiled heating element. On the outside of this a zinc oxide based semiconductor layer is found with an outer electrode which is porous and catalytically active. Again the resistance of the semiconducting oxide layer is the measured quantity. This device is interesting because it combines the properties of a semiconducting oxide with those of catalytically active materials such as porous activated alumina. The change from tin oxide to zinc oxide is reputed to give improved specificity and improved resistance to poisoning. However, the device is poisoned by H_2S. According to the surface reaction model outlined above, the catalyst acts by selectively increasing the rate of some of the surface oxidation reactions.

This concept of combining alternative semiconducting oxides with alternative catalysts is an interesting one which is believed may well lead to improvements in the performance of gas detectors.

Other semiconducting oxides which show surface conductivity changes with gas adsorption include γ-Fe_2O_3[83], $LaNiO_3$[84] and $Ln_{1-x}Sr_xCoO_3$[84,85]. Specific catalytic activity has been reported for a variety of conducting perovskite oxides[86,87]. Although any change in resistivity accompanying this catalysis has not been reported, it is likely that there will be changes in surface conductance which could form the basis of new gas detectors.

5. CATALYTIC GAS DETECTORS

For purposes of environmental monitoring, rather than process control, most inflammable gases can be detected by so-called catalytic gas detectors. Such a detector, termed a pellister bead, is illustrated in figure 13. The bead comprises a finely wound platinum coil on which is deposited porous oxide materials, usually by dipping into an aqueous suspension of the material and then repeated flash firing of the resistance element. These materials might be alumina, thoria, perhaps tin oxide, or combinations of these. Either during or after the formation of the porous oxide, the material is catalytically activated by formation of surface deposits of materials such as platinum or palladium.

In a typical detector head, two such beads are to be found. They are both identical except that one of the beads has been poisoned so it has no catalytic activity. Both beads are heated to temperatures of about 600°C by passing a small current through them, and when the catalytically active bead comes into contact with inflammable gases, in the presence of great amounts of oxygen (usually from the atmosphere) these inflammable materials are burnt. This is an exothermic reaction and the catalytic bead heats up. This causes a change in the resistance of the platinum heating element. The two heating elements form one arm of a Wheatstone bridge, and the changing resistance of the catalytic bead causes an imbalance in the bridge and a signal to be detected which is roughly proportional to the percentage of combustibles in the environment.

Unfortunately such devices do not yet provide differentiation between combustible gases, they are subject to poisoning, and need recalibration periodically due to aging. This aging can arise from gradual poisoning of the catalytic bead, or degradation of its pore structure arising from its continued use at relatively high temperatures. However, the devices are quite successful and find wide application in monitoring of fuel dumps, oil rigs and coal mines. They are extremely small and cheap and it is likely that advances in preparation technique and catalyst technology will result in improvements in performance.

6. SEMICONDUCTOR JUNCTION DEVICES

The devices to be described in this section are in many ways similar to the surface conductivity devices treated in section 4.2. Adsorption of a gas modifies the electron distribution in the surface region of a semiconductor. The modification can arise either directly as a result of electron exchange with the gas molecules being detected, or indirectly by electron exchange with oxygen, the surface concentration of which is a function of the gas being detected. The devices previously described addressed this modulation of the electron distribution by means of a measurement of surface conductivity. The devices described in this section probe this modulation by means of measurement of conduction across the interface (Schottky diodes) or by means of measurement of the capacitance of the surface space charge layer (MOS capacitors and gas-sensitive field-effect transistors).

6.1 Shottky Diodes[88]

In figure 14 the electron energy levels across a metal-semiconductor junction are represented schematically; this structure forms a diode. The reverse current through the diode is an exponential function of the barrier height, $\Delta\Phi$ (figure 14). It is the barrier height which is the gas-sensitive parameter detected, probed by the variation of reverse current through the diode structure. The barrier height can be modulated by adsorption and electron exchange of gases with the metal resulting in changes in the work function of the metal. In order for this to work the metal layer has to be extremely thin since the charge exchanged with the adsorbate needs to be a significant fraction of the conducting electron population. Alternatively, adsorption at the metal-semiconductor interface could provide electronic surface states and thus modulate the barrier height by changing the charge distribution in the semiconductor or could provide states from which electrons could tunnel through the barrier. Such adsorption could be brought about by having a porous metal contact or by the gas dissociating on the metal to form a species (e.g. H) which could diffuse through the metal to the junction.

To date, PbS-Si[89], Pd-CdS[90] and Pd-TiO$_2$[91] diode structures have shown usable sensitivity, particularly to H$_2$, as has a 3-layer MOS structure[88] - Pd/SiO$_2$/Si - in which the intermediate layer of SiO$_2$ is 15-30 Å and the Pd 300 Å thick. Formation of the oxide layer prevents the reaction of Pd and Si to Pd$_2$Si. Three factors appeared to be involved in the hydrogen sensitivity of this device, namely a change in metal work function, the formation of a dipole layer at the Pd-SiO$_2$ interface and the appearance and removal of surface states at the Si-SiO$_2$ interface[88].

6.2 MOS Capacitors and Field Effect Transistors[92,93]

In these devices (figure 15) a modulation of the thickness of a space-charge layer in the surface of a semiconductor is measured. Useful responses have been obtained from Pd-SiO$_2$-Si structures,

with the SiO_2 layer 100–1000 Å thick (much thicker than in the
Schottky diodes described above), operated at around 150°C. The
principle of the response of the devices is believed[92] to be
that hydrogen molecules in the ambient are dissociated on the
catalytic metal surface. Some of the adsorbed hydrogen atoms so
formed diffuse through the thin metal film and are adsorbed at the
metal–insulator interface, and an equilibrium is reached between
hydrogen atoms and the gas and insulator interfaces. Hydrogen
atoms at the metal–insulator interface are polarized and give rise
to a dipole layer. This dipole layer at the interface contributes
a voltage drop, ΔV, which is added to the externally applied
voltage. The characteristics of the MOS structures are therefore
shifted along the voltage axis by an amount, ΔV, dependent on the
surface concentration of adsorbed hydrogen atoms at the gas
interface of the metal film. This concentration is a function not
only of the hydrogen pressure in the ambient but also of the
pressure of oxidising gases such as O_2 which could remove
adsorbed hydrogen by a surface reaction.

When the palladium film was non–porous, the devices responded
only to hydrogen or to gases which could dissociate to hydrogen
atoms (H_2S or NH_3) or to oxidising gases (O_2, Cl_2) in the
presence of H_2. When the metal layer was made porous, on the
other hand, the devices became sensitive to any gas which could
adsorb at the Pd–gas interface and create a dipole there, CO for
example.

It is clear that devices like these offer powerful tools for
studying gas–solid interactions. Although only simple
constructions have been made so far, it is clear that there is
great scope for materials research to extend the range of useful
responses.

7. CONCLUSIONS

It is possible to draw out from this article a number of
important features which can help in directing future research:

(I)   In connection with vehicle exhaust oxygen sensors, the
      optimum combination of increased fuel efficiency and
      reduced pollution from petrol engines in Europe may be
      found in operating the engines at lean mixture. However,
      this is a very demanding problem. In figure 16 the
      dotted area in the lean region in Figure 2 is shown
      expanded, and the vertical scale shows the signal output
      of various types of sensor at 600°C, calculated from the
      oxygen level variation across the area shown in figure 2.
      Inspection of figure 16 allows some important conclusions
      to be drawn.

      (i) Of the various devices, the Nernst sensor has the
           largest output change with air–fuel ratio.

450

(ii)   An imaginary oxide with a $p(O_2)^{-\frac{1}{2}}$ response
approaches the Nernst sensor in signal output.
Oxides with $p(O_2)^{-\frac{1}{4}}$ or $p(O_2)^{-1/6}$
responses are very inferior.

(iii)   Whilst the variation in Nernst sensor output
resulting from a change of temperature of $\pm$ 10°C
is quite small in comparison with the voltage
change due to oxygen level variations, in the case
of TiO_{2-x} such a temperature variation would
swamp the signal from $p(O_2)$ variations, on
account of the high activation energy of this
material.

From a technological point of view, considering size of
response and degree of temperature stabilisation
required, the Nernst sensor would appear to be an
attractive proposition for operation in the lean region.
However, with the geometry of existing devices it is very
difficult to achieve even this degree of temperature
uniformity, ignoring stabilisation, which would itself
also be difficult with this geometry.   Furthermore, the
device relies heavily on the catalytic activity of its
electrodes, and since fuel in Europe is unlikely to be
lead-free, its durability would be open to severe doubt.

The alternative approach, using a semiconducting oxide,
is not, however, without its technological difficulties.
Not least of these is the identification of a suitable
material.   Clearly, from figure 16 an oxide with a
$p(O_2)^{\pm\frac{1}{2}}$ resistance response would be desirable, and
it is relatively easy to calculate that to get equivalent
insensitivity to temperature variations as the Nernst
sensor, the semiconducting oxide needs an activation
energy around 0.1 eV.   This particular combination of
properties is not available from any published material
and this clearly represents an important area of
research.

(II)   On a broader front, there is a great variety of devices,
many of which are being actively researched.   It is
perhaps true that the devices themselves are often
demonstrated, and then the problems associated with their
implementation are given less attention.   These problems
include lack of sensitivity and more particularly
selectivity.   The devices are often unstable over long or
sometimes short times, and they are more particularly
often deficient in terms of their durability,
susceptibility to poisons, etc.   The theory describing
the devices is sometimes sketchy, with reliance placed on
empiricism to make progress.   There is thus scope for

451

considerably more research, of a multidisciplinary
nature, to improve understanding and develop devices with
the rugged characteristics necessary for practical use.

ACKNOWLEDGEMENTS

We would like to thank Mr. K.T. Harrison for painstaking
preparation of several of the figures and diagrams, and for
maintaining a careful watch on the literature. Dr. B.C. Tofield
contributed enthusiastically to discussions on particular points.

8. REFERENCES

1. For more details see P. McGeehin, J. Brit. Ceram. Soc. 80 (2),
   (1981), 37.

2. In this section we draw upon the unpublished work of K.T.
   Harrison.

3. W.L. Worrell, Ceramic Bull., 53 (1974), 425.

4. M. Gauthier, A. Belanger, Y. Meas and M. Kleitz in "Solid
   Electrolytes - General Principles, Characterization,
   Materials, Applications", (P. Hagenmuller and W. Van Gool,
   Eds.), p. 497, (1978), N.Y., Academic Press.

5. "Solid Electrolytes and Their Applications", (E.C. Subbarao,
   Ed.,), (1980), N.Y., Plenum Press.

6. B.C. Tofield, D. Gilling, P. McGeehin, A. Hooper and D.E.
   Williams, unpublished work.

7. J.A. Little and D.J. Fray, Trans. Inst. Min. Metall., Sect. C,
   88 (1979), C229.

8. R.J. Lauf and C.S. Morgan, ORNL Rept. TM8154 (1982).

9. T.A. Ramanarayanan, M.L. Narula and W.L. Worrell, J.
   Electrochem. Soc., 126 (1979), 1360.

10. S-F. Chou, Diss. Abs., 40 (1980), 3336B.

11. T.A. Ramanarayanan and W.L. Worrell, J. Electrochem. Soc., 127
    (1980), 1717.

12. W.L. Worrell and Q.G. Liu, Electrochem. Soc. Ext. Abs., 81
    (2), (1981), 1104.

13. K.T. Jacob and D. Bhogeswara Rao, J. Electrochem. Soc., 126
    (1979), 1842.

14. M. Gauthier, R. Bellamare and A. Belanger, J. Electrochem.
    128 (1981), 371.

15. M. Gauthier, D. Fauteux and A. Belanger, Electrochem. Soc. Ext. Abs., 82 (1), (1982), 1178.

16. T.A. Ramanarayanan, J. Electochem. Soc., 128 (1981), 2487.

17. H.L. Tuller and A.S. Nowick, J. Electrochem. Soc., 122 (1975), 255.

18. G.H.J. Broers, H.T. Cahen, A. Honders and J.H.W. Dewit, J. Appl. Electrochem., 10 (1980) 229.

19. F.J. Esper and K.H. Friese, Ber. Dt. Keram. Ges., 55 (1978), 314.

20. W.J. Fleming, SAE 770400 (1977), Society of Automotive Engineers (USA).

21. C.T. Young and J.D. Bode, SAE 790143 (1979), Society of Automotive Engineers (USA).

22. W.J. Fleming, J. Electrochem. Soc. 124 (1977), 21.

23. J. Schlatter et al, Science 179, (1973), 798.

24. W.J. Fleming, General Motors Research Publication GMR-3107 (1979).

25. W.J. Fleming, General Motors Research Publication GMR-3148 (1979).

26. H. Maurer, K. Müller, E. Linder, F. Rieger and G. Stecher, (Robert Bosch GmbH), UK Pat. Appl. 2054868A (1980).

27. M.P. Murphy and G.W. Hildebrand, (General Motors), UK Pat. Appl. 2017927A (1979).

28. M.J. Verkerk and A.J. Burggraaf, J. Electrochem. Soc., 128 (1981), 75.

29. Hitachi Ltd., UK Pat. 1563964 (1977).

30. M. Croset, J.P. Schnell, G. Velasco and J. Siejka, J. Appl. Phys., 48 (1977), 775.

31. C.F. Bauer, L.B. Welsh, K.J. Youtsey and F.R. Szofran (Universal Oil Products Ltd.), US Patent 4040929 (1977).

32. K. Ikezawa, H. Takao, K. Matoba, S. Ishitani and S. Kimura, (Nissan Ltd.), UK Pat. Appl. 2027729A (1979).

33. Thompson-C.S.F. Ltd., European Patent EP G012647 (1982).

34. M. Tohda, H. Takao and S. Kimura, (Nissan Ltd.), UK Patent Appl. 2066478A.

35. J.E. Anderson and Y.B. Graves, J. Electrochem. Soc., 128 (1981), 2181.

36. J.E. Anderson and Y.B. Graves, J. Appl. Electrochem., 12 (1982) 335.

37. J.E. Anderson and Y.B. Graves, J. Appl. Electrochem., 12 (1982), 463.

38. D.M. Haaland, J. Electrochem. Soc. 127 (1980), 796.

39. Y.L. Sandler, J. Electrochem. Soc., 118 (1971), 1378.

40. H. Okamoto, H. Obayashi and T. Kudo, Solid State Ionics, 1, (1980), 319.

41. H. Okamoto, H. Obayashi and T. Kudo, Solid State Ionics, 3/4 (1981), 453.

42. H. Shinohara, Y. Otsuka, S. Matsumoto, T. Furutani and H. Wakizaka, (Toyota Ltd.), US Patent 4265930 (1981).

43. H. Obayashi and H. Okamoto, (Hitachi Ltd.), US Patent 4210509 (1980).

44. M.L. Jeunehomme, (Bendix Autolite Corp.), UK Patent Appl. 2004067A (1978).

45. D.E. Williams, P. McGeehin and B.C. Tofield, AERE Report R-10638 (1982).

46. A.O. Isenberg, US Patent 4226692 (1978).

47. J. Salardenne, Vide Couche Minces, 37 (1982), 51.

48. R.A. Rapp and D.A. Shores in "Physicochemical Measurements in Metals Research, Part 2", (R.A. Rapp, Ed.), p. 123, (1970), N.Y., Wiley.

49. C.Y. Yang and H.S. Isaacs, J. Electroanal. Chem., 123 (1981), 411.

50. L.W. Niedrach, J. Electrochem. Soc., 127 (1980), 2122.

51. H. Shinohara, Y. Otsuka, S. Matsumoto, T. Furutani and H. Wakizaka, (Toyota Ltd.), US Patent 4299627 (1981).

52. Nissan Ltd., UK Patent 2051379 (1980).

53. S. Kimura, H. Takao, S. Ishitani, K. Ikezawa and K. Sone, (Nissan Ltd.), US Patent 4207159 (1980).

54. T. Fujishiro, (Nissan Ltd.), UK Pat. Appl. 2050625A (1980).

55. R.E. Hetrick and W.A. Fate, (Ford Motor Co.), US Patent 4272329 (1981).

56. K. Muller, H. Maurer, E. Linder, F. Rieger, K.H. Friese, H. Riese, H. Dietz, H. Ziene, F. Esper and G. Holfelder (Robert Bosch GmbH), US Patent 4334974 (1982).

57. Heijne, US Patent 3907657.

58. (a) R.E. Hetrick, (Ford), US Patent 4272331 (1981), and
    (b) R.E. Hetrick, W.A. Fate and W.C. Vassell, SAE 810433 (1981), Soc. of Automotive Engineers (USA).

59. R.E. Hetrick and W.C. Vassell, Appl. Phys. Lett. 37 (1980), 494.

60. R.E. Hetrick and W.C. Vassell, J. Electrochem. Soc. 128 (1981), 2529.

61. W.A. Fate, R.E. Hetrick and W.C. Vassell, Appl. Phys. Lett. 39 (1981), 924.

62. P. Kofstad, "Nonstoichiometry, Diffusion and Electrical Conductivity in Primary Metal Oxides", (1972), N.Y., Wiley-Interscience.

63. E.M. Logothetis, K. Park, A.H. Meitzler and K.R. Land, App. Phys. Lett. 28 (1975), 209.

64. K. Park and E.M. Logothetis, J. Electrochem. Soc. 124 (1977), 1443.

65. (a) T.Y. Tien, H.L. Stadler, E.F. Gibbons and P.J. Zaconanidis, Ceramic Bulletin 54 (1975), 280; and
    (b) M.J. Esper, E.M. Logothetis and J.C. Chu, SAE paper 790140 (1979), Society of Automotive Engineers (USA).

66. Ford Motor Co., European Patent G001-510 (1982).

67. E.M. Logothetis, Ford Technical Report SR-80-01 (1980).

68. W.J. Fleming, SAE 810432 (1981), Society of Automotive Engineers (USA).

69. T. Nitta, J. Terada and F. Fukushima, IEEE Transactions on Electron Devices, ED-29 (1982), 95.

70. A.C. Jason in "Humidity and Moisture, Vol. 1", (A. Wexler, Ed.), 372, (1965), N.Y., Reinhold Publishing Corp.

71. A. Mountvala, G. Onoda, E. Onesto and A. Pincus, American Ceramic Society Bulletin 50 (1971), 170.

72. A.N. Willis and M. Silarajs, (Ambac Industries), US Patent 4197089 (1980).

73. T. Oyabu, J. Appl. Phys., 53 (1982), 2785.

74. H. Windischmann and P. Mark, J. Electrochem. Soc. 126 (1979), 627.

75. J.G. Firth, A. Jones and T.A. Jones, Ann. Occup. Hyg. 18 (1975), 63.

76. S-C. Chang, IEEE Trans Electr. Dev. ED-26 (1979), 1875.

77. M. Nitta, S. Kanefusa, Y. Taketa and M. Haradome, Appl. Phys. Lett. 32 (1978), 590.

78. M. Nitta and M. Haradome, IEEE Trans. Electr. Dev. ED-26 (1979), 219.

79. H. Pink, L. Treitinger and L. Vité, Jap. J. Appl. Phys., 19 (1980), 513.

80. M. Nitta and M. Haradome, IEEE Trans. Electr. Dev. ED-26 (1979), 247.

81. D. Baresel, P. Scharner, G. Huth and W. Gillert, (Robert Bosch GmbH), US Patent 4194994 (1980).

82. L. Heyne, Electrochem. Soc. Ext. Abs. 78(1) (1978), paper 78.

83. Y. Nakatani, S. Nakatani, M. Sakai, M. Matsuoka (Matushita), US Patent 4241019 (1980).

84. H. Obayashi, Y. Sakurai and T. Gejo, J. Solid State Chem. 17 (1976), 299.

85. J.M. Parry and P. Raccah, US Patent 4221827 (1978).

86. R.J.H. Voorhoere, J.P. Remeika and L.E. Trimble, Mat. Res. Bull. 9, (1974), 1393.

87. O. Prakash, P. Gauguly, G.R. Rao and C.N.R. Rao, Mat. Res. Bull. 9 (1974), 1173.

88. J.N. Zemel, B. Keramati and C.W. Spivak, Sensors and Actuators, 1 (1981), 427.

89. J.N. Zemel, J.J. Young and H. Rahnamai, Crit. Rev. Solid State Sci., 1 (1975), 1.

90. M.C. Steele and B.A. MacIver, Appl. Phys. Lett. 28 (1976), 687.

91. N. Yamamoto, S. Tonamura, T. Matsuoka and H. Tsubomura, Surf. Sci., 92 (1980), 400.

92. I. Ludström, Sensors and Actuators, 1 (1981), 403.

93. I. Lundström and D. Söderberg, Sensors and Actuators, 2 (1981/2) 105.

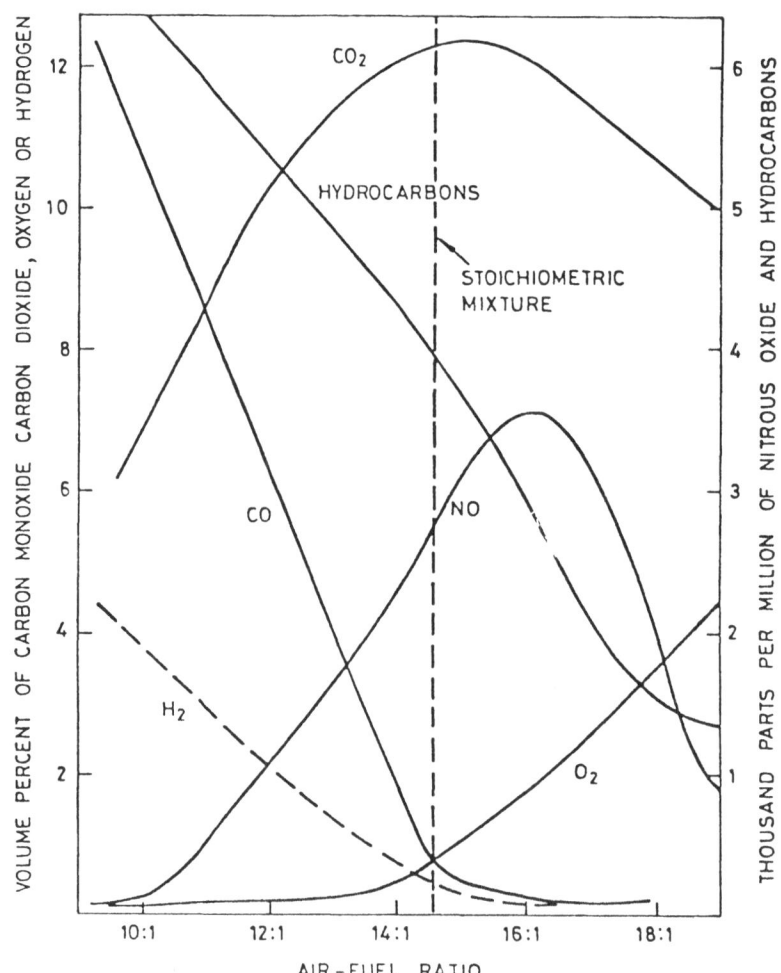

Fig. 1   Diagrammatic representation of the variation in the
          levels of the products of combustion arising in poorly
          equilibrated exhaust gases as a function of engine input
          air-fuel ratio.

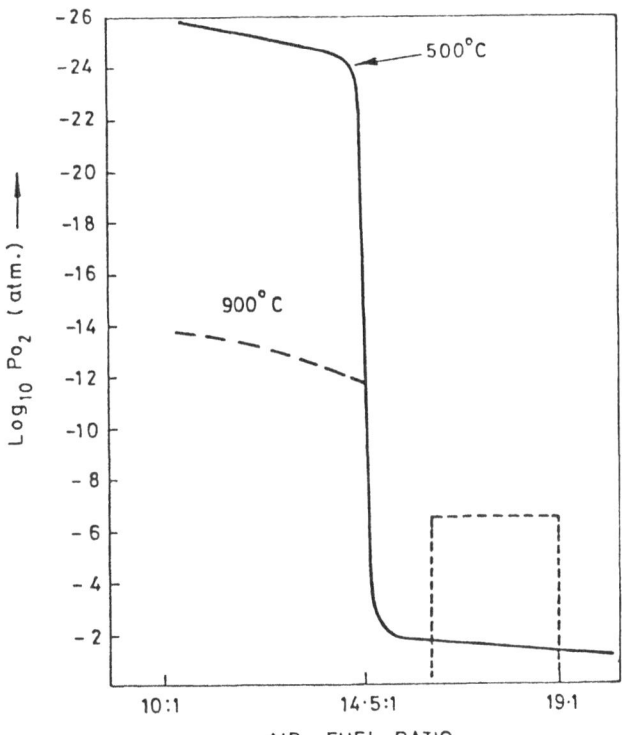

Fig. 2   Variation in the equilibrium oxygen partial pressure in
           vehicle exhausts as a function of air-fuel ratio, showing
           the characteristic S-shaped curve. Boxed: lean mixture
           control range.

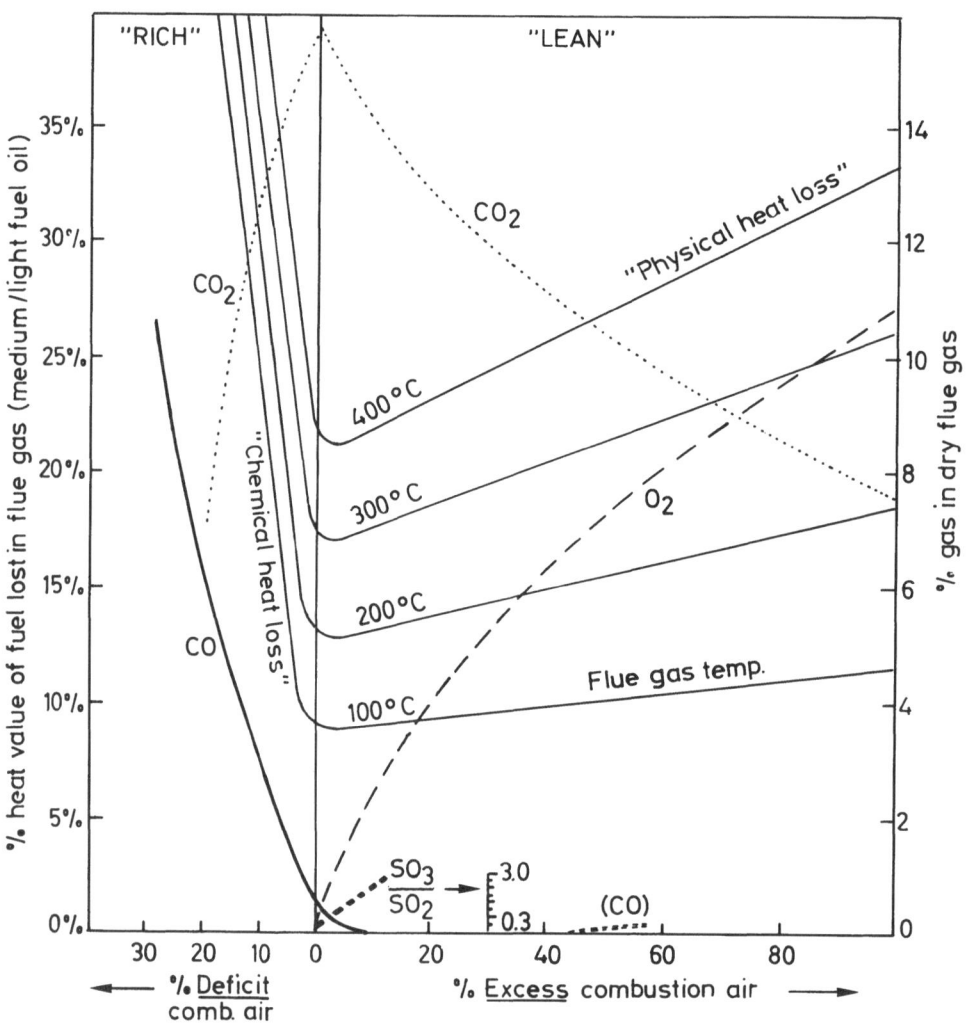

Fig. 3   The percentage of heat of the fuel lost in flue gases for
a medium/light fuel oil and the volume percentage of the
constituents of the flue gas plotted against the
proportion of air supplied to the burner.

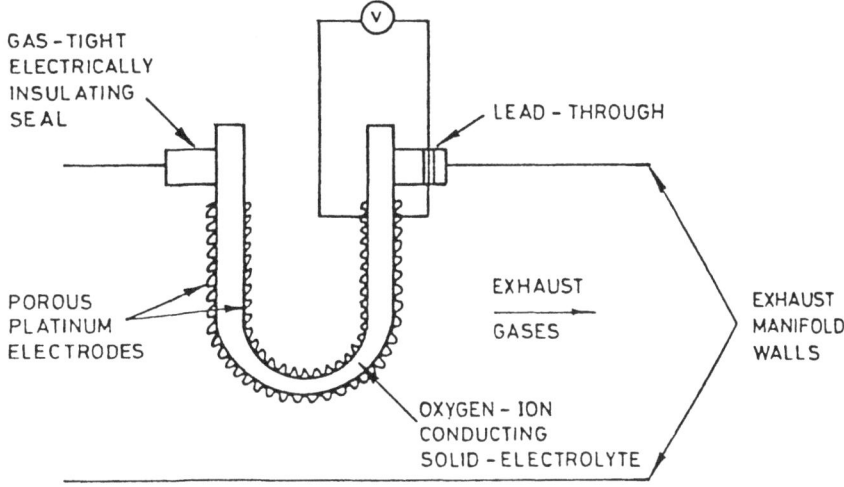

Fig. 4   Schematic drawing of a Nernst sensor placed in a vehicle exhaust manifold.

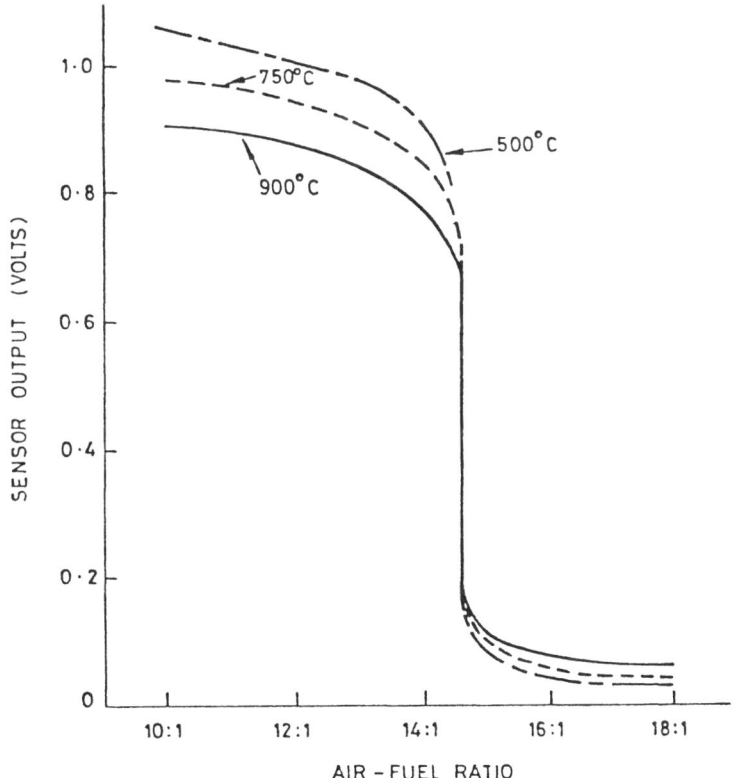

Fig. 5   Variation of the voltage output of the Nernst sensor as a function of air fuel ratios at various temperatures.

461

SOLID ELECTROLYTE

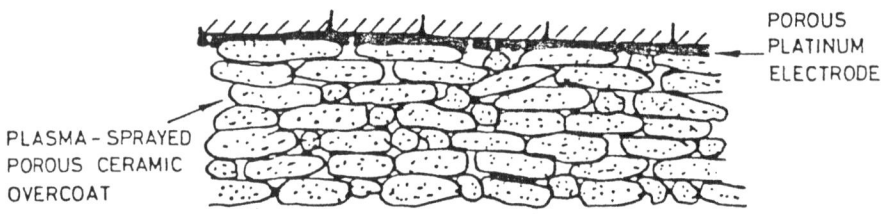

POROUS
PLATINUM
ELECTRODE

PLASMA - SPRAYED
POROUS CERAMIC
OVERCOAT

EXHAUST GAS

Fig. 6   Surface structure of Nernst vehicle exhaust sensor.

(a) Combustion control detecter

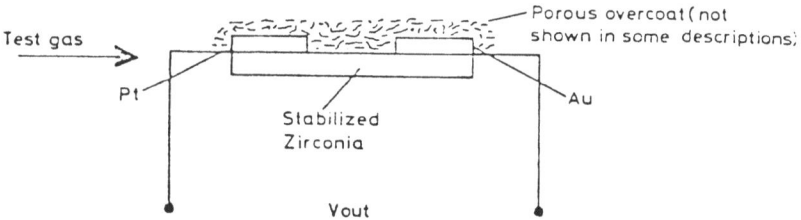

Test gas

Porous overcoat (not
shown in some descriptions)

Pt

Au

Stabilized
Zirconia

Vout

(b) Usual form of solid electrolyte oxygen gauge.

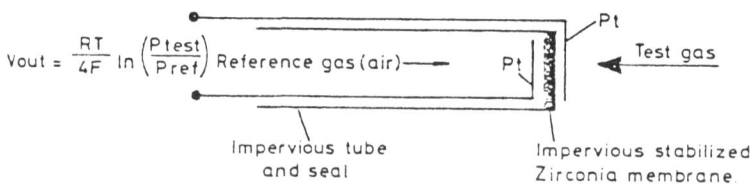

$V_{out} = \frac{RT}{4F} \ln \left(\frac{P_{test}}{P_{ref}} \right)$ Reference gas (air)

Pt

Pt

Test gas

Impervious tube
and seal

Impervious stabilized
Zirconia membrane.

(c) Carbon monoxide detector

Porous separator supporting an
oxidation catalyst.

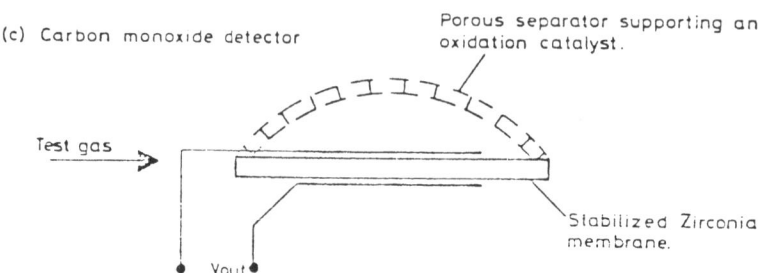

Test gas

Stabilized Zirconia
membrane.

Vout

Fig. 7   Solid electrolyte mixed potential devices.

462

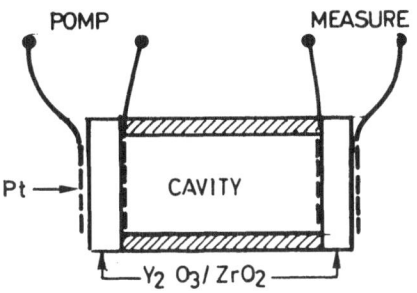

Fig. 8   Sealed cavity reference electrode.

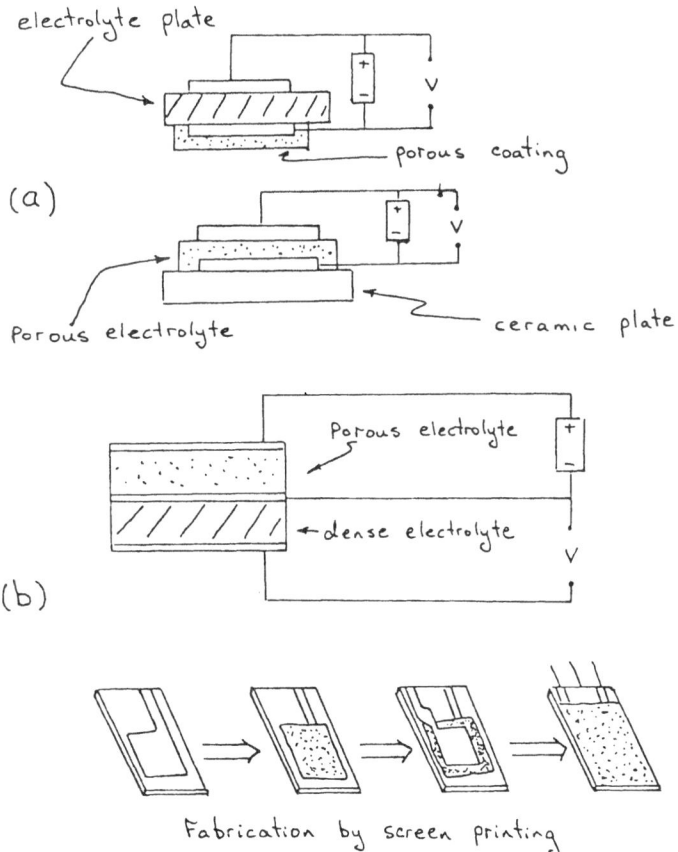

Fig. 9   Dynamic reference electrode systems: (a) 2-electrodes,
           (b) 3-electrodes.

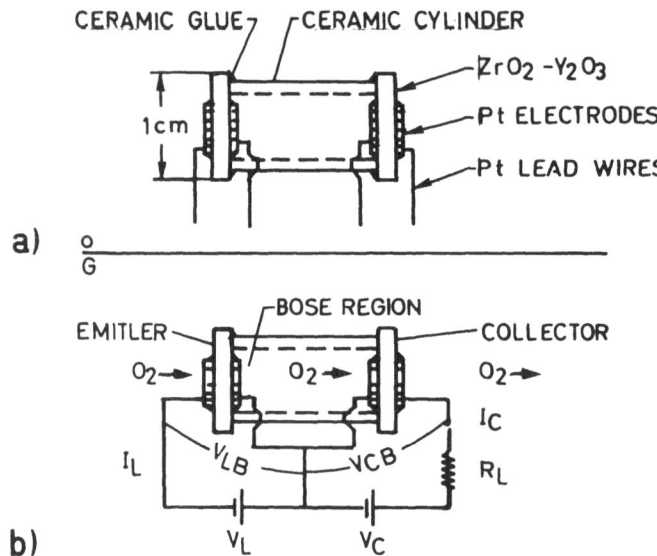

Fig. 10   (a) Solid electrolyte transistor structure.
           (b) Transistor wired in the common base configuration.

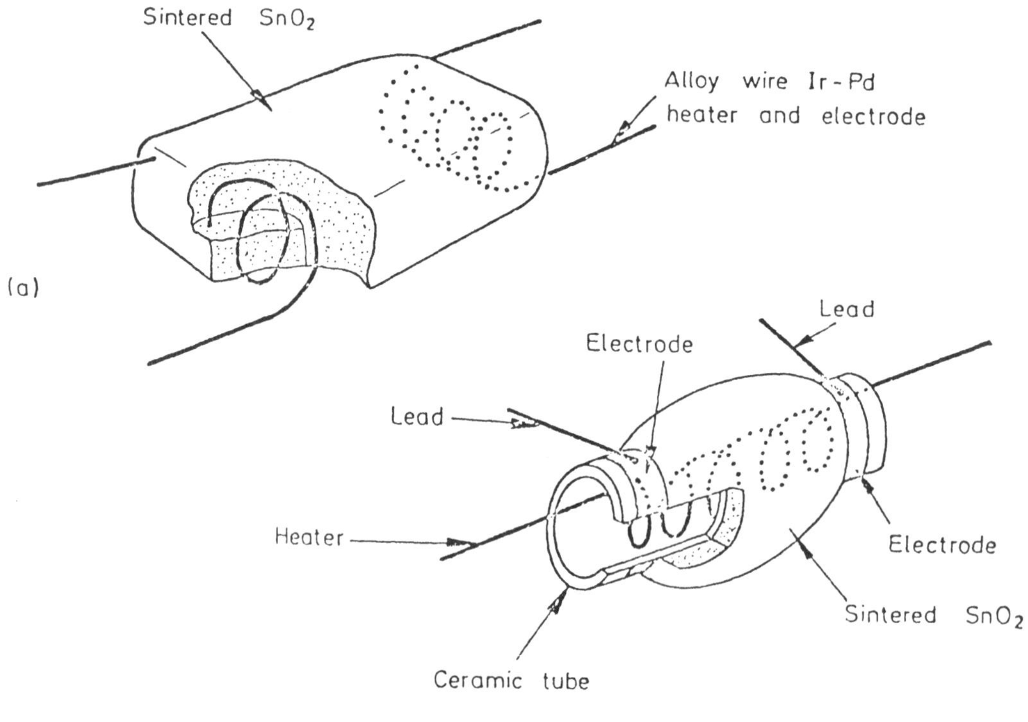

Fig. 11   Early form of Taguchi sensor.

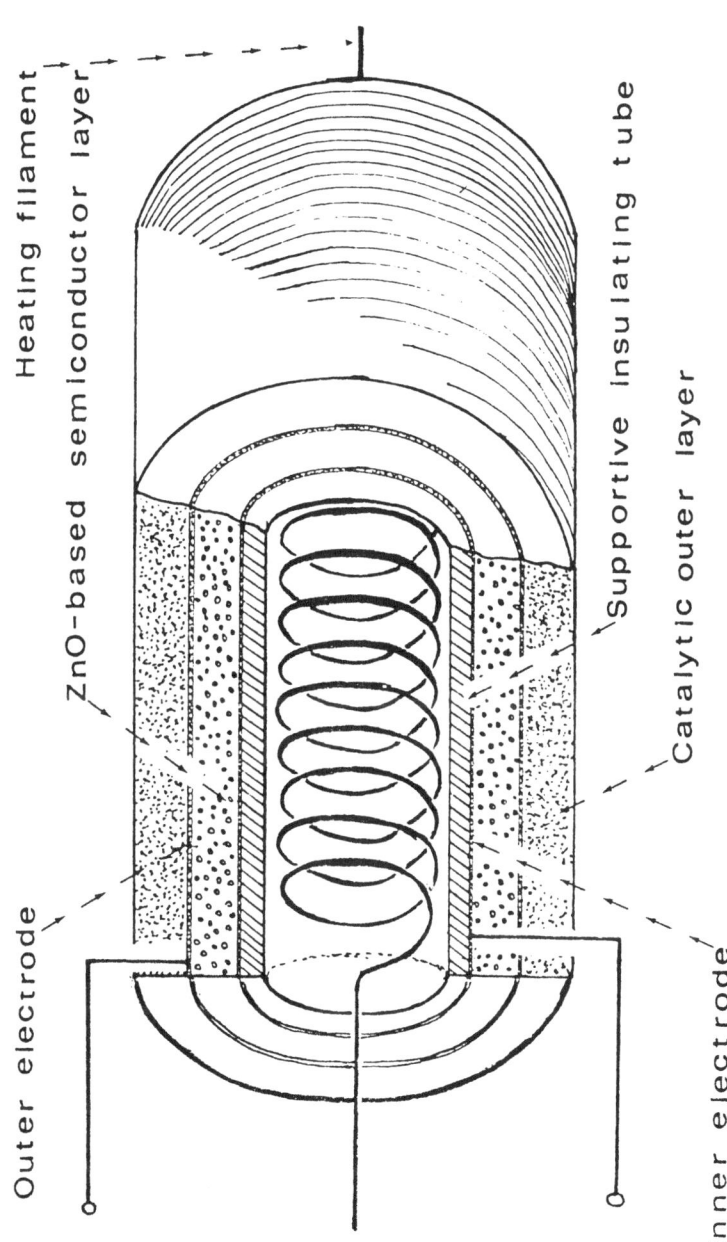

Heating filament

ZnO-based semiconductor layer

Outer electrode

Supportive insulating tube

Catalytic outer layer

Inner electrode

Fig. 12   Improved form of Taguchi sensor.

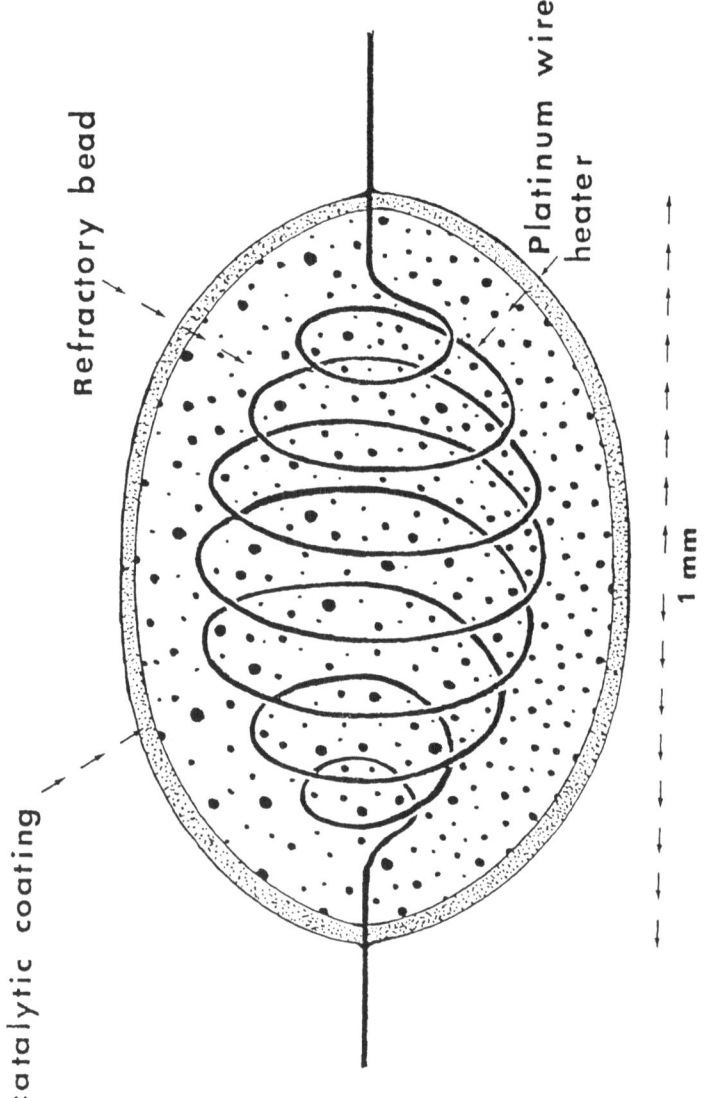

Refractory bead

Platinum wire heater

catalytic coating

1 mm

Fig. 13   Pellistor bead for detecting catalytic gases.

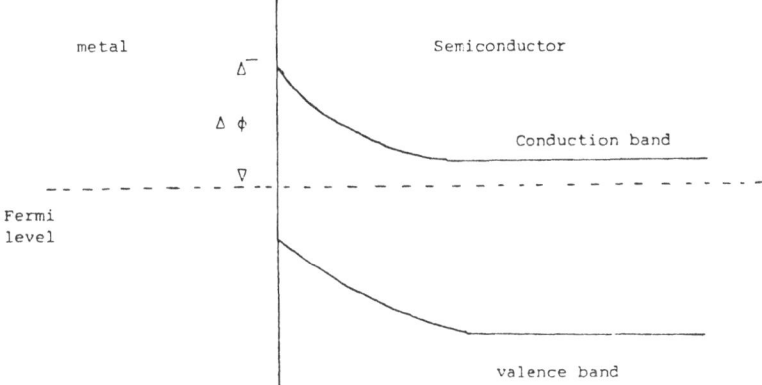

Fig. 14   Schematic diagram of the electron energy levels across a
           metal-semiconductor junction.

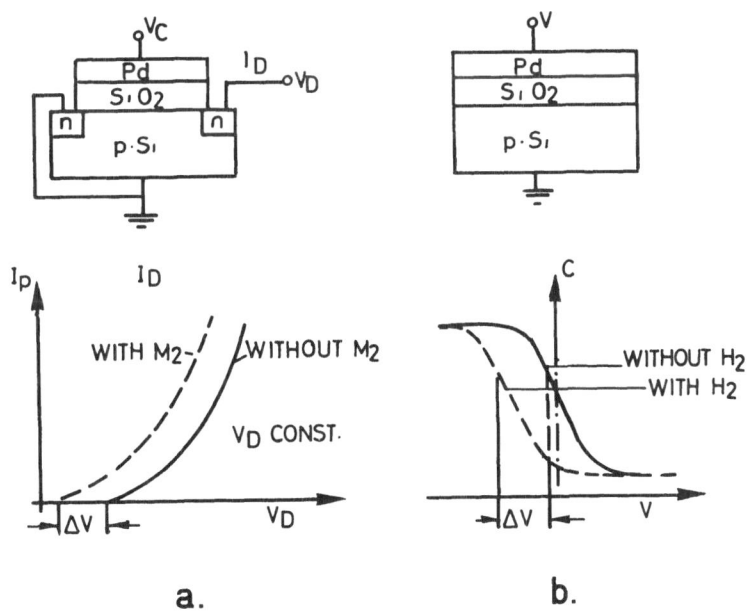

Fig. 15   Schematic diagrams of gas sensitive MOS capacitor and
           field-effect transistor.

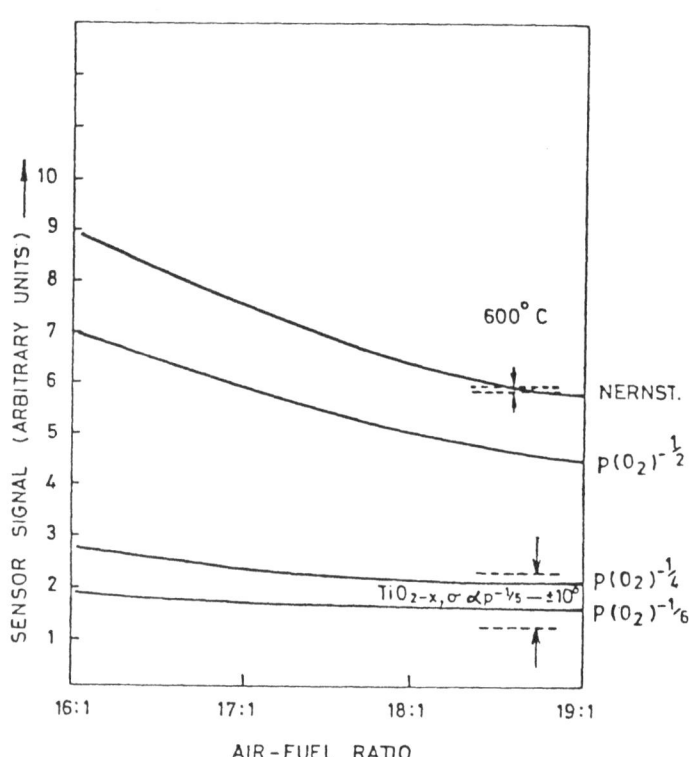

Fig. 16   Signal output and its sensitivity to device temperature
fluctuations, for the Nernst sensor and a variety of
semiconducting oxides with different characteristics, in
the lean mixture region.

SENSORS, THIN FILMS, ELECTRODES

Olivier de Pous

BATTELLE, Geneva Research Centres
CH - 1227 Carouge/Geneva

ABSTRACT

Over the last ten years, a spectacular breakthrough has been observed in the microprocessor industry. This development was initially aimed to achieve better control of industrial processes and is now being applied on a large scale in the fields of domestic appliances, safety systems, automobiles. Therefore, it looks attractive to make a survey of existing technologies which are used in sensor manufacture and to identify the R & D efforts which have to be made in order to satisfy the future market demand.

In each case the problems to be solved are :

- the reliability of the sensor output

- the non-linearity of the signal

- the inter-dependence of different parameters in multivariant systems (e.g. the effect of the temperature)

- the stability of the sensor (need for calibration, reference and standards)

- the resistance to the environment (corrosion, passivation, etc.)

Thick and thin film technologies are currently involved in the manufacture of sensors. The future demand for sensors of high reliability will require a significant multidisciplinary R & D effort. Following a recent evaluation, the size of the existing market is in the range of 500 mio $. Priority has to be given to the development of optical devices.

1.   INTRODUCTION

Following W.G. Wolber [1], an electronic sensor is a simple parameter measuring instrument which transduces a physical parameter
into a corresponding electrical signal with significant fidelity. The
basic concept of solid state electronic sensors is to use the
variation of transport parameters when the material is submitted to
the variation of the surrounding conditions such as temperature,
mechanical strain, radiation (UV, visible, IR, XR, etc.), gas
composition, magnetic field, electric field, etc.

The variation of transport parameters such as the nature of the
electric charge carrier, the carrier concentration, the carrier
mobility, etc., will be detected by measuring the material conductivity or the output voltage delivered by a galvanic cell formed by
the material submitted to different chemical potentials.

In most cases the parameter to be measured generates an electrical
signal or induces a modification of the electric current applied to
the sensor. In both cases two electrodes are applied on each side
of the sensing material. These electrodes could be single metal or
ceramic layer with a sufficient electric conductivity. In somes cases
the electrode should present a specific property such as being
transparent (In_2O_3, SnO_2, etc.) and could have a catalytic effect
(Pt, Pd, ReO_2). Sensors can be classified either by their function
or by considering their level of integration.

2. SENSOR FUNCTION

In Table 1, we have listed the different parameters to be measured and the relevant mechanisms which are involved in the sensing operation.

TABLE 1 - LIST OF PARAMETERS TO BE MEASURED

CLASS	PARAMETER	PRINCIPLE
Radiation	UV Visible IR Ultrasonic Microwave	luminescent photo diodes pyro-electric magnetostriction piezo-electricity magnetostriction
Strength	Pressure Acceleration Vibration	piezo-electricity semi-conductivity
Temperature	Temperature Heat flow	pyro-electric, ferro- electric, thermo-electric, thermistor, PTC, resistor, posistor, semi-conductivity
Atmosphere	Gas consumption Gas flow Liquid composition Liquid flow Moisture	semi-conductivity heat conductivity ionic conductivity
Magnetic field		hall effect
Electric field	over voltage overload	ZnO varistor semi-conductivity

3. GRADE OF INTEGRATION

Over the past ten years a spectacular breakthrough has been observed in the microcomputer industry. This development was initially aimed at achieving better control of the industrial processes and is now applied on a large scale in the field of domestic appliances, safety systems, automobile industry, etc.

471

Compatibility with a microcomputer depends on the level of a possible integration of the sensing components to the electronic circuit. Conventional non-integrated electronic sensors deliver an analog signal. The analog to digital conversion requires an interface unit (ADC) which represents a significant increase of the total cost of the system. Two ways are open as measurement techniques; the use of a pulsed code or the frequency modulated output. In the second case, the time varying signal is directly related to the amplitude of the varying signal.

A typical example of such a three stage system is the temperature control of an industrial furnace (Figure 1).

The signal is delivered by the two wires of the thermocouple. This analog signal is converted into a digital output directly used by the microcomputer in charge of controlling the electric power supply through a thysistor unit.

The cost for the complete monitoring system using a non integrated sensor can be in the range of US$ 2 000 to 3 000, for a programmable unit.

In hybrid technology, analog chips or digital chips are bound to the electronic circuit (Figure 2). The sensor is then considered as a hybrid integrated sensor. In some cases a thin film made by CVD, PVD or other technologies such as thick films made by silk screen printing (serigraphy) could be preferred to the chips technology. A typical system of this type are humidity sensors for which the sensible element can be made of tin oxide as a bulk component or as a film deposited by sputtering. This sensing element is for example located on the ceramic support of the ADC unit.

The average cost of such a complete monitoring system using an hybrid sensor varies from US$ 500 to 1 000.

A cheaper system can be manufactured by using the semiconductor of the integrated circuit as the sensing element itself (Figure 3). Many systems can be used as MOS, MIS or FET transistors. A very satisfying system can be realised such as OPFET developed for IR detection [2].

Therefore the cost of the complete monitoring system using an integrated electronic sensor can be as low as US$ 50 to 500, depending on the level of sophistication. This cost makes it possible to consider the market extension on a worldwide basis, particularly for domestic appliances.

Many people are using the appelation "Smart Sensor". A smart sensor is a microcomputer aid instrument in which the micro-computer is both dedicated to the measurement and integral to the sensor. A wider degree of design freedom can be then achieved, and such a system will be able to treat multi-sensors systems or non-linear systems with the maximum of efficiency.

4.   SENSORS AND CERAMICS

Ceramic technology is related to sensor development in many ways:

- The ceramic (inorganic material produced by sintering) can constitute the active part of the sensor (thermistor, varistor, photochromic material, electro-chromic material, solid electrolyte, et.); this active component can be:
  . a single-wired component
  . active chips
  . an active thick film made by silk screen printing
  . an active thin film made by PVD or CVD.

- The ceramic can be used as a support for the active element (thin film casting).

- The ceramic can also be used as protective coating (CVD, PVD, sol-gel, etc.)

Many sensors required in the field of energy optimisation are exposed to high temperature or corrosive atmosphere. The ceramic represents a significant advantage over other materials. In this analysis, we will consider only the use of ceramic as a sensing element.

4.1 SOLID ELECTROLYTE

Materials with a preponderant ionic electric conductivity are known as solid electrolytes. These are used in form of a membrane (Figure 4) (many times in tubular shapes) coated on each side with porous electric conductors. A disequilibrium of the chemical potential on each side of such a membrane will deliver a current in the external circuit (galvanic mode). The output is delivered in the form of voltage if the impedance of the measuring system is significantly higher than the internal impedance of the sensing system. The output voltage is then directly related to the diffe-rence of chemical potential (or gas partial pressure P_A and P_B) on each side of the solid electrolyte membrane (Figure 5)

$$E = \frac{RT}{NF} \log \frac{P_A}{P_B} \quad ;$$

if the partial pressure P_B is fixed and if the temperature is known, univariant dependance between E and P_A exists, this will constitute the analog signal delivered by the sensor.

The selection of the solid electrolyte can be made by considering the ionic conductivity as a function of the temperature (Figure 6).

Three fields of application of oxygen sensors are currently investigated.

The first application concerns the automobile industry.

The sensor being placed on the exhaust gas system, the air/fuel ratio can be continuously controlled. Moreover, the presence of a catalytic electrode such as platinum can promote the combustion of carbon monoxide and of unburned hydrocarbon. The sensor output, as shown in Figures 7 and 8, will permit to control the presence of the unburned component. Such a sensor, initially developed on the industrial scale by BOSCH, is used for the control of the air/fuel ratio in the vicinity of the stoichiometric value [3 to 12].

The research efforts in this field concern:

- the development of a cheap solution to eliminate the effect of the operating temperature which could vary from 300 to 1 000°C;

- the development of an electrode which is resistant to the poisoning resulting from the presence of lead content in the gas. In fact, the present market of BOSCH Type sensor is limited to the U.S. or Japan and to cars using unleaded gasoline. The resistance to lead remains a critical improvement to be achieved so as to extend the market of oxygen sensors in Europe.

  Whatever the future of leaded or unleaded gasoline this market remains very attractive, and it could represent a few million pieces per year and about US$ 2 000 mio. Nevertheless, the use of solid electrolyte sensors could be limited by the presence of competitive techniques such as TiO_2 semi-conducting sensors or Knock sensor system [13], [14].

The extension of the use of turbo chargers with partial recycling of exhaust gas will also require permanent control of the exhaust gas composition. This constitutes a new promoting parameter of the development of exhaust gas sensors. The final result should be a decrease of the gas consumption ranging from 5 to 20% with respect to the pollution level specifications (Figure 9).

The second application concern the control of industrial as well as household gas and oil burners. The temperature problem is also important in particular for the regulation of flame quality of burners operating in multimode condition. This could represent a significant decrease of energy consumption as compared to burners operating in a discontinuous manner. The poisoning of the sensor owing to the presence of sulfur compounds in the gas and the erosion remain the main problems to be solved. The development of less sophisticated systems which will not require a permanent oxygen reference should be welcome with respect to the sensor durability and cost. This will be particularly required for small household heating units. Any drift of the sensor output could lead to an increase of the gas consumption. Then the reliability of the sensor remains the priority of any development.

The third application concerns the steel industry.

During the elimination of oxygen in steel manufacturing, each batch is controlled by using a solid electrolyte sensor similar to those used for combustion control. For each treatment (approx. 30 minutes), two or three sensors are used (consumable tip), to detect the point where the oxygen level is lower than the specifications. The development of sensors which could stay in the melt for 10 to 20 minutes should be suitable for a continuous monitoring of the deoxidation treatment which would make it possible to minimize the time required for such a treatment and then the energy consumption of the process.

The same requirement exists for controlling the production process of other metals. For example, sensing elements able to control the alumina content (or the oxygen concentration) in liquid cryolite are required in order to optimize the electrolyte bath composition in the aluminium production and then to minimize the electric consumption.

By using an oxygen solid electrolyte with very high conductivity such as $CeO_2Gd_2O_3$, new applications can be considered such as the level of oxygen in liquid sodium cooling loop of a nuclear plant. This allows control of the absence of corrosion in the system [15].

Other solid electrolytes could be used for specific applications. For example beta alumina can be used for monitoring the sodium content in liquid aluminium [16]. In presence of protonic hydrogen, beta alumina can also be used as a detector for hydrogen and hydrocarbon.

Three general problems are related to the use of galvanic cells with solid electrolytes.

1. A stable reference is required on the opposite side of the atmosphere to be controlled. The presence of air cannot be considered in some complex systems as for example in nuclear technology. Scientists are looking at two solutions:

    - the use of a solid state reference (metal + oxide)

    - the use of reference partial pressure supplied by an oxygen pump; this oxygen pump is also made of a solid state electrolyte on which a well-defined current is applied [17].

    In both cases the temperature dependance of the reference itself constitutes an additional problem to be solved for example by the microcomputer itself.

2. The performance of a solid electrolyte system is strongly dependant on the nature and quality of the electrodes used for the signal collection. Response time, selectivity, durability are mainly related to the electrode. This point will be discussed in the section, concerning the electrodes.

3. Existing systems which are used today are composed of thick electrolytes coated on each side by thin electrodes. This geometry corresponds to the maximum ohmic losses in the sensor itself. Thin and thick film technologies should make it possible to develop a structure with thick porous electrodes (at least on the reference side) and thin electrolyte membranes. Such sensors should be able to operate at lower temperatures with a response time compatible with the industrial requirements (Figure 10).

4.2 SEMI-CONDUCTING OXIDE SENSORS

Many elementary oxides (SnO_2, ZnO, TiO_2, etc.) present a wide range of semi-conduction (Figures 11, 12, 13). A clear review of this conductivity has been made by G. PARIS [18]. This is also the case of ionic conductors at high temperatures. Although semi-conducting oxides are classified as n or p types, many species are involved in the conductivity mechanism, such as:

-   interstitial cation
-   atomic vacancy
-   oxygen anions
-   oxygen vacancy
-   electron.

In oxide, the relative influence of each type of carrier is highly affected by the surrounding partial pressure of oxygen. Oxides containing atoms which can adopt different valencies or containing two kinds of atoms, such as perovskite constitute a family of more complexe compounds. The conduction mode can be for example affected by the charge transfer from a divalent to a trivalent atom.

In the family of semi-conducting oxides, a wide range of products exist. The industrial development of these materials as possible sensing components would require a systematic research effort in order to evaluate the behaviour of each material as a function of:

- the shaping technique
  . double-wired component
  . chips component
  . thick film component
  . thin film component

- the morphology of the ceramic
  . grain size
  . grain boundary

- the chemical composition
  . stoichiometry
  . doping element

- the surface morphology
  . porosity
  . chemical reactivity

Semi-conductive oxides are generally produced in polycrystalline form, then the reliability of the experiments carried out by different scientists is difficult to assess. The weakness of the exact description of the manufacturing technique does not allow reporting published experiments. Our literature review has not made it possible to make a clear analysis of the published data in consequence of their dispersion.

It seems therefore a priority to combine the know-how of ceramists in the production of high-quality polycrystalline materials with a well-defined grain structure with the know-how of a physicist in the measurement of the physical properties by using a selected technique accepted by the majority of the scientific community. Precision on sample geometry, sample history, surface preparation and measurement procedures such as temperature, atmosphere, should allow a more complete evaluation of the industrial possibility of semi-conducting oxides as sensing elements to be made.

RESISTIVE SENSOR

The most simple semi-conductive oxide system concerns the measurement of the resistivity of polycrystalline material. Constant voltage being applied to the component, the sensor output is given by the current intensity from which the resistance of the component can be measured (Figure 14).

A list of materials which have been considered for the manufacture of resistive sensors is given in Table 2. As for the solid electrolyte, the sensor output varies strongly with the temperature. In addition, under heating the size of crystallites can increase and the behaviour of the sensor will be strongly affected. In order to get a stable sensor, it seems to be necessary to carry out the measurement on annealed components at a temperature significantly higher than the maximum operating temperature of the sensor.

The temperature dependance of resistive sensors could be adjusted by using in series a suitable thermistance (with a negative or a positive coefficient). This thermistance should not be sensitive to the parameter to be controlled.

The cheapest way to make this type of sensor remains the thick or thin film technology applied on a ceramic substrate.

As an example of applications, we would like to discuss the possibilities offered by the use of SnO_2. Work has been carried out by universities and large groups such as SIEMENS and MATSUSHITA [19 to 25].

Looking at Figure 15, it seems that the results which have been obtained on sputtered SnO_2 films are better than those obtained on sintered SnO_2. In Figure 16, we can observe that the maximum output can be achieved for ultra fine grain structure with a crystallite size below 25 nm. Under heating, the grain size increases and the sensor output decreases. This type of temperature limitation is expected to exist for any other semi-conductive oxide.

TABLE 2 -   LIST OF SEMI-CONDUCTIVE OXIDES USED FOR GAS DETECTION

MATERIAL	GAS	RELATIVE EMPHASIS %
SnO_2	NO_x CH_4 H_2 H_2S H_2O CO	30
TiO_2 $\left\{\begin{array}{l} SnO_2 \\ MgCr_2O_4 \end{array}\right.$	O_2	15
Tungstate $\left\{\begin{array}{l} Mo \\ Mn \end{array}\right.$	H_2O	4
Al_2O_3	H_2O	4
Fe_2O_3	CH_4	2
CoO		2
TaN	C_2H_4	2
MnO_2	H_2O	2
Ta_2O_5	H_2O	2
$ZrOCl_2$	H_2O	2
OTHERS		35

Such sensors can be used for the detection of many gases. In particular, H_2, H_2O, hydrocarbon and NO_x can be detected in air. As several of these compounds can be simultaneously present in the gas stream, it seems interesting to evaluate the possibility of manufacturing selective sensors.

As shown in Figure 17, the temperature of the substrate during the sputtering does not influence the sensor selectivity. Following the data published by Matsushita, some can be achieved with the sensor operating temperature (Figure 18). For example, H_2O will be detected up to 100°C, C_2H_5OH in the range of 250°C and C_4H_{10} at 380°C. This temperature dependance selectivity is related to the surface exchange mechanism of the oxide layer. In certain cases the other gas components can modify the sensor response; in oxygen for example, the signal resulting from the presence of few ppm of NO_x will depend on the level of humidity of the gas stream (Figure 19).

Effectively, in absence of selectivity, the possible interaction of the different gases present in the gas stream and more precisely in the water content could be a limiting parameter to the industrial development. Solid electrolytes in this aspect can be considered as more feasible owing to the selectivity resulting from the pre-dominance of specific ionic conductivity.

Concerning the application of semi-conducting oxides as sensing elements, the main difficulty concerns the long-term reliability and the possible deviation resulting from a long time exposure to the reducing atmosphere. This type of sensor should nevertheless be suitable for control of small gas impurities in air or for control of pollution in water.

Research efforts have to be made in:

- the optimisation of the manufacturing procedure
- the temperature control process
- the increase of the operating temperature
- the increase of the number of tests carried out in real conditions.

The development of less performant but more stable sensors using semiconducting oxide could be preferred for many industrial applications.

As a conclusion, semi-conducting oxide sensors and more precisely SnO_2 sensors could be suitable for the detection of minor gas in air stream and for the determination of humidity.

A very cheap system can be considered, but the selectivity has to be clearly demonstrated in non-artificial atmosphere. The manufacturing process has to be optimized and the long-term reliability clearly demonstrated.

The field of application could be very broad. Semi-conducting oxide sensors can be used for the monitoring of continuous industrial processes, traffic pollution, fluid level control, humidity sensor for food conservation or exhaust gas control. The development carried out by Ford by using TiO_2 resistive sensors [13] is a clear demonstration of a possible development (Figure 20), in automobile field. Nevertheless it is not suitable for the control of oxygen content in lean gas mixture.

TRANSISTOR SENSOR

A second way is open by the use of semi-conductive oxides in form of a a coating deposited on silicon chips. This constitutes the base of many sophisticated sensors fully adapted to integrated electronics. Examples are MOS, MIS, FET components. In order to illustrate these applications, we have selected the control of humidity level [26] and the pH of liquid [27].

The basic design is shown in Figure 21. As an example, we can observe that the output voltage is related to the nature of the ceramic coatings, in the case of a ISFET type sensor (Figure 22, 23).

SiO_2 coating gives the higher voltage, but silicon nitride or alumina layers give a more linear signal and a wider range of measurements (Figure 23). The use of non oxide ceramic coatings such as BN, TaN is also commonly used in such an application [28].

The main advantage of such a system results from the full adaptability and integration of the sensing component to the microprocessor unit. The main disadvantage results from the possible contamination of the system.

Concerning the selectivity, the problems are similar to the problem for pure resistive sensors.

Fully integrated sensor developments are limited by:
- the maximum working conditions of the component itself and of the surrounding electronics,

- the contamination of the semi-conductor as a result of direct contact with the atmosphere to be controlled.

Therefore, this technology can be applied mainly for electro-optic detections of radiation for which a protective layer can be applied. For gas detections, it will be difficult to get a sufficient reliability on a long-term basis.

4.3 FERRO-ELECTRIC SENSOR

For the manufacture of sensors it should be necessary to consider the possibility to use ferro-electric material as sensitive element. Typical properties of these materials are related to the variation of the dielectric constant of the ceramic with the temperature (Figure 24) and the variation of the material polarisation with the temperature (Figure 25).

Today, many ceramic products present a ferro-electric effect. This temperature dependance can be used to control any thermal process and is then directly suitable for the monitoring of related energy. Many of these materials are commonly considered as pyro-electric materials. Conventional materials are $LiTaO_3$, $SrBaNb_2O_6$ and lead zirconate. One of the major advantages of a poly-crystalline ceramic over single crystal is that the production of large blocks and plates is possible at low cost. The electric conductivity in the range of 10^9 ohm^{-1}/m is such that FET bias resistors commonly used in detector devices can be omitted, which allows a reduction of the cost of the sensor.

The main application of ferroelectric material is detection of IR radiation. This makes it possible to manufacture intruder alarms or fire alarms. Selective gas detectors can be made by using the absorption of IR issued from an infra-red source. This system will be quite performant but more expensive than conventional semi-conductor devices. Pyroelectric ceramic can be also used for the manufacture of radiometers suitable for temperature control for example in solar energy technolgy. Other applications not related to energy are the thermal imaging system and the laser detectors.

5.   ELECTRODES

Electrode manufacture in sensing technology presents the following problems:

- in harsh environment, the electrode must be stable and resistant to corrosion,
- in a radiation system, the electrode must be transparent (doped SnO_2 or In_2O_2 are commonly used for such applications)
- in gas phase detection, the electrode must have a high conductivity and high porosity.

The manufacture of electrodes for gas sensors remains one of the key points of development. In many cases the electrode has a critical effect on the sensor selectivity.

The material can be noble metal (Pt, Pd, Ag, An, etc.) but also electro-conductive oxide (Figure 26). Different morphologies of the electrode are possible such as dense layers (Figure 27 A) columnar structure (Figure 27 B), sandwich structure (Figure 27 C), dendritic structure (Figure 27 D), cauliflower structure (Figure 27 E) or sponge structure (Figure 27 F).

Many techniques can be used for making them; CVD, sputtering, vacuum deposition, silk screen printing (serigraphy). Here also, the development would require an optimization of the procedure technology.

6.   CERAMIC SENSOR AND ENERGY

In Table 3 are listed the different needs directly related to the problem of energy production, transportation and consumption.

Following a recent analysis [29] performed in the US by the Engineering Societies Commissions on Energy (ESCOE) on behalf of the US Department of Energy, eight most important areas of research have been identified:

1.   smarter sensor (fully integrated)
2.   instrument performance (long term reliability)
3.   control on flame quality (multimode operation)
4.   fuel analysis (use of low grade fuel)
5.   control of toxic gas and gas stock safety
6.   internal combustion engine control (exhaust gas sensor)
7.   harsh environment (resistance to temperature, errosion and corrosion)
8.   heating, ventilating and air conditioning (HVAC sensor).

It also appears from this study that in the following four areas, the needs cannot be met with existing technologies and new approaches are needed:

- flame quality in multimode combustion burner
- fuel analysis for the expansion of low quality fuel consumption
- control of NO_x , SO_2 and particulate emission level
- reliable system working in harsh environment defined by the conjunction of corrosion (H_2S, SO_2, etc.), high temperature and high level of radiation.

	NUCLEAR	HYDRAULIC	FOSSILE	GEOTHERMIC	SOLAR
PRODUCTION	Temperature Radiation Pressure Corrosion Gas flow	Pressure Gas flow	Temperature Gas composition Particle size	Temperature Flowmeter	Radiation Flowmeter Temperature
	ELECTRIC		GAS	HEAT	
TRANSPORTATION	Overvoltage Insulator fault		Pressure Gas composition Corrosion	Temperature Flowmeter Corrosion	
CONSUMPTION INDUSTRIAL	Watt hour Billing meter		Fluidized bed level Combustion Control Temperature Pollution	Temperature Flowmeter Heat transfer	
HOUSEHOLD CONSUMPTION			Housing temperature Air conditioning Ventilating Cooking		

TABLE 4 -; SENSORS FOR FUTURE RESEARCH AND DEVELOPMENT SURVEY OF USERS OPINION MADE BY THE JAPAN INDUSTRY DEVELOPMENT ASSOCIATION (technocrat Vol. 15 No. 5 May 1982)

Market distribution 1 000 sensors		Machinery	Precision instrument	Measuring instrument	Computer technology	Broadcasting communic.	Household appliance	Transport engine	Others
Photo sensitive	190	10	31	39	63	37	8	2	-
Pressure	150	8	8	38	31	15	10	31	-
Displacement	110	13	5	26	21	10	10	23	-
Temperature	110	8	8	33	22	17	13	8	-
Gas sensors	100	5	9	26	16	14	19	12	-
Humidity	90	4	2	24	24	10	16	8	-
Magnetic	60	5	2	17	14	14	7	-	-
IR sensor	60	-	6	18	14	12	4	4	-
Speed meter	30	-	-	-	-	-	-	-	-
Ultrasonic	20	2	3	3	6	3	2	-	-
Flow meter	20	-	-	-	-	-	-	-	-
Acceleration	10	-	-	-	-	-	-	-	-
Others	40	-	-	-	-	-	-	-	-

7. MARKET TRENDS

The worldwide sensor market today represents a total turnover in the range of US $ 500 M. In Table 4 the distribution of the demand based on a recent inquiry performed in Japan is reported [30].

Looking at the priority with respect to the potential business opportunities, we can define the following items:

1. photo-electric (opto-electronic) devices for computer technology, measurement instruments and telecommunications

2. pressure control devices for measurement instruments, computer technology, engine control and transportation system

3. displacement for measuring devices applied to numerical machines, robotics and engine control

4. temperature measurements and control for industry but also for the large market that domestic appliances represent (heating, cooling, ventilation, air conditioning)

5. gas sensors as measuring instruments (pollution control), household appliances (safety) and combustion engine control (decrease of gas consumption)

6. humidity sensor represents also an attractive market for food conservation, industrial process control, and air conditioning. Many of these operations are energy consuming processes.

In the past, military and space requirements were the main demanding sectors. Today, the worldwide demand results from the conjunction of several driving forces:

- economic parameters in terms of productivity, energy saving, nutrition problem, etc.
- social parameters in terms of health, safety
- technical parameters in terms of the expansion of the microcomputer technique and their applications.

Today, this market is covered to the extent of 50 to 60% by large companies and 40 to 50% by small companies.

This constitutes an open field for joint venture development. Nevertheless, the problem encountered by investors is the multiplicity of a proposed system and the absence of complete feasibility of the majority of systems proposed by scientists.

The automobile market remains one of the most attractive fields. In 1979, 15% of the new US cars were equipped with one sensor. In 1983, we can expect that 100% of the new US cars will contain at least one sensor. Reduction from five to 10% of the gas consumption could be achieved using electronic control of the combustion.

This market will be opened only if a sensor with a sufficient reliability at lower cost (US $ 40) can be manufactured. For comparison, current electronically controllable carburettors cost around US $ 200.

It should be of interest to engage upon a sectorial analysis of the energy which could be saved by the progressive introduction of sensor technology.

Ceramic materials which have a good reputation of stability, temperature and corrosion resistance, remain as a bulk or as a thin film an ideal material for the design of a reliable sensor.

BIBLIOGRAPHY

[1] W.G. WOLBER, K.D. WISE
IEEE Trans. Elec. Devices (1979), Vol. EP.26 N° 12,
p 1864-1874

[2] M. OKUYAMA, Y. MATSUI, H. NAKANO, Y. HAMAKAWA
Ferroelectrics (1981) Vol 33, p 235-241

[3] D.J. ROMINE (1980), US Patent 4 234 542

[4] T. YASUO, K. KENJI, K. TAKAYUKI
Natl. Tech. Rep. Matsushita Elec. (1979) Vol 25, N° 5
p 1042-1052

[5] R.T. DIRSTINE, W.O. GENTRY, R.N. BLUMENTHAL
Am. Ceram. Soc. Bull. (1979) V58, N° 8, p 778-783

[6] E. HAMANN, H. MANGER
SAE Prep. N 770401 (1977)

[7] R. WILHELM, D. EDDY
Am. Ceram. Soc. Bull. (1977), V.56, N° 5, p 509-512

[8] J.E. ANDERSON, Y.B. GRAVES
J. Electrochem. Soc. (1981) V. 128, N° 2, p 294-300

[9] D.M. HAALAND
J. Electrochem. Soc. (1980), V.127, N° 4, p 796-804

[10] G. VELASCO, J.P. SCHNELL, M. CROSET
Proc. Int. Conf. passive component (1982) Paris, p 262-267

[11] W.J. FLEMING
SAE spec. public (1977), SP-418, p 39-51

[12] J.D. BODE
(1978) US Pat. 4080276

[13] M.J. ESPER, E.M. LOGOTHETIS, J.C. CHU
SAE spec. publ. (1979), p 19-27, Paper No 790140

[14] J.P. DOUGHERTY
SAE spec. publ. (1979), Paper No 790139

[15] P. ROY, B.E. BUGBEE
Nucl. Technol. (USA) (1978) Vol 39, 216

[16] BATTELLE GENEVA DEVELOPMENT

[17] R.E. HETRICK, D.K. HOHNKE, E.M. LOGOTHETIS
AIP. Conf. Proc. (1981) Vol 66, p 140-146

[18] G. PARIS
Conductibility ionic and electronic of oxydes 2n thesis
private communication

[19] R.B. COOPER, G.N. ADVANI, A.G. JORDAN
J. Electronic Water (1981), Vol 10, N° 3, p 455-472

[20] H. PINK, L. TREITINGER, L. VITE
Japanese Journal of Applied Physics (1980), Vol 19, N° 3,
p 513-517

[21] H. OGAWA, A. ABE, M. NISHIKAWA, S. HAYAKAWA
J. Elect. Chem. Soc. (1981), Vol 128, N° 9, p 2020-2025

[22] J. Appli. phys. (1982), Vol 53, N° 4, p 2785-7

[23] G.N. ADVANI, Y. KOMEN, J. HASENKOPF, A.G. JORDAN
Sens. Actuators (1981), Vol 2, N° 2, p 139-47

[24] P. TISCHER, H. PINK, L. TREITINGER
Proc. Conf. Solid State Devices (1980), Vol 11, p 513-17

[25] H. WINDISCHMANN, P. MARK
J. Electrochem.. Soc. (1979), V.126, N° 4, p 627-633

[26] T. NITTA, Z. TERADA, S. HAYAKAWA
(1977) US Patent 4 015 230

[27] I.R. LAUKS, J.N. ZEMEL
IEE Trans Electron Devices (1979), Vol ED26, N° 12,
p 1959-1964

[28] T. KIMURA, K. YAMAMOTO, T. SHIMIZU, S. YUGO
Thin solid films (1980), Vol 70, p 351-362

[29] US Engineering Societies Commissions of Energy - DOE (1981)

[30] TECHNOCRAT (1982), Vol 15, N° 5, p 45-46

[31] R.W. WEST, J.M. HONIG
Electrical conductivity in ceramics and glass (1974), Vol 2,
p 343-452

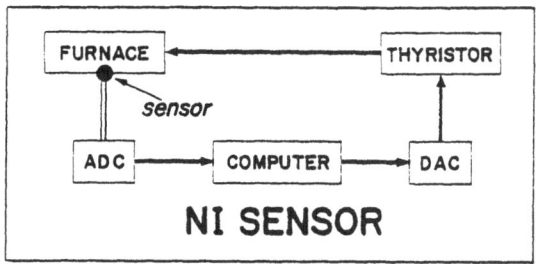

Figure 1 - Typical system using a non-integrated sensor device

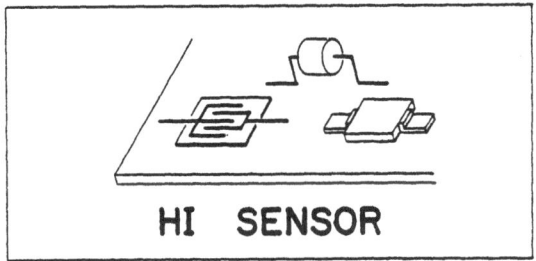

Figure 2 - Hybrid sensor on electronic circuit (thick film, double-wired, or chips)

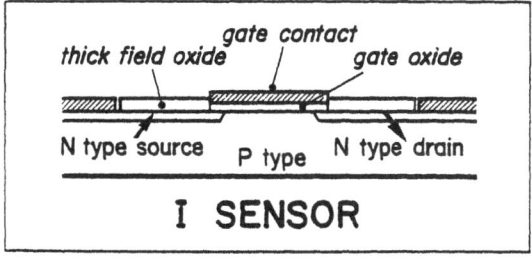

Figure 3 - Example of integrated sensor

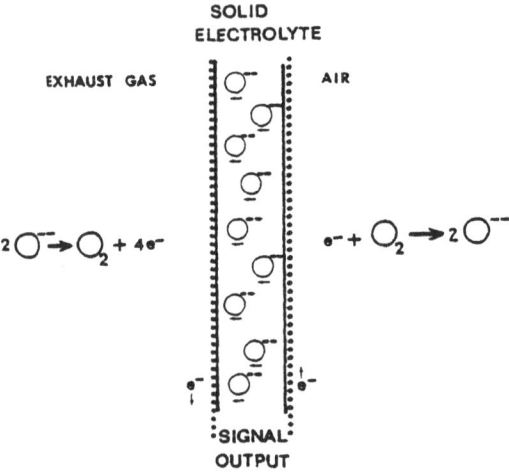

Figure 4 - Schematic representation of solid electrolyte membrane for oxygen determination

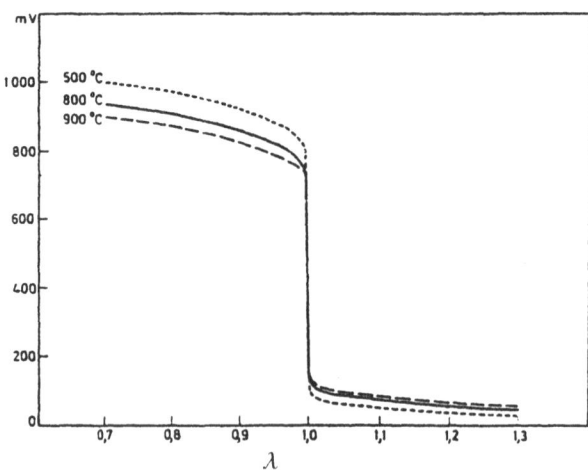

Figure 5 - Theoretical output of solid electrolyte sensor as a function of air/fuel ratio (λ)

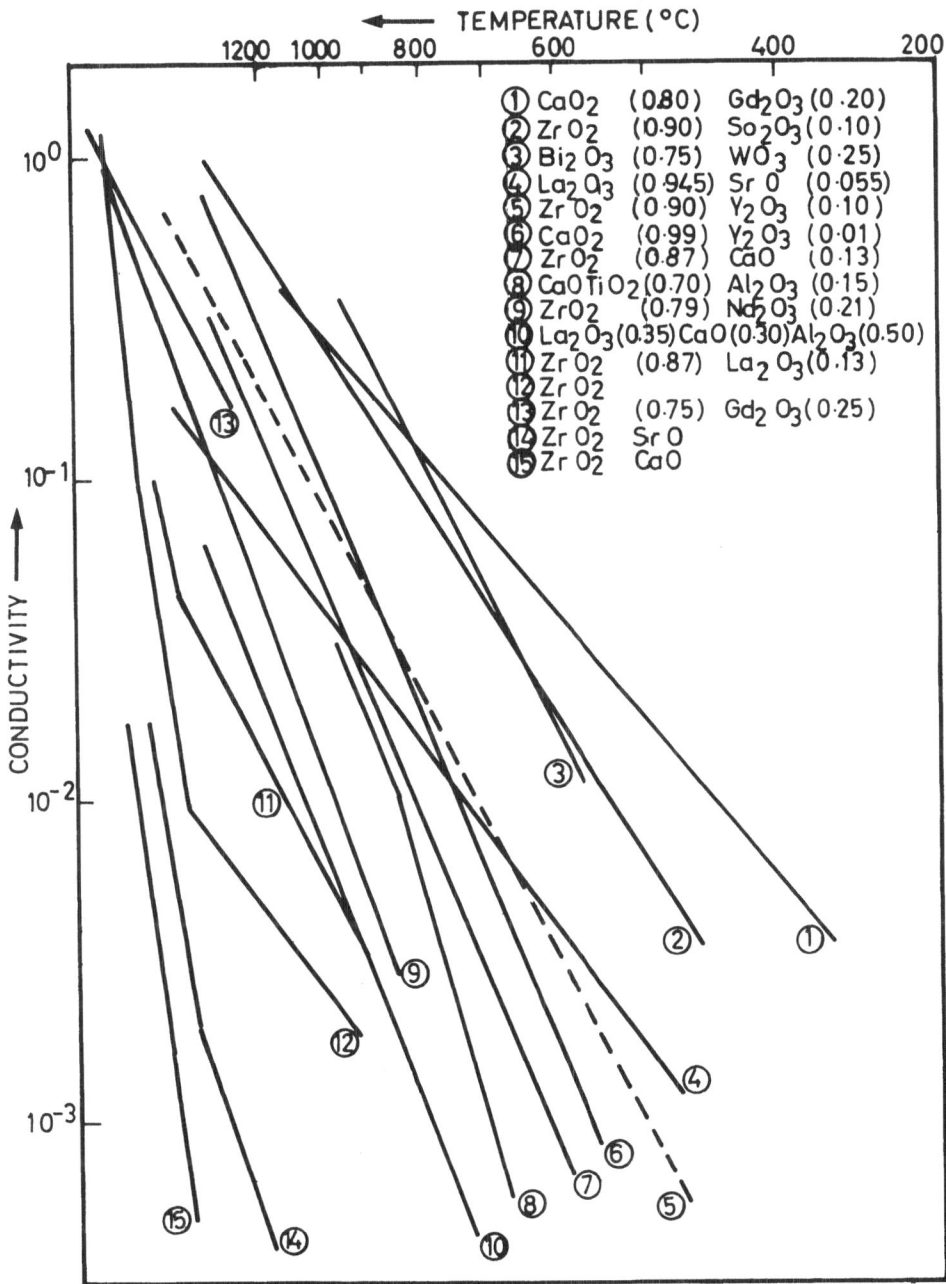

Fig.6 Ionic conductivity (O^{--}) of solid electrolytes as a function of temperature

Legend shown in figure:

① CaO_2 (0.80) Gd_2O_3 (0.20)
② ZrO_2 (0.90) So_2O_3 (0.10)
③ Bi_2O_3 (0.75) WO_3 (0.25)
④ La_2O_3 (0.945) SrO (0.055)
⑤ ZrO_2 (0.90) Y_2O_3 (0.10)
⑥ CaO_2 (0.99) Y_2O_3 (0.01)
⑦ ZrO_2 (0.87) CaO (0.13)
⑧ $CaOTiO_2$ (0.70) Al_2O_3 (0.15)
⑨ ZrO_2 (0.79) Nd_2O_3 (0.21)
⑩ La_2O_3 (0.35) CaO (0.30) Al_2O_3 (0.50)
⑪ ZrO_2 (0.87) La_2O_3 (0.13)
⑫ ZrO_2
⑬ ZrO_2 (0.75) Gd_2O_3 (0.25)
⑭ ZrO_2 SrO
⑮ ZrO_2 CaO

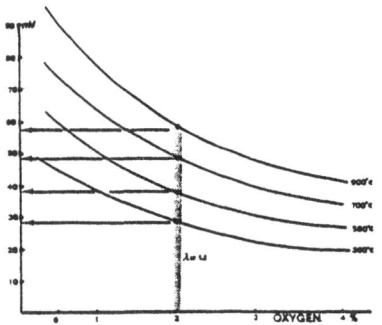

Figure 7 - Signal of the lambda Bosch sensor as a function of temperature

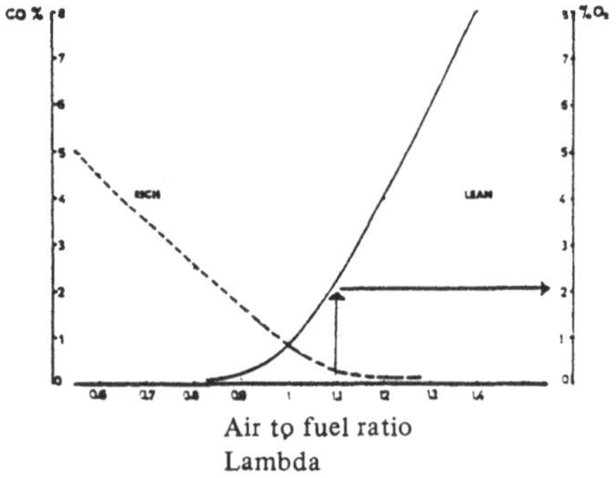

Figure 8 - Typical exhaust gas composition

Air to fuel ratio
Lambda

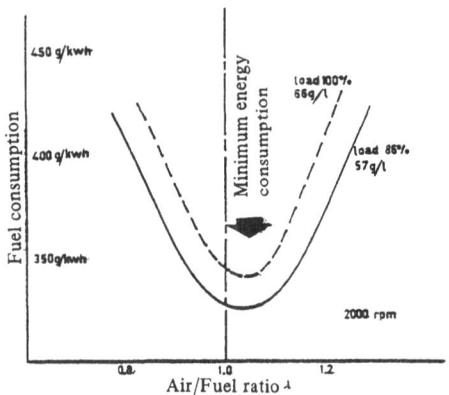

Figure 9 - Fuel consumption as a function of the excess-air factor

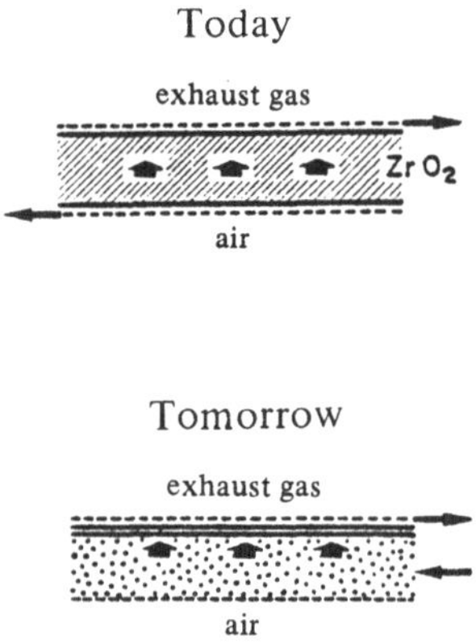

Figure 10 - Thin and thick ZrO_2 sensor concepts

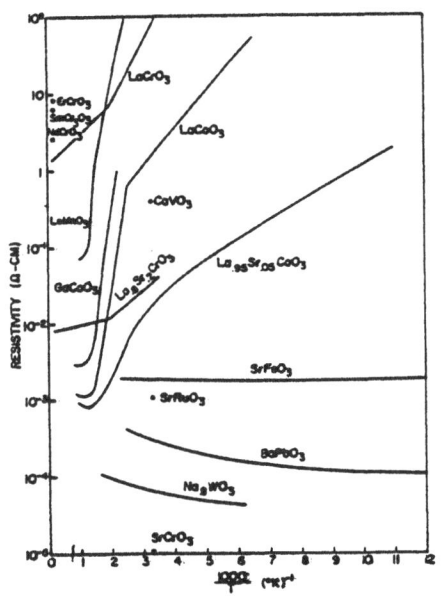

Figure 11 - Resistivity of selected rutile structure oxides [31]

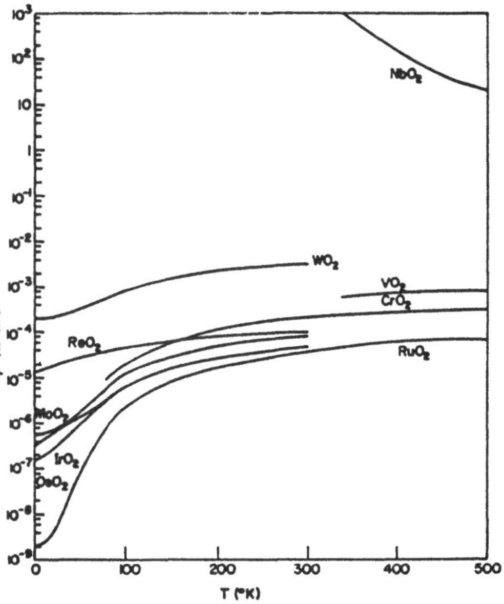

Figure 12 - Resistivity of highly conducting monoxides [31]

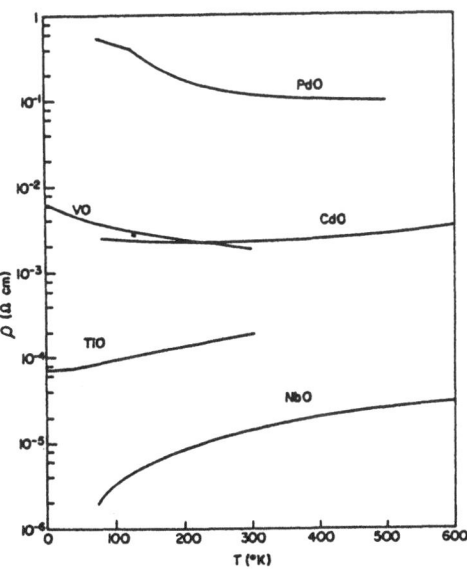

Figure 13 - Resistivity of selected Perovskite structure oxides [31]

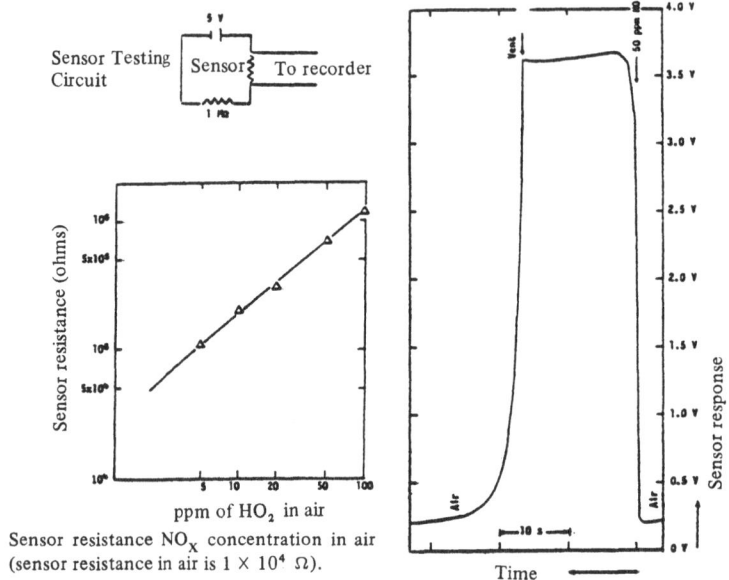

Sensor resistance NO_x concentration in air
(sensor resistance in air is 1×10^4 Ω).

Figure 14 - NO_x detection in using thin SnO_2 resistive layer [19]

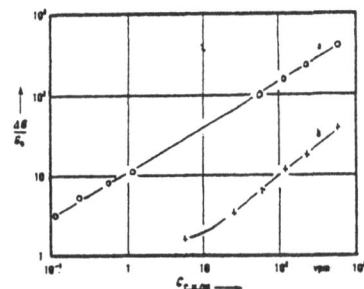

Sensitivity of an SnO_2 (a) film on quartz
glass compared with sintered SnO_2 (b) vs
ethanol concentration.

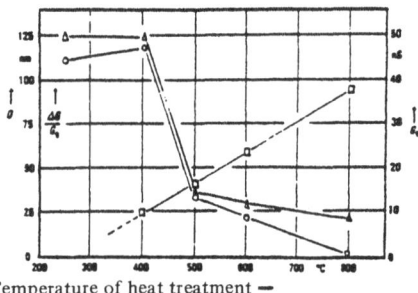

Heat treatment of SnO_2 films in air: ○
sensitivity to 0.05%C_2H_5OH; △ conductance
at 400°C; □ size of crystallites.

Figure 15 - Comparison of thick
film and bulk SnO_2 sensor [20]

Figure 16 - Effect of size of
crystallite on sensor response
[20]

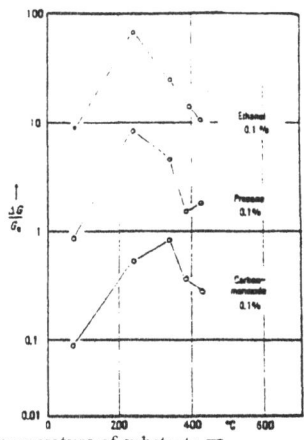

Temperature of substrate →

Relative conductance change $\Delta G/G_e$ at $500°C$ vs substrate temperature during the deposition.

Figure 17 - Effect of the substrate temperature during the sputtering deposition [20]

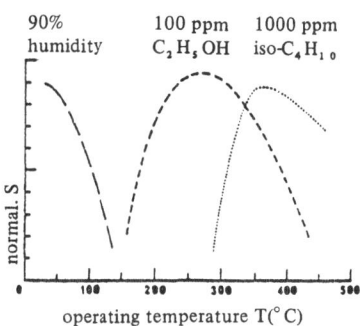

90% humidity   100 ppm C_2H_5OH 1000 ppm iso-C_4H_{10}

normal. S

operating temperature T(°C)

Figure 18 - Sensor selectivity as a function of the sensor temperature [21]

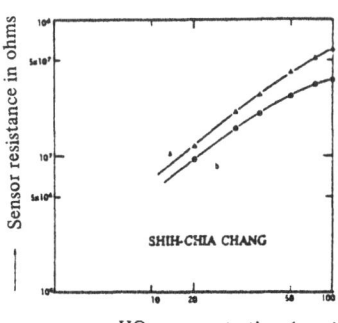

Sensor resistance in ohms

SHIH-CHIA CHANG

— HO_2 concentration (ppm)

Water vapor effect. a: Carrier gas is dry oxygen, sensor resistance $\sim 7 \times 10^5 \, \Omega$. b: Carrier gas is water-vapor saturated oxygen, sensor resistance $\sim 4.5 \times 10^5 \, \Omega$. Sensor temperature $\sim 250°C$.

Figure 19 - Detection of NO_2 in dry and wet atmosphere [22]

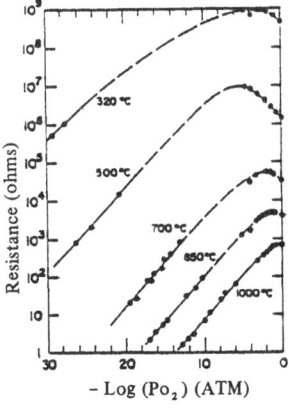

Resistance (ohms)

320 °C
500 °C
700 °C
850 °C
1000 °C

- Log (P_{O_2}) (ATM)

Figure 20 Principle of the TiO_2 Ford sensor for the analysis of exhaust gases [13]

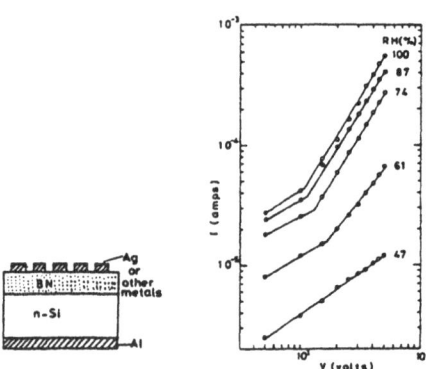

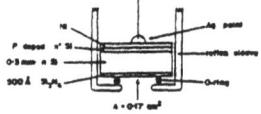

Experimental *p*H response (Steady state-dashed line and initial response-dotted line) compated with theoretical membrane potential difference (solid lines).

Si/Si₃N₄ electrode structure

Figure 21 - Example of sensor for the measurement of relative humidity [28]

Figure 22 - Example of ion sensitive field effect transistor for measurement of pH [27]

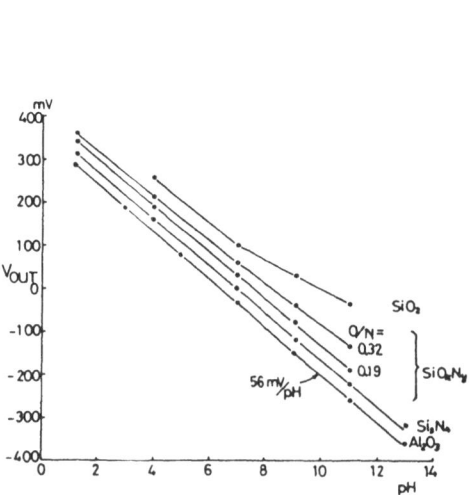

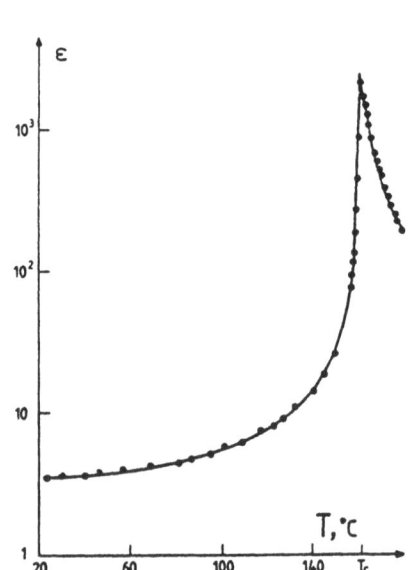

Figure 23 - Effect of the nature of the gate oxide in the performance of ion sensitive field effect transistor

Figure 24 - Variation of the dielectric constant of pyroelectric material with the temperature

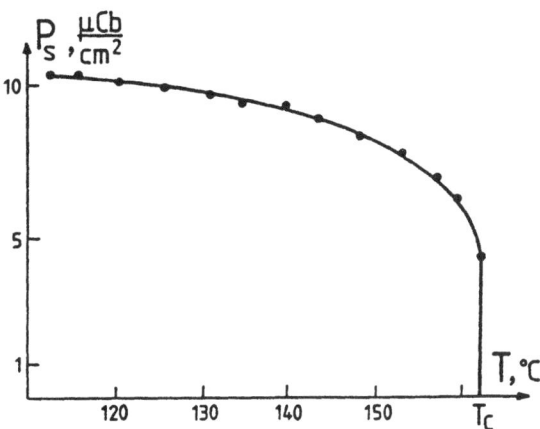

Figure 25 - Variation of the reversible polarisation of pyroelectric material with the temperature

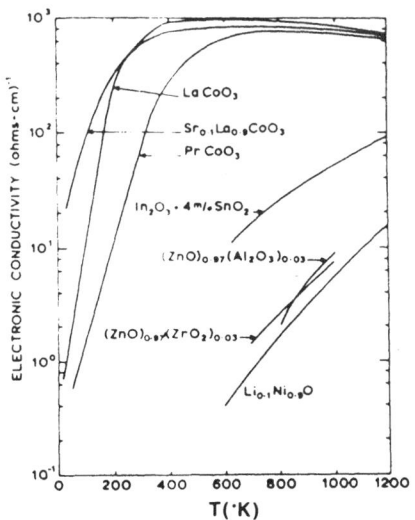

Figure 26 - Oxide with high electric conductivity suitable for the manufacture of electro-conductive electrodes [31]

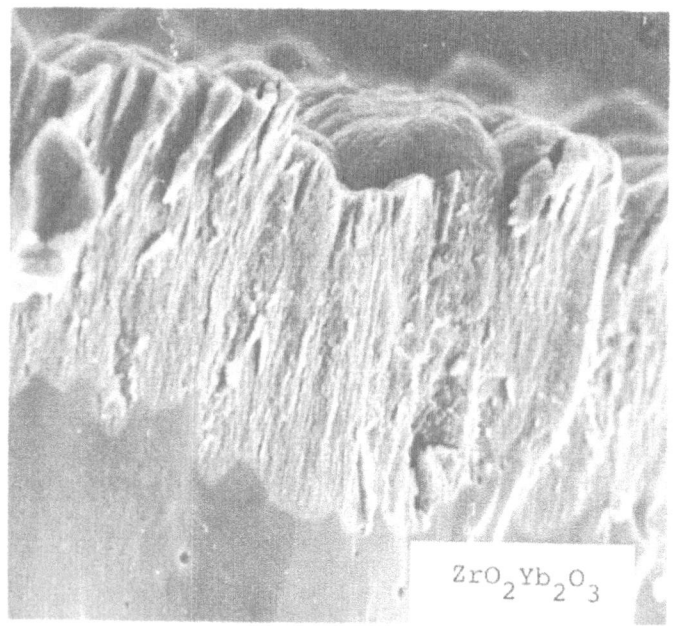

Figure 27 A - Dense solid electrolyte layer

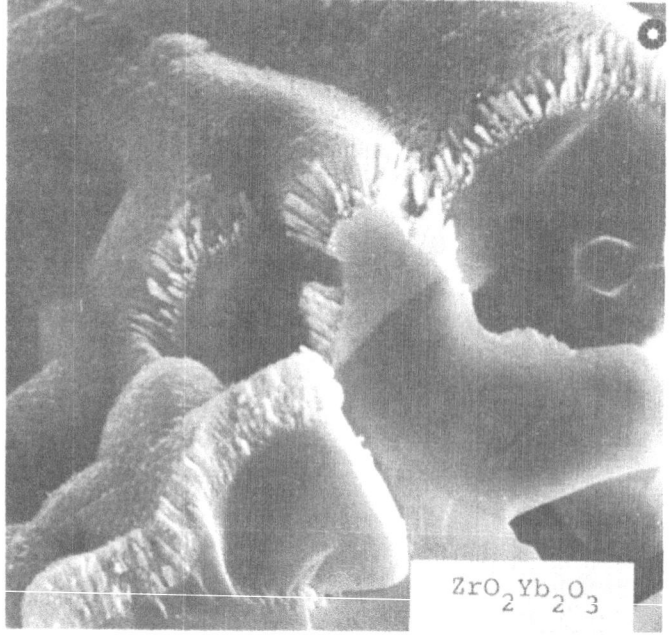

Figure 27 B - Solid electrolyte with columnar structure

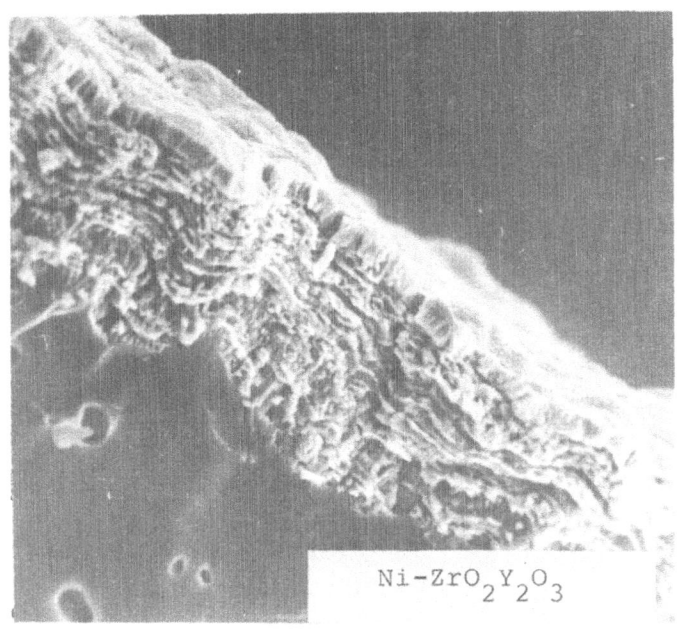

Figure 27 C - Sandwich electrode with metal and oxide layer

Figure 27 D - Dendritic electro-conductive oxide layer

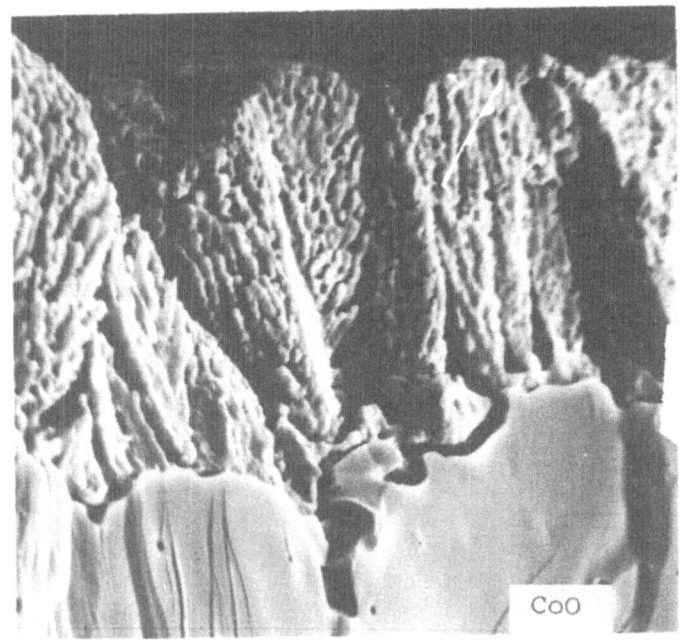

Figure 27 E - Electro-conductive oxide layer with cauliflower structure

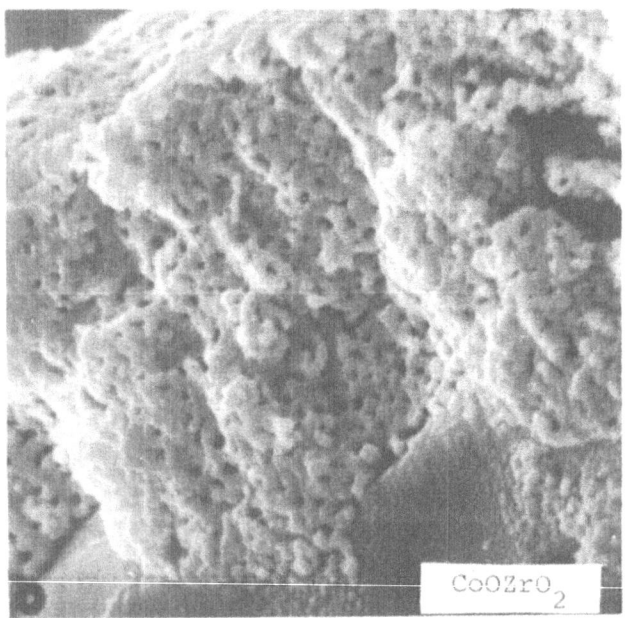

Figure 27 F - Electro-conductive oxide layer with sponge structure

PRODUCTION OF ZIRCONIA FIBERS FROM SOL-GELS FOR WATER ELECTROLYSIS
APPLICATION

E. Leroy, C. Robin-Brosse, J. P. Torre

Laboratoire des Céramiques Techniques,
Lafarge Réfractaires
BP 59, 78193 TRAPPES CEDEX, FRANCE

ABSTRACT.
    Zirconium acetate solutions were concentrated in such a
manner that short fibers could be obtained from them by
extrusion and air-jet drawing. The chemical and structural
evolution of these fibers when heated was followed by
DT-TGA, X-ray diffraction, IR analysis and SEM observations.
Special attention was given to the influence of the firing
atmosphere and of the addition to the starting solutions of
compounds known to inhibit the allotropic transformations of
zirconium oxide. When correctly controlled this evolution
was found to lead to fine-grained metastable tetragonal
zirconia without major flaws.

1.   INTRODUCTION.
    The most critical part of the hydrogen production devices
which are based on water electrolysis is the porous diaphragm
separating hydrogen from oxygen. This diaphragm must exhibit
a high enough open porosity to allow ionic exchange within
the liquid phase, while at the same time having a low pore
size in order to prevent gas bubbles of about 10 microns
diameter from passing across it. It must also resist the
attack of potash at 150° C to 200° C and exhibit no
mechanical property degradation after thousands of hours
operating time in such an aggressive environment.
    Zirconia is a suitable material in terms of corrosion
resistance and a zirconia fiber mat might thus be used as a
diaphragm, provided that fibers with a good mechanical
behaviour can be produced. To our knowledge, the only
commercially available zirconia fibers at the present time
are the ZIRCAR fibers manufactured by the UNION CARBIDE

CORPORATION. The fabrication process of these fibers is based on the impregnation of organic viscose fibers with a zirconium salt solution [1, 2]. During a suitable heat treatment, the zirconium salt is transformed into zirconia. The final fibers exhibit the typical polylobate morphology of the initial viscose fibers, which probably limits their mechanical properties. Other oxide fibers are however obtained with a more appropriate circular cross section. Alumina fibers, for example, manufactured by different techniques (notably by sol-gel processing), do attain a high strength level. Promising results were obtained several years ago with zirconia fibers produced from sol-gels[3, 4, 5, 6] but seemingly without any present commercial development.

The work to be reported in the present paper was undertaken in an attempt to identify and further clarify the basic mechanisms involved in the production of zirconia fibers from sol-gels.

2. FUNDAMENTALS OF FIBER PRODUCTION FROM SOL-GELS.

A metallic salt in an aqueous solution generally precipitates when the solution is concentrated beyond a precise level which is called "the solubility limit" of the salt. Some salts behave in a different way. They seem to exhibit no solubility limit and their aqueous solutions can be highly concentrated, changing continuously into viscous liquids and finally into homogeneous solids. This phenomenon is called "sol-gel process" and can be compared to polymerization.

When a given polymerization stage is reached (which is difficult to measure quantitatively in a precise fashion) the solution acquires properties which enable it to give fibers. Those properties are notably a high viscosity (depending on the fiber forming technique) and a low surface energy. Fibers can thus be obtained from aluminium oxychloride, acetate or nitrate solutions which become further converted to alumina by heat treatment [7, 8]. Zirconium oxychloride or acetate might similarly be used for zirconia fiber production.

As surface energy is difficult to measure in the case of viscous solutions, viscosity is the only parameter which can be used to follow the polymerization progress.

Figure 1 shows the typical variations of viscosity versus concentration observed when concentrating a solution giving a sol-gel at a constant temperature. The formation of fibers generally requires viscosity values corresponding to the beginning of the high-sloped part of the curve (for example η_1 on figure 1). It is thus possible to directly concentrate the solution so that the desired viscosity value is

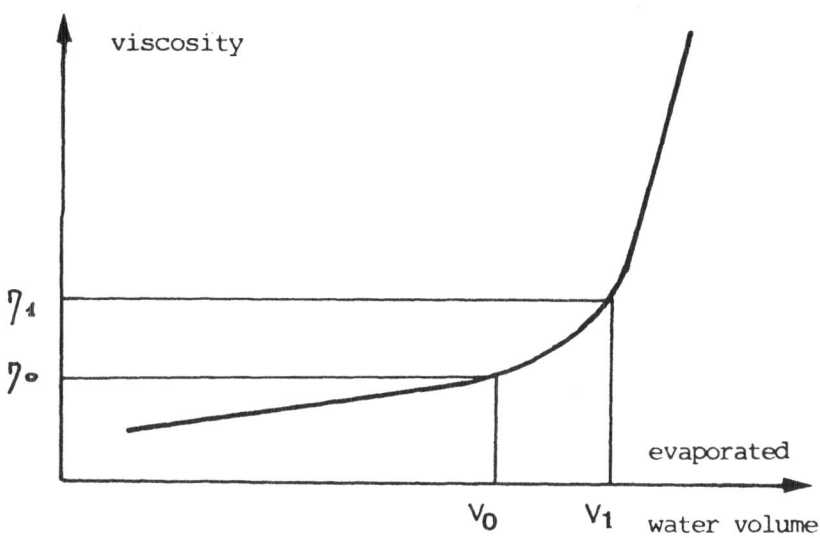

figure 1: variations of viscosity vs concentration

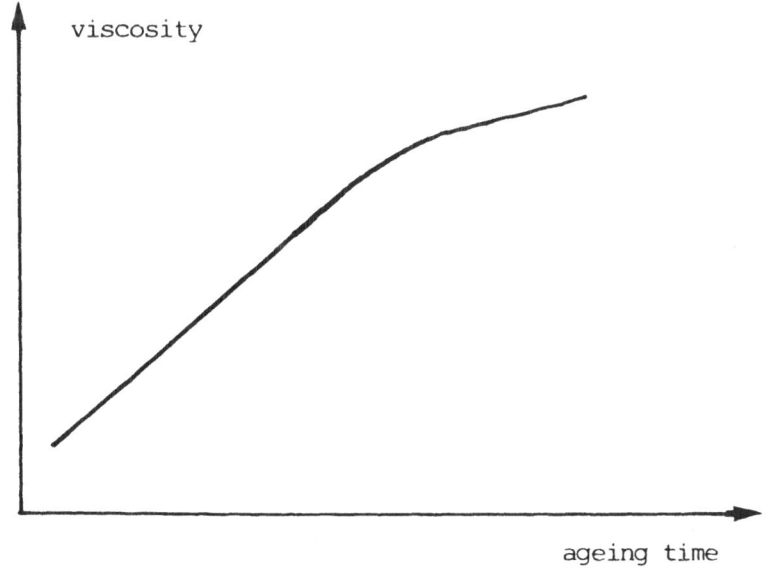

figure 2: variations of viscosity vs ageing time
after concentrating

reached, but it is very difficult to adjust the corresponding concentration. Moreover bad quality fibers will be
produced because polymerization has not progressed far
enough.

A better procedure therefore is to stop concentrating
the solution at a lower viscosity level (e.g. η_o) and to
keep it then at a given temperature for some time.
Polymerization will then progress, resulting in an increase
of viscosity as a function of the ageing time (figure 2).

Two major phenomena occur when the fibers are heated :
first the transformation of the metallic salt into oxide,
during which gaseous species are released (for instance
chlorine or carbon compounds when starting from a metallic
oxychloride or acetate) ; then the sintering of the very
fine particles resulting from this transformation. Both
phenomena are generally not independent : sintering starts
at temperatures where the transformation is not complete.
The main difficulty hence consists in completely eliminating
elements such as chlorine or carbon (the presence of which
appears to be detrimental to the mechanical behaviour of the
fibers) without damageing the fibers and without promoting
grain growth which also results in a decrease of the
mechanical properties.

3. EXPERIMENTAL PROCEDURE.

The starting product was a commercial zirconium acetate
solution * containing 22 equivalent weight per cent zirconia.
This was concentrated at 50° C under vacuum in a 10 l
evaporator. The concentration was continuously followed by
measuring the amount of evaporated water (figure 3).
Once concentrated at the desired value, the solution was
cooled to 20° C and kept at this temperature for ageing.

Fibers were obtained by extrusion and drawn by air jet
as schematized on figure 4. The caracteristics of the air
jet were 50 m^3 h^{-1} flow rate and 1.6 atm pressure resulting
in a 140 m.s^{-1} speed. The chemical and structural evolution
of the fibers when heated was followed by DTA and TGA, X-ray
diffraction, IR absorption and SEM observations.

4. RESULTS AND DISCUSSION.
4.1 - Fiber forming

Figure 5 shows the variation of viscosity versus
evaporated volume of water and concentration for 2.5 l
starting zirconium acetate solution concentrated at 50° C.
Figure 6 demonstrates the variations of viscosity as a
function of ageing time at 20° C which resulted from
stopping the concentration of the solution at different
stages. These stages were determined by the fraction of
* MAGNESIUM ELEKTRON LTD, U. K.

504

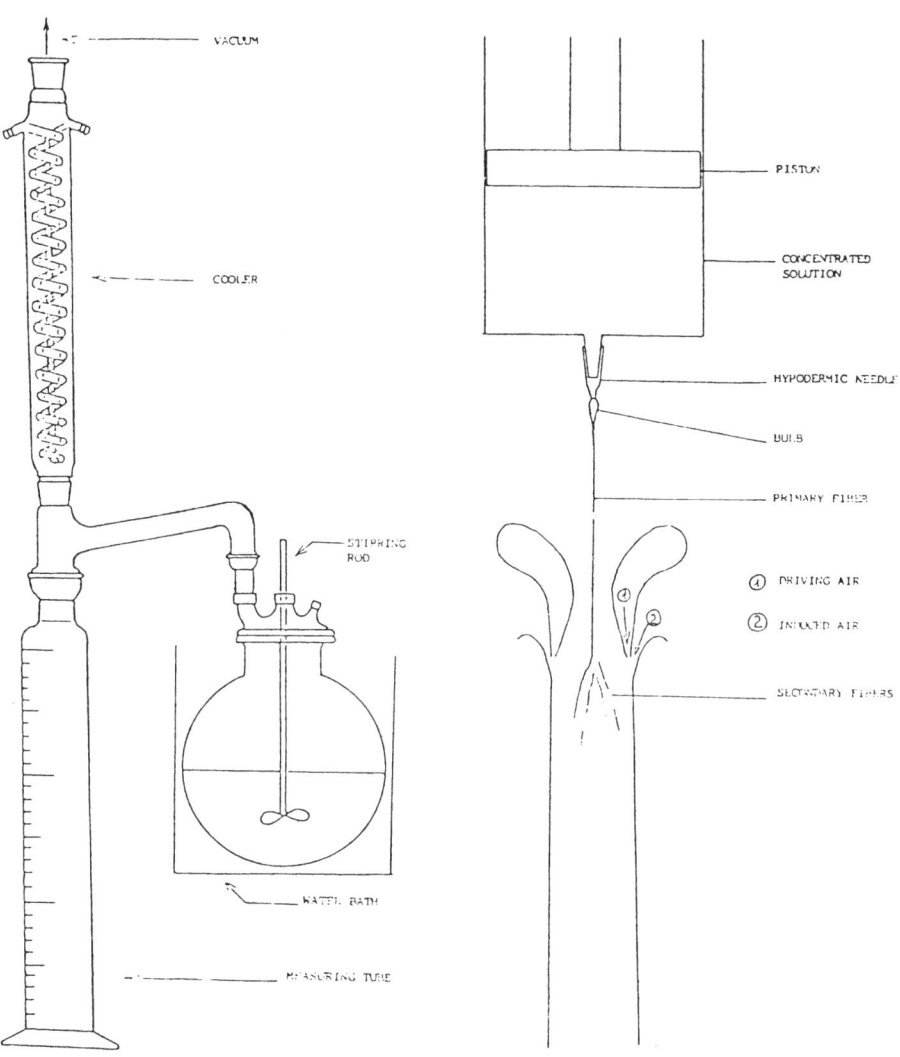

figure 3: schematics of
evaporator

figure 4: schematics of fiber
forming device

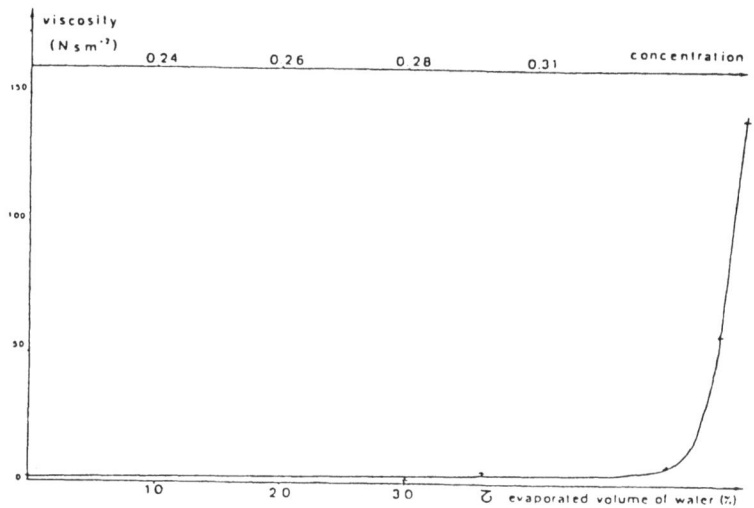

figure 5 :variation of viscosity vs evaporated volume
of water and concentration .(50°C ,2.5 l
zirconium acetate solution)

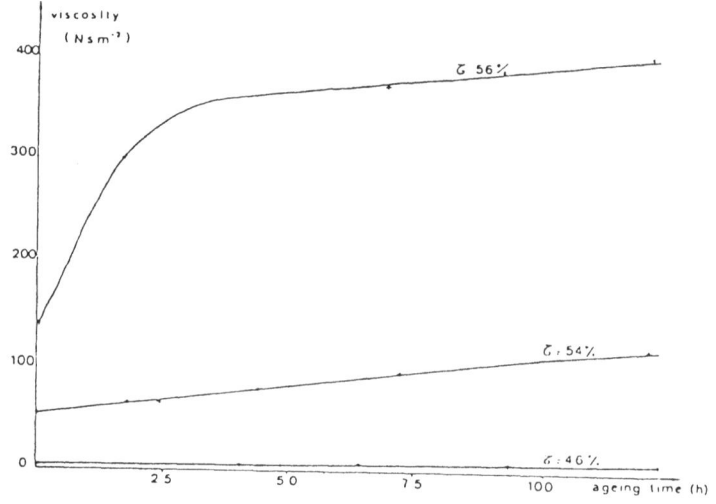

figure 6 : variation of viscosity as a function
of ageing time for different concentrations

water evaporated. The fiber forming technique used in this work was shown to give the best results for viscosity values between 20 and 50 Nsm^{-2}. The fibers studied further were thus obtained from solutions concentrated so that about 50 % water was evaporated. This resulted in a low polymerization rate and hence a long ageing time which might be shortened for industrial purposes by increasing temperature. This is beyond the scope of the present paper.

The extrusion nozzle diameter could not be reduced under 0.3 mm because of the low extrusion pressure used (1 MNm^{-2}). Nozzles with lower diameters were rapidly blocked up. This might be avoided by adding organic binders to the solution. Discontinuous fibers with about 10 um diameter and 20 mm maximum length could however be drawn using the 0.3 mm nozzle under the air-blowing conditions given above. Figure 7 shows the micrographic aspect of the fibers.

4.2 - Fiber transformation
Figure 8 shows the DT-TGA curves recorded when heating the fibers at 60° C h^{-1} up to 1000° C in air. The DTA curve exhibits sharp exothermic peaks with the exception of one endothermic peak, corresponding to the early stage of transformation and accompanied by an important weight loss. This early stage of transformation consists in water volatilization. At temperatures up to the first exothermic peak (300° C), which is also accompanied by an important weight loss, the X-ray diffraction patterns reveal no crystallization (figure 9a). At 300° C blurred diffraction lines appear at the same Bragg angles than those for tetragonal zirconia (figure 9b). IR spectra of fibers both unfired and kept during 30 hours at 300° C and of zirconia powder are shown on figure 10 (for the analysis samples were embedded in optical grade potassium bromide pellets). It can be seen that absorptions at less than 900 cm^{-1} similar to those of zirconia have appeared as a result of heat treatment at 300° C. They seem to have developped at the expense of those at 1070, 1400 and 1530 cm^{-1} which are due to zirconium acetate major bonds. A weak absorption at 3400 cm^{-1} characteristic of the O-H bond is also visible at this stage. All these results clearly suggest that transformation of zirconium acetate into badly crystallised zirconium oxide and hydroxide occurs at 300° C.

Hence at temperatures above 300° C the chemical evolution of the fibers consists in transformations of zirconium oxide and hydroxide and residual acetate. At the temperature of the second exothermic peak (500° C) the fibers suffer no weight loss. X-ray diffraction analysis reveals that well crystallized tetragonal zirconia is formed at this stage (figure 9c). A new weight loss takes place at

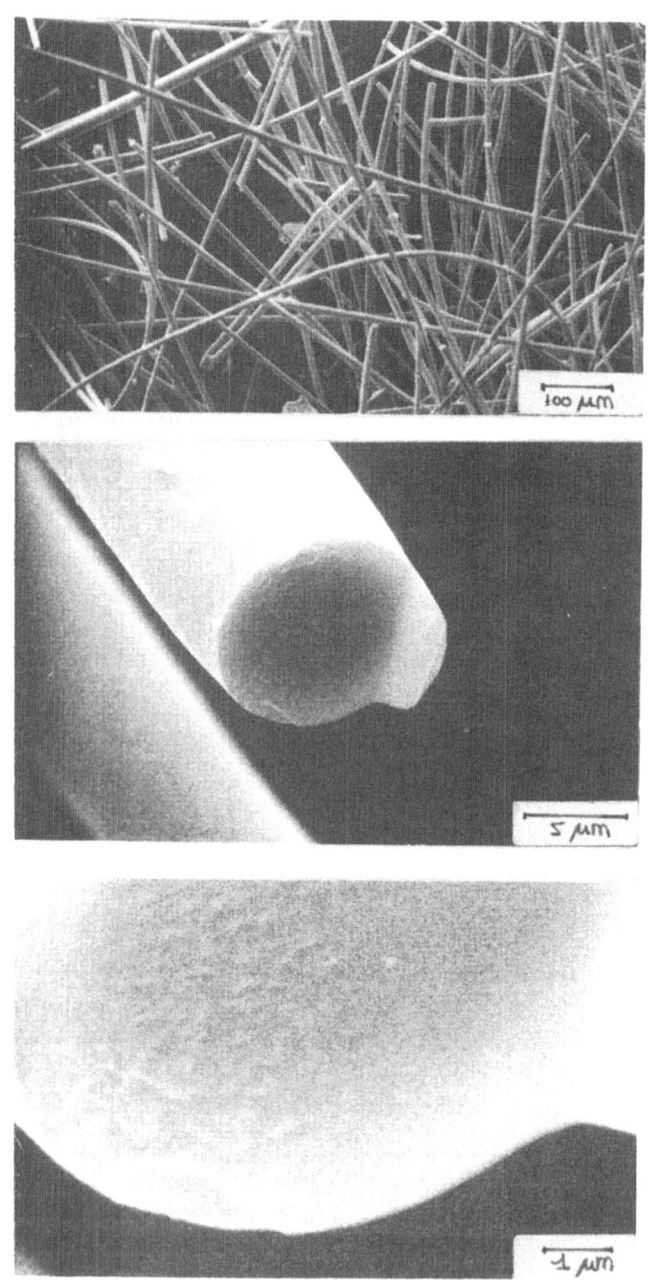

figure 7 : micrographic aspect of as-drawn fibers

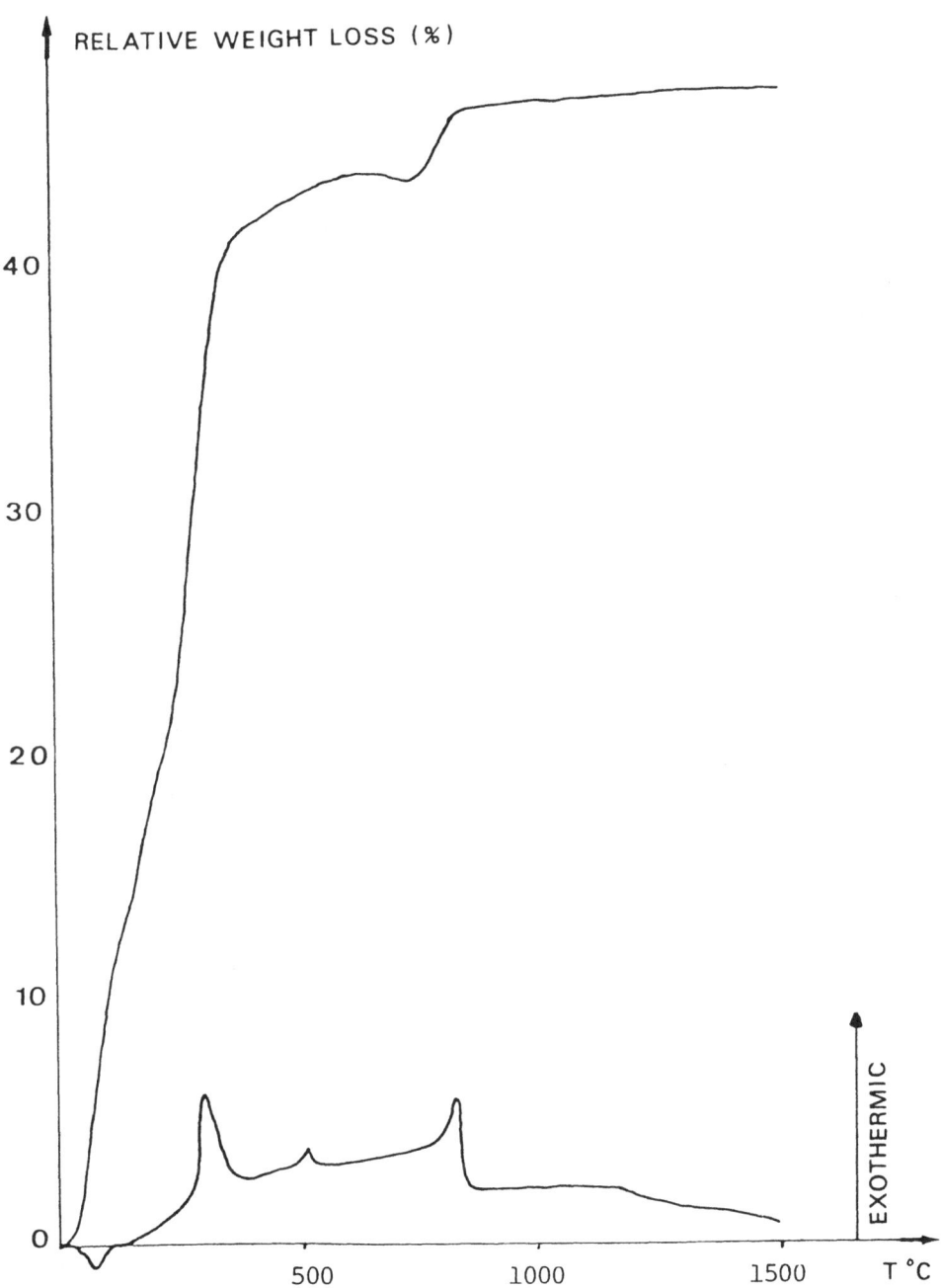

figure 8 : DT-TGA curves

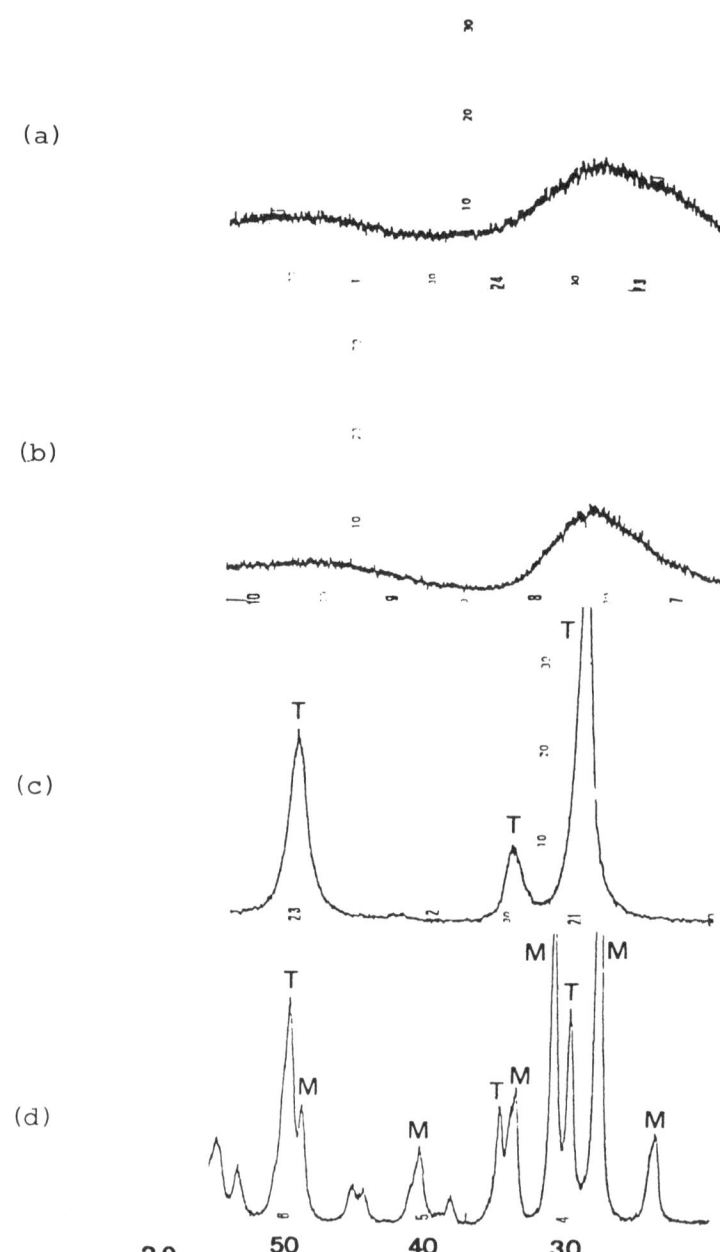

figure 9 : X-ray diffraction patterns of fibers heated at
60°C h^{-1}

(a) : 250° C   (b) : 300° C
(c) : 500° C   (d) : 750° C

750° C corresponding to the third exothermic peak. Simultaneously the fibers turn clearer, suggesting that final volatilization of carbon compounds occurs then. The dark colour visible below this temperature suggests on the other hand that free carbon is formed during the decomposition of acetate, in agreement with observations reported elsewhere [9, 10]. Figure 9d finally shows that tetragonal zirconia is partly transformed into monoclinic phase at 750° C. This transformation does not result in a peak on the DTA curve because it is known to take place continuously in a wide temperature range [11].

4.3 – Microstructure control
    Micrographs of fibers heated up to 750° C show that they have suffered an important damage during the heat treatment as illustrated by figure 11a. This damage can be related to both the rapid elimination of carbon compounds at high temperature and to the tetragonal to monoclinic zirconia transformation which is accompanied by a large volume increase. It is still visible on fibers kept during 30 hours at 300° C before the temperature was raised to 750° C at 60° C h^{-1} (figure 11b). This can be explained by the IR analysis which shows that noticeable amounts of carbon compounds are still present in the fibers after 30 hours at 300° C (figure 10). However the 300° C treatment had a profitable effect : less monoclinic zirconia is found in the final fibers (i. e. fired at 750° C) as demonstrated by the X-ray diffraction patterns (figures 9d and 12a). This reveals that substantial densification has occurred at low temperature. The transformation of tetragonal zirconia into monoclinic phase at higher temperatures has then been limited by the compressive stresses applied on each tetragonal grain by its neighbours and due to the volume increase associated with the transformation [12, 13, 14].
    As a primary practical conclusion of the present work it appears that fibers with good mechanical properties should only be obtained provided that both densification and elimination of carbon compounds are enhanced at low temperature. The preceding results clearly suggest that the carbon compounds cannot be easily removed from the fibers at temperatures below 750° C in air. But it has been pointed out elsewhere [8] that using water steam and/or ammonia as firing atmosphere helped removing acetate anions from aluminium acetate fibers at much lower temperatures. For instance aluminium acetate is completely transformed into aluminium hydroxide and ammonium acetate between 300° C and 500° C in the presence of ammonia. In addition this transformation takes place without decomposition of the acetate anions and therefore without production of free carbon, known to inhibit densification. Similar results

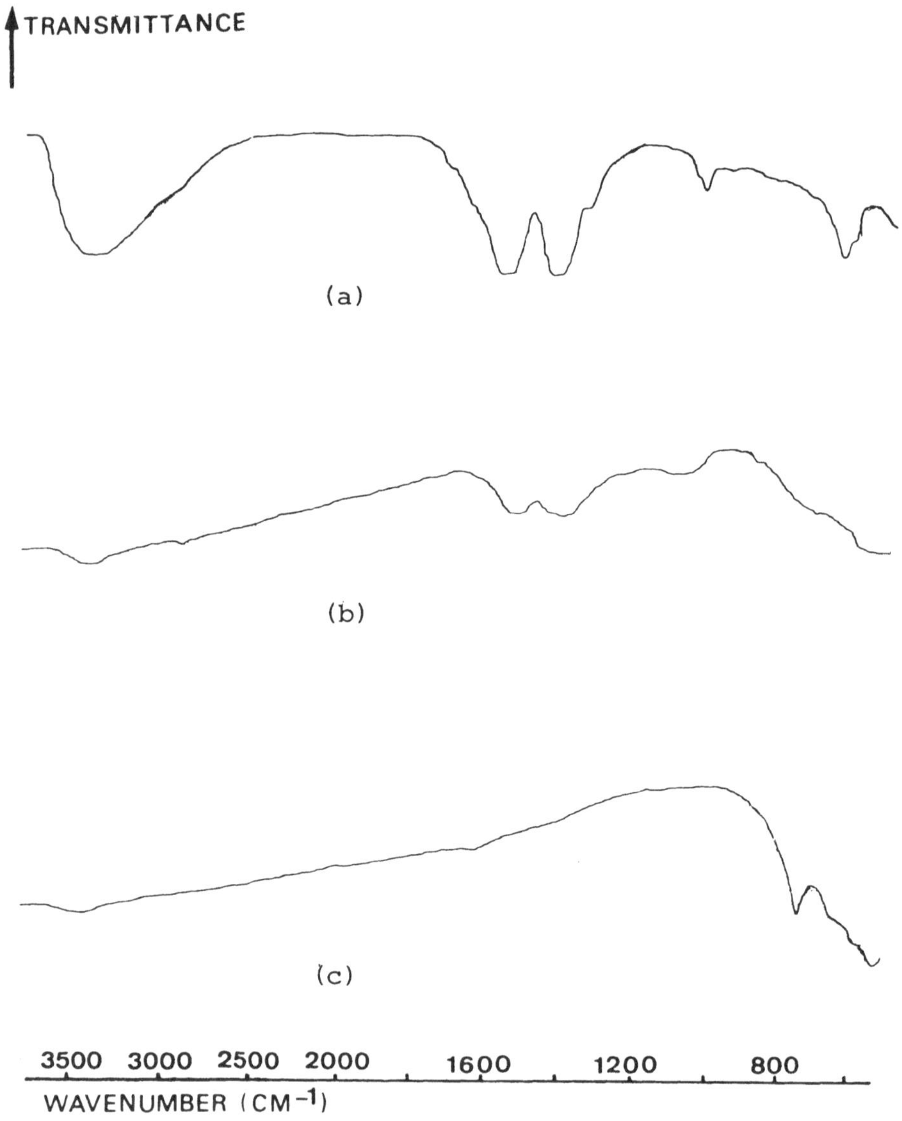

figure 10 : IR spectra

        (a) : unfired fibers

        (b) : fibers fired at 300° C

        (c) : zirconia powder

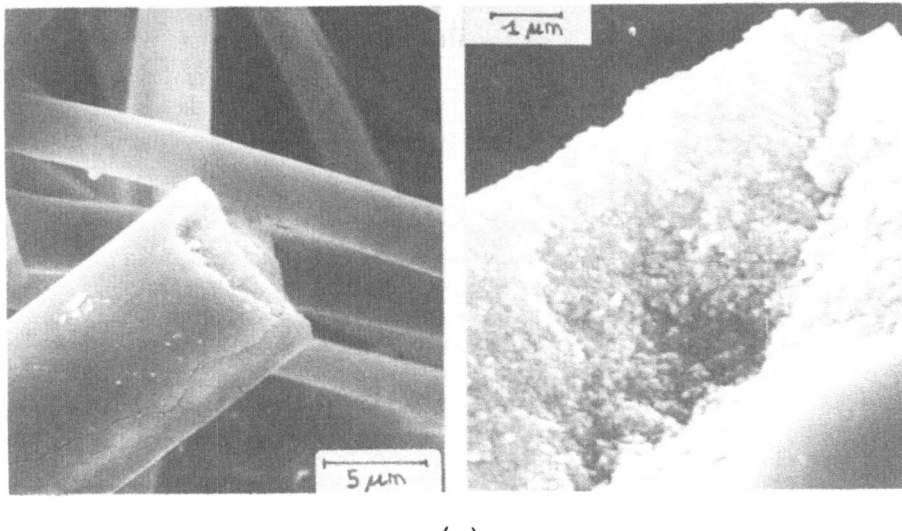

(a)

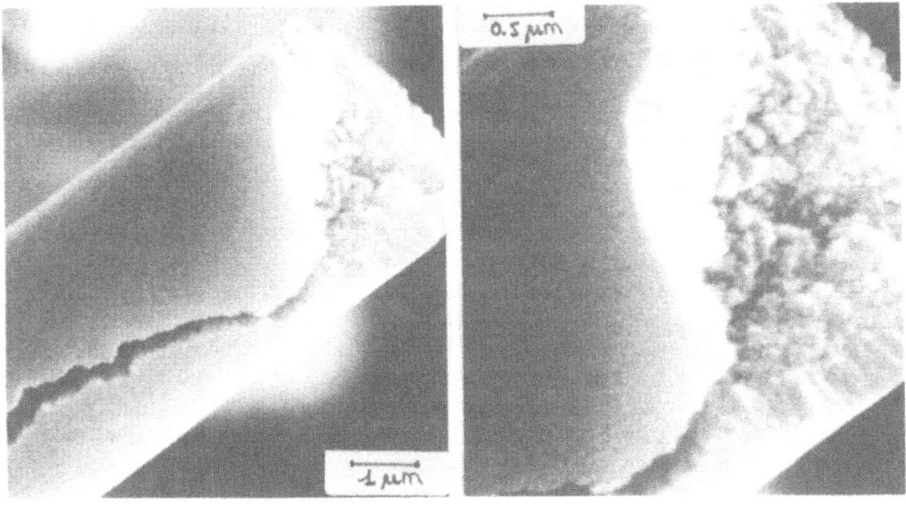

(b)

figure 11 : micrographic aspect of fibers heated up to 750° C

    (a) heated at 60° C h⁻¹

    (b) kept at 300° C during 30 h

could be expected for the present zirconium acetate fibers. Complementary experiments were made in order to check this point.

The starting acetate fibers were kept during 30 hours at 300° C and heated up to 750° C at 60° C h^{-1} in ammonia. After firing, the fibers were perfectly white, which was never observed when similar treatments were conducted in air. Figure 12b shows the typical X-ray diffraction pattern of the final fibers. It can be seen that the amount of monoclinic phase is smaller than in fibers fired in air (figure 12a). Finally the micrographic observations did not reveal any visible damage (figure 13a). All these results suggest that complete elimination of carbon compounds occurred at low temperature without decomposition, thus promoting densification and stabilization of the tetragonal phase. However this did not completely prevent the monoclinic phase from being formed.

The effect of condensed additives known to inhibit the allotropic transformations of zirconia was finally investigated. The fiber forming technique used in the present work was shown to be applicable to zirconium acetate solutions containing a few per cent magnesium acetate or cerium nitrate. The resulting fibers were kept during 30 hours at 300° C and heated up to 750° C in ammonia. The final X-ray diffraction pattern and microstructures are shown on figures 12c and d and 13b and c. Only traces of monoclinic phase are detectable in both cases and micrographs reveal fine-grained structures without major flaws.

5. CONCLUSION.

The various stages of the chemical and structural evolution of fibers formed by extruding a concentrated zirconium acetate solution and subsequent heating have been identified. When correctly controlled, this evolution leads to fine-grained metastable tetragonal zirconia fibers. The fibers produced were unfortunately too short to allow direct measurements of their mechanical properties. Producing optimized fibers for practical use would in any case require a further effort in order to improve the fiber forming conditions.

ACKNOWLEDGEMENTS.

Financial support for this work was provided by Gaz de France, Direction des Etudes et Techniques Nouvelles.

REFERENCES.

1 - B. H. HAMLING (UNION CARBIDE CORPORATION) :
    brevet français 2 000 691 du 12/09/1969
    (Appl. U. S. 700 031 du 24/01/1968)

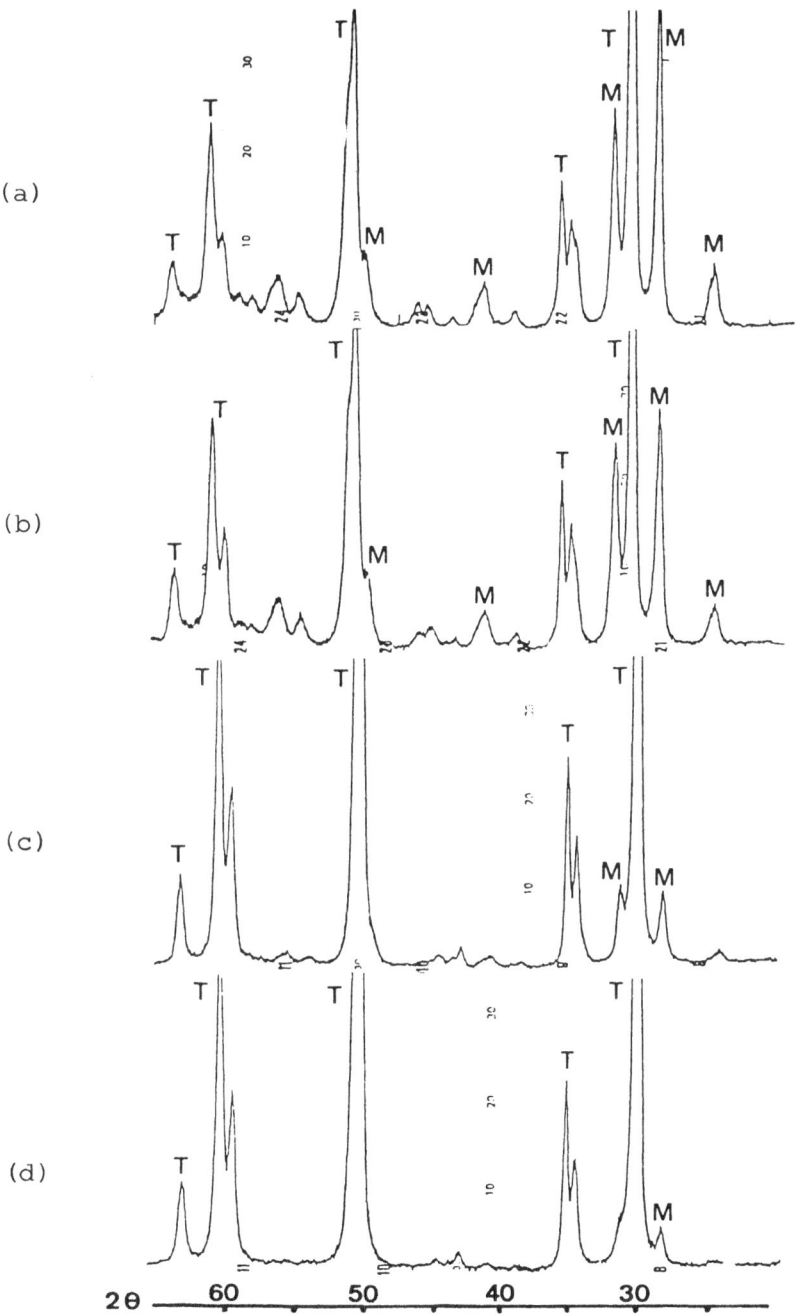

(a)

(b)

(c)

(d)

2θ   60   50   40   30

figure 12 : X-ray diffraction patterns of fibers fired at 300°C
          + 750° C

          (a) : in air   (b) : in ammonia

          (c) : in ammonia (Mg-doped)   (d) : d° (Ce-doped)

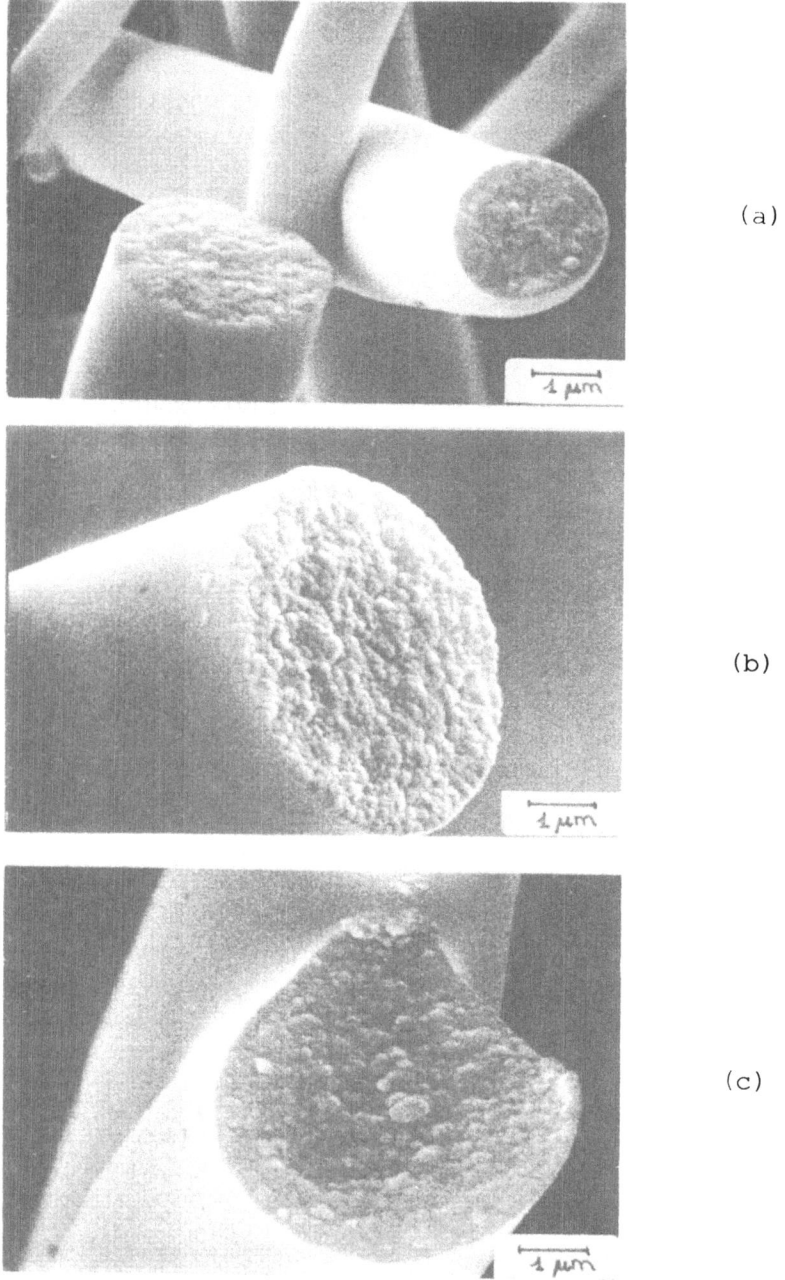

figure 13 : micrographic aspect of fibers fired at 300° C
+ 750° C in ammonia

(a) : undoped   (b) : Mg-doped
            (c) : Ce-doped

2 - B. H. HAMLING (UNION CARBIDE CORPORATION) :
U. S. patent 3 385 915 (05/28/1968)

3 - M. G. MORTON, J. D. BIRCHALL, J. E. CASSIDY
(I. C. I.) : British patent 1 360 200 (21/05/1971)

4 - M. G. MORTON, J. D. BIRCHALL, J. E. CASSIDY
(I. C. I.) : British patent 1 360 199 (21/05/1971)

5 - M. G. MORTON, J. D. BIRCHALL, J. E. CASSIDY
(I. C. I.) : British patent 1 360 198 (21/05/1971)

6 - M. G. MORTON, J. D. BIRCHALL, J. E. CASSIDY
(I. C. I.) : British patent 1 360 197 (21/05/1971)

7 - S. J. HARRIS (I. C. I.) : British patent
1 402 544 (23/06/1972)

8 - S. G. P. R. : British patent 1 228 243 (10/04/1968)

9 - L. N. KOMISSAROVA, M. V. SAVEL'EVA, V. E. PLYUSHCHEV :
Zh. Neorg. Khim., 1975, 8, 56-62

10 - V. B. FEDORUS, V. E. MATSERA, Z. A. ZALTSEVA,
L. I. KOPYLOVA : Poroshkovaya Metallurgiya,
1981, 218, 38-40

11 - R. C. GARVIE, R. H. HANNINCK, R. T. PASCOE :
Nat. Phys. Sci., 1975, 258, 703-704

12 - T. K. GUPTA : Science of Sintering, 1978, 10, 205-216

13 - T. MITSUHASHI, M. ICHIHARA, U. TATSUKE :
J. Amer. Ceram. Soc., 1974, 57, 97-101

14 - D. L. PORTER, A. G. EVANS, A. H. HEUER :
Acta Met., 1979, 27, 1649-1654

DISCUSSION

R. Metselaar: Have any of these fibres been tested already
for water electrolysis?

J.P. Torre: Fibres are being tested at the present time;
no result is available so far.

CERAMIC FABRICATION

R.J. Brook

Department of Ceramics, University of Leeds,

Leeds LS2 9JT, U.K.

The ability to exploit the favourable properties of a ceramic depends on the ability to fabricate in a reproducible manner a material with a carefully designed microstructure. The requirements in terms of porosity, grain size, grain boundary character are often determined within narrow limits by the application and the objective is therefore to select the methods of powder preparation, of forming, and of heat treatment which yield the desired structure. The more common fabrication routes are discussed and some problem areas requiring research attention are identified: these include particularly the control of agglomeration in powder processing and the control of coarsening mechanisms during sintering. The control of reproducibility during manufacture, though less clearly a research area, remains a problem of the greatest importance.

Introduction

A common feature to all the applications of ceramics in the advanced energy technologies is the need to fabricate articles which in fact provide the properties which are in theory available. Fabrication is consequently a key issue and one in which technological progress brings important advances in a wide range of dependent applications.

The topic is very broad and detailed matters of fabrication procedure are often specific to the particular material and application in question. Furthermore many of these details are proprietary information insofar as they confer technological advantages to manufacturers. As a consequence it is the objective of the present paper to provide a brief summary of the most general fabrication

518

routes that are available, indicating both the main research
matters and some of the relevant literature.

Requirements

Many of the applications of concern to the advanced energy
technologies (heat engines, MHD, slag resistant refractories,
battery electrolytes, monitors) require high strength, impermeable
materials, often with very specific requirements relating to composition and purity. Thus a common wish in the case of materials
such as Si_3N_4, SiC, ZrO_2 is that they be dense (minimum porosity),
fine grain size, and with clean grain boundaries free from any
second phase.

In addition, there is a clear requirement for materials with
reproducible and reliable properties; since certain key properties
of ceramics (notably fracture strength) depend on the existence of
isolated flaws or faults in the material[1], accidents during fabrication, even if resulting in highly localised damage, can have a more
serious consequence on the performance and acceptability of a material for applications. This aspect is of great importance when
moving work from a laboratory (where it is not unknown to emphasise
the most promising results stemming from a piece of work) to the
manufacturing plant (where the matter of routine reproducibility
becomes central).

Fabrication

With the exception of glasses and glass ceramics, the common
route for the fabrication of ceramic materials involves the consol-
idation of powder by heat treatment. There are three broad categories of process, all of which depend upon the sequence:

Powder $\longrightarrow$ Shaping $\longrightarrow$ Debonding $\longrightarrow$ Firing $\longrightarrow$ Finishing
+ binder   and forming   (removal of   (removal of
                               binder by   porosity
                               heating)   by heat
                                           treatment)

The three categories are (i) vitrification in which compositions
are chosen which yield enough liquid at the firing temperature to
fill the porosity. Since this results in a material with an inhomo-
geneous microstructure and with poor high temperature mechanical
properties (creep resistance), it is not generally promising for the
applications considered here;

(ii) liquid phase sintering in which compos-
itions are chosen which yield a small quantity of liquid at the firing

temperature which coats the solid grains. Densification then results from grain reshaping, material transport occurring through the liquid phase. This is a common and efficient process (Si_3N_4 + MgO, MgO + LiF, ZrO_2 + Bi_2O_3), but problems in terms of creep resistance at high temperature will be found unless the liquid is removed from the grain boundary subsequent to firing;

(iii) solid state sintering in which the powder is densified in the absence of any liquid. This is a very attractive process in terms of the properties that can be achieved since the final microstructures can be dense, small grain size, and of high purity.

From the point of view of the advanced energy technologies, the preferred general route in terms of final properties is therefore solid state sintering and much research effort[2-12] has been deployed on the selection of the processing conditions (particle size, firing temperature, use of trace additives) that yield the desired result of high density and fine grain size. Since these conditions are not always easily identified, a set of related processes in which pressure is applied during heat treatment have been developed (hot pressing, hot isostatic pressing, explosive forming, hot forging). These are effective but expensive.

The second route is that of liquid phase sintering followed by heat treatment to remove the liquid. Again much effort has been given to the development of this route, particularly for the nitrogen ceramics. Brief accounts of some of the considerations of these processes are given in the following sections.

Solid State Sintering

In this process where all densification is achieved by a reshaping of the grains of the powder preform as a result of diffusion, careful design of the processing conditions are required if suitable microstructures are to be achieved. To enhance the driving force (surface energy reduction) and to accelerate the kinetics (short diffusion distances), fine powders are required[13] (typically $\leqslant$ 0.1µm)

Considerable attention has been given to powder preparation and characterisation[14-18], the ideal objective being[19] fine powders, which are free from agglomeration, which contain a narrow distribution of particle sizes, which comprise spherical shaped particles, and which have closely controlled composition. Generally processing is seen as an optimisation with the attainment of this ideal form not only being technologically difficult but also being expensive. Nonetheless there has been a steady trend in the advanced ceramics to use progressively better controlled and more expensive powders, the increased cost being amply justified by the increased product performance and reliability.

520

A major question and problem is that of agglomeration, i.e. the bonding together of the fine powder crystallites into larger porous particles which are sufficiently strong to survive pressing and forming operations. These particles then produce low density faults in the fired product; examples where the importance of agglomeration has been demonstrated are ZrO_2[20] and Al_2O_3[21]. The preparation of agglomerate-free powders is a key research issue and the approaches have ranged from thorough beneficiation of powders, e.g. ZrO_2[22], to sophisticated procedures for powder formation, e.g. Si, Si_3N_4, SiC[23].

For forming and shaping, the most common methods involve mixing the powder with a binder followed by isostatic pressing, die pressing, injection moulding or extrusion. These techniques have now been taken to the stage where good finish in complicated shapes can be achieved. For systems where high volume production is involved, dry bag isostatic pressing[15] has many merits; powders either with or without binder (PVA, PEG, soluble waxes) are pressed to approximate shape, machined in the green condition, and fired. If the powder properties are constant, accurate prediction of the firing shrinkage can be made and good dimensional tolerance after firing can be achieved. Isostatic pressing is often the most reliable technique where fine powders are being processed.

During the firing stage, the objective is to use a high temperature treatment to remove the porosity from the powder preform with a minimum of associated grain growth. This stage involves a balance between the two processes by which the powder can reduce its surface energy, namely densification (replacement of two free surfaces by a grain boundary) and coarsening (reduction of surface area as the particles grow); the first process is desired, the second is not. The design of the firing conditions therefore represents the selection of powder size, temperature and composition so that the ratio of densification to coarsening is maximised.

The appropriate control of these variables requires that the mechanisms (controlling species and controlling diffusion pathway) of densification and of coarsening be known, a state of affairs which does not usually hold. Since, however, the rewards of suitable process control can be striking, e.g. in Al_2O_3 lamp envelopes, this remains a very significant research area. More particularly, since the objective is to minimise the extent of coarsening, there is a need to develop procedures for studying coarsening processes (surface diffusion and evaporation/condensation) and methods for controlling them.

Among the variables that are available during firing, the control of the temperature/time cycle is perhaps the most easily achieved. Depending on the material, advantages can be found in short, high temperature cycles (zone sintering or flash firing), a procedure that has proved effective for β-Al$_2$O$_3$[24], or in longer lower temperature routes such as controlled rate sintering[25] where the temperature is varied to achieve a constant densification rate. The relative enthalpies for the densification and coarsening processes are significant in determining the effectiveness of such methods[26].

In respect to particle size, advantage is generally expected from the use of fine powders as noted above; an exception can arise, however, where surface diffusion is responsible for coarsening[27]. The use of additives, as in the classical example of MgO in Al$_2$O$_3$[28] is a particularly effective means of controlling solid state sintering; the complications of possible additive functions are severe[26] but the great potential effectiveness ensures that this remains a particularly significant research area.

Liquid Phase Sintering

The requirements for powder uniformity and careful forming that have been noted for solid state sintering apply equally to the liquid phase process. The densification/coarsening ratio is, however, raised by the presence of an additive which produces a liquid which acts as a medium for rapid material transport and grain reshaping at high temperature.

An important consideration for the use of ceramics in advanced energy applications is the removal of the second phase from the boundary once densification has been completed. Among the techniques that have been proposed are the crystallisation of the liquid to form a creep resistant polycrystalline boundary phase rather than a glass, or the reaction of the boundary phase with the major crystalline phase to yield a single phase solid solution (these procedures have been advocated[29-30] for silicon nitride/sialon types of ceramics), or the evaporation of the boundary phase (as in MgO-LiF[31]). Despite the complications and uncertainties of such methods, the great effectiveness of liquid phase sintering in yielding a high density product has ensured its widespread use.

Research Needs

The research needs in ceramic fabrication are both numerous and significant; they are numerous because the specific needs vary substantially from one application to another, they are significant because the rewards from successful processing are important in terms of improved properties and performance in applications. A careful review of the needs in respect to energy applications is available[19] in the literature.

It is possible, however, to identify some central features and problems that apply widely across a range of applications. Thus the trend to better characterised powders and to carefully designed and synthesised powders is a clear one, well justified by the promise of greater reproducibility and performance in the finished ceramic. Key research areas in classical processing are, as noted earlier, agglomerate control in powders, and the control of the rheology of powder/binder systems during forming. In the firing stage, the promising areas include the study and control of coarsening mechanisms and the whole question of sintering additives.

Furthermore, pending solution to these problems, there will be continuing interest in those special methods which, though expensive, avoid some of the deficiencies of the classical route. These include hot isostatic pressing and other hot forming methods where the ability to ensure a high degree of flaw removal is particularly attractive.

References

1. A.G. Evans: J. Am. Ceram. Soc. 1982, 65 (3) 127-137.
2. J.E. Burke and J.H. Rossolowsky: 'Sintering', Chapter 10 in Treatise on Solid State Chemistry, Volume 4: Reactivity of Solids, 1976, ed. N.B. Hannay, Plenum Press, New York.
3. J.E. Burke: 'Sintering and Microstructure Control', Chapter 18 in Chemical and Mechanical Behaviour of Inorganic Materials, 1970, eds. A.W. Searcy, D.V. Ragone and U. Colombo, Wiley-Interscience, New York.
4. R.L. Coble and J.E. Burke: 'Sintering of Ceramics', in Progress in Ceramic Science, Volume 3, p.197, 1963, ed. J.E. Burke, Pergamon, Oxford.
5. A.L. Stuijts: 'Synthesis of Materials from Powders by Sintering', in Annual Reviews of Materials Science, Volume 3, p.363, 1973, eds. R.A. Huggins, R.H. Bube and R.W. Roberts, Annual Reviews Inc., Palo Alto.
6. F. Thümmler and W. Thomma: 'The Sintering Process', in Metallurgical Reviews 12, p.69, 1967, Metals and Metallurgy Trust, London.
7. H.E. Exner: 'Principles of Single Phase Sintering', in Reviews on Powder Metallurgy and Physical Ceramics, Volume 1, p.7, 1979, ed. F.U. Lenel, Freund Publishing House, Tel Aviv.
8. Y.E. Geguzin: 'Physik des Sinterns', VEB Deutscher Verlag für Grundstoffindustrie, 1973, Leipzig.
9. G.C. Kuczynski, N.A. Hooton and C.F. Gibbon (eds.): 'Sintering and Related Phenomena', 2nd Notre Dame Conference, 1967, Gordon and Breach, New York.
10. G.C. Kuczynski (ed.): 'Sintering and Related Phenomena', 3rd Notre Dame Conference, Mat. Sci. Res. 6, 1973, Plenum Press, New York.
11. G.C. Kuczynski (ed.): 'Sintering and Catalysis', 4th Notre Dame Conference, Mat. Sci. Res. 10, 1975, Plenum Press, New York.

12. G.C. Kuczynski (ed.): 'Sintering Processes', 5th Notre Dame Conference, Mat. Sci. Res. 13, 1970, Plenum Press, New York.
13. C. Herring: J. Appl. Phys. 1950, 21 (4) 301-303.
14. Hayne Palmour III, R.F. Davis and T.M. Hare (eds.): 'Processing of Crystalline Ceramics', Mat. Sci. Res. 11, 1978, Plenum Press, New York.
15. F.F.Y. Wang (ed.): 'Ceramic Fabrication Processes', Treat. on Mat. Sci. and Tech. 9, 1976, Academic Press, New York.
16. G.Y. Onoda, Jr., and L.L. Hench (eds.): 'Ceramic Processing before Firing', 1978, Wiley, New York.
17. D.W. Johnson, Jr.: Bull. Am. Ceram. Soc. 1981, 60 (2) 221.
18. K.S. Mazdiyasni: Ceramurgia Int. 1982, 8 (2) 42-56.
19. H.K. Bowen: Mat. Sci. Eng. 1980, 44 (1) 1-56.
20. W.H. Rhodes: J. Am. Ceram. Soc. 1981, 64 (1) 19.
21. D.W. Johnson, Jr., D.J. Nitti and L. Berrin: Bull. Am. Ceram. Soc. 1972, 51 (12) 896.
22. C.E. Scott and J.S. Reed: Bull. Am. Ceram. Soc. 1979, 58 (6) 587.
23. W.R. Cannon, S.C. Danforth, J.H. Flint, J.S. Haggerty and R.A. Marra: J. Am. Ceram. Soc. 1982, 65 (7) 324.
24. I. Wynn Jones and M.J. Miles: Proc. Brit. Ceram. Soc. 1971, 19 161.
25. M.L. Huckabee, T.M. Hare and H. Palmour III: Mat. Sci. Res. 1978, 11 205.
26. R.J. Brook: Proc. Brit. Ceram. Soc. 1982, 32 7-24.
27. C. Greskovich and J.H. Rossolowski: J. Am. Ceram. Soc. 1976, 59 336.
28. R.L. Coble: J. Appl. Phys. 1961, 32 793.
29. K.H. Jack: J. Mat. Sci. 1978, 6 1135.
30. F.F. Lange: Int. Met. Rev. 1980, 1 1-20.
31. R.W. Rice: Proc. Brit. Ceram. Soc. 1969, 12 99.

DISCUSSION

G. de With comments and has a question. The question is: are there any serious steps taken to use the rapid firing cycle, as advertised by you for alumina, to be used for large structures?

R.J. Brook: There is a size limitation in the use of fast firing, namely that the sample must be small enough to allow heating to the desired temperature in a short time. Generally samples which have their shortest dimension (plate thickness, tube wall thickness) less than 1 cm are capable of being heated in a matter of minutes (the time required depends on the square of this dimension). Thus a car of refractory bricks cannot be fast fired in the meaning of fast used here; a long, thin-walled tube, however, is suitable.

H.M. Verhoog: In the past we have used a technique to fabricate dense lime crucibles from calcium carbonate powder. Are methods using decomposing powders still valid?

R.J. Brook: For many systems, decomposing powders can still be valid. The argument being advanced here is that for the achievement of full density by solid state sintering, highly refined powders in terms of granulometry and purity are very advantageous. For applications where the presence of liquid phase at high temperature is tolerable, liquid phase sintering or viscous flow sintering can be employed and the powder requirements for these are less demanding. The use of decomposing powders may then become attractive.

H. Hausner: You showed a schematic drawing, in which an agglomerate caused a porous (less densified) area. In the microstructure of sintered ZrO_2 very dense areas could be seen, which in my opinion are caused by agglomerates too. Therefore agglomerates may act in different ways, depending on the original density of the agglomerates.

R.J. Brook: I agree. The agglomerate is a structure formed from powder crystallites which are bonded together sufficiently strongly to avoid breakdown or collapse during forming (i.e. die pressing, isostatic pressing). As a consequence, the agglomerate is a region within the green compact which has a density different from that of the regions consisting of non-agglomerated crystallites. This density can in principle be either greater or less than the non-agglomerated regions: if it is less (as can occur with Al_2O_3 powders), then a low density fault occurs at the side of the agglomerate in the sintered product; if it is greater, then the agglomerate can give rise to local high density regions as you indicate.

Panel Discussion

Prof. Mocellin for mechanical/structural features:

Session one was dealing with mechanical and structural features.
It seems that one could organize the seven papers into three
clusters. A first cluster dealt with materials that are available
as of today. Papers by Prof. Gugel and Dr. Claussen would belong
to this cluster. Following a logical scheme: if you have
materials available you want to worry about the kind of qualities
that they offer to you, particularly in view of the possible
applications you might think you could have for these materials.
There we come to a second cluster of papers, like the ones
presented by Dr. Davidge and Dr. Goebbels, who discussed
properties and structural quality of materials, aiming at giving
information and data for the design. For the next logical step of
dealing with materials we have a set of three papers, more or
less discussing behaviour in real or more or less simulated
service conditions, and I suggest that those papers by
Dr. Walzer, Le Doussal and Palin be regrouped in this cluster.
Now I would like to take just a few minutes to discuss in more
detail the various clusters as I have presented. The first one,
dealing with the materials themselves (ceramic materials that are
available): if you consider those that were available 5-7 years
ago (first generation materials) they were really of two kinds:
reaction-bonded, hot-pressed materials, both silicon nitrides and
silicon-carbides. It appears that these first generation
materials have found some applications, which is encouraging, but
there are still big problems with reproducibility in fabrication.
In the past several years, it seems fair to say, we have
witnessed the appearance of a second generation of materials,
those that were available or nearly available in 1981-82. Among
those, together with the first generation ones, we can mention
sintered silicon carbide, sintered silicon nitride, hot
isostatically pressed materials. I do not know as yet how easily
available these are, but at least it appears that the process is
feasible.
We also have some sialons which, as we heard, have found some
applications as cutting tools; and zirconia-containing systems
are also there, as have been discussed by Dr. Claussen.

For this second generation of materials it seems to appear that the processing conditions might in some respects be more extreme or delicate to control than for the first generation materials, in particular with respect to e.g. nitrogen pressure for sintered silicon nitride.

It was a pleasure to have heard Prof. Brook's talk earlier this afternoon about problems with raw materials. I was a bit surprised not to hear more about properties of raw materials and their associated problems. A question also raised: should we prefer silicon nitride materials or should we prefer silicon carbides? It might be a good thing to go back to this question as there might be some interesting points of discussion.

The second cluster of papers had to do with properties and structural quality. One can draw a parallel between the papers of Dr. Davidge and Dr. Goebbels. The ultimate goal of both these papers was the provision of design, or construction information, or both. These designer, of course, deal with components. We heard about the problems of a.o. proof testing, inspection, elimination of bad components by non-destructive evaluation techniques. The central matter here was the flaw population that is present within the component. This flaw population was dealt with from two points of view: property measurement in relation to the flaws present (Dr. Davidge) and application of non-destructive evaluation techniques to investigate and characterize them. We heard we have to worry about flaws that can be as small as about 1 micron in contrast to flaws of about a 100 microns for metallic materials.

These flaw populations have great influence on the strength and the slow crack growth of the material. They can be identified or approached by very modern techniques such as microfocus X-rays, ultrasonic vibrations etc.

I would like to stress here that there is a strong relationship between flaw population and microstructure. A few of the problems that were dealt with concerned multiaxial strength, Weibull statistics, multiple flaw populations etc. In parallel one has non-destructive evaluation concepts and defect characteristics that can be brought in line with the first ones. The message was that one has to put in a lot more effort in the high temperature behaviours particularly because changes in the flaw population are taking place then. From the non-destructive evaluation techniques I gathered that these techniques need developing, perhaps putting together a handbook of acoustic "defect signatures" would be quite valuable.

The third cluster of papers, discussed three different examples
of working problems. However, they all have in common the very
complex situation one encounters when combining mechanical/
thermal/chemical effects. We heard a number of interesting
comments about the potential of ceramics in automotive power
plants, and questions were raised about thermal shock, erosion,
corrosion behaviour etc. Coming to cracking in alumina polyphase
ceramics, the main message of that paper was, I think, that it
attempted to show how to combine relatively modern mechanical
testing and acoustic emission techniques to ceramics.
Finally the paper about refractory structures. These are highly
complex: there are many engineering properties that are involved
and need to be known. It was suggested that it would be
worthwhile to have a more comprehensive data bank on such
parameters as thermal expansion or conductivity for which we may
need improved theoretical models. Attention also should be
directed to the now available standards which may not always be
pertinent ?

Dr. Popper for thermal/physical features:

Mr. Palm discussed the use of ceramics in diesel engines. For
those like myself who had been working for a long time in the
field of materials for the gas turbine it was gratifying to
hear that there may be a spin-off from our work, i.e. the
application of ceramics in the diesel engine. He did not cover
in detail all the components for which ceramics may be used,
but concentrated on those which provided heat insulation in
the engine and thereby would give the greatest benefit from
the point of view of efficiency. This meant that he also
wanted materials which had properties different from those
required for the gas turbine, i.e. low thermal conductivity.
He guessed correctly that the ideal material, which is
required for some parts of the diesel engine, does not exist.
It is useful for an engineer to state what he really would
like to have; it stimulates the materials people to endeavour
to give something which at least partially fulfills the
requirements. Since the requirements are incompatible, e.g.
high thermal coefficient of expansion so as to be able to cast
iron around the ceramics, and at the same time have properties
which result in low thermal stresses, one may have to consider
two-phase materials or a construction using one material in

different forms, e.g. a low density thermal insulator with a dense coating. In the end engineers may have to compromise and use the best of the existing materials and techniques.

The next two papers were concerned with heat exchangers. The first, using silicon nitride for a domestic heater, was given by Dr. Krauth, Rosenthal Technik. This was a badly-required success story for silicon nitride, which has now been around for over 25 years. It is the first time, I think, that a large-scale application of this material was close to production. Many speakers mention the lack of reproducibility in ceramic materials. I venture to guess that when you have a large-scale production many of the problems regarding reproducibility will disappear. Today suppliers are asked to make a variety of components for experimental purposes which leads to inconsistent furnace loadings, etc., which, particularly in a material like silicon nitride, leads to problems.

Dr. Roettenbacher used silicon infiltrated silicon carbide in his high temperature heat exchanger. It is another example of how long it takes for a material to find a suitable application, since I did the first development on this in 1958.

The paper by Professor Rietjens on MHD was a great surprise to me. I thought MHD was dead and buried for ever! In the past I could not foresee how the materials requirements for MHD could ever be solved. It appears that by using coal instead of nuclear energy the materials problems have become much more amenable, and it will be very interesting to hear about the performance of the high-power plant now being constructed in Russia.

I would like to go back to the problem of the reliability which was mentioned frequently. I thought there was a tremendous gap between the industrial needs and the academic approach demonstrated by Prof. Brook. I cannot see that people working on the feasibility of many applications of ceramics can wait until entirely novel types of materials, e.g. laser-produced sub-micron particles of uniform size, can be produced industrially. The problem now is that one in a hundred or one in a thousand tubes used in the sodium alumina battery fails. The reason for the failure is the result of a flaw in the production. In many cases suitable raw materials are not available and will not become available until a commercial need for them is established. Most of the early work on silicon nitride had to be done on silicon which was produced for metallurgic purposes.

I feel that this problem can only be overcome by the injection
of a suitable amount of money such as has been achieved in the
German materials programme. Not only is raw material
unsuitable but also the equipment used in the production. This
invariably is the same equipment which is used for the
production of traditional ceramics where nobody is bothered
about a flaw of the order of 100 microns occurring in one out
of a hundred articles.

In answer to Prof. Brook's comments, I am not against the
academic efforts to produce the most suitable raw materials,
but I think that in many cases applications cannot wait for
the approach he suggested.

Prof. Majdic for chemical features*:

The first paper surveyed Coal Gasification processes which are
under development or are already in operation, in which
temperatures may range from 750°C to 1300°C, and for which,
temperatures up to 1700°C are envisaged. These temperature ranges
include therefore, the regimes of both solid (particulate) and
liquid slag formation. The composition of slags is known, and the
temperature of slag decomposition is established as 1100°C –
1200°C, which leads in some cases, to drastic changes in the
corrosive loading of the refractory.

The development of technology has proceeded in different ways,
one example being the achievement of rapid decomposition of slags
in molten iron or liquid slag baths, characterised by their high
heat content.

Often mentioned as known deterioration mechanism are:

- Erosion by particulates in gas streams - requiring the use of
  hard and dense materials

- Corrosion, by diffusion through porous material and by
  fluxing. Investigation has shown that from many refractory
  materials, only two were found where the loss rate by surface
  corrosion could be reduced from 0.2 mm/h to 0.01 mm/h. Both
  materials were chromia-rich refractories.

---

* Editor's translation from verbal presentation.

530

Technology constantly aims to increase efficiency, by increasing temperature and pressure, ultimately to temperature regimes beyond the capability of refractories. Beyond this limit, water cooling is required to provide a wall surface of frozen slag, whose thickness is controlled by the equilibrium conditions of thermal conductivity and cooling rate. The current target for refractory lifetimes is one year, but in many processes this is not nearly achieved.

In another, more material-oriented paper, further deterioration mechanisms were ennumerated, viz: alkaline attack, carbon deposition from decomposed carbon monoxide and reduction by SO_2. The author believes that difficulties arise in that laboratory experience cannot be transferred directly or completely into operational service and therefore proposes that laboratory testing environments should simulate as near as is possible, plant processing conditions. Particular importance is attached to dynamic tests where refractories are in continuous contact with fresh slag and not with refractory-enriched slag.

The influence of slag viscosity may also be considered. However at constant composition and under isothermal conditions, viscosity must result from variation in composition of the slag. The influence of these changes may be small, but in some instances, vital.

There has been a proposal to apply steel fibre re-inforcement to refractory concrete. In this application, it should be remembered that, exposed iron surfaces will lead to enhanced CO decomposition and carbon dumping. Our own studies on fibre orientation effects confirm that thermal transport is enhanced with fibres arranged parallel to the thermal gradient.

For further research it has been proposed to optimise the microstructure of those materials which until now have given the best results. Professor Muan has pointed out the importance of thermodynamic considerations. I would like to add that these considerations should not only be applied to the major phase, but also to the binding phase which has less resistance to chemical attack. This raises the question of the wisdom of pursuing the option of direct-binding in basic materials, and indeed whether direct-bindings exist at all in technical products.

The section dealing with the physical chemistry of SiC, Si_3N_4 and sialons showed that the intrinsic thermodynamic instability of these materials is counteracted by the very slow kinetics of decomposition.

Prof. Metselaar for electrical and electronic features:

You all have the lectures in mind as they were given today. We can group them into three different topics. First I will discuss the two lectures on battery sytems by Dr. Steele and by Dr. Wicker. It is clear, I think, especially from the overview of Dr. Steele, that there is a need for high power battery systems. This topic is under development for many years now and it seems that the most promising work, anyhow most of the work, is spent on the sodium sulphur battery and I think, having heard the two lectures, we can pose the question: why is the sodium sulphur battery, after exactly 25 years now, still the battery of the decade ?

You heard that there are a number of problems involved, all concentrating on the lifetime of the electrolyte, the sodium electrolyte and in all cases this is the beta-alumina. The question should be asked whether this is a problem in the ceramic parts or the chemistry parts, e.g. stability of the material - is this still a problem or is it purely the ceramic parts? We can only find out by experimenting in a careful way, and I'd like to stress the words of Prof. Brook in this respect, that we should try to develop that aspect as far as possible. Otherwise production will remain in this phase.

Some other points I would like to raise: it has been said that there is still a need for better electrolytes working at a lower temperature. That would be very useful indeed, but personally I do not believe that after the many years of work directed especially to the finding of this kind of electrolyte a real breakthrough will occur. If a new electrolyte would be found, we would still have to develop the technology involved in the production of such an electrolyte - and that would take a long time again.

We have not heard much about the fuel cells in comparison to the batteries, but maybe there, looking at the state of the art, as far as I know could be a good chance too to utilise fuel cells; it is a very interesting field.

If I may pass to the second topic, the sensors: we had a very

good overview from Dr. McGeehin. The problems as posed both
by Dr. McGeehin and Dr. De Pous were very clearcut:
temperature dependence of these materials (I think this is an
inherent problem), choice of reference, electrodes, ageing of
the materials, in some cases lack of sensitivity. These are
problems which have to be solved. It is interesting that
zirconia still is the material which is in the picture most
of the time - we heard it at many different places during this
colloquium, not only at this session.

Finally, the lecture we heard on electrolysis.
Here again we encounter the problems which we heard of so many
times: production of these tubes in larger quantities, reliable,
with long lifetimes, in this respect it may be the low temperature
application which Dr. Torre mentioned, which might be simpler
to realise, but we cannot say much about it after his lecture
as it is still in the very beginning.

Post-Colloquium Workshop Conclusions

The colloquium was followed by a one day workshop in which experts from EEC countries and Commission services (see list of participants) discussed R & D priorities in the field of ceramics in Europe.

Important topics for research were identified within the following broad areas :

(a) Materials development and materials properties research,
(b) Ceramic fabrication, processing and quality control,
(c) Ceramic component design and component performance evaluation,
(d) Information collection and transfer.

In the area of Materials Research and Development, the following topics were judged to be important:

- The study of the interdependence between properties and microstructure. The relationship is also critical for the optimisation of properties by process control.

- The development of new powder preparation methods as well as the processing of powder and their characterisation are important, because new and improved materials are expected to be developed through the availability of new powder sources.

- There is still plenty of scope to improve existing materials and to develop new areas. With respect to the latter, the field of ceramic based composites needs a renewed effort.

- The workshop recommended strongly the development of reference materials and the standardisation of testing methods. The development of new testing techniques is also necessary.

The area of <u>Ceramic Fabrication, Processing and Quality Control</u>
concerns the transfer of laboratory scale achievements in
materials development to the mass production of components. In
order of importance, the following topics are considered worthy
of R & D efforts :

- Development of low cost non-destructive testing to monitor
  mass production.

- Development of methods for the control of fabrication
  parameters and study of the relationship with microstructure.

- Development of low cost raw material sources and powder
  processing methods in order to achieve greater consistency in
  the starting materials.

- Adaptation of processing equipment to modern materials and to
  new fabrication methods.

- Development of near net shape production techniques.

- Study of the causes of the discrepancies observed, in
  practice, between the properties of laboratory produced parts
  and those of mass produced parts.

- Study of microstructure/property relationships in order to use
  these for process and quality control during production.

- Study of surface treatments and coatings to strengthen ceramic
  components or, as required in some applications, to leakproof
  them.

In the area of <u>Component Design and Component Performance</u>
<u>Evaluation</u> the following were deemed to be important areas of
R & D :

- Design methodology of ceramic components should continue to be
  improved.
- Design is recognised to be particularly important in joining
  problems. Techniques for joining ceramics to ceramics and to
  metals need to be further developed.
- Development of life time prediction methods is needed. This
  means testing for long times on a statistical basis under

application conditions to explore reliability. Failure
analyses should be included and the results fed back to
producer and materials developer.
- Methods for non-destructive evaluation of components are
  needed. In these methods it is necessary not only to learn to
  detect defects through interpretation of the signal but also
  to develop accept/ reject criteria.
- Efforts to test components and component assemblies on a pilot
  scale should be mounted.

Finally, in the area of <u>Information Collection and Transfer</u> the
workshop recognised the need for a ceramics data bank not only to
collect data on the properties of materials and finished products
but also those of the raw materials.
The cause of advanced ceramics will benefit from the transfer of
information between technical disciplines, e.g. between ceramics
and solid state chemistry, between ceramicists and refractories
specialists. There is still a need for more general publicity
aimed at improving the image of ceramics as engineering
materials.
In order to advance the field and widen the scope of
applications, close cooperation and more exchange of information
is needed between potential users and ceramic suppliers.

LIST OF PARTICIPANTS

AMBS H.	Thyssen AG, Duisburg, G
AURIOL A. °	CERAVER, Paris, F
BABINI G.N.	National Research Council, Faenza, I
BARRENBERG H.	Fried. Krupp GmbH, Essen, G
BAUDRAN A. ° (Chairman)	Société Française de Céramique, Paris, F
BLACKSTONE R.	Netherlands Energy Research Foundation, ECN, Petten, NL
BLOEM P.J.C.	N.V. Kema, Arnhem, NL
BÖHMER M.	DFVLR, Köln, G
BONNET C.	Société Européenne de Propulsion, St. Médard en Jalles, F
BOONE D.H.(Author)	University of California, Berkeley, CA, USA
BOURDEAU P.	CEC, DG XII, Brussels, B
BRADBURY B.A.	Champion Spark Plug Company, Ltd., Wirral, UK
BRESSERS J.	CEC, JRC, Petten, NL
BRIGGS J. °	J. Morgan Thermic Ltd., Stourport-on-Severn, UK
BRONSVELD P.M.	University of Groningen, Groningen, NL
BROOK J.R. (Author)	The University of Leeds, Leeds, UK
BROUSSAUD D.	Ecole Nationale Supérieure des Mines de Paris, Evry, F
BÜHL H.	Daimler-Benz AG, Stuttgart, G
BULLOCK E.	CEC, JRC, Petten, NL
BURESCH W.E.	KFA Jülich GmbH, Jülich, G
BURGGRAAF A.J. (Author)	Twente University of Technology, Enschede, NL
BUTLER E.P.	Imperial College, London, UK
BUTTER J.A.M.	Hoogovens BV, Ijmuiden, NL
CARLSSON R.	Swedish Institute for Silicate Research, Göteborg, S
CASTEELS F.	SCK/CEN, Mol, B
CLAUSSEN N. (Author)	Max-Planck-Institute für Metallforschung, Stuttgart, G
CLUBLEY M.H.	Magnesium Elektron Ltd, Manchester, UK
COCHET-MUCHY B.	Produits Chimiques Ugine Kuhlmann, Courbevoie, F
COEN-PORISINI F.	CEC, JRC, Ispra, I
COHEUR L.	SCK/CEN, Mol, B
DAL MASCHIO R.	Facoltà di Ingegneria dell'Università, Padova, I

° Workshop Participant

DAVIDGE R.W. (Author)	AERE, Harwell, UK
DE POUS O. (Author)	Battelle Institute, Carouge/Genève, CH
DESPLANCHES G.	Regienov, Rueil Malmaison, F
DE WITH G.	Philips Research, Eindhoven, NL
DIDERRICH E.	Centre de Recherches Métallurgiques, Liège, B
DIETRICHS P.	TH, Aachen, G
DOENITZ W.	Dornier System GmbH, Friedrichshaven, G
DRAMAIS R.	Société Belge des Produits Réfractaires, BELREF, Saint-Ghislain, B
DÜRRFELD R. (Author)	Ruhrkohle, Öl und Gas GmbH, Bottrop, G
DWORAK U.	Fa. Feldmuehle AG, Plochingen, G
EHRENTREICH J. °	CEC, DG III, Brussels, B
EISENBACH B.	Magnetisal, G
ENGMAN U.	Studsvik Energiteknik AB, Nyköping, S
ERBEN E.	M.A.N. Neue Technologie, München, G
ESNOULT M.	S.E.P.R. Recherches, Le Pontet, F
FALCE J.	Creusot-Loire, Le Creusot, F
FANTOZZI G.M.	INSA, Villeurbanne, F
FLICK W.	Fa. Dynamit/Nobel, Niederkassel, G
FRIDEN, R.A.	Volvo Truck Corporation, Göteborg, S
GAROT W.	Silicon, Den Haag, NL
GARVELINK E.A.	Netherlands Energy Research Foundation, ECN, Petten, NL
McGEEHIN P. (Author)	AERE, Harwell, UK
GOEBBELS K. (Author)	Fraunhofer-Institut IZFP, Saarbrücken, G
GRATHWOHL G.	University of Karlsruhe, Karlsruhe, G
GREEN J.M.	Billiton Research BV, Arnhem, NL
GUGEL E. ° (Author)	Annawerk Keramische Betriebe GmbH, Rödental, G
HALPIN M.K. (Chairman)	Institute for Industrial Research and Standards, Dublin, Ireland
HAMBURG F.W.	Netherlands Energy Research Foundation, ECN, Petten, NL
HAMPSHIRE S.	The National Institute for Higher Education, Limerick, Ireland
HANSRANI S.P.	British Steel Corporation, Grangetown, UK
HAUSNER H. ° (Chairman)	Technische Universität Berlin, Berlin, G
HEIDER W.	Sigri Elektrographit GmbH, Meitingen, G
HERAUD L.	Société Européenne de Propulsion, Saint-Médard-en-Jalles, F
HERMANSSON L.	Swedish Institute for Silicate Research, Göteborg, S
HIND D.	G.R.-Stein Refractories Ltd, Worksop, UK

° Workshop Participant

HOLMBERG B.	Research Institute of National Defence, Stockholm, S
HOLLUND B.	B & W Diesel AS, Copenhagen, DK
HOWLETT S.P.	Doulton Industrial Products Ltd., Stone, UK
JEANNEZ Y.	Lafarge Réfractaires, Vénissieux, F
JOHN R.C.	Koninklijke Shell, Amsterdam, NL
KAPTEIJN J.	Technische Hogeschool, Delft, NL
KEIZER K. (Author)	Twente University of Technology, Enschede, NL
KEMENY G.	CEC, JRC, Petten, NL
KEMPER C.A.L.	FDO Technische Adviseurs, Amsterdam, NL
KENNEDY C.R.	Exxon Research & Engineering Company, Florham Park, USA (Author)
KENNEDY P.	UKAEA, Salwick Preston, UK
KIEHL J.P. (Author)	Lafarge Réfractaires, Vénissieux, F
KNOCH H.	Elektroschmelzwerk Kempten GmbH, Kempten, G
KOLASKA H.	Krupp Vidia Fabrik, Essen, G
KOSLOWSKI H.	Renault, Rueil Malmaison, F
KRAUTH A. ° (Author)	Rosenthal Technik AG, Selb, G
KREFELD R.	CEC, JRC, Petten, NL
KRÖCKEL H. °	CEC, JRC, Petten NL
KRUITHOF J.	DAF-Trucks, Eindhoven, NL
LANG J.	Université de Rennes, Rennes, F
LARGUIER P.	Peugeot SA, Vélizy, F
LARKER H.T.	ASEA AB, Rodertspors, S
LE DOUSSAL H. (Author)	Société Française de Céramique, Paris, F
LIM K.H.	UHDE GmbH, Dortmund, G
LIVEY D.T.	AERE, Harwell, UK
LOHSCHEIDT K.	Thyssen A.G., Düsseldorf, G
LUX B.	Technische Universität Wien, Wien, A
MAJDIC A. (Panel member)	Forschungsinstitut der Feuerfest-Industrie, Bonn, G
MÄNTYLÄ T.A.	Tampere University of Technology, Finland
MARKOV A.	Martin & Pagenstecher, Krefeld-Linn, G
MARRIOTT J.B.	CEC, JRC, Petten, NL
MARTINEZ G.	Société Bertin et Compagnie, Tarnos, F
MARYNISSEN G.H.	Elbar B.V., Lomm, NL
MEETHAM G.	Rolls Royce Ltd., Derby, UK
MERZ M. °	CEC, JRC, Petten, NL
METSELAAR R. (Panel member)	University of Technology, Eindhoven, NL
MOCELLIN A. (Panel member)	Ecole Polytechnique Fédérale de Lausanne, Lausanne, CH
MOORTGAT G.	Centre de Recherches de l'Industrie Belge de la Céramique, Mons,B
MOROCUTTI O. °	CEC, DG XII, Brussels, B
MORRELL R.	National Physical Laboratory, Teddington, UK

° Workshop Participant

MÖRTL G.	Österreichisch-Amerikanische Magnesit AG, Radenthein, A
MUAN A. (Author)	Pennsylvania State University, Pennsylvania, USA
MÜLLER-ZELL A.	Hutschenreuther AG, Selb, G
MURRAY, P.R.	Shell Research, Amsterdam, NL
NEUMANN W.	Österreichisches Forschungszentrum Seibersdorf Ges.mbH, Wien, A
NICHOLAS M.G.	AERE, Harwell, UK
NOSBUSCH H. °	CEC, JRC, Brussel, B
NUTTER K.M.	Smith Industries Ltd., Rugby, UK
PALIN F.T. (Author)	British Ceramics Research Association, Stoke-on-Trent, UK
PALM B. (Author)	Saab-Scania AB, Södertälje, S
PALMONARI C. (Chairman)	Centro Ceramico, Bologna, I
PALSGRAAF A.	Kon. Mij. De Schelde, Vlissingen, NL
PETER K.	Dr. Otto, Buchum, G
PIEPER J.H.A.	Billiton Research B.V., Arnhem, NL
PILAVACHI P.	CEC, DG XII, Brussels, B
PLOSS G.E.	Lonza-Werke GmbH, Waldshut-Tiengen, G
POPPER P. ° (Panel member)	British Ceramic Research Association, Stoke-on-Trent, UK
QUELL P.	KFA Jülich GmbH, Jülich, G
RAE A.W.J.M.	Anzon Ltd., Wallsend UK
RANC R.	C.E.A./C.E.N. Grenoble, F
REHN S.J.	United Turbine AB, Malmö, S
RIEHL, M.	Confédération des Industries Céramiques de France, Paris, F
RIETJENS L.H.Th. (Author)	Technische Hogeschool Eindhoven, Eindhoven, NL
ROETTENBACHER R. (Author)	Dornier System GmbH, Friedrichshafen, G
SCARINCI G.	Facoltà di Ingegneria, Università di Padova, Padova, I
SCHIFFELEERS P.J.M.	Van Doorne's Bedrijfswagenfabriek DAF B.V., Eindhoven, NL
SCHINDLER S.R.	Rosenthal Technik AG, Selb, G
SCHMIDT W.	Hutschenreuther AG, Selb, G
SCHUSTER K.	CEC, JRC, Petten, NL
SIEBELS J.E.	Vokswagenwerk AG, Wolfsburg, G
SISKENS, C.A.M.	Institute of Applied Physics, TNO, Geldrop, NL
SNELL N.A.	Rolls Royce Ltd., Filton Bristol, UK
STARR F.	British Gas, London, UK
STEELE B.C.H. (Author)	Imperial College of Science & Technology, London, UK
STEIN H.	Didier-Werke AG, Wiesbaden, G
STEINBRECH R.W.	Universität Dortmund, Dortmund, G
TAYLOR R.J.	Anzon Ltd., Tyne & Wear, UK

° Workshop Participant

TENAGLIA A.	Centro Ceramico, Bologna, I
THEVENOT F.	Ecole Nationale Supérieure des Mines, Saint-Etienne, F
THOMPSON D.P.	University of Newcastle upon Tyne, UK
TITEUX M.P.	Produits Chimiques Ugine Kuhlmann, Levallois-Perret, F
TORRE J.P. (Author)	Lafarge Réfractaires, Trappes, F
TURBERFIELD K.C.	AERE Harwell, UK
VAN DE VOORDE M. °	CEC, JRC, Petten NL
VAN DER BIEST O. °	CEC, JRC, Petten, NL
VAN DER GIESSEN D.	Estel Technical Services BV, Ijmuiden, NL
VAN KONIJNENBURG J.T.°	Estel Hoogovens BV, Ijmuiden, NL
VAN RHIJN A.A.	CEC, DG III, Brussels, B
VAN WESTEN P.J.	CEC, JRC, Petten, NL
VERHOOG H.M.	Estel Hoogovens BV, Ijmuiden, NL
VERINGA H.J.	Netherlands Energy Research Foundation, ECN, Petten, NL
VINCENZINI P. °	National Research Institute for Ceramics, Faenza, I
VINSON J.	Raychem Ltd, Swindon, UK
VIRTAMO J.T.	Neste Oy, Espoo, Finland
WALZER P. (Author)	Volkswagenwerk AG, Wolfsburg, G
WEBB R.S.	Magnesium Elektron Ltd., Manchester, UK
WICKER A. (Author)	Centre de Recherches de la Compagnie Générale d'Electricité, Marcoussis, F
WILLIAMS D.E.	AERE, Harwell, UK
WILLMANN G. (Author)	Dornier System GmbH, Friedrichshafen, G
WOLFF L.R.	Technische Hogeschool Eindhoven, Eindhoven, NL
WURM J. °	CEC, DG XII, Brussels, B
YVARS M.	CEN Saclay, Gif-sur-Yvette, F
ZAALBERG R.J.	Thomassen International BV, Rheden, NL
ZIEGLER G.	Deutsche Forschungs-und Versuchsanstalt für Luft- und Raumfahrt (Author) e.V., Köln, G
ZILKENS E.R.N.E.	Lonza Benelux BV, Rotterdam, NL

° Workshop Participant